PROGRAMMING AND CUSTOMIZING THE 8051 MICROCONTROLLER

Other Books in the McGraw-Hill Microcontroller Series

MYKE PREDKO • *Programming and Customizing the PIC Microcontroller*
SCOTT EDWARDS • *Programming and Customizing the BASIC Stamp Computer*
MYKE PREDKO • *Handbook of Microcontrollers*

PROGRAMMING AND CUSTOMIZING THE 8051 MICROCONTROLLER

MYKE PREDKO

McGraw-Hill

New York San Francisco Washington, D.C. Auckland Bogotá
Caracas Lisbon London Madrid Mexico City Milan
Montreal New Delhi San Juan Singapore
Sydney Tokyo Toronto

Library of Congress Cataloging-in-Publication Data

Predko, Myke.
Programming & customizing the 8051 microcontroller / Myke Predko.
p. cm.
Includes index.
ISBN 0-07-134195-1—ISBN 0-07-134192-7 (soft)
1. Intel 8051 (Computer) 2. Digital control systems. I. Title.
II. Title: Programming and customizing the 8051 microcontroller.
TJ223.M53P74 1999
629.8'95—dc21 98-48076
CIP

McGraw-Hill

A Division of The ***McGraw·Hill*** *Companies*

5 6 7 8 9 0 DOC/DOC 0 3

P/N 134196-X (HC) P/N 134193-5 (PBK)
PART OF ISBN 0-07-134195-1 PART OF ISBN 0-07-134192-7

The sponsoring editor for this book was Scott Grillo, the editing supervisor was John Baker, and the production supervisor was Pamela Pelton. It was set in Times New Roman by Lisa M. Mellott through the services of Barry E. Brown (Broker—Editing, Design and Production).

Printed and bound by R. R. Donnelley & Sons Company.

CONTENTS

ACKNOWLEDGMENTS

There's an old saying that purports; "The hardest job there is, is looking for one."

Well, I don't ever remember finding a job to be as much work as writing a book. Like my previous books, this one would not be possible without the support, enthusiasm, and help from many individuals.

It would be impossible to recognize everyone on the 8051 Internet list servers that I belong to. I appreciate the suggestions and answers to my questions as I worked through the book. The list server community has always demonstrated what a powerful source of help and information the Internet can be. Thank you all for your ideas, suggestions, and occasional jokes.

I would like to thank Frank Taylor of Dallas Semiconductor for his help on this book. Frank's energy about the HSM products was catching and helped spur me on to see opportunities to use the HSM to simplify many of the projects presented in this book.

Thanks, as always, goes to Ben Wirz for his support, ideas, and help on the 51Bot. It was nice to have you to lean on.

Thanks also go to Phillipe Techer for his hard work on UMPS. I'm thankful for your efforts in adding additional devices (many of which were used in this book) and listening to my comments and suggestions (and implementing many of them). UMPS has grown into a remarkably mature and powerful tool in the year I've been working with it.

I have to recognize the many talented and conscientious people at Celestica, my regular employer. Thank you for not paging me too often as I proofread this book.

To my children—Joel, Elliot, and Marya—thank you for your consideration with me as I created this book, and thank you for not trashing the projects presented here before I had a chance to photograph them.

The biggest thank you of all goes to my wife, Patience, for her love, support, and encouraging words, even when the current project I was working on was burning and I felt like I should have become a plumber. We still haven't gotten to that beach yet, but when we do, it will be so much more sweeter.

Myke Predko • Toronto, Canada • June, 1998

ABOUT THE AUTHOR

Myke Predko is the author of *The Microcontroller Handbook* and *Programming & Customizing the PIC Microcontroller*, also from TAB Books, and New Product Test Engineer at Celestica in Toronto, Ontario, Canada, where he works with new electronic product designers. He has also served as a test engineer, product engineer, and manufacturing manager for some of the world's largest computer manufacturers. Mr. Predko has a patent pending on an automated test for PC motherboards as well as patents pending on microcontroller architecture design. He is a graduate of the University of Waterloo in electrical engineering.

INTRODUCTION

In my day job, working for Test Engineering at Celestica, I am amazed at the reluctance of many new engineers to design applications (both products and test interfaces) with an embedded microcontroller. This is disappointing for me because I think that the projects could be developed significantly faster and much more easily using a microcontroller.

The purpose of this book is to introduce the engineer or technician to the Intel developed 8051 embedded microcontroller and provide enough information and tools (with the CD-ROM that comes with this book) for someone who has never worked with the 8051 before to develop and debug their own applications. Along with the information, I have provided a number of experiments and example applications to show how the 8051 can be interfaced to a number of different common electronic devices.

An embedded microcontroller (as I define it) is a chip which has a computer processor with all it's support functions (clocking and reset), memory (both Program Storage and RAM), and I/O (including bus interfaces) built into the device. These built in functions minimize the need for external circuits and devices to be designed in the final application. See Fig. I-1.

Often, interface and control circuits are attempted with discrete logic or PLDs because of the assumptions that microcontrollers:

- are more expensive than discrete logic
- require a long learning curve

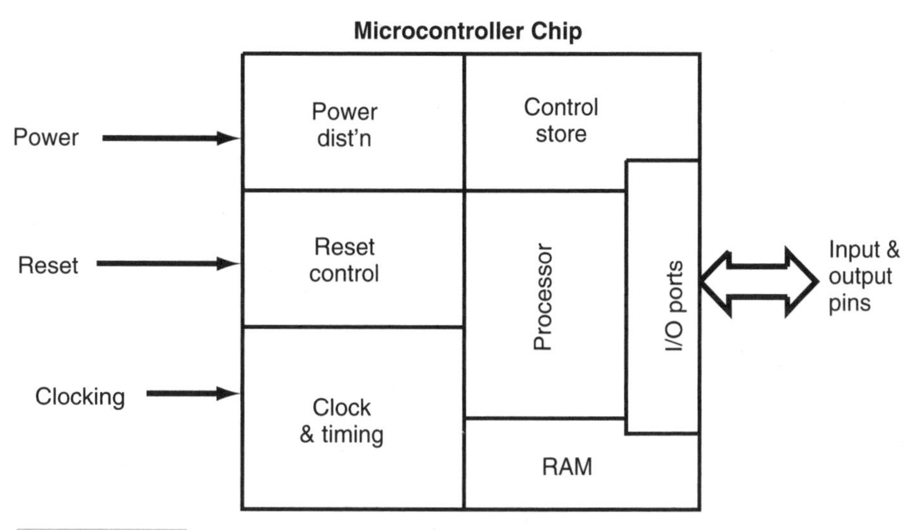

FIGURE I-1 **Microcontroller block diagram.**

- require a lot of expensive support tools
- do not have a lot of support available (especially for test/process engineers and hobbyists, who's useage ranges from single quantities to tens of devices and are not high volume consumers)
- do not have easily obtainable extended environment devices (for use in harsh environments and Environmental Stress Screening)
- are difficult to actually implement a "single device solutions" (additional circuitry will be required to complete the application)

These assumptions are largely false. Improvements in microcontroller technology (including the inevitable drop in cost that all electronic devices experience) has meant that it is often more cost effective, faster and more efficient to develop an application using a microcontroller rather than discrete logic. The purpose of this book, and the others in McGraw-Hill's "Programming and Customizing" series is to introduce you to microcontrollers and give you the knowledge and information needed to develop your own applications.

Most microcontrollers do not require a substantial amount of time to learn how to efficiently program them, although many of them have quirks which you will have to understand before you attempt to develop your first application. While this book will probably not answer all the questions you will have about developing 8051 applications, it should give you enough information to plan out how you will do the application as well as look for additional resources needed to complete the project.

Along with microcontrollers getting faster, smaller and more power efficient they are also getting more and more features. Often, the first version of a microcontroller will just have memory and simple digital I/O, but as the device family matures, more and more part numbers with varying features will be available.

With all the 8051 manufacturer's products taken into account, there are over two hundred different 8051 part numbers, each with different features and capabilities. For most applications, you will be able to find a device within the family that meets your specifications with a minimum of external devices, or with an external bus which will make attaching external devices easier, both in terms of wiring and programming.

This book is organized in the way that I find to be most effective in learning a new device. Initially, I want to get a feel for the device and what it's capabilities are. I usually do that by first understanding the packaging, timing and I/O capabilities.

Next, I look at the processor's architecture and try to get an idea of how it is organized in my mind. This is an interesting exercise because often my idea of how the processor works is completely different than what's presented in the manufacturer's datasheets. Once I have a block diagram of my own completed, I can then compare it to the operation of the various instructions. If I cannot plot the data paths for an instruction, then I modify my block diagram until it makes sense and works for every instruction.

I love to play around with intelligent digital electronics, seeing how circuits interrelate and looking for shortcuts in software and hardware interfacing. In the "Experiments" chapter, I'll go through the different experiments that I created so that I could understand how the device works. Along with this, I have also presented a number of example applications to show how the 8051 can interface to the outside world.

As part of the experiments and example applications, I have tried to make sure the book contains examples of how to interface to:

- LEDs
- LCDs
- Buttons
- Analog Voltage I/O
- Matrix Keyboards
- Memory Mapped Devices
- Serial Devices
- Motors, Relays and Servos
- Critically Timed Applications

The software presented in this book, while only written in assembly language will show off many of the features of the 8051's processor architecture and give you some tricks that you can use in your own applications. I will describe high level languages and give you examples of an RTOS and monitor (debugger) programs.

For many microcontrollers, programmers can be built very cheaply, or even built into the final application circuit eliminating the need for a separate circuit. Also simplifying this requirement is the availability of microcontrollers with SRAM and EEPROM for Control Store which will allow program development without having to remove the microcontroller from the application circuit. The 8051 is not a device which fits the above mold. I will present how 8051s are programmed (I have included a programmer circuits and software driver) as well as explained the algorithms used to program the device (and protect the code once the device is in a product and shipped in the field).

The Internet has meant an incredible improvement in the support available to everyone using electronic devices. Along with manufacturer web sites which contain datasheets available for downloading, there are also user supported FAQs Frequently Asked Questions documents listing common questions and their answers) and list servers in which mail sent to a central address is distributed to a number of people. The Internet also provides a method to contact manufacturers and distributors allowing fast and easy ordering as well as answers to your questions. I can honestly say that through the Internet, I see individuals getting the same level of support regardless of whether they are buying one chip per year or one million.

The Intel 8051

When you first think of Intel, you probably think of the PC processors made by the company and while you probably know that Intel has a very complete line of products, microcontrollers are probably not something that jumps immediately to mind. Despite this, the Intel 8051 microcontroller processor architecture is one of the most popular microcontroller architectures available on the market today.

The 8051 is unusual because of the number of different manufacturers that are currently building products based on licensed versions of the processor architecture. As I write this, there are over twenty companies that currently make 8051 architected products available, while in comparison, most other microcontroller architectures only have one or two manufacturers.

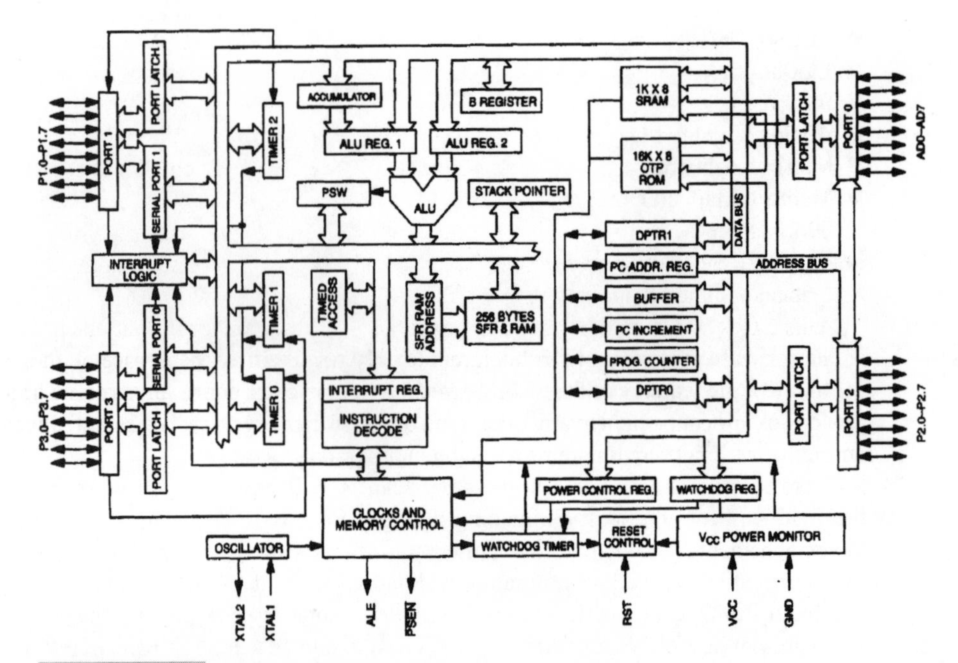

FIGURE I-2 DS87C520 block diagram.

This large number of manufacturers has meant that the 8051 is available with a startling number of features and configurations.With literally hundreds of different part numbers, each with different features and packages to choose from, developing single chip applications for the 8051 first starts off with looking at parts catalogs. While the original (or "baseline") 8051 does not have many features, there are many of different versions now available that will simplify just about any application.

The 8051's processor architecture is unusual in it's flexibility; along with having many features for efficiently handling bits (which is necessary for a microcontroller), it is also very well designed to take advantage of high-level languages and large data spaces (the 8051 can interface with up to 64K Bytes of Control Store and 64 KBytes of RAM). In the experiments and example applications, I will try to show what is possible with the 8051 processor architecture.

The 8051 is also somewhat unusual in how it interfaces to other devices. I will go through a number of different interfaces, both using the parallel digital I/O pins and enhanced features. I will concentrate on the "standard" I/O, but by going through catalogs and datasheets, you should be able to find an 8051 derivative that exactly meets your requirements.

Additional Information and Resources

While the process of this book is to introduce you to the 8051 microcontroller, it is not a vehicle for teaching electrical engineering and computer science basics. It is assumed that you will have some experience in electronics and assembly (including binary numbering systems) and high-level language programming techniques. Without a basic understanding of how computer processors work or digital and analog electronics, you may find this book to be confusing and difficult to understand.

Included with this book is a CD-ROM, containing the code used to provide the application presented in the book for each different microcontroller along with a set of datasheets in adobe ".PDF" format for various members of device family. Also included on the CD-ROM is a demonstration version of one of the most impressive assembler/simulator tools that I have ever seen. This program can be used for developing and testing out the experiments and applications presented in this book.

This CD-ROM should not be considered the final resource for the various products presented in this book. Chip Manufacturers often work at a breakneck pace, with new products and errata sheets released on a weekly basis. When doing any type of application using a microcontroller (or any other type of part for that matter), you should always make sure the information you are using is the most current possible.

I am a great believer in the Internet and it's ability to help the engineer get technical information. For all the products presented in this book, this is very true and along with web sites, where updated information can be found, I have also included information for subscribing to list servers along with information on contacting the part manufacturers directly.

Conventions Used in This Book

kHz = thousands of cycles per second
K = 1,000 ohms
MHz = millions of cycles per second
µF – microfarads
ms – milliseconds
µs – microseconds

0x0nn, $nn, 0nnh and H'nn' – Hex Numbers
0b0nnn, %nnn, 0nnnb and B'nnn' – Binary Number
nnn and 0nnnd – Decimal Number

AND and & – Bitwise "AND"
OR and | – Bitwise "OR"
XOR and ^ –Bitwise "XOR"

_Label – Negative Active Pin. In some manufacturer's data sheets this is represented with a leading "!" character or with a bar over the entire label.

1

MICROCONTROLLERS

CONTENTS AT A GLANCE

When I differentiate microcontrollers, I tend to do it at a very basic level. Rather than looking at individual I/O features, I go down into the processor architecture and work my way up. With this level of knowledge, when I'm first presented with a job to do, I immediately have a good idea of what's involved as well as having some good ideas of what is the best way to accomplish the task. Getting "under the covers" and understanding all the aspects of a device is necessary to working at this level.

Before introducing the 8051, there are some aspects of microcontrollers that you should understand. Now, you might feel that you are only going to write your applications in a common high-level language, like C, which is available on a wide variety of devices, and assume you don't have to understand the intricacies of the architecture of the device that you are writing software for. However, I feel that, to effectively create applications for a device, even if you are doing it at a high level that insulates you from the hardware, you have to understand all possible characteristics and features of the device.

Different Types of Microcontrollers

Creating applications for microcontrollers is completely different than any other development job in computing and electronics. In most other applications, you probably have a number of subsystems and interfaces already available for your use. This is not the case with a microcontroller, where you are responsible for:

- Power distribution
- System clocking
- Interface design and wiring
- Systems programming
- Application programming
- Device programming

These work items might seem obvious, but having to do them all is really quite profound in modern computing system development. In no other aspect of electronics are all these requirements found. It's like the early days of computing where to get an application running, you would be starting from scratch.

I personally find this very exciting because it means that I have the freedom to develop the "best" application for a set of requirements. The process is also made more enjoyable (if that's the right word) by learning how to work with the features built into the devices that are designed to simplify the task of directly connecting to other devices. Often, very useful applications can be created using a microcontroller and a few passive components.

When you look through manufacturer's catalogs, you'll probably find that anybody who makes chips makes a microcontroller. These devices range from being startling simple to unimaginably complex (rivaling the complexity of the Pentium processor in your PC) with just about anything in between. The devices can be completely self-contained or require a large amount of support chips to execute properly. Understanding the differences is imperative if you want to choose an appropriate device to meet your requirements.

Before selecting a particular device for an application, it's important to understand what the different options and features are and what they can mean with regard to developing your application. I want to go through the basic features of microcontrollers and introduce how the 8051 has implemented these features and why the 8051 is a device well suited to a large variety of different tasks.

EMBEDDED MICROCONTROLLERS

When you are first presented with the built-in features of the 8051, chances are that you will think of it in terms of a completely self-contained device, with all the necessary hardware built into the chip (Fig. 1-1).

When all the hardware required to run the application is provided on the chip, it is referred to as an *embedded microcontroller*. All that is typically required to operate the device is power, reset, and a clock. Digital I/O pins are provided to allow interfacing with external devices.

This complete hardware on a chip is extremely useful for some applications. For example, if you wanted to wait for a button to be pressed and then wait for a second before turning on an LED, the circuit shown in Fig. 1-2 could be used.

The DS87C520 is an 8051 derivative that has 32 I/O pins, some of which can also be used for serial I/O and interrupt input to the device.

Embedded microcontrollers are now replacing some very common devices like 555 timers because they are actually cheaper to use in applications and they are much more precise and easier to control.

EXTERNAL MEMORY MICROCONTROLLERS

Sometimes, the program memory (which I call *Control Store*) is insufficient for an application or, during debug, a separate ROM (or even RAM) would make the work easier. Some microcontrollers (including the 8051) allow the connection of external memory. (See Fig. 1-3.)

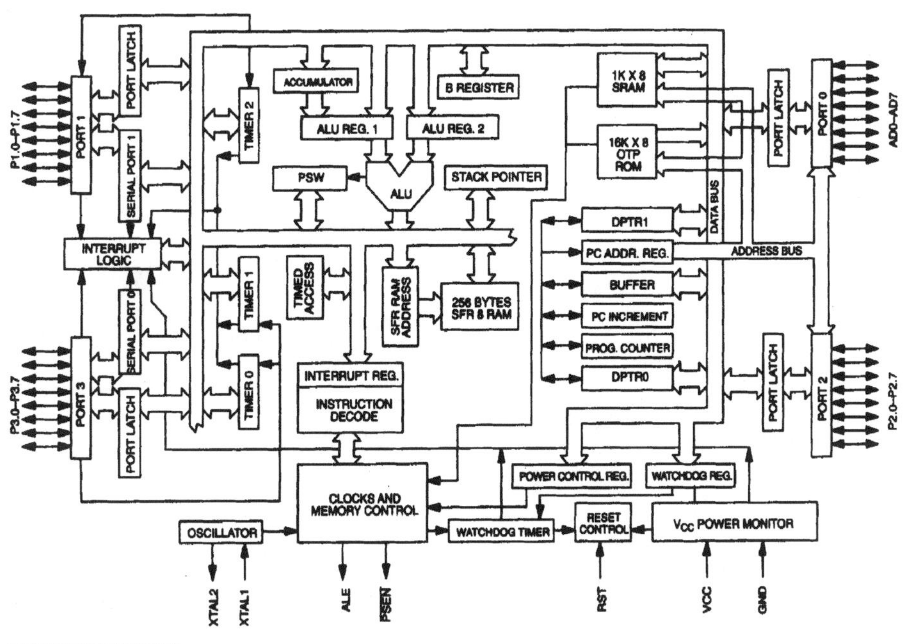

FIGURE 1-1 DS87C520 block diagram.

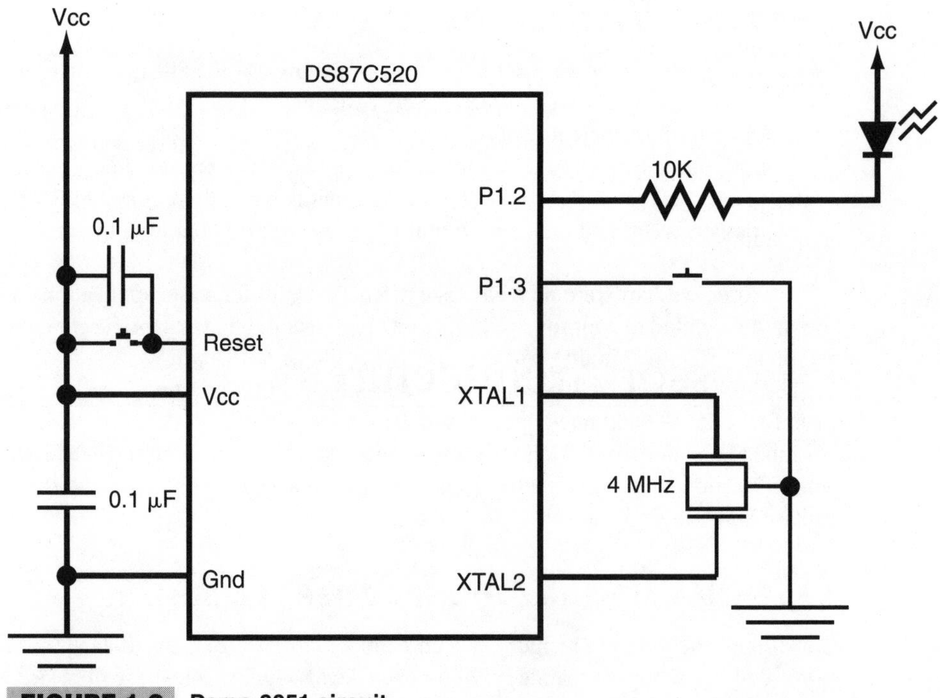

FIGURE 1-2 **Demo 8051 circuit.**

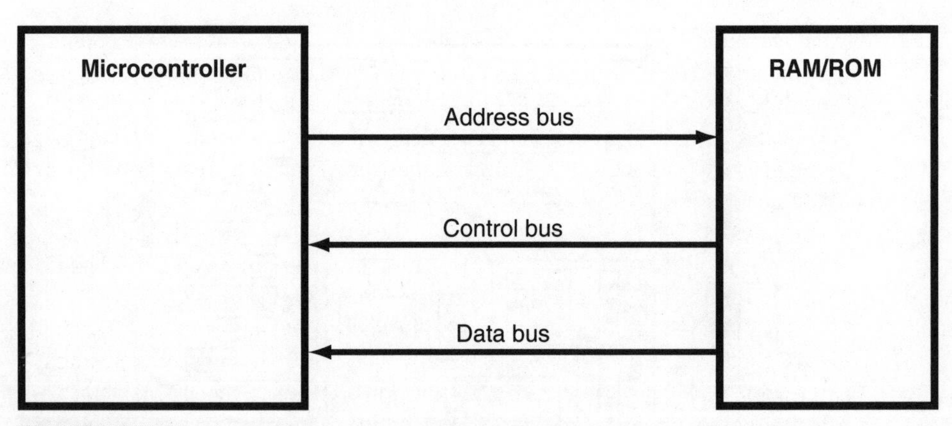

FIGURE 1-3 **External memory wired to an MCU.**

When you get into the higher-end microcontrollers (16- and 32-bit processors), you'll discover that they typically *only* have external memory, which will make the term "microcontroller" seem somewhat poorly defined when compared against the terms "embedded microcontroller" and "microprocessor," both of which seem to be much more crisply defined.

An external memory microcontroller seems to primarily differ from a microprocessor in the areas of built-in peripheral features. These features could include memory device selection (avoiding the need for external address decoders or DRAM address multiplexers), timers, interrupt controllers, DMA, and I/O devices like serial ports.

The 8051 can work very effectively as an embedded device or with external memory. As I will show later in this book, the external memory bus can be used to address memory-mapped I/O devices, which can expand the capabilities of the 8051 significantly. It's also important to note that, when external memory is used with the 8051, all of the built-in hardware interfaces can still be accessed, which makes the 8051 very flexible for different applications.

Processor Architectures

When discussing processor architectures, I always feel like I am about to open up a can of worms because, for different types of architectures, there are some very strong proponents with some very valid reasons for having their preferences. To make matters worse, it is hard to say definitely what type of processor—Harvard or Princeton and CISC or RISC—an 8051 is because it has features of all these different types of processors.

I've always felt that the 8051's processor architecture is a very thoughtful and practical combination of the differing philosophies, and this will really be demonstrated as I go through the processor's architecture and show how applications can be created for it.

HARVARD VERSUS PRINCETON

Many years ago, the United States government asked Harvard and Princeton Universities to come up with a computer architecture to be used in computing tables of Naval artillery shell distances for varying elevations and environmental conditions.

Princeton's response was a computer that had common memory for storing the control program as well as variables and other data structures. It was best known by the chief scientist's name "Von Neumann" (Fig. 1-4).

The *memory interface unit* is responsible for arbitrating access to the memory space between reading instructions (based upon the current program counter) and passing data back and forth with the processor and its internal registers.

It might at first seem that the memory interface unit is a bottleneck between the processor and the variable/RAM space (especially with the requirement for fetching instructions at the same time); however, in many Princeton architected processors, this is not the case because the time required to execute a given instruction can be used to fetch the next instruction (this is known as *pre-fetching*) and is a feature on many Princeton architected processors.

In contrast, Harvard's response was a design that used separate memory banks for program storage, the processor stack, and variable RAM. (See Fig. 1-5.)

The Princeton architecture won the competition because it was better suited to the technology of the time. Using one memory was preferable because of the unreliability of current electronics (this was before transistors were in widespread use). A single memory interface would have fewer things that could go wrong.

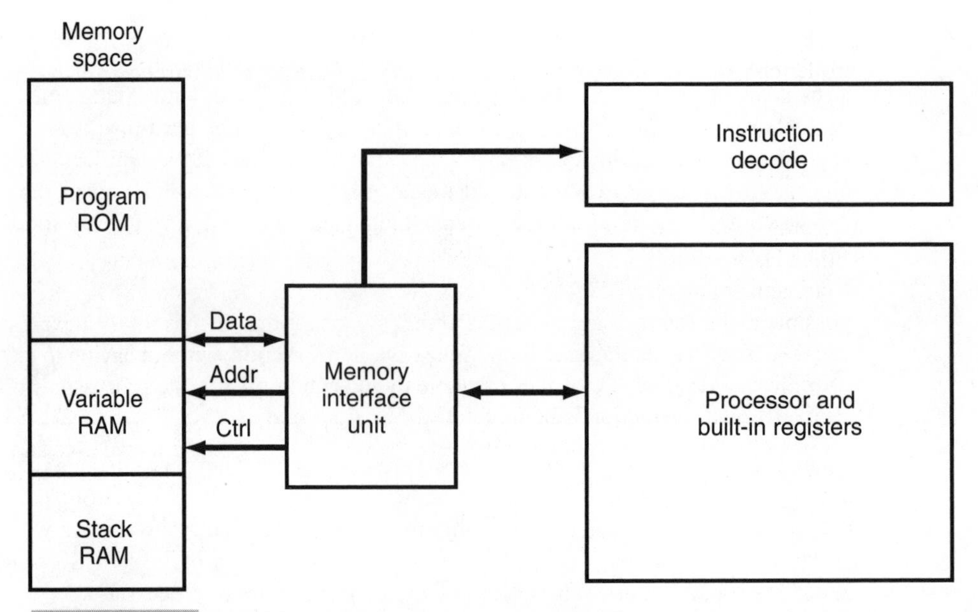

FIGURE 1-4 **Princeton architecture block diagram.**

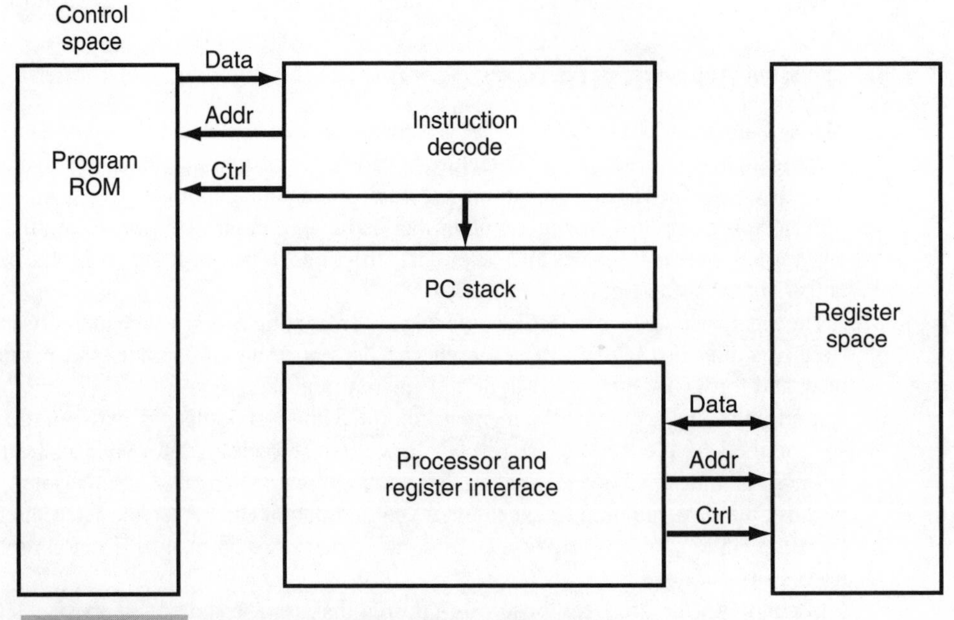

FIGURE 1-5 **Harvard architecture block diagram.**

The Harvard architecture was largely ignored until the late 1970s when microcontroller manufacturers realized that the architecture had advantages for the devices they were currently designing.

What are the advantages of the two architectures?

The Von Neumann architecture's largest advantage is that it simplifies the microcontroller chip design because only one memory is accessed. For microcontrollers, its biggest asset is that the contents of RAM (random-access memory) can be used for both variable (data) storage as well as program instruction storage. An advantage for some applications is the program counter stack contents that are available for access by the program. This allows greater flexibility in developing software, primarily in the area of real-time operating systems (which will be discussed at greater length elsewhere in the book).

The Harvard architecture executes instructions in fewer instruction cycles than the Von Neumann architecture. This is because a much greater amount of instruction *parallelism* is possible in the Harvard architecture. Parallelism means that fetches for the next instruction can take place during the execution of the current instruction, without having to either wait for either a "dead" cycle of the instruction's execution or stop the processor's operation while the next instruction is being fetched.

For example, if a Princeton architected processor was to execute a read byte and store in the accumulator instruction, it would carry out the instruction sequence illustrated in Fig. 1-6.

In the first cycle of the instruction execution, the instruction is read in from the memory space. In the next cycle, the data to be put in the accumulator is read from the memory space.

The Harvard architecture, because of its increased parallelism, would be able to carry out the instruction while the next instruction is being fetched from memory (the current instruction was fetched during the previous instruction's execution). See Fig. 1-7.

Executing this instruction in the Harvard Architecture also takes place over two instructions, but the instruction read takes place while the previous instruction is carried out. This allows the instruction to execute in only one instruction cycle (while the next instruction is being read in).

This method of execution (parallelism), like RISC instructions, also helps instructions take the same number of cycles for easier timing of loops and critical code. This point, while seemingly made in passing, is probably the most important aspect that I would consider in choosing a microcontroller for a timing sensitive application.

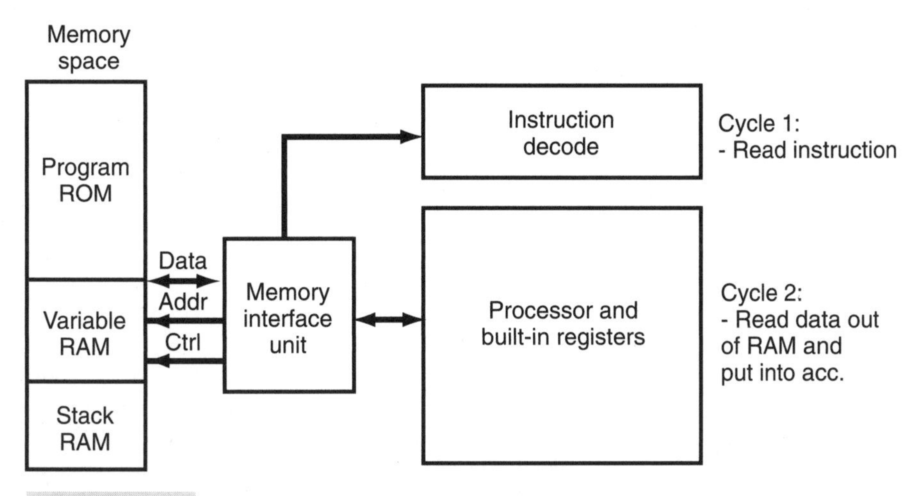

FIGURE 1-6 `mov Acc, Reg` **in the Princeton architecture.**

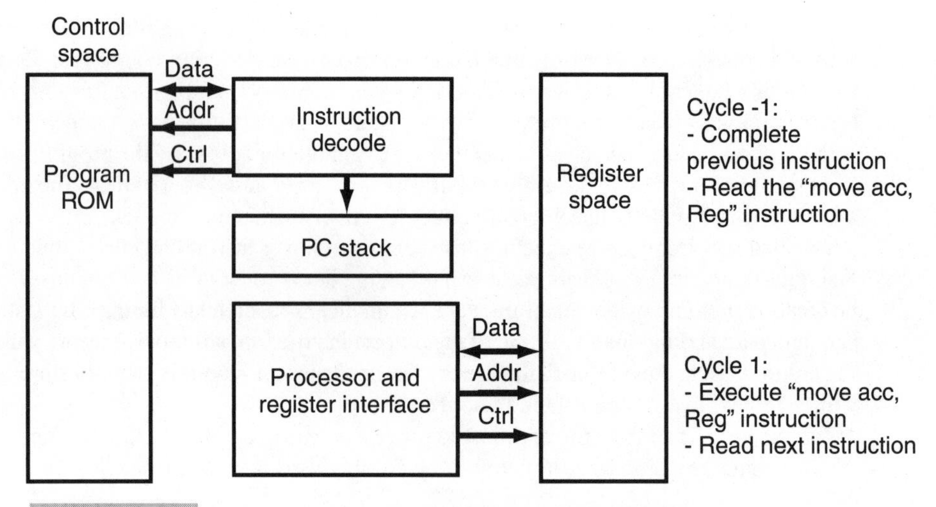

FIGURE 1-7 `mov Acc,` *`Reg`* **in the Harvard architecture.**

For example, the 8051 executes every instruction, except ones that modify the program counter, in 12 clock cycles (one *instruction cycle*). This makes critical timing operations quite easy to do. Often, a simulator or a hardware emulator will be required to accurately time a function, rather than trying to figure out manually how many cycles will be used from the code.

CISC VERSUS RISC

Currently, many processors are called "RISC" ("Reduced Instruct Set Computers") because there is a perception that RISC is faster than "CISC" (or "Complex Instruction Set Computers"). This can be confusing because there are many processors available that are identified as "RISC-like" but that are actually CISC processors. In some applications, CISC processors will execute code faster than RISC processors or execute applications that RISC processors cannot.

What is the real difference between RISC and CISC? In CISC processors, there tends to be a large number of instructions, each carrying out a different permutation of the same operation (accessing data directly, through index registers, etc.) with instructions perceived to be useful by the processor's designer.

In a RISC system, the instructions are at as bare a minimum as possible to allow the user to design their own operations, rather than let the designer do it for them. Later, I show how a stack "push" and "pop" would be done by RISC system in two instructions, which allows the two simple constituent instructions to be used for different operations (or "compound instructions" such as this).

This ability to write to all the registers in the processor as if they were the same is known *orthogonality* or *symmetry* of the processor. This allows some operations to be unexpectedly powerful and flexible. This can be seen in conditional jumping. In a CISC system, a conditional jump is usually based on status register bits. In a RISC system, a conditional jump can be based on a bit anywhere in memory. This greatly simplifies the operation of flags and executing code based on their state.

However, for a RISC system to be successful, more than just reducing the number of things that are done in an instruction has to be done. By careful design of the processor's architecture, the flexibility can be increased to the point to where a very small instruction set, able to execute in very few instruction cycles, can be used to provide extremely complex functions in the most efficient manner.

The 8051 combines many of the features of a RISC processor (namely the largely orthogonal register set) and the large instruction set of a CISC processor. This combination makes the creation of complex program functions surprisingly easy in many cases. When you have a good understanding of how the 8051 is architected and how instructions execute within the processor, you can develop code that is very fast and takes up very little "control store" space.

Microcontroller Memory Types

Memory is probably not something you normally think about when you create applications for a personal computer. In a microcontroller, understanding how much memory you have and how it's architected is critical, especially when you are planning on how to implement the application code. In a microcontroller, memory for different purposes is typically segregated and arranged to allow the device to execute most efficiently.

CONTROL STORAGE

In a PC, when you execute an application, you read the application from disk and store it into an allocated section of memory. In a microcontroller, this is not possible because there is no disk to read from. The application that is stored in nonvolatile memory is always the only software the microcontroller will execute. Having the program always available in memory makes the writing of it somewhat different than PC or workstation applications.

Control store is known by a number of different names including *program memory* and *firmware* (as well as some permutations of the various names). The name really isn't important. What is important is understanding that this memory space is the maximum size of the application that can be loaded into the microcontroller and that the application also includes all the low-level code and device interfaces necessary to execute an application.

A PC program could be as simple as:

```
// Print a simple message

main(){

  printf( "Hello World\n" );

}
```

and the actual code created could be:

```
MAIN:                               ; Program execution starts here

  mov  AH, 9                        ; Print the message
  lds  DX, Msg
  int  021h

  mov  AH, 04Ch                     ; End the program
  xor  AL, AL                       ;  with a return code of "0"
  int  021h

Msg    db   "Hello World", 00Dh, 00Ah, 0
```

This code simply calls the operating system routine that will print the message stored at location "Msg" and then returns execution to the processor.

If you were to execute this program in a microcontroller, this code would be considerably longer because *you* have to write the display subroutine (not to mention that you would have to figure out what kind of device to output to and initialize it). The software would have to make sure any other hardware in the device or connected to it is properly initialized. At the end of the application, the microcontroller has to figure out what to do to stop the application.

The extra code is put into control store along with the application code. For devices like an LCD, it's not inconceivable that the support routines require more control store space than the application.

Rather than scaring you, I hope I've peaked your interest. Because this aspect of a microcontroller is what really gets me excited about doing applications and projects on a microcontroller.

If you've ever traced through a program executing on a PC and wanted to understand how something as simple as displaying a character on the screen is accomplished, chances are you would have to go through many subroutines and lots of code before you would find out actually how the video hardware was updated. When it finally comes down to it, writing to the display is writing 8 bits to a video memory location addressed by a pointer. It could take you literally days to understand exactly how this was done.

In a microcontroller, because you have total control over how an application executes, writing to the video memory could simply be an indexed write from the current cursor in one instruction.

This opportunity for very simple and efficient operations and functions means that the control store can be very small. In modern PCs, memory to load and execute programs is measured in megabytes. In a microcontroller, control store is measured in just a few kilobytes.

While I might have caused you some concern when I said you would be providing all the interface code, hopefully you feel better because the code will not be all that difficult or complex.

With this understanding of how applications execute in a microcontroller, you can look at how it is actually implemented in the device. Earlier in this section, I mentioned that the control store was nonvolatile.

Nonvolatile memory is memory that does not lose its contents even when power is taken away from it. Normal, or *volatile*, memory circuits lose their contents when power is lost. Volatile memory is most commonly known as *random-access memory* (RAM) and can be read from or written to by the microcontroller's processor. The nonvolatile control store is often known as *read-only memory* (ROM) because during execution, the processor can only read from it, not write new information into it.

In the 8051, there are five different types of control store available in different devices and applications: none (external ROM), mask ROM, PROM, EPROM, and EEPROM/ Flash. While these five types of memory all provide the same function, memory for the processor to read and execute, they each have different characteristics and are best suited for different purposes.

"None" probably seems like a strange option, but in the 8051, it is a very legitimate one. With no internal control store, the device has to be connected to an external ROM chip. (See Fig. 1-8.)

To reduce the number of pins required to interface with the ROM, I have drawn an 8051 (Fig. 1-8) with a single external ROM that does not require any ROM address decoding. When the "_PSEN" line is asserted along with "ALE," a byte from the ROM is read into the 8051 and executed as if it were located in the 8051's control store.

One of the first 8051 devices available, the 8031 was designed without any control store at all. The 8031 could only run with an external ROM. The 8031 was originally designed as a debug device because early erasable ROM technology was very expensive to add to a microcontroller. By using an external ROM and an 8031, a fairly cost effective application development tool could be created.

You can see this philosophy of using an 8051 without control store in many early 8051 designs where the "P0" and "P2" ports are left unconnected to allow for an 8031 with an external ROM.

Today, the external ROM feature is primarily used when more application control store is required or somebody wants to start working with 8051s cheaply. When I say cheaply, I mean to say that mask ROM or previously programmed OTP (which is defined later) parts can often be found for less than a dollar. By adding a (surplus) EPROM and a 74LS373, you can have an 8051 with 16 I/O pins to experiment with.

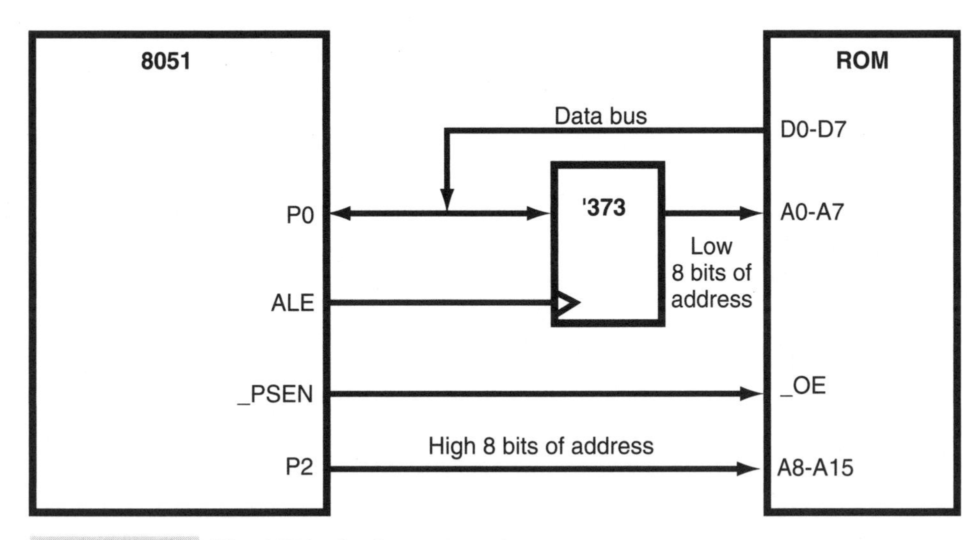

FIGURE 1-8 **The 8051 wired to external memory.**

There are microcontrollers available with read-only memory control store. When the chips are built, they are completed except for etching the last metal layer. When an order comes in for a batch of microcontrollers with a ROM with a customer specified application, these chips are pulled from stock and the last metal layer is then etched using a mask made from the customer-supplied program. This is known as *mask ROM programming*.

With the program put into the chip, customers will have devices that they can use in their products without having to worry about programming them. Often the ROM contents cannot be read out by others trying to reverse engineer the product.

There are some downsides to this method of buying microcontrollers. The first two are the cost and lead time required to get the chips in. While the actual piece price is less than a device with a customer (or "field") programmable control store, the nonrecurring expenses (NRE) cost of getting the mask made would only be cost effective in lots of 10,000 or more microcontrollers. As well, the lead time for getting mask ROM devices built is typically on the order of 6 to 10 weeks.

For certain applications, such as for the automotive market, the downsides of mask ROM microcontrollers are not significant or a problem. Here, the parts are ordered well in advance of their use, and a large, guaranteed order is assured.

A major concern about mask programmed parts is what happens if the application built into them has a problem or it has to be changed.

If it's an 8051, it's good news for the hobbyist because it can still be used in external memory mode, although it's bad news for the company that had the parts made—they will have to sell their stock to a surplus house for pennies on the dollar. Actually, in the case of the 8051, it's better news for the company left with the stock, because with other microcontrollers that can't be used with external memory, there isn't anybody who would want the parts for any price.

The most popular way of purchasing microcontrollers is to have a programmable control store. When programmable ROM (PROM) was first introduced, it consisted of an array of fuseable links that could be burned out by the application of high current to the link during the programming process. As soon as something better came along, PROM control store was dropped because the programming of the parts wasn't anywhere close to having 100% yield and over time, the blown fuse links could "grow back." This lack of reliability with the PROM contents really made it less than desirable.

The first reprogrammable control store or *erasable PROM* (EPROM) control store microcontrollers (Fig. 1-9) were introduced in the early 1980s. These devices used ultraviolet light erasable memory cells that could be loaded with a program and then erased so that another application could be programmed into them.

The EPROM memory cell consists of a MOSFET-like transistor with a floating gate surrounded by silicon dioxide above the substrate of the device. Silicon dioxide is best known as glass and is a very good insulator. To program this bit, the control gate above the floating gate is raised to a high enough voltage potential to cause the silicon dioxide surrounding it to break down and allow a charge to pass to the floating gate. With a charge in the floating gate, the transistor is turned on (is conducting) all the time, until the charge escapes (which will take a very long time that is measured in tens of years).

Before programming, all the floating gates of all the cells are uncharged and the act of programming the control store will load a charge into some of these cells. By convention, if an unprogrammed memory cell is read, a "1" will be returned. After the cell is pro-

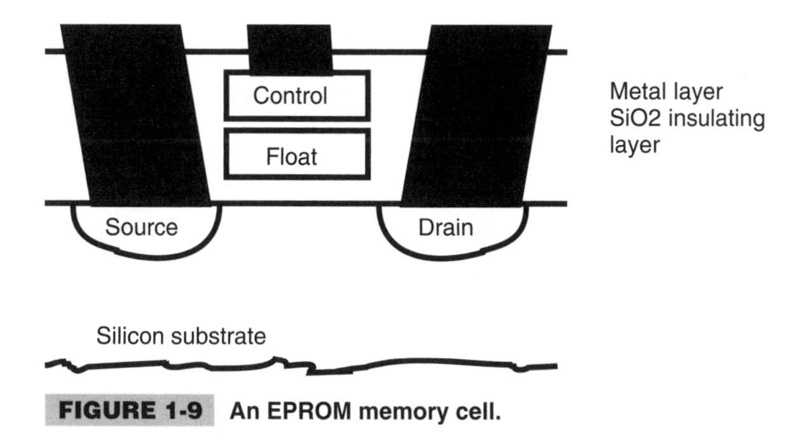

FIGURE 1-9 An EPROM memory cell.

grammed a "0" is returned. So, to see if a control store is ready to be programmed, each byte is read out of it and compared with 0FFh (all 8 bits set).

To erase a programmed EPROM cell, UV light energizes the trapped electrons in the floating gate to an energy level where they can escape. In some manufacturers' devices, you might discover that some EPROM cells are protected from UV light by a metal layer over them. The purpose of this metal layer is to prevent the cell from being erased. This is often done in memory-protection schemes in which critical bits, if erased, will allow reading out of the software in the device. By placing the metal shield over the bit, UV light cannot reach the floating gate and the programmed cell cannot be erased.

This might seem like an unreliable method of storing data, but EPROM memories are typically rated as being able to keep their contents without any bits changing state for 30 years or more.

Microcontrollers with EPROM control store can be placed in two types of packages. If you've worked with EPROM before, you probably have seen the ceramic packages with a small window built in for erasing the device. EPROM microcontrollers are also available in packages with no window and are known as *one-time programmable* (OTP).

OTP devices might seem odd when you consider that the advantage of the EPROM is its ability to be erased and reprogrammed. By taking away the window for UV light to access it, you might as well go with a PROM or mask ROM device.

OTP devices actually fill a large market niche. Windowed ceramic packages can cost 10 times what cheap plastic packages cost, and in most microcontroller applications and products, the device will never be reprogrammed. So, by using OTP packaging, the part can still be field programmed, will be electrically identical to the part used to develop the application, and is very cost effective for quantities less than the break-even point for mask ROM.

Using OTP parts has a significant advantage for the card manufacturer. If the manufacturer is building several products with the same device (which is not unheard of for microcontrollers), then keeping unprogrammed parts in stock and only burning the number required for the current build run can be a real advantage in terms of inventory carrying costs. Rather than keeping five or six of the same devices, at a minimum chip manufacturing build quantity in stock, the manufacturer just needs to keep the minimum number of parts on hand to satisfy their current orders for the various products.

An improvement over UV erasable EPROM technology is *electrically erasable PROM* (EEPROM). This nonvolatile memory is built with the same technology as EPROM, but the floating gate's charge can be removed by circuits on the chip and no UV light is required.

There are two types of EEPROM in use in microcontrollers. The first type is simply known as "EEPROM" and allows each bit (and byte) in the control store array to be reprogrammed without affecting any other cells in the array. This type of memory first became available in the early 1980s and found its way into microcontrollers in the early 1990s. EEPROM has been very successful when implemented in small, easy-to-access packages.

In the late 1980s, Intel introduced an improvement to EEPROM that was called *Flash*. The difference between Flash and EEPROM is Flash's use of a bussed circuit for erasing the cells' floating gates rather than making each cell independent. This reduced the cost of the EEPROM memory and reduced the time required to program a device. Rather than having to erase each cell in the EEPROM individually, the Flash erase cycle, which takes as long for one byte, erases all the memory in the array.

In the current marketplace, there is some confusion about the terms "EEPROM" and "Flash" with some microcontroller manufacturers labelling their EEPROM devices as "Flash" microcontrollers. It's not really a point to get very concerned over. EEPROM is not Flash; however, except for the difference in the programming time, there really is no practical difference between the two technologies.

8051 instructions can be 1, 2, or 3 bytes long. The control store memory array is only 8 bits (1 byte) wide. This means that a multi-byte instruction will require multiple control store read operations before it can execute. This means a multi-byte instruction will take longer to execute than a single-byte instruction.

Planning your application to use single byte instructions wherever possible will actually speed up execution of the application as well as make the control store in the device able to be loaded with more instructions (or a longer application). Later in the book, I will show you some strategies for choosing the most efficient instructions.

In this section, I have introduced you to a number of different types of control store that are available in different versions and different manufacturers' 8051s. There is a fairly simple convention that is used to identify what type of control store a device has.

I use the generic term "8051" to describe the devices I'm writing about in this book. Many other people use the generic term "8x51" where "x" is the control store type and is defined as shown in Table 1-1.

TABLE 1-1 DEFINING "8x51"

"x" VALUE	CONTROL STORE TYPE
0	None
3	Mask ROM
7	EPROM
9	EEPROM/FLASH

VARIABLE AREA

If you've spent some time programming PC applications, you've probably never worried about the space that variables and data structures take up. Most modern PC languages will allow just about unlimited direct storage. So, when you first looked at the 8051's variable storage (or *scratchpad RAM*) capacity, you probably wondered how complex applications could be written for the device.

Creating complex applications with limited variable RAM isn't that hard, although large arrays cannot be implemented without external memory. Throughout the book, I will present some very substantial applications without requiring external memory.

In a microcontroller, there are four types of internal variable data storage: bits, registers, variable RAM, and the program counter stack. In this section, I will focus on the first three types and discuss the program counter stack in the next section.

All variable storage in the 8051 is implemented as *static random-access memory* (SRAM), which will retain the current contents as long as power is applied to it. This is in contrast to the ROM used by the control store, which does not loose its contents when power is taken away. SRAM can be referred to as *volatile memory.*

Each bit in an SRAM memory array is made up of the six transistor memory cell shown in Fig. 1-10. This memory cell (probably known to you as a *flip flop*) will stay in one state until the write transistor is enabled and the write data is used to set the state of the SRAM cell.

The PMOS/NMOS transistor pair on the write side of the flip flop will hold this value as a voltage level because it will cause the PMOS/NMOS transistor pair on the read side to output the complemented value. This complemented value will then be fed back to the write side's transistors, which complements the value again, resulting in the actual value that had been set in the flip flop.

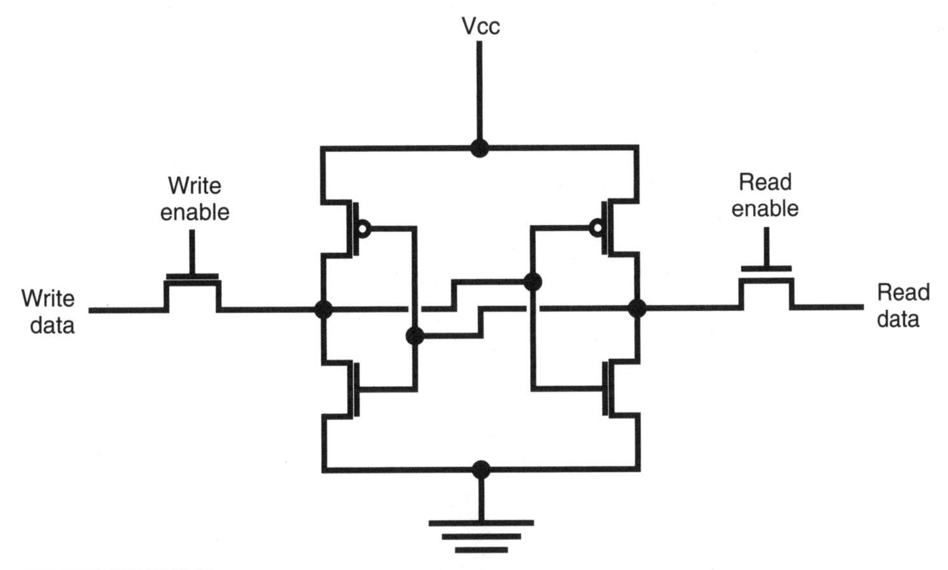

FIGURE 1-10 **Static RAM (SRAM) memory cell.**

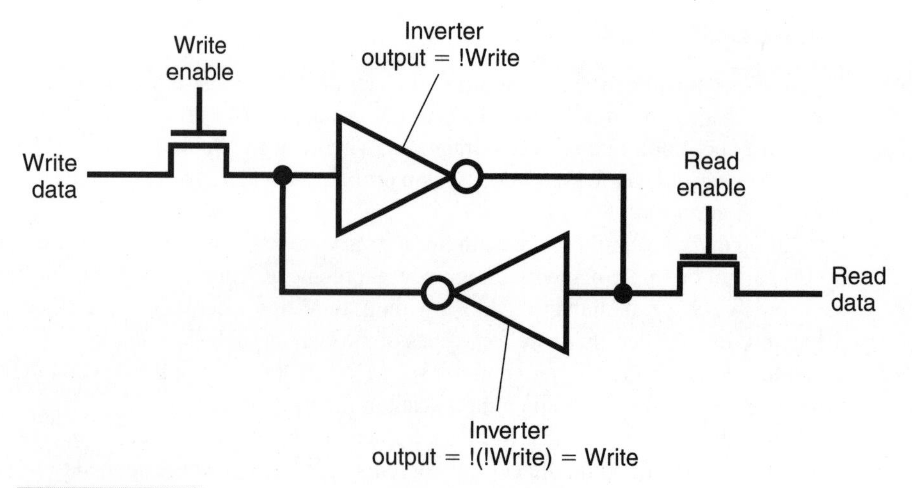

FIGURE 1-11 **SRAM memory cell block diagram.**

What this circuit really is is a pair of inverters feeding back to each other. (See Fig. 1-11.) Once a value has been set in the inverters' feedback loop, it will stay there until changed.

Reading data is accomplished by asserting the read-enable line and inverting the value output (because the read side contains the inverted write side's data).

This method of implementing a flip flop is well suited to a microcontroller because it uses very little power (current only flows when the state is changed) and is quite fast. It is not very efficient in terms of silicon space; the six transistors required for each memory cell actually take up quite a bit of surface area.

Bits might seem to be not that useful a data type if you are used to writing PC applications; however, in a microcontroller, they can really help make the application much more efficient by allowing fast manipulation of pin states, flags, or state variables. In PCs and workstations, these functions can be carried out in byte-level logical statements.

For example, setting a bit in a byte can be accomplished by the statement:

```
byte = byte | ( 1 << bit );          // Set the specified bit
```

It will take multiple instructions to execute this instruction. In a processor that is able to manipulate bits, this can be executed in a single instruction that takes up less control store and runs in a lot fewer instruction cycles.

Working with bits will probably not be something that comes easily to you. If you've been working with "typical" processors and hardware, you will probably be used to thinking in terms of byte-wise instructions. The 8051 can manipulate bits very easily and efficiently, although I often find myself thinking in terms of byte operations when I can use the bit instructions instead.

While instructions might make modifying individual bits seem like single bits are being modified, in some processors, an entire group of bits can be read, the desired bit modified, and the entire group of bits written back. Depending on how the bits are written back, this might cause a problem. In the 8051, all accessible bits are the only data bits that are manipulated; however, in other microcontrollers, this is not true. Before using bit set and reset instructions, make sure you understand how the processor is architected and operates on the bits.

When 8 bits are grouped together, they are known as a *byte*. The byte is really the basic unit of storage in most simple microcontrollers like the 8051. Memory bytes can be arranged as *registers* or *RAM* to the microcontroller's processor. The difference between registers and RAM might appear arbitrary; however, in many cases, they can be very crisply defined.

Registers are RAM locations that can be accessed very easily by the processor itself. Typically, there are only a few registers that are accessed by the processor, and often these require special instructions or addressing modes. Registers can also pull double duty, performing processor hardware tasks as well as just saving data bytes.

RAM can typically be accessed in a very arbitrary way with instructions that either specify absolute addresses or indexed addresses. RAM's only function is to save data and is not used for any additional functions. The first 128 addresses of the 8051's memory space is devoted to RAM.

RAM in the 8051 is known as scratchpad RAM because the external memory capabilities of the 8051 allow the addition of up to 64K of RAM for variable storage. This large addressing capability is often used in high-level languages as the primary variable storage area, and the first 128 addresses is used as temporary or scratchpad variables. Further confusing how the 8051 is architected is the availability, in the 8052 (an enhanced 8051), of an additional 128 scratchpad RAM bytes starting at address 080h of the 8051's variable address space and can only be accessed by index registers.

PROGRAM COUNTER STACK

Processor stacks are a simple and fast way of saving data during program execution. This is a bit of a dry explanation of stacks and does not really tell you what *you* would use them for.

Stacks save data in a processor the same way you save papers on your desk. As you are working, the work piles up in front of you, and you do the task which is at the top of the pile (Fig. 1-12).

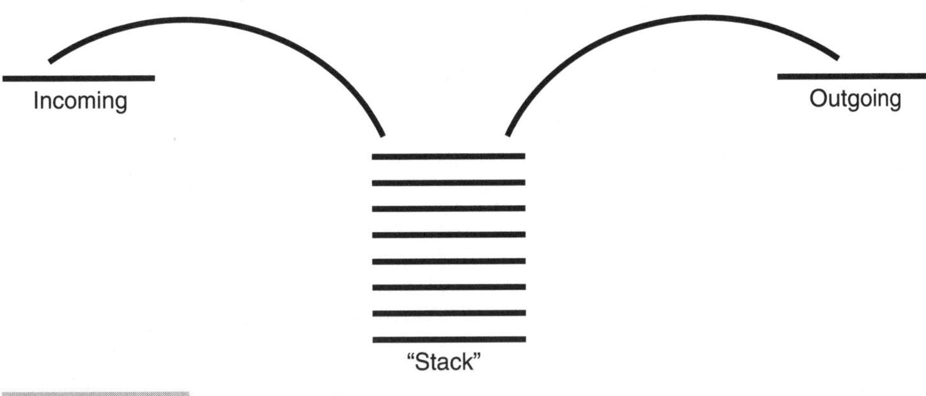

FIGURE 1-12 **Stack data flow.**

A stack is known as a *last in/first out* (LIFO) *memory*. This should be pretty obvious because the first work item on the stack of paper will be last one you get to.

Looking back over this example, I guess you can see that a good strategy for getting somebody to do some work for you in the shortest amount of time is to place the work on their pile just before they are about to start going through it.

In a computer processor, a stack works in exactly the same manner. Data most recently put onto the stack (this is known as a *push*) is the first item pulled off the stack (this is a *pop*). Pushing and popping the data on and off the stack is useful in the case of program execution where multiple subroutines are called from within each other:

```
main() {

    :

  i = Factorial( j );          // Get the factorial of "j"

    :

}  //  End main

int Factorial( int i ) {

  if ( i == 1 )
    return 1;
  else
    return i * Factorial( i - 1 );

}  // End Factorial
```

In this example, each time "Factorial" is called, it saves the current location of the program counter and jumps to itself again and passes one less than the value it was given (except in the case where 1 is passed to it and then it just returns 1). When the routine has finished, the `return` statement pops the return address off the stack and executes where it left off, just after the subroutine call.

If you wanted to find the factorial of 5, the "Factorial" subroutine would be called five times with the last call simply returning 1. Using a LIFO stack to save the return addresses, the program will always return to the correct place under the correct circumstances (to execute properly, execution has to return to the last saved value).

The "Factorial" subroutine is a special type and is known as *recursive* because it calls itself repeatedly. Recursive subroutines and subroutines that are called from other subroutines are known as *nested subroutines*. Before using nested subroutines, you should make sure you understand how much stack space is required and how much is available. Calling subroutines from within themselves can use up more stack space than has been made available for the application with disastrous results.

For operations like calculating a factorial of an integer, this is an excellent technique for simply calculating a value without having to create a loop. However, caution must be used because, if your routine cannot stop calling itself, you will find that all the stack space in your processor will be used up very quickly. If "Factorial" was implemented on an 8051, you would find that, each time it is called, 2, 3, or more bytes of stack memory are used. With only 128 bytes available, you might find that you have run out of stack space very quickly.

Along with return addresses for subroutines, stacks can also store data. In the factorial example, the parameter passed to the "Factorial" subroutine could be pushed onto the stack before the subroutine is called. There are other methods of passing parameters to a subroutine, but this method is particularly efficient because none of the processor's internal registers are used for passing and saving the data. This leaves a maximum number of registers free for data processing inside of the routines.

If you are passing parameters to a subroutine via the stack, care should be taken that, after execution returns to the calling routine, the stack is popped until it is at the value before the subroutine call. Failing to do this could cause the stack to be used up or change value in such a way as a subroutine cannot return to its caller.

The 8051 initializes the *stack pointer* (SP) *register* to start storing data at address 008h of the scratchpad RAM. The algorithms used are:

```
push( data ) {

  SP++;                       // Point to the next address in memory
  Stack[ SP ] = data;         // Store the data

} //  End push

int pop() {

int i;

  i = Stack[ SP ];            // Get data pointed to by the SP
  SP--;                       // Decrement the SP

  return i;

} //  End pop
```

When you first look at the 8051 and its documentation, you'll see that the stack pointer is set to 007h. The first actual address written to is 008h. If you follow the "push" routine shown earlier, you can see that the stack pointer is incremented before the memory write takes place.

HARDWARE INTERFACE REGISTERS (I/O SPACE)

Being able to interface with processor status and control registers, as well as peripheral registers, is important in any computer system. In a microcontroller, providing an efficient method of interfacing to the peripheral registers is critical because complicated interfaces and procedures will make the device more difficult to develop applications for and will take up valuable control store space.

Understanding how the registers are addressed in a processor architecture is an issue that affects how programs will execute. In Von Neumann (Princeton) architected processors, the register space can either be a part of the memory space or in a separately addressed I/O space (Fig. 1-13).

These two different ways of implementing the hardware registers are known as *memory mapped I/O* when the hardware interface registers are in the same memory space as the ROM and RAM. A separate I/O space can be implemented with its own addressing structure.

As an aside, the first computer processor architecture that used a separate I/O and memory space was the IBM System/360 introduced in 1964. Except for some Harvard architecture prototypes, all computers up to that time had a single address space in which ROM, RAM, and I/O were all addressed in the same space and used the same instructions for access.

The advantage of having a separate I/O space is that invalid program operations (i.e., when the computer is running amok) will be less likely to affect the operation of the hardware. However, it does come at the price of being more complex than simply placing the I/O registers in the memory map. By placing the hardware interface registers within the memory map, the hardware required for addressing is simpler and no special instructions are required for passing data back and forth.

For Harvard architectures (Fig. 1-14), there are three different ways of implementing the I/O space in the processor.

Because of the two separate memory spaces already available, there is the option of putting the I/O registers with the control store, with the variable RAM, or in it's own Space. In Harvard-architected microcontrollers, putting the I/O registers into the control store space might not be practical because there usually isn't any hardware or instructions available for directly reading or writing this address space. Adding a separate I/O space is implemented in some microcontrollers (such as the Atmel AVR) but does make the device more complex due to the requirements of additional hardware to address the space and implement the special instructions. In most Harvard-architected microcontrollers, including the 8051, the I/O registers are located in the variable memory space.

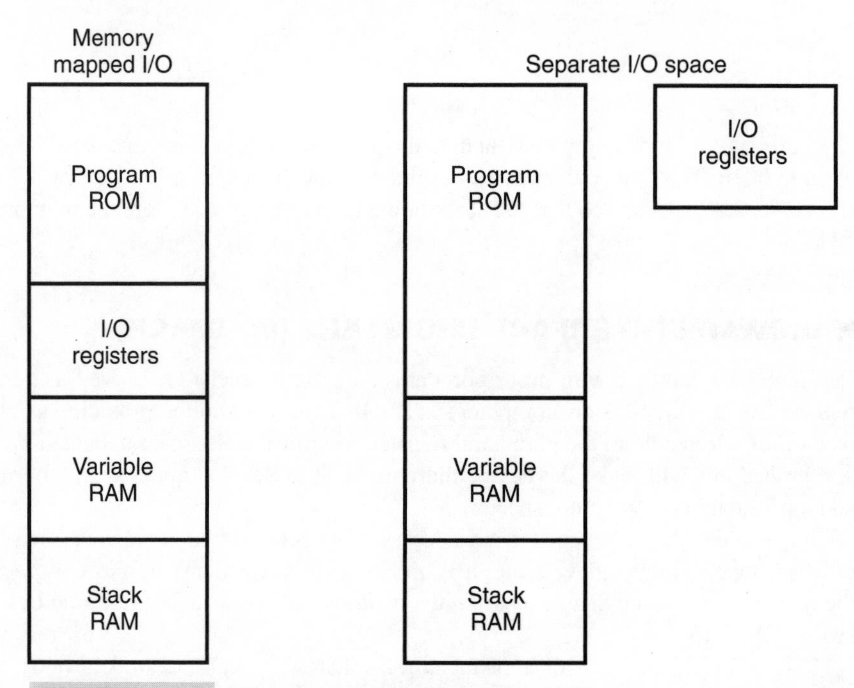

FIGURE 1-13 **Princeton architecture I/O registers.**

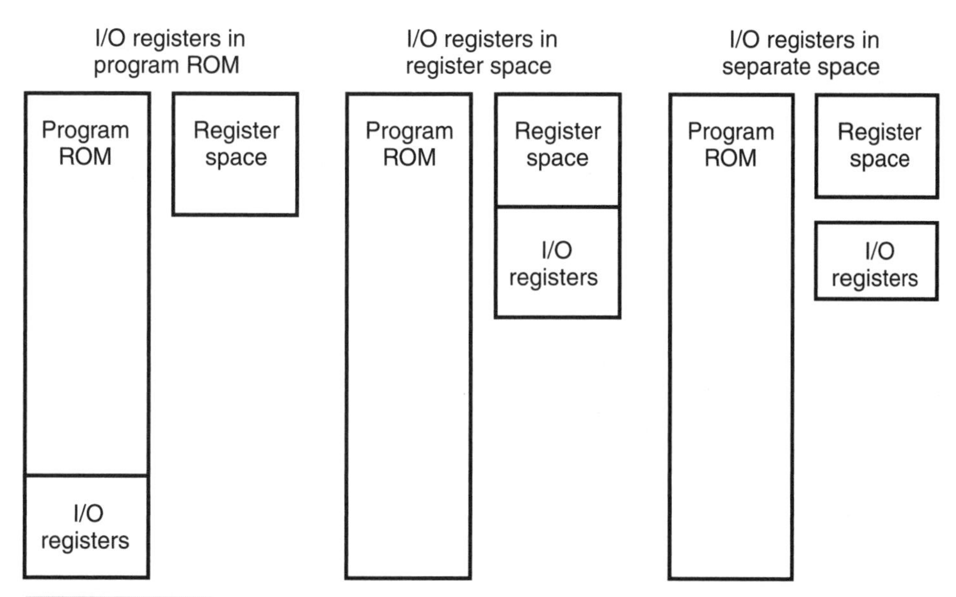

FIGURE 1-14 **Harvard architecture I/O registers.**

Microcontroller Features

At the beginning of this chapter, I introduced you to the idea that a microcontroller, like the 8051, consisted of a computer processor, all required memory, power, and reset functions. Along with these concepts, there are a number of features including how a microcontrollers interfaces with the outside world and actually executes its programs that you should understand. Not only are these features available in the 8051, but they are also available for all microcontrollers and computer processors.

CLOCKING

Providing operating resources to a microcontroller is generally pretty easy, with input power, reset, or clocking that would seem downright primitive compared to more conventional computer systems.

Maybe the word "primitive" is the wrong one to use; maybe describing microcontrollers as being *robust* with respect to their power, reset, and clocking would be more accurate. The reason for this built-in robustness is the variety of different environments microcontrollers will be expected to work in. Later in this book, I will show how the 8051 can be connected to a variety of power, clocking, and reset sources and still run. In this section, I want to introduce you to some of the issues related with how a microcontroller is clocked and how programs are timed.

Most microcontrollers are capable of running from a clock that is not running at all to several tens of MHz. On most microcontrollers, there is built-in circuitry to allow a simple connection of a crystal or other hardware, such as ceramic resonators or an external clock

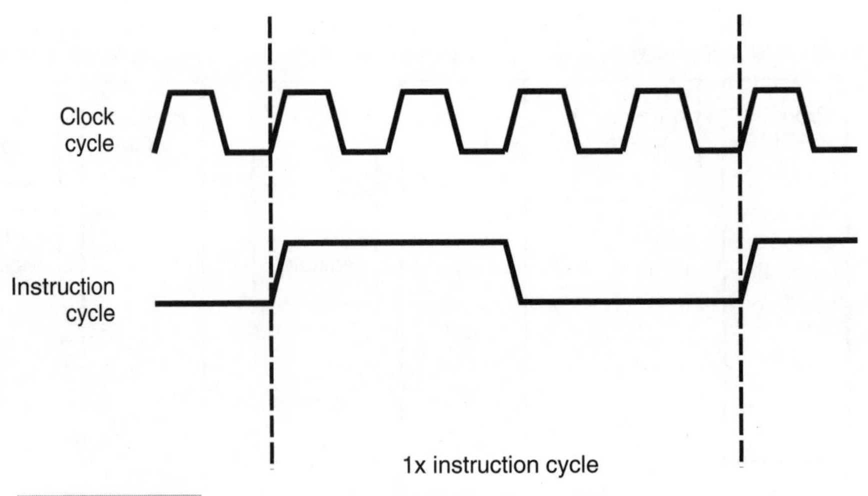

FIGURE 1-15 **Instruction cycle versus clock cycle.**

source. The microcontroller might also have an internal *ring oscillator* built in, allowing it to run without any external components other than power.

As you would probably expect, the microcontroller's clock is used to sequence the execution of the instructions. For each instruction, a specific number of *instruction cycles* is required for the instruction to execute.

Each instruction cycle is made up of a number of clock cycles (Fig. 1-15). In the "true" 8051, each instruction cycle takes 12 clock cycles to execute. In some 8051 part number data sheets that list the instructions, you'll see that most instructions take one or two cycles. This might lead you to believe that, if you run the 8051 at 2 MHz, you can expect the device to run at over one MIPS (million instructions per second) when, in actuality, the 8051 is probably executing at somewhere around 150,000 instructions per second.

Requiring 12 cycles for each instruction is somewhat unusual in the microcontroller world, with some of the newer microcontrollers currently on the market capable of running at one clock cycle per instruction cycle.

Some 8051 manufacturers have redesigned the 8051 processor core (instruction execution unit) so that instructions will run faster (i.e., each instruction cycle run in few clock cycles). In this book, I will present a number of applications developed for the Dallas Semiconductor HSM (high-speed microcontroller) devices. These 8051-compatible microcontrollers require 4 clock cycles per instruction cycle with some instructions taking a different number of instruction cycles. The instruction execution speed is approximately 2.5 times that of a "standard" 8051.

As well as the Dallas Semiconductor HSM parts, Intel has the MCS-151 and MCS-251 and Philips Semiconductor has the "XA" architectures, each of which run the 8051 instructions in fewer clock cycles.

The variance in instruction execution has lead to the development of some techniques for making sure that applications can be run on a variety of different 8051 types. These techniques center around the use of keying off the clock cycle and not the instruction cycle. This subtle, but important, difference will come back to haunt you if you try to run an application written and debugged on one manufacturer's 8051 compatible device and attempt to port it to another's.

As one last comment on clocking microcontroller applications, I want to note that you should design the application to run on the slowest microcontroller clock speed possible. The primary reason for doing this is to minimize the power consumed by the microcontroller. This, of course, leads to its own philosophies of coding. However, in many applications where the microcontroller is powered by a battery, by keeping the microcontroller's clock running as slowly as possible, you will minimize the power consumed by the microcontroller (from mAmps to tens or hundreds of uAmps) and minimize the amount of electrical noise generated by the application.

This feels like it goes against the grain with me. I've worked on PCs for a number of years, and the goal of working with PCs is to always increase the speed the computer works at. In the microcontroller world, the "need for speed" isn't as great and might not be appropriate in some applications.

I/O PINS

When I was in university, I designed my first computer system around a Z-80. I originally wanted to provide a simple input-and-output (I/O) capability on the computer system (like all students, I had visions of making the house completely electronic and basically not have to get out of bed). To start off, I wanted to provide eight simple inputs and outputs that I could control from the system's software.

Rather than buying a commercial I/O device (like an Intel 8255 or Zilog PIO), I decided to design my own out of discrete logic. My solution was, to my thinking, reasonably efficient; it consisted of a '374 (with its output control always enabled) controlling the output drivers of another '374. I used a '244 for reading back the state of the pin. (See Fig. 1-16.)

This method of providing I/O is almost exactly how virtually all microcontrollers except the 8051 usually implement I/O. The major difference between my method and how parallel I/O is implemented in microcontrollers is that each bit's output driver enable is usually controlled by a single bit in the control register. In my implementation, only one bit in the control register controls the output driver of all eight bits.

As well as being individually set as output or input, many microcontroller I/O pins are dedicated to specialized functions that further complicates them. These specialized functions include serial ports, analog I/O, external device buses, etc.

In some applications, you might want to output a signal on a *dotted-AND bus*. This type of bus utilizes a number of transistors pulling down line to ground using open collectors (in the case of bipolar logic) or open drain (in the case of NMOS or CMOS logic). In Fig. 1-17, the bus is pulled up with a number of transistors, controlled by control signal "Cn," any of which can pull the line low.

When any of the transistors in this circuit are turned on, the bus pin will go low. Only when all the transistors are turned off will the bus be high. Because this bus circuit will only be high when all the transistors are off, or their functions are high, this is known as a "dotted AND".

Many microcontrollers provide pins with this characteristic output to allow them to be used on busses which require these features. Many bus protocols, such as I2C, which I will introduce later in the book, require open collector circuits for correct operation.

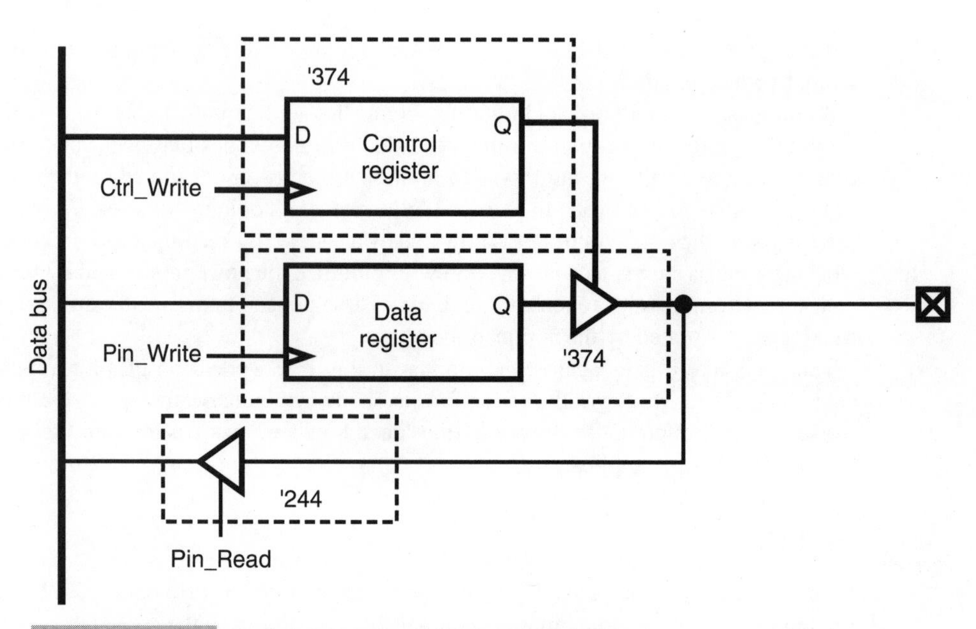

FIGURE 1-16 **Standard bidirectional parallel I/O pins.**

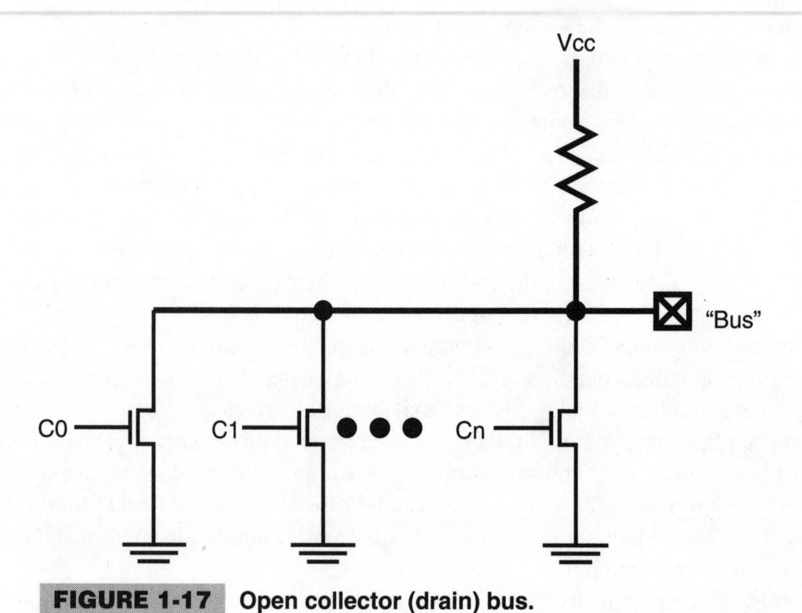

FIGURE 1-17 **Open collector (drain) bus.**

The open collector circuit's operation can be simulated by turning off the output driver except when a low voltage is to be output (which will pull down the bus). However, this method requires somewhat more complex software than if just a straight open collector was provided as the output.

The 8051 is unusual in that all I/O pins are designed as open-drain outputs when used as parallel I/O pins. As I will present later in the book, most of the 8051's I/O pins have a weak built-in pull up. (See Fig. 1-18.) Using open drain I/O pins exclusively in the 8051 means that you have to plan for adding external pull ups for some applications and only sinking current, never sourcing it.

In this type of I/O pin, note that there isn't an output control. If the pin is to be used for input, then a "1" has to be loaded into the output data register so that the 8051 does not drive the pin state.

INTERRUPTS

For people just starting out with microprocessors or microcontrollers and just learning assembly language, interrupts are usually lumped into the category of things that are best left alone until expert status is attained. This attitude arises from how basic assembly language is taught, with interrupts being "filler" material at the end if the "core" material has been presented before the end of the semester. In computer architecture courses, the mechanics of interrupts are presented, but no working code or examples are given. The result is that many college and university students graduate without really understanding how interrupts work or even having gone through the effort of creating the code to handle them.

This attitude is further reinforced when somebody develops applications for the IBM PC or other workstations. Using interrupts in the PC is something of a challenge and best left to the experts because you have to interface not only with the hardware (and you might have multiple devices from different applications requesting the same physical interrupt) but also with the operating system (which seems almost impossible to interface with).

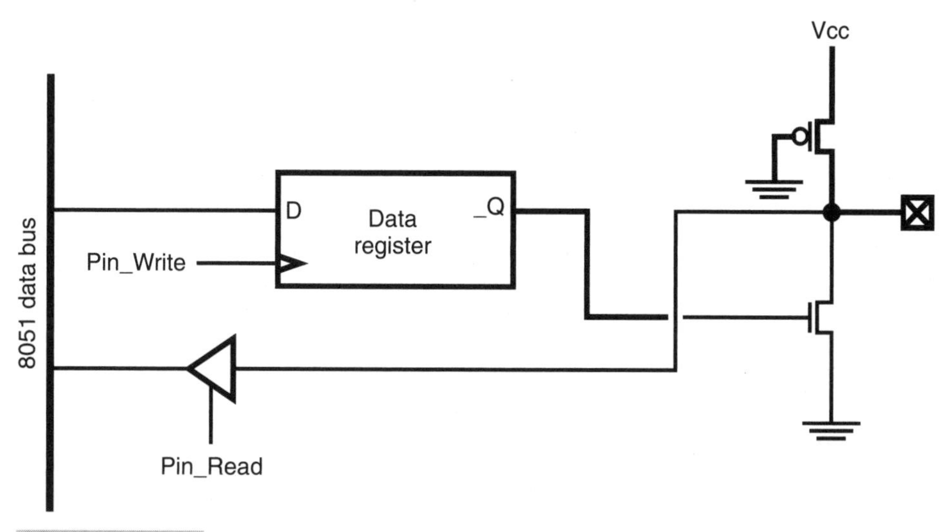

FIGURE 1-18 8051 parallel I/O pins.

With this kind of background, it's not surprising that, when many new engineers develop applications for computer systems (including microcontrollers), they shy away from using interrupts within the application.

This is very unfortunate because interrupts can be very useful in applications, allowing the code to respond much more quickly to input stimulus, and can make the code simpler as well as shorter than an application "polling" the inputs and responding when a change is detected. All of these attributes make using interrupts attractive for working in devices like the 8051, which has a limited amount of control store and will run at a set speed.

Before reviewing how interrupts are implemented in a computer system, I want to review how interrupts work in your own life. You are constantly interrupted, and understanding how you handle the requests that are made of you is analogous to how they are handled in a computer system.

Interrupts are defined as requests because they can be refused (or *masked*). If they are not refused, then when an interrupt request is acknowledged (either by you or a computer), a special set of events or routines are followed to handle the interrupt (Fig. 1-19).

These special routines are known as *interrupt handlers* or *interrupt service routines* and are located at a special location in memory (both yours and the computer system's). The mainline code does not resume executing until the interrupt has finished being serviced and the processor (or your brain) indicates that the originally executing function can resume.

In real life, suppose you were watching an episode of classic "Star Trek" on your VCR and Captain Kirk, Mr. Spock, and Dr. McCoy have just beamed down to a uncharted

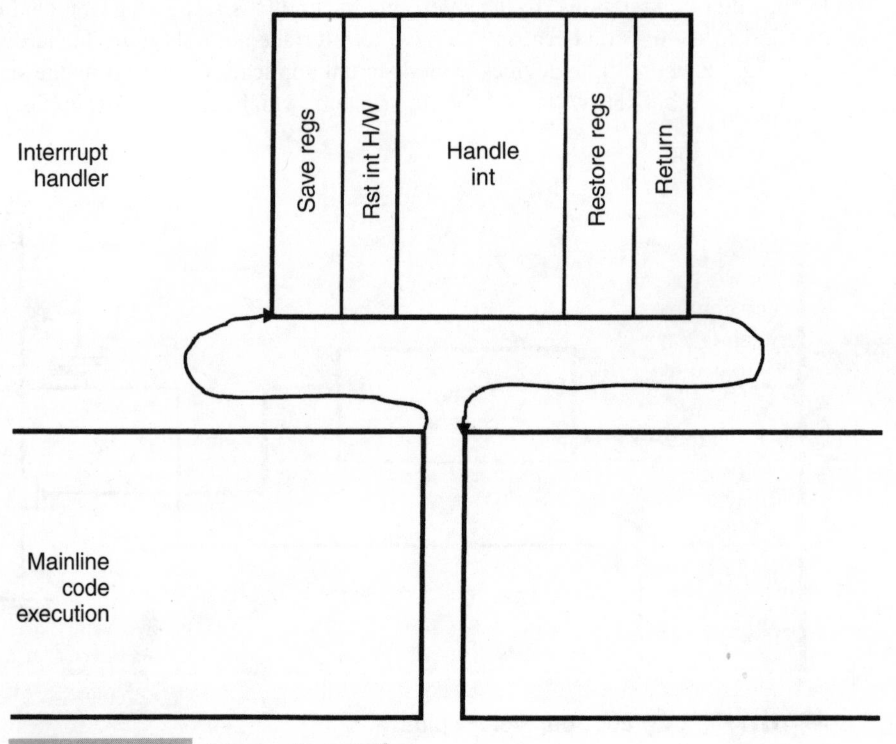

FIGURE 1-19 Interrupt execution.

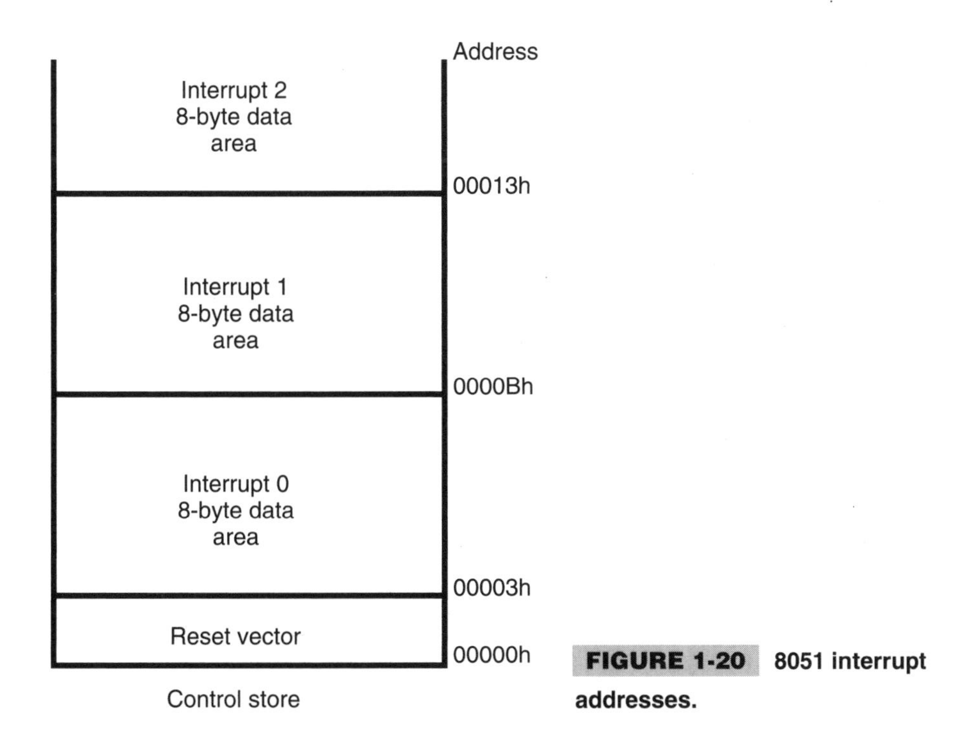

FIGURE 1-20 8051 interrupt addresses.

planet with two security guards. Following the "Red Shirt Rule," one of the security guards is immediately hit by some kind of ray, glows bright green, and whistles as he fades from view. In response Captain Kirk whips out his phaser and.... The telephone rings.

How do you handle this interruption?

There are three different ways that you could respond, each one analogous to how interrupts are handled in a computer system.

The first method of handling the interrupt is to ignore it—the TV show is more important than any call you could be getting. After the show, you can check your answering machine and see who called. (In this case, I highly recommend that you say you were washing your hair and couldn't hear the phone ring; I doubt very many people would be sympathetic to your feeling that a "Star Trek" episode is more important than talking to them.)

The next way of responding to the interrupting phone call is to answer the phone, take the caller's name and number and tell them that you will call them back at the end of the show. This is also carried out in a computer system implementing a fast interrupt handler: just setting a flag or incrementing a counter to indicate that an interrupt occurred to keep the impact of the interrupt on the speed of the mainline's execution to a minimum.

The 8051 is ideal for implementing such a type of interrupt. In some cases, I lament the lack of a zero flag, but the 8051 is uniquely suited to allowing very simple interrupt handlers that run very quickly. These features include a separate vector for each type of interrupt along with space (8 bytes) devoted to each vector to allow a small interrupt handler to be located without having to jump to other locations. (See Fig. 1-20.)

The third method of handling an interrupt is to process it as it comes in. In the "Star Trek" example, this would be analogous to stopping the video tape and taking the call. When the call was finished, then you could finish off watching the episode.

When an interrupt request is acknowledged by a computer system, the following actions are taken to service the interrupt.

1. Save the context register information.
2. Reset the hardware requesting the interrupt.
3. Reset the interrupt controller.
4. Process the interrupt.
5. Restore the context information.
6. Return to the previously executing code (the mainline).

Context registers are the registers used by the interrupt handler during the operation of the code. For example, when an interrupt was requested and acknowledged, if the mainline was about to run the instruction:

```
addc     A,@R0
```

the context registers that would need to be protected by the interrupt handler before it changed those registers would be:

- The accumulator
- The status register (which saves the carry flag)
- The contents of "R0," which is pointing to the data to add to the accumulator.

If any of these three registers are changed without being saved and restored, then the result of the instruction will be incorrect.

The context information is typically saved on the stack as a series of `push` instructions. Normally, all registers in a processor are saved; however, in an 8051, the range of registers accessed can be hard to generically bound, so some care is required to only save those registers that are modified within the interrupt handler.

Once the context registers are saved, I reset the hardware responsible for making the interrupt request, followed by the interrupt controller hardware. In the 8051, interrupt requesting hardware and the interrupt controller are often reset by just the execution of the `reti` (return from interrupt) instruction.

Ideally, processing the interrupt should be as fast (i.e., taking up as few instruction cycles) as possible. By keeping the interrupt handler's execution very short, the probability that repeating interrupts will be missed or another interrupt not be serviced in a timely manner is reduced.

Interrupts can occur at different priority levels. Going back to the "Star Trek" example, what would happen if you were interrupted by the phone *and* a knock at the door at the same time? You might deal with the door first, ask the person to wait for a second while you answered the phone.

Occasionally, interrupts in computer systems might be requested at the same time, and you will have to design your application to not only service both interrupts, but you also might want both interrupt handlers to execute at the same time (which is known as *nested interrupts*).

In the 8051 and most other computer processors, different interrupt sources are given different priorities. A higher priority interrupt will be serviced before a lower priority one.

Nesting of interrupts means that interrupts are re-enabled inside an interrupt handler. If another interrupt request codes in while the first interrupt handler is executing, processor execution will acknowledge the new interrupt and jump to its vector. To avoid this, you might want to disable (mask) interrupts during the handlers that you think should have the highest priority and should never be interrupted.

Before allowing nested interrupts, the requesting and interrupt-control hardware have to be reset. By doing this right at the start of the interrupt handler, nested interrupt requests are allowed to execute very early in the current interrupt handler's execution.

TIMERS

One of the most critical functions required by all computer systems is a timer. Along with providing real-time information interrupts to the processor, timers used in microcontrollers (and microprocessor systems) also provide a number of other tasks helpful in the operation of the application.

A basic computer timer (Fig. 1-21) consists of a counter that can be read from or written to by the processor and is driven by some constant frequency source. A typical enhancement is to provide an interrupt request from an overflow. The counter's clock source is typically either the microcontroller's clock or an external source.

The counter itself is usually 8 or 16 bits wide. Typically, the value in the counter can be read from or written to during the processor's operation. When the counter overflows ("OF" in Fig. 1-21), an interrupt request is made of the processor.

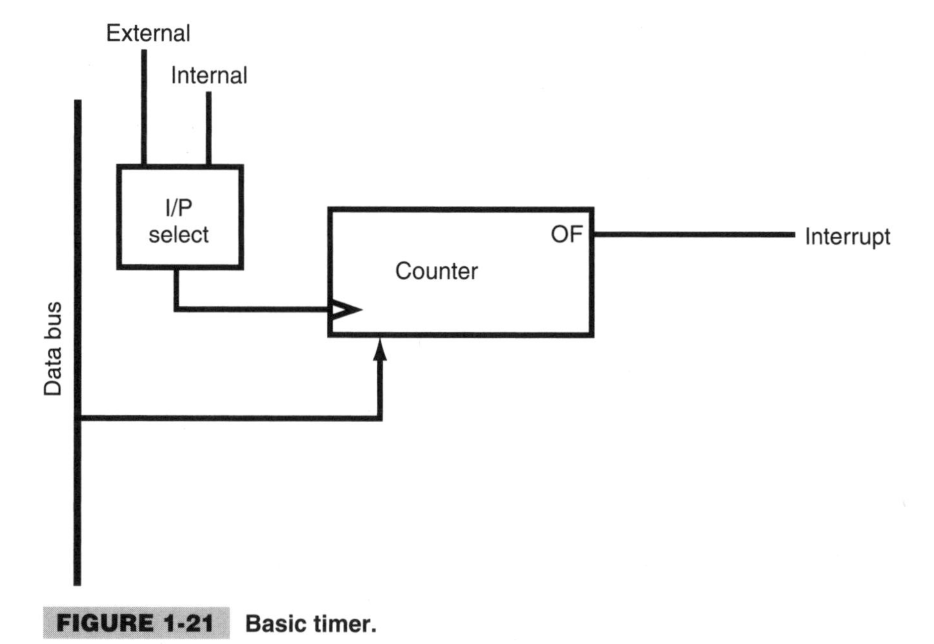

FIGURE 1-21 Basic timer.

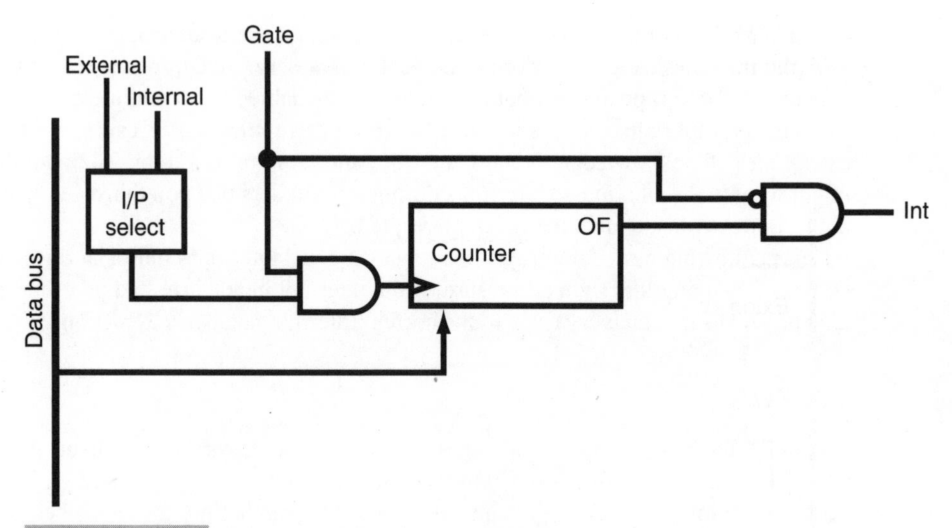

FIGURE 1-22 **Timer used for pulse measurement.**

From this basic circuit, enhancements are often used with the clock, including circuitry to measure incoming pulse widths, provide repeating interrupt at a specific interval, or output a pulse width modulated signal. These enhanced features are usually implemented as enhancements to the basic circuit shown in Fig. 1-21.

To measure pulse widths, an external control or *gate* input is used to mask the clock source except when the pulse is active. (See Fig. 1-22.) When the gate becomes high (the pulse is active), the counter's clock source is allowed to increment the counter. When the pulse becomes inactive (low), the counter's clock source is masked and an interrupt is requested to allow the application to read the counter value that is proportional to the width of the pulse. In some circuits, you might see that the counter is reset when the pulse becomes active, so the value is correct for each pulse.

If the count overflows during the active pulse, the processor is also notified via an interrupt request, and it can increment a word counter used for sensing the pulse width.

To add a specific delay interrupt, the contents of the counter are reloaded with a set value when the counter overflows, rather than be reset back to zero. (See Fig. 1-23.) To determine the interval between overflows (in which the interrupt request is made and the timer is reloaded), the following formula is used:

$$\text{Interval} = \frac{(\text{CountMax} - \text{Reload})}{\text{ClockFreq}}$$

CountMax is the overflow value of the counter (for an 8-bit counter, it would be 256, and for a 16-bit counter, it would be 65,536). The *ClockFreq* is the frequency in which the counter is updated (this is important to understand because, in the 8051 using the internal clock, this is the clock input to the 8051 divided by 12).

This circuit is often used for driving peripherals that need a constant clock. In the 8051, this circuit is used to drive the asynchronous serial ports' data rates.

A PWM signal (Fig. 1-24) consists of a repeating signal with a pulse of a given *duty cycle* (the fraction of the cycle that the pulse is active). This fraction can be anywhere from 0% to 100% of the cycle. PWM controls are often used for controlling the speed of a DC motor.

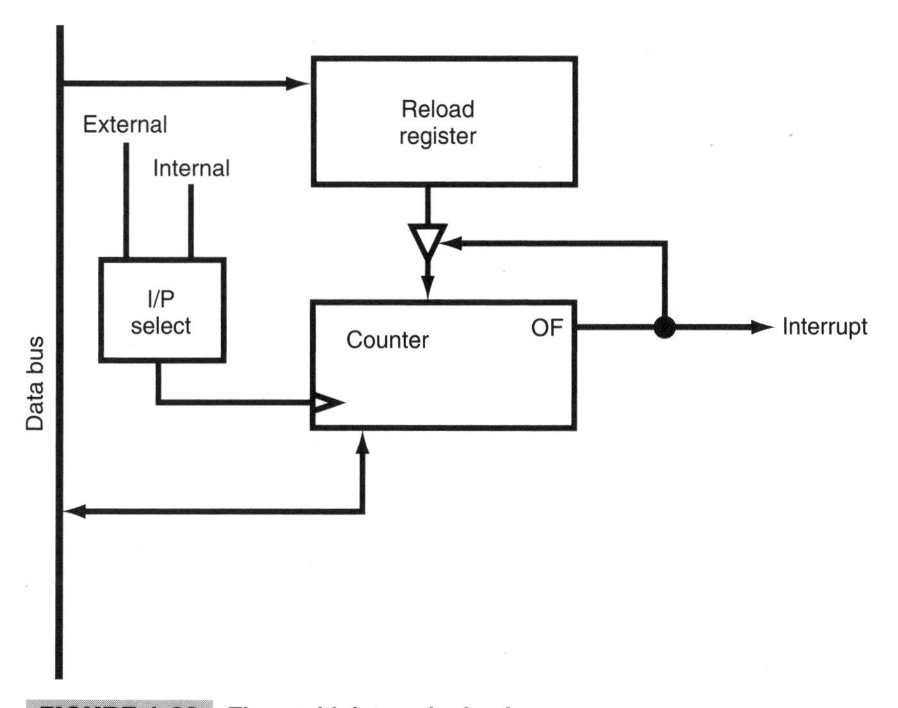

FIGURE 1-23 **Timer with interval reload.**

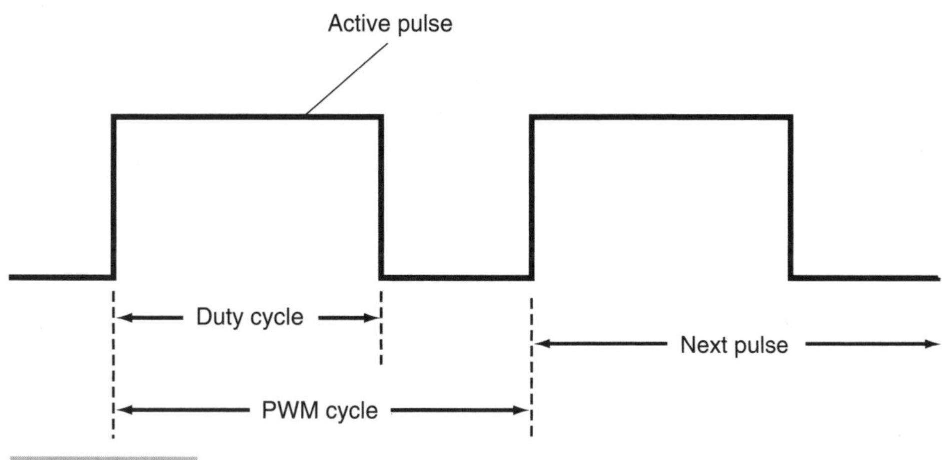

FIGURE 1-24 **PWM signal output.**

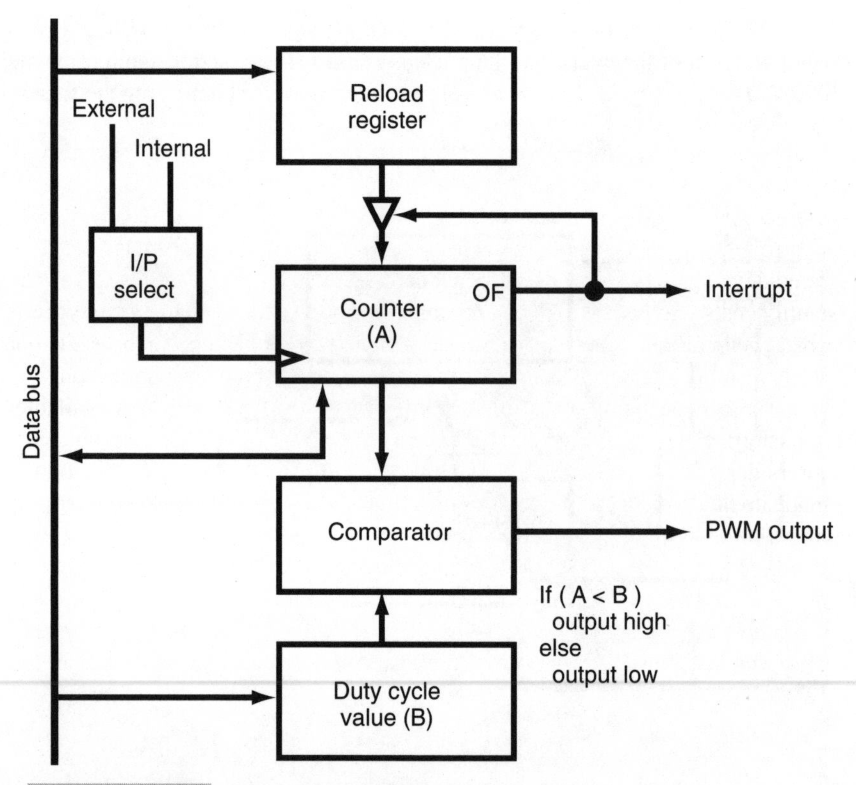

FIGURE 1-25 **Timer with PWM signal output.**

To output a pulse-width-modulated signal, the timer circuit in Fig. 1-23 is enhanced to have the current clock value compared to a set value (Fig. 1-25).

When the timer is less than the set value (the duty-cycle value), the comparator will output a high value. When the counter is equal to or greater than this set value, a low value will be output. To program this counter, the PWM value must be greater than the reload value to ensure that an active signal will be output.

The "standard" 8051's timers do not support the pulse input measuring or pulse-width-modulated outputs, but does support a number of operating modes that will allow you to implement these features in your applications very easily.

PERIPHERALS

So far, I've described a microcontroller as a device that consists of a computer processor, program memory, RAM, I/O pins, interrupts, and timers. This list is what I would consider to be "basic" for a microcontroller. Starting from this list of features, many microcontroller manufacturers add additional peripherals to create devices suited for particular applications.

These peripherals include interfaces to microcontroller network devices to simplify wiring applications and not depend on using an external memory interface. These interfaces allow a microcontroller to be simply wired to digital devices—such as EEPROM memory, thermometers, analog-to-digital converters (ADCs), etc.—and other microcontrollers.

Rather than including a new bus with the microcontroller, many peripherals are built directly onto the chip. Right now, there are literally hundreds of different 8051 part numbers, each devoted to providing different features for different applications. More are being designed all the time.

Microcontrollers with enhanced features might appear to be more costly than basic microcontrollers and external hardware peripherals, but there are advantages in using an integrated device in terms of costs (the raw card used for the application will be smaller and assembly costs will be less), wiring complexity, power requirements, quality (fewer parts means fewer things to go wrong), and performance. These factors can be easily quantified to help you make the best decision for the parts to be used in your application.

Along with there being a real plethora of differently featured 8051s available, if you have custom requirements, you could have a custom 8051 designed for you. Most manufacturers design their microcontrollers and other chips using *macros* (i.e., an 8051 could be made up of a processor, EPROM, RAM, timer, power, and serial I/O macros rather than having a designer create each function from scratch). If you need a microcontroller that has a custom set of peripherals, you can get it designed and built in a few weeks more than what's required to get a mask programmed device built.

Alternately, you could choose an ASIC technology that has the 8051 core and other peripheral macros available, which would allow you to add the microcontroller to an ASIC that is already being designed. Doing this could radically improve the final end product.

The 8051

When you first think of Intel, you tend to think of it as a company totally devoted to PC systems. Along with developing the first microprocessor (the 4004) in the 1970s, the company was involved right from the beginning with developing microcontrollers. The microcontroller development effort resulted in the 8051 architecture, which was first introduced in 1980 and has gone on to be arguably the most popular microcontroller architecture available.

The 8051 is a very complete microcontroller with a large amount of built-in control store (ROM and EPROM) and RAM (respectably large even for today), enhanced I/O ports, and the ability to access external memory. The original 8051 was also quite fast, with a 20-MHz maximum clock speed.

Whereas most of the other microcontrollers measure their success in terms of total units sold, the 8051 can claim an unusual honor as being, far and away, the device that is manufactured by the most companies. This is actually quite an advantage for the part because each manufacturer has enhanced the 8051 in terms of features or speed and has made the architecture very accessible to a wide range of applications.

As I go through this book, I will primarily focus on what I will call the "original," "classic," "stock," or "true" 87C51 and the Dallas Semiconductor "High Speed Microcontroller" processor enhancement to the 87C52. The 87C51 is a version of the 8051, with the "7" indicating that it has EPROM control store and the "C" indicating that it is a CMOS part.

The standard features of the 87C51 are:

- 24-MHz clock speed
- 12-clock cycle per instruction cycle
- 4 Kbytes of control store
- 128 bytes of RAM
- 32 I/O lines
- Two 8/16-bit timers
- Multiple internal and external interrupt sources
- Programmable serial port
- Interface for up to 128 Kbytes of external memory

8051 Suppliers

Right now, there are more than 20 vendors building their own versions of the 8051. These versions range from being pin-, code-, timing-, and feature-compatible with the 87C51 to having significant enhancements to the original design in terms of speeding up program execution or adding different features to the part. For some devices, the enhancements actually consist of providing *fewer* features to the 8051 reference standard to make more cost-effective parts for applications not requiring advanced functions. The original 8051 might seem rather featureless when compared to the varied part numbers of the PICMicro and 68HC05; however, when the range of features from all manufacturers is tallied, the 8051 has at least as diverse a feature set as any microcontroller.

Even though they are the originator of the 8051 architecture, Intel has not made a lot of changes to the original device. The most significant changes made to the 8051 are the MCS-151 and MCS-251 microcontrollers. These pin- and code- compatible devices offer significant performance improvements over the original 8051 and 87C51 devices.

Atmel, which is a newcomer to the microcontroller field, offers "shrunk" 8051 compatibles (along with its line of AVR microcontrollers), which are available in a 20-pin package. This smaller version of the 8051 has made the architecture more accessible to smaller/cheaper applications. The Atmel versions of the 8051 also use EEPROM technology for control store and nonvolatile data storage, which makes a useful part for education and hobbyists to experiment with the 8051.

I will focus a lot of attention on the Dallas Semiconductor "High-Speed Microcontrollers" (HSM) devices for the example applications. With the HSM, Dallas Semiconductor, more than any other 8051 vendor, has lead the way to improved speed in the operation of the architecture. The HSM 8051s compatible microcontrollers execute code two to three times faster than the "stock" parts.

One of the most interesting microcontrollers available is the Dallas Semiconductor encrypted (or soft program) microcontrollers. These devices will allow the download of an application and store it in an encrypted format in an external SRAM. This chip is very useful in remote-sensor applications where the microcontroller or data might not be physically secure.

Philips Semiconductor has the most feature-rich collection of 8051-based microcontrollers of any vendor. When developing an application that will require external peripherals, I highly recommend that you check out Philips' catalog of parts because there is a good chance that there will be a Philip's part number that will do exactly what you are looking for without any external devices. Along with the wide variety of "standard" 8051 parts, Philips also has the XA architecture, which is an enhancement to the original 8051 architecture to allow 16-bit data processing.

One of the interesting developments of the early 1980s was the creation of piggy-backed microcontrollers and microprocessors. With these devices, a standard PROM or ROM was literally plugged into the back of the processor chip. The advantage of this method was that the processor chip would not have to be removed from a product when the code was updated. The combination would be cheaper because they wouldn't be built with hybrid technologies (logic and EPROM/ROM require different manufacturing processes). Another advantage of this method is fewer pins and board space used for external memory. These advantages are reflected in some of Oki's 8051 products, which are built without control store but will take an EPROM/ROM chip to plugged into the top of the chip.

2

8051 PROCESSOR ARCHITECTURE

CONTENTS AT A GLANCE

The 8051 processor architecture is Harvard-based with external memory read/write capabilities built in as part of the architecture. The 8051's architecture is really a "typical" MCU architecture. If you compare the architecture to the AVR or the PIC, you shouldn't have any problems with understanding how the 8051 works relative to these parts. However, if you are expecting the 8051 to work like the 8080 or 8086 (i.e., Intel microprocessors), you are in for quite a shock.

The CPU

The basic 8051 architecture, from a high level, isn't all that different from the other Harvard architectures presented in this book. With the architecture diagram shown in Fig. 2-1, the basic 8051 should look like the Harvard architecture I presented earlier in the book.

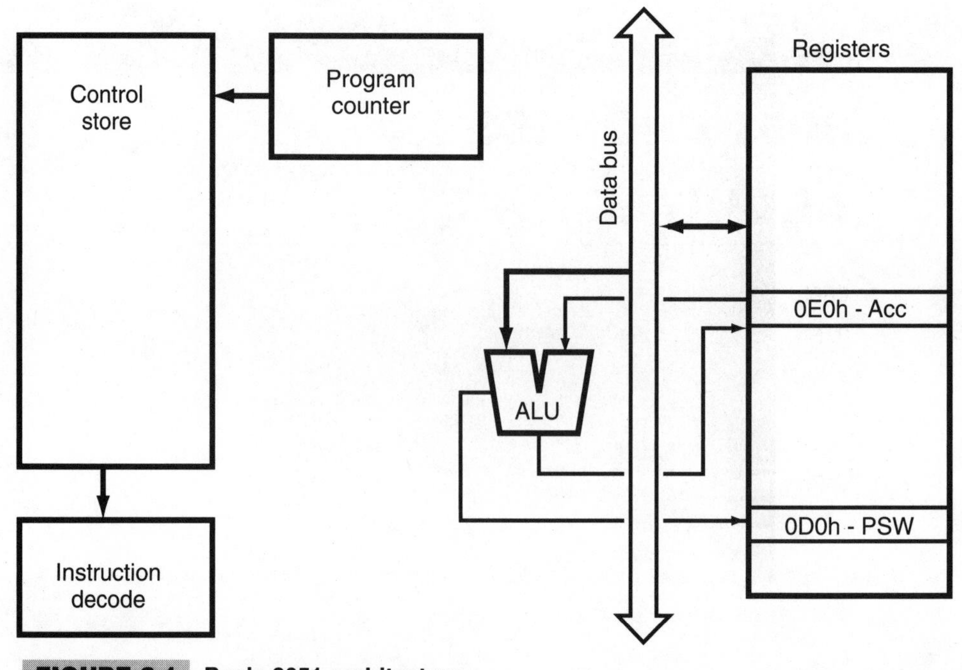

FIGURE 2-1 Basic 8051 architecture.

For the most part, the 8051 isn't a difficult device to program. However, there are a few quirks that you need to understand before you will feel comfortable with the design. The two areas that I will expand upon are the memory organization in general (including the control store and register boxes in Fig. 2-1) and how registers are implemented specifically.

To explain how the registers work, I'm going to apply a magnifying glass to the register box from Fig. 2-1. (See Fig. 2-2.) I have broken up the first 256 addresses by how the registers are accessed by *direct addressing* (which is when the register address is specifically given in the instruction) and *indirect addressing* (when the register address is located within an index register).

You might note that I have marked the direct addresses 020h to 02Fh differently in Fig. 2-2. These 16 bytes can be accessed as 128 bits as well as bytes. Some of the bits in the special-purpose registers (in address 080h and above) can be accessed directly, although for others they could be loaded into the bit addressable area (addresses 020h to 02Fh), modified, then stored back into the special-purpose (I/O) register area if individual bits were to be modified. The 128 bits can also be accessed as 16 regular RAM bytes.

If you directly address 080h to 0FFh, you will be accessing the special-purpose (I/O) register area of the 8051. This register area is reserved for processor and I/O peripheral registers. (See Fig. 2-3.)

The shaded areas are addresses that might or might not have hardware registers for controlling peripheral functions. The areas that aren't shaded are the standard registers that are available in all 8051s. I have refrained from extending the standard register definitions to anything more than an absolute minimum because, with all the different

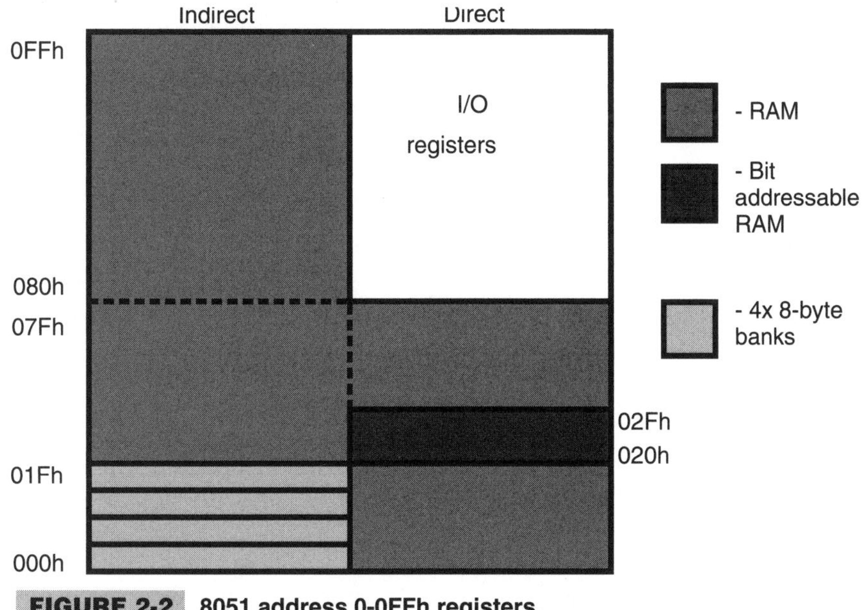

FIGURE 2-2 8051 address 0-0FFh registers.

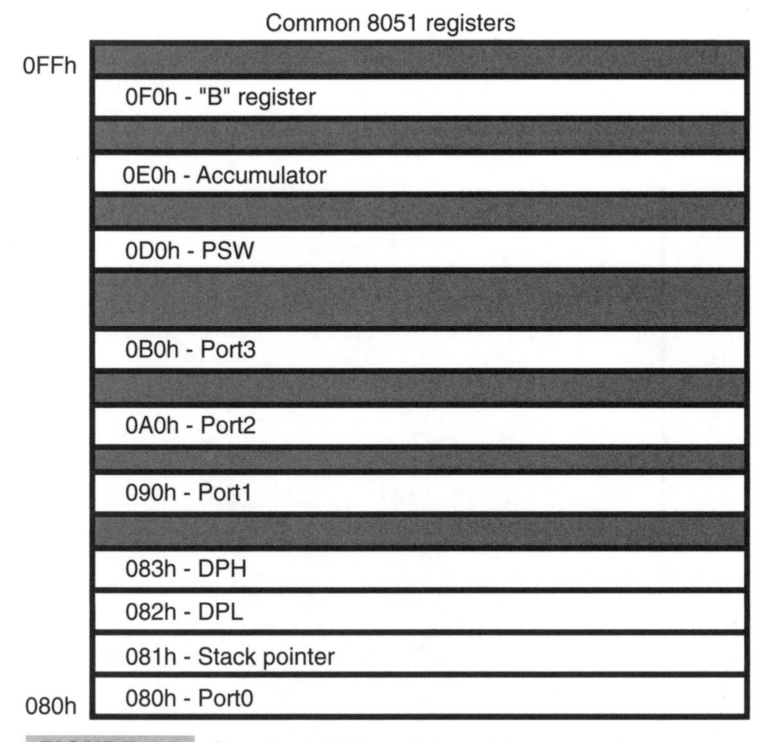

FIGURE 2-3 Standard 8051 register addresses.

8051 manufacturers, there are a number of standard peripherals and their associated registers that are not available in every device.

The special function registers I have identified are: the stack pointer, index pointer (DPL and DPH), I/O port addresses, the status register (program status word or PSW), and accumulators. I will present more information on these registers later in the book.

All direct addressing can only access the first 256 addresses in the memory space (which is marked as "Registers" in Fig. 2-1). To access RAM bytes at addresses above this 256 address space, you'll have to use an index register (such as DPTR, which is made up of DPL and DPH).

Indirect (or *indexed*) *addressing* uses either the stack pointer (the SP register identified earlier) or an index register (DPTR, for example). The special function registers (at address 080h to 0FFh) cannot be accessed by the indirect addressing. The 8052 enhancements have put in 128 bytes of RAM that can only be accessed here by the stack pointer or an index pointer.

With RAM put into the indirect space at the same addresses as the "special function registers" something funny starts to happen: You can access 384 different RAM bytes and I/O registers in the first 256 addresses of the 8051. This is 256 addresses of RAM (which can only be fully accessed by using indirect addressing instructions) and 128 addresses of special function registers (which can only be accessed using direct addressing instructions). When I first read this in the Dallas Semiconductor's HSM documentation, I really ended up shaking my head.

With this information on the register area, I can now expand the basic architecture block diagram to that shown in Fig. 2-4.

In describing the special purpose registers, I think I left a few open questions. The first is: What are all the special purpose registers that I identified? I will go through the processor-specific registers here, but leave the I/O specific registers for later in this section.

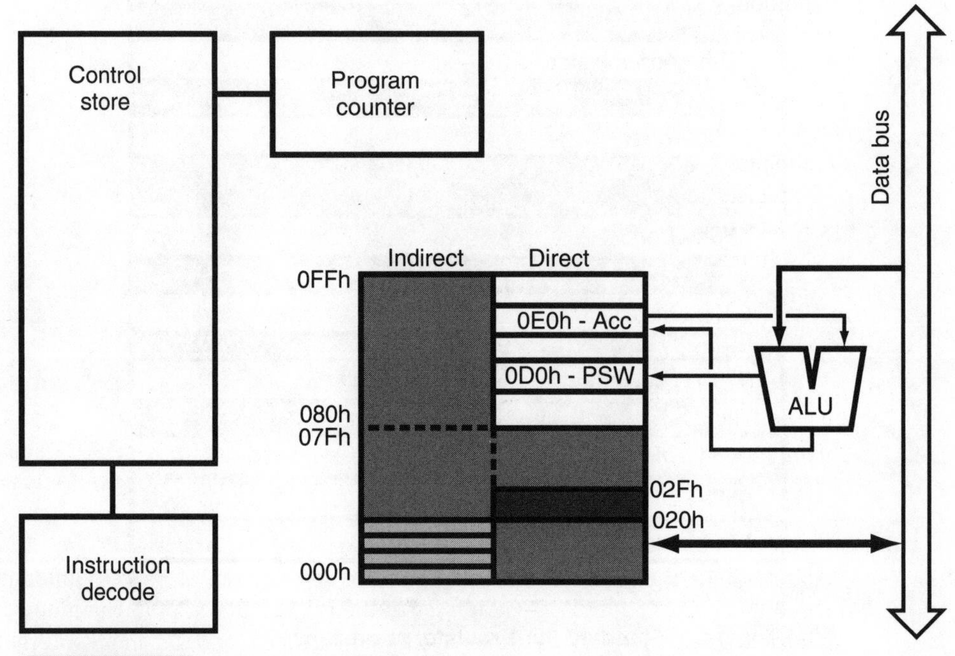

FIGURE 2-4 **8051 architecture with register addressing.**

TABLE 2-1 DEFINITION OF THE 8051 PSW

BIT	PSW BIT FUNCTION
7	C (carry flag)—Set when addition > 0FFh or when subtraction < 0
6	AC (auxillary carry flag)—Set when the low nybble affects the high nybble
5	F0 (user flag 0)
4	RS1 (register bank select 1)
3	RS0 (register bank select 0)
2	OV (overflow flag)—Set after addition or subtraction, cleared by all others
1	F1 (user flag 1)
0	P (parity flag)—Set when the accumulator has an odd number of bits

The accumulator registers (A and B at addresses 0E0h and 0F0h, respectively) are used to store temporary values and the results of arithmetic operations. If you flip through the arithmetic and bitwise instructions quickly, you'll see that B is just about never used. It is only used for multiplication and division. The accumulators can be accessed either as part of an instruction or as addresses 0E0h or 0F0h for ACC (which is the special function register identification for the accumulator) and B, respectively.

The program status word (most commonly known as PSW at address 0D0h) is the status register for the 8051. It is defined in Table 2-1.

The CY and AC bits are the full carry and half carry flags. CY is set when an addition or subtraction causes a carry to or a borrow from the next highest byte of the number (which allows easy manipulation of 16 bit or larger values). The AC, or half carry, flag is set when an addition or subtraction makes the lower nybble change the value (by one) of the upper nybble. These two status flags are in all the microcontrollers presented in this book.

Looking at the PSW definition, there is one flag that you would expect to see but don't. That is the zero flag (which is set when the result of an arithmetic, bitwise, or shift/rotate operation is equal to zero). Checking for zero is handled differently in the 8051. The `jz` instruction tests the contents of the accumulator (A) and carries out the jump if the contents of the register is equal to zero. I will expand on this operation in chapter 3.

The lack of a zero flag is something I've pondered about, wondering whether it's good or not. In many ways, it is a positive (most notably in helping to keep interrupt handlers short and not unnecessarily pushing and popping the PSW on and off the stack). I guess it seems foreign to me not to have one and takes some getting used to (and learning how to think through developing applications for it).

The RS0 and RS1 bits are used to select which of the four 8-byte banks is currently being used. These 8-byte banks are used for providing single-byte arithmetic instructions. By providing a very small area to operate out of, smaller and faster instructions can be used in your application. One aspect of the 8-byte banks that you should be aware of is that the two least-significant bytes of the bank (identified as R0 and R1) can be used for index addressing in the 256-address RAM area.

Another item I have not addressed so far is the operation of the stack pointer and how it can only access the first 256 addresses of the RAM register area. Upon power up, it is set to 007h, but when using an 8052, I like to change it to 080h and just forget about worrying how much space is required for the stack.

This might seem a bit strange, and there are a few things I really haven't explained here. The stack's `push` and `pop` instructions *increment* the stack pointer (rather than decrement the stack pointer as is done in many other devices). This means that, to set up the stack pointer to give it maximum space in a stack area, you have to give it the lowest address of the stack area, rather than the highest addresses, as you would with most other devices.

I set the stack pointer address to 080h because the RAM is only accessible by the index registers (which the stack pointer is one). Giving the stack pointer the bottom of the accessible area, I can put arrays at the top of the first 256 addresses and not really have to worry about whether or not the arrays or the stack will write over each other's data area.

The DPTR is a 16-bit index register that can access up to 64K different addresses in external memory. It is primarily designed for transferring data to/from the external memory area, and I'll discuss it in more detail in the next section.

With the special registers, the current block diagram looks like the one shown in Fig. 2-5. In this diagram, you can see how the stack pointer (SP), DPTR, A, and PSW registers are all accessible from the direct register area and how they interact with the 8051's register data and address busses.

With this, the block diagram of the 8051 is just about complete. The only change that I would make is adding a multiplexor selection between reading an immediate value in the instruction or using a value in the Register space. This is shown in Fig. 2-6.

With Fig. 2-6, I consider *my* block diagram of the operation of the 8051 to be complete. If you compare it to the block diagram contained in the Dallas Semiconductor's 87C520 data sheet (Fig. 2-7), you'll see just about no relation between the two diagrams. These differences

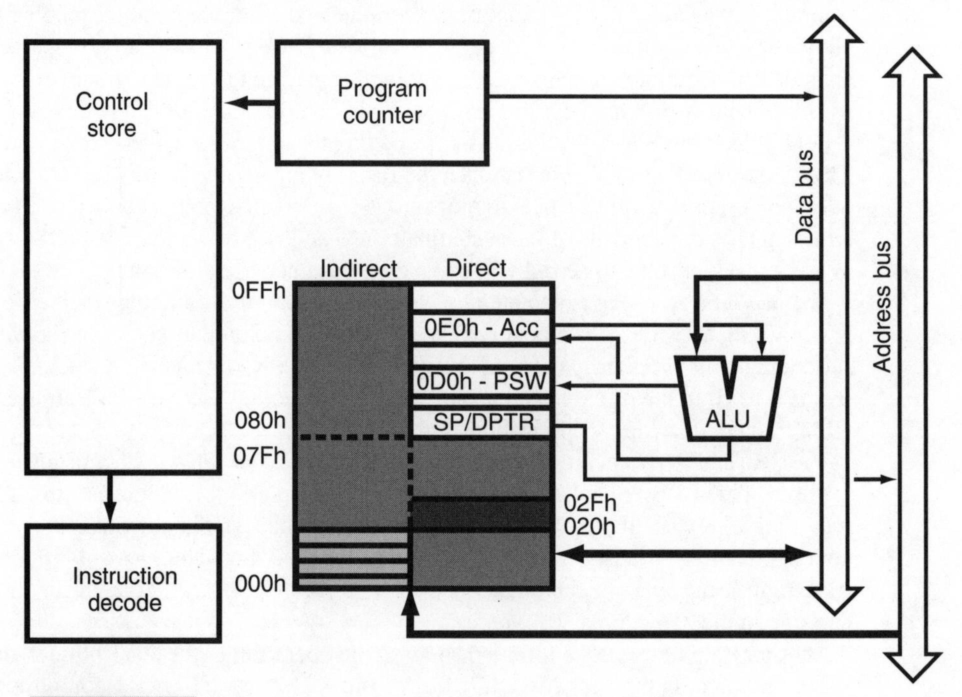

FIGURE 2-5 **8051 architecture with register addressing.**

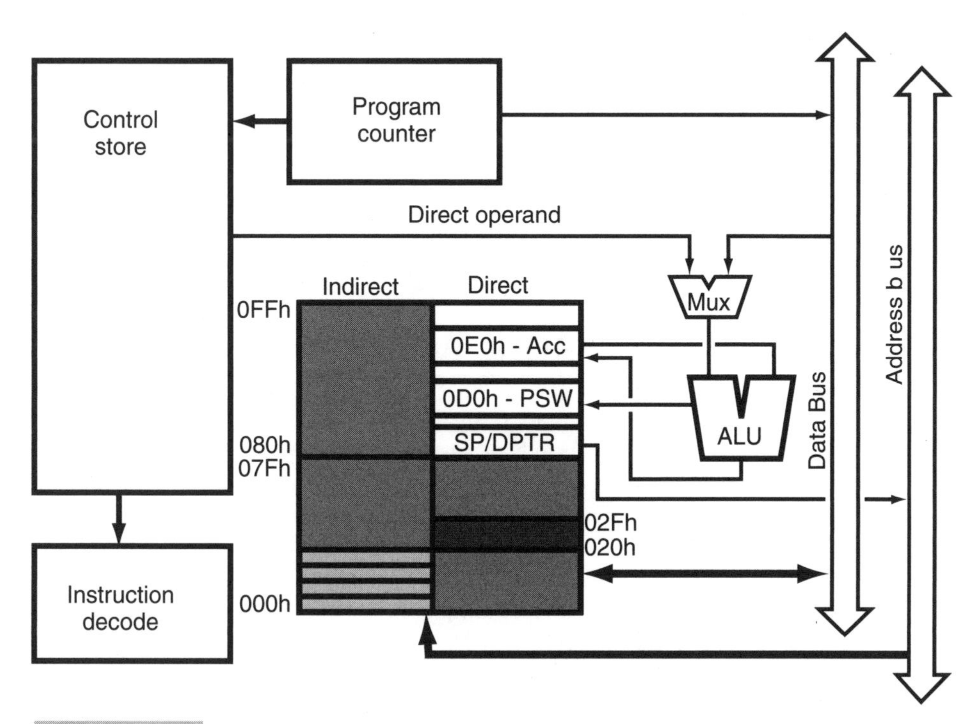

FIGURE 2-6 8051 architecture with immediate addressing.

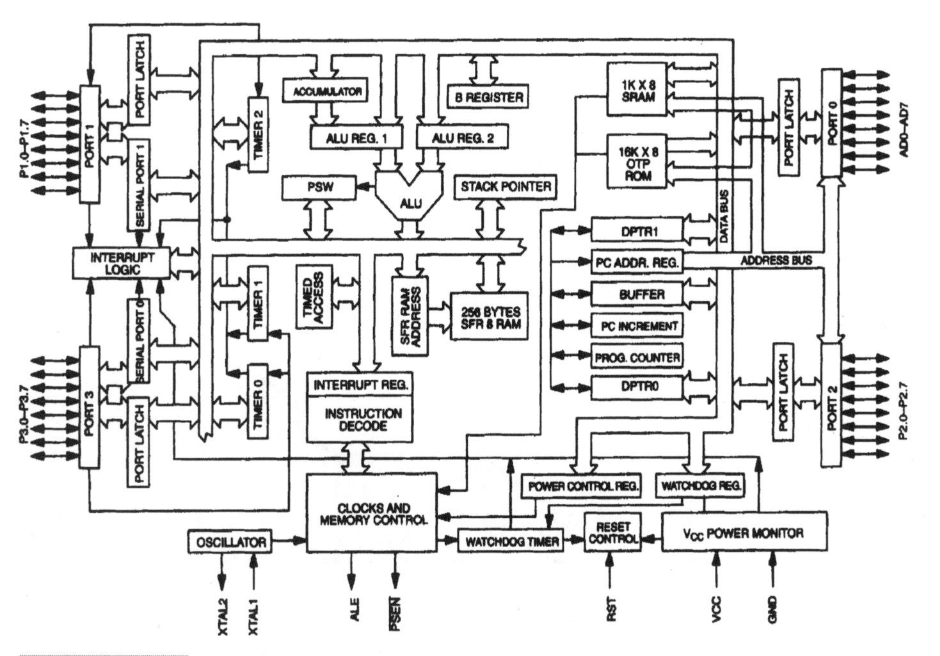

FIGURE 2-7 DS87C520 block diagram.

really come from how I try to understand how the devices work. I tend to focus on instruction execution and try to understand what has to happen inside the processor for the instruction to execute properly. My block diagrams tend to simplify how the peripheral hardware interfaces to the processor (although I will expand upon the interfaces later in each section).

Is this the right way to learn how the different microcontrollers and their processors work? I don't know, but I do know that it is a very effective way for me to learn how they work and visualize how data moves inside the devices.

8051 Addressing Modes

The 8051's addressing modes are designed from the perspective that data-access priority should reflect how the data is to be accessed. In most applications, very few variables are accessed a lot, some quite a bit, and most quite infrequently. In designing the 8051, Intel used this philosphy to specify the single-byte/12-clock-cycle instruction cycle for determining how data would be accessed. Accessing the current register banks only takes 1 byte and 12 cycles. This is followed by the 2-byte/24-clock instructions into the first 256 register addresses and the 3-byte, multiple instruction cycle access to the memory above the first 256 addresses.

This can be shown graphically in Fig. 2-8. This diagram is a good one to remember when developing complex applications in assembler. As you are writing the code, you should be asking yourself: What is the distribution of the instructions accessing the variables and is it optimized to execute in the fewest number of cycles and require the fewest number of control store bytes?

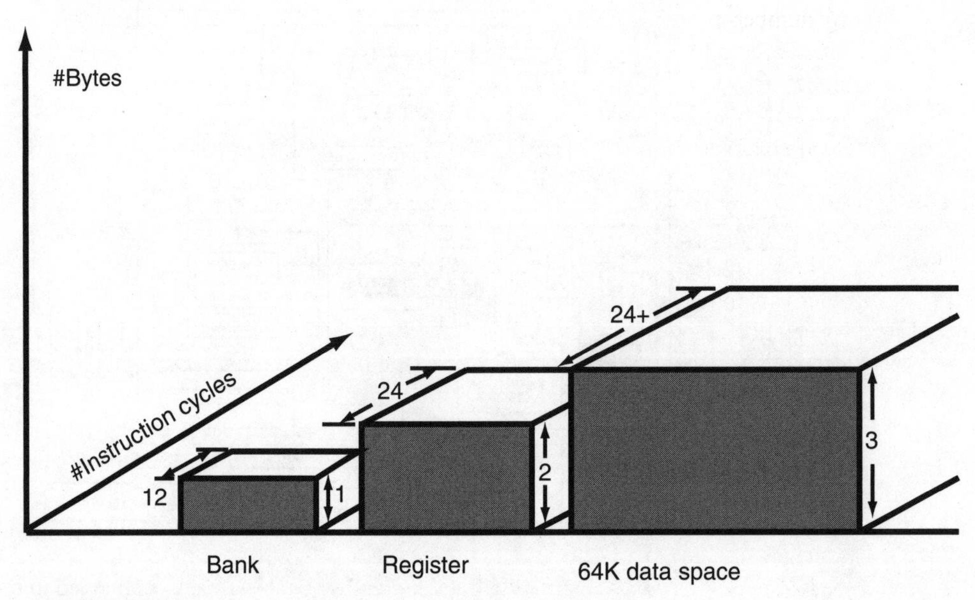

FIGURE 2-8 **8051 addressing mode comparison.**

The first addressing mode really isn't an addressing mode at all. *Immediate addressing* passes the value to be executed as part of the instruction. The immediate value is specified by placing a "#" character in front of the immediate value:

```
add  A, #77                    ; Add "77" to the accumulator
```

This instruction will add decimal 77 to the contents of the accumulator and store the result back into the accumulator.

Bank addressing allows you to access a byte in the current register bank. This is the most efficient (both in terms of clock cycles and control store) method of accessing data. Most register instructions execute in one register cycle and only require one byte to execute the instruction. The 8 bytes are known as R0 though R7.

Direct memory addressing differs from register addressing in that any byte within the first 256 addresses can be accessed by specifying an 8-bit address. When using this mode, there are a few things to watch out for. The first RAM addresses (080h to 0FFh, if the device you're using has RAM at these locations) cannot be accessed by direct addressing (instead you'll have to use an indirect mode). If you specify an instruction like:

```
mov  A, 088h
```

You'll load the accumulator with the contents of the TCON register (at 088h) rather than the contents of a RAM byte.

The second thing to note is something that always irritates me when I'm developing an 8051 application. I always forget to put in the "#" character for immediate addressing, and I inadvertently wind up with an instruction that uses direct addressing. This seems to be my number-one syntax problem when developing 8051 code, and it means I have to be especially careful when I'm simulating an application to make sure the correct addressing mode is being used.

Register indirect addressing uses R0 or R1 as an 8-bit index register to access a byte in the first 256 addresses of the data space:

```
orl  A, @R0
```

The register indirect is identified by the "@" character before either R0 or R1. Using any other bank registers will result in an error.

The DPTR register can also be used as a 16-bit index register. This addressing mode can also be enhanced with an offset for accessing data structures in data space memory. This mode is known as *register indirect with displacement.*

When you want to read a table out of control store (i.e., read a string of ASCII characters), the `movc A, @A+PC` instruction can access an offset from the current address. This offset has to be arithmetically generated before the instruction can be used.

The remaining addressing modes are used for changing the current program counter. You will see similar modes in the other microcontrollers. These modes are expanded upon in chapter 3. Actually, the different data addressing modes are also explained in more detail in chapter 3 as well.

External Addressing

At the start of this section, I introduced the 8051 as a device that could access up to 64K of program memory and 64K of variable SRAM. Looking through what I've shown earlier in this chapter and looking at the data sheets, you're probably wondering how this is accomplished. The 8051 is designed with a built-in external memory interface that uses the P0 and P2 I/O port bits for accesses. (See Fig. 2-9.)

Anytime an address (either control store or data space) is greater than what's in the 8051, the external I/O is activated and the 8051 tries to access external memory. If you're familiar with other microcontrollers, you might wonder how this is possible with a "typical" microcontroller's I/O pin design. You might want to skip ahead and look in chapter 4 to see what I'm talking about and see how different bit states are output.

When accessing external memory, first the least significant 8 bits of the address are output, followed by either reading or writing data.

The external control store byte read waveform looks like Fig. 2-10, which is actually how the Intel 8085/8088/8086 microprocessors access external memory. For RAM, the _RD (read) and _WR (write) pins are used instead of _PSEN (which is used to access external control store). In the Dallas Semiconductor HSM 8051s, the length of time used for reading/writing data can be extended or the cycles stretched. This is analogous in the "PC universe" as adding wait states. Stretching the accesses might be required when going from a "true" 8051 to a HSM part because of the latter's faster instruction cycles or to allow "slower" devices to be accessed without violating their timing specifications.

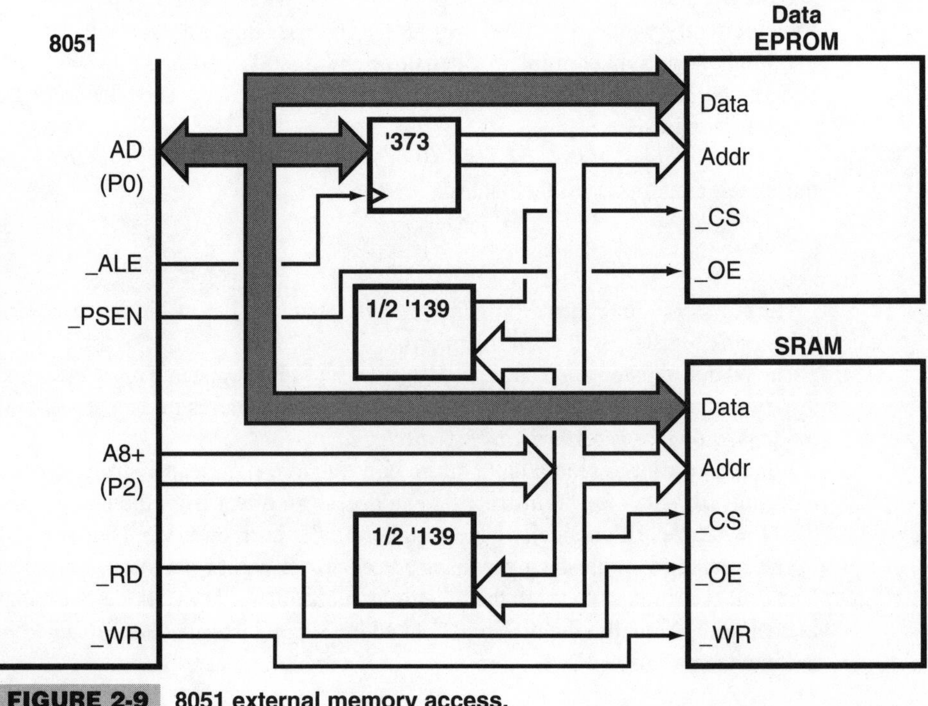

FIGURE 2-9 8051 external memory access.

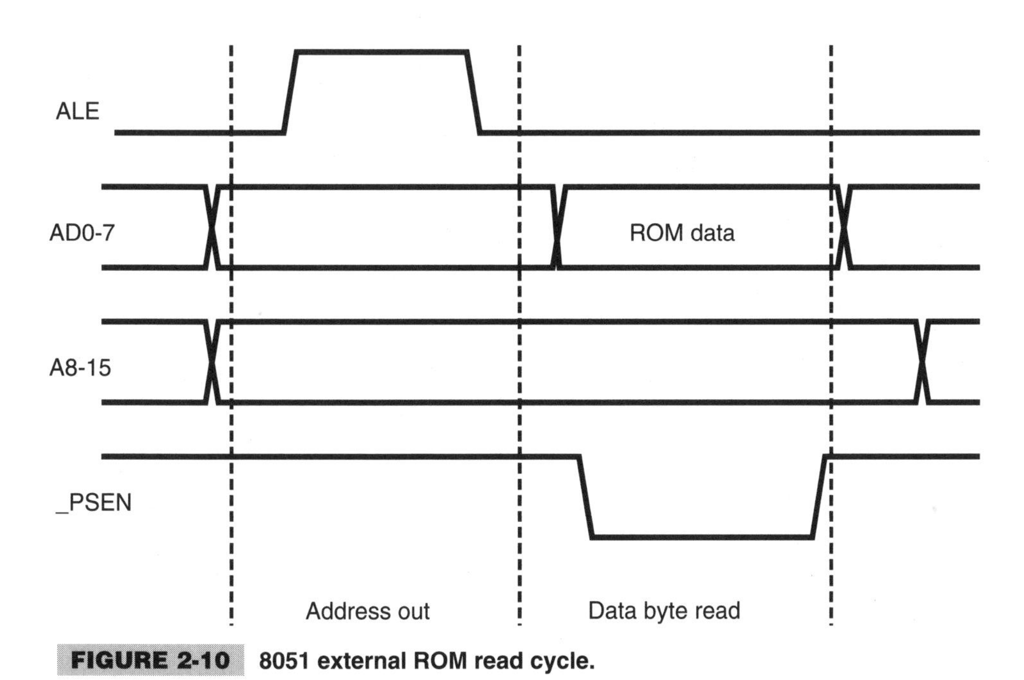

FIGURE 2-10 **8051 external ROM read cycle.**

Now, if you want to provide external memory (or devices) on the 8051's external busses but don't want to use the entire 64K address space, you could put these devices at high memory, which means that the I/O pins are always high and can be used for functions other than providing memory addresses.

For example, an application could be designed where 2K of SRAM was required along with two pins also required for bit I/O purposes, the 8051 and SRAM could be wired as shown in Fig. 2-11.

In this scheme, the P2.5 and P2.6 pins not used by the external RAM and are available for bit I/O because they will not change state during SRAM access. If the bit in the address is set, then its external I/O state will not change. This means that the SRAM pins should always be set high (or left at their initial value of all high) to prevent problems that could prevent certain addresses from being accessible.

Earlier, I mentioned that I/O devices could be interfaced with the 8051, and this can be done very simply by using one of P2's bits for the _CS or clock pin on the device. The last example 8051 application shows a memory mapped I/O solution using similar wiring. (See Fig. 2-12.)

For the example circuit in Fig. 2-12, writes to Device 1 could be carried out by writing to address 0FE00h, Device 2 can be read at 0FD00h, and the ROM (Device 0) could be read starting at 0FC00h. This scheme allows each device to be accessed without affecting any of the others or requiring extensive "glue" logic to determine whether or not reads or writes should be carried out. The lower 8 address bits are used for selecting registers or addresses within the peripheral devices.

One really interesting thing that can be done with the 8051 is to use one memory for *both* control store and RAM. This is done by ANDing _PSEN and _RD together (Fig. 2-13).

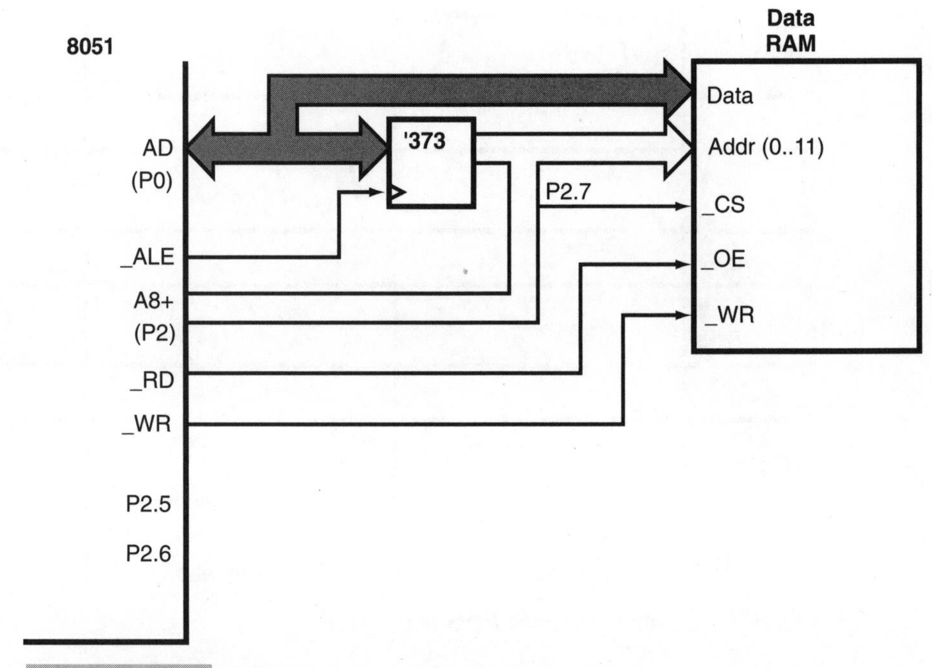

FIGURE 2-11 8051 2K external SRAM access.

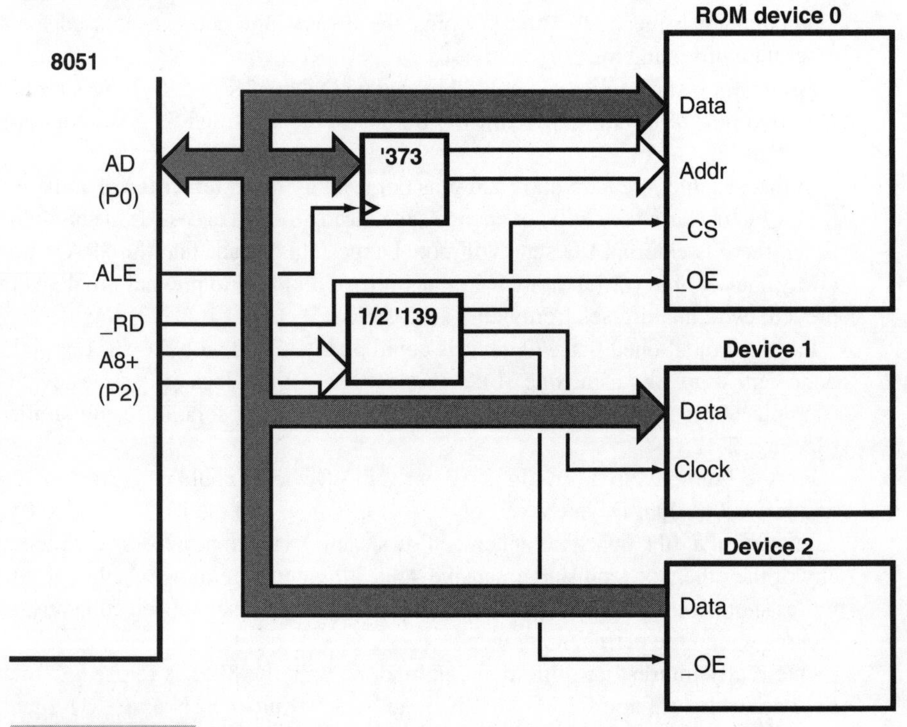

FIGURE 2-12 8051 external memory mapped I/O with a data ROM.

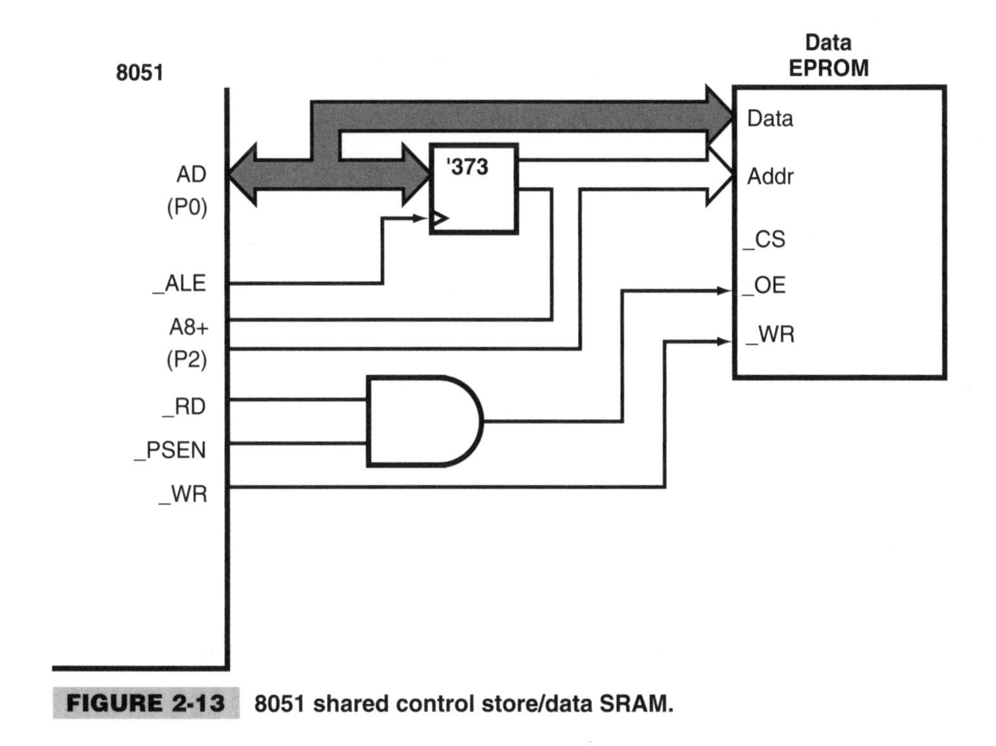

FIGURE 2-13 **8051 shared control store/data SRAM.**

By doing this, I've now created an 8051 application that can write to its control store (an immediate use would be for creating an 8051 experimental debug tool for downloading code into the SRAM). This circuit is used for the emulator presented later.

It also poses an interesting philosophical question: By creating this circuit, have I created a Von Neumann architected processor out a Harvard Architecture?

Interrupts

The "bottom" of the 8051's control store memory map can be represented as shown in Fig. 2-14. Address 0 is the reset (and watchdog timer overflow) execution vector address. When the 8051 is reset, execution jumps to this address. When an interrupt request is acknowledged, execution jumps to the appropriate address (i.e., interrupt 0 goes to address 03h, interrupt 1 goes to address 0Bh, and so on).

The exciting thing is, while other microcontrollers operate in a similar fashion, they don't provide memory space to allow an application to handle each individual interrupt directly in line. The 8 bytes of control store available for each interrupt handler can either be used to house the entire interrupt handler or jump to a more substantial handler.

I always look for features built into devices that make my life easier. By "easier," I'm trying to say that I have to do less thinking. Having 8 bytes available for instructions makes simple interrupt handlers (i.e., reset timer interrupt and increment a real-time clock value) very easy to implement.

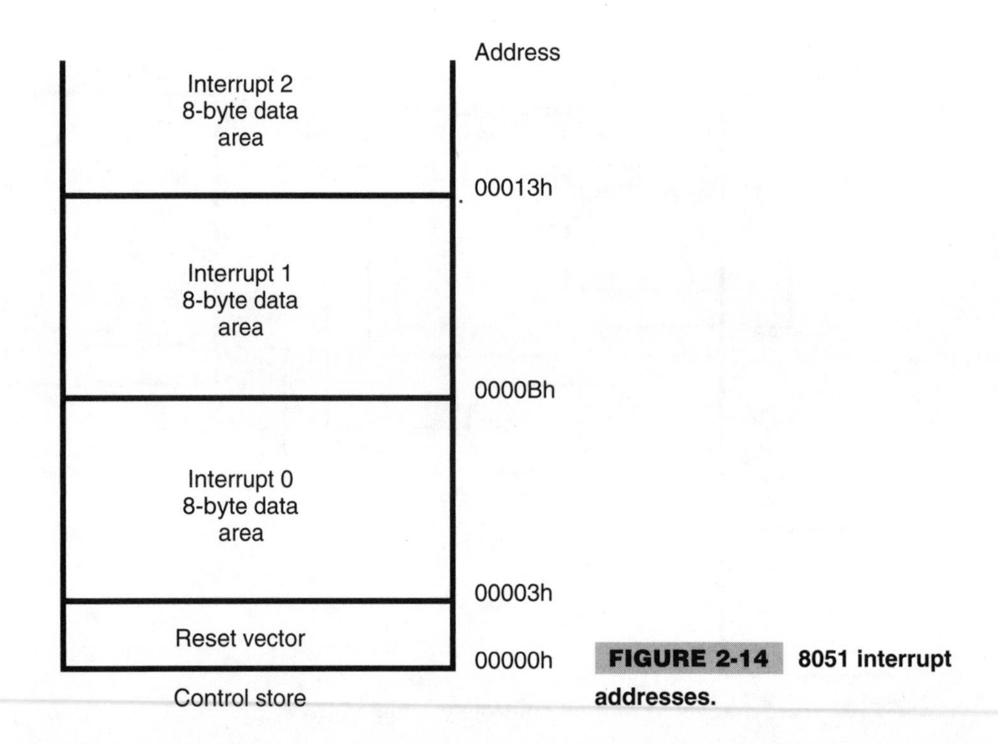

FIGURE 2-14 8051 interrupt addresses.

If you're thinking that the 8 bytes are probably not sufficient to save the context registers before handling the interrupt, you're probably correct if you're thinking about your experiences with other processors. However, because there is no zero flag in the 8051's PSW register, surprisingly complex operations can be carried out in 8-byte interrupt handlers. Later in the book, I will show you many examples of this.

An important feature that makes writing short interrupt handlers easier is the automatic interrupt controller hardware reset available for the basic 8051 interrupts.

8051 Instruction Execution

The 8051 processor core is designed differently than most other microcontrollers. The 8051 contains a microcoded processor that is in contrast to other microcontrollers that use a hardwired one. (See Fig. 2-15.)

What is a microcoded processor? This is really a processor within a processor, or a state machine that executes each different instruction as the address to a subroutine of instructions. When an instruction is loaded into the instruction-holding register, certain bits of the instruction are used to point to the start of the instruction routine (or microcode) and the microcode instruction decode-and-processor logic executes the microcode instructions until an instruction end is encountered.

As a quick aside, I should point out that having the instruction-holding register appear wider, in Fig. 2-15, than the control-store memory is not necessarily a mistake. The control store in the 8051 is only 8 bits wide, and quite a few of the instructions are more than 8 bits. This could mean that, before an instruction can be executed, the entire instruction might have to be loaded into the holding register (which can take additional time).

A hardwired processor uses the bit pattern of the instruction to access specific logic gates (possibly unique to the instruction), which are executed as a combinatorial circuit to carry out the instruction. (See Fig. 2-16.)

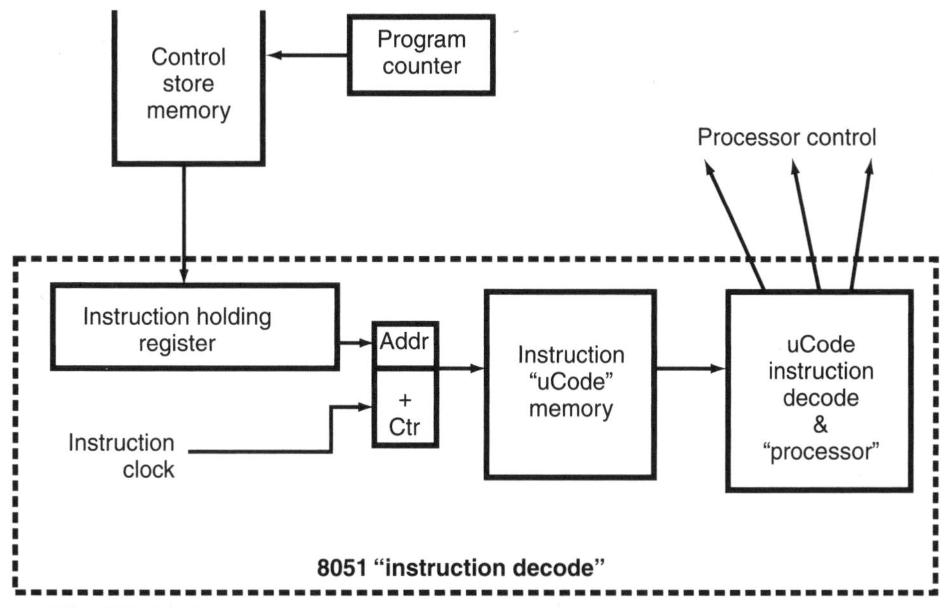

FIGURE 2-15 8051's microcoded processor.

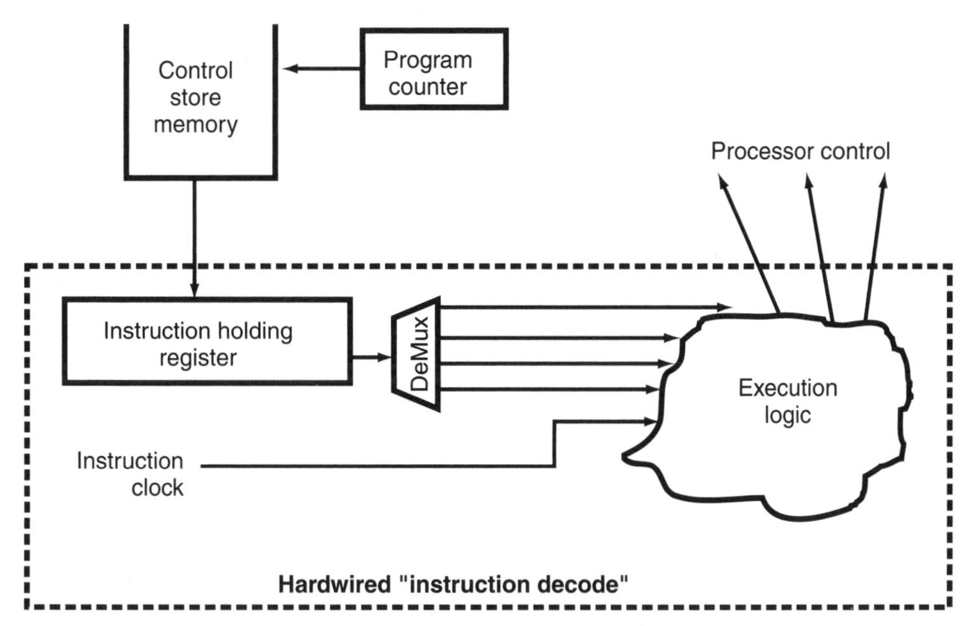

FIGURE 2-16 Hardwired instruction processor.

Each of the two methods offer advantages over the other. A microcoded process is usually simpler than a hardwired one to design, and the actual design can be implemented faster and with less chance of having problems at specific conditions.

A great example of the quick-and-easy changes that a microcoded processors allow was a number of years ago when IBM wanted to have a microprocessor that could run 370 assembly language instructions. Before IBM began to design their own microprocessor, they looked around at existing designs and noticed that the Motorola 68000 had the same hardware organization as the 370 (although the instructions were completely different). IBM ended up paying Motorola to rewrite the microcode for the 68000 and came up with a new microprocessor that was able to run 370 instructions, but at a small fraction of the cost of developing a new device.

A hard-wired processor is usually a lot more complex because the same functions have to be repeated over and over again in hardware (how many times do you think that a register read or write function has to be repeated for each type of instruction?). This means that the processor design will probably be harder to debug and be less flexible than a microcoded design, but it will probably run a lot faster.

The "true" (original) 8051 uses a microcoded processor that requires at 12 to 24 clock cycles for each instruction to execute. Later in this book, I will present three 8051-compatible devices that have redesigned processor cores aimed at improving the execution speed of the 8051's instruction set.

3

8051 INSTRUCTION SET

CONTENTS AT A GLANCE

You might be surprised to discover that, in the 8051, instructions execute very similarly to a "classic" von Neumann architected processor; data has to be first stored in an accumulator before operating on it. I would strongly urge you to look at chapter 2 before reading through the instruction set to understand such concepts as register banks and addressing before you will be able to really understand how the different instructions operate.

One thing that you will notice in the "Data Movement Instructions," "Arithmetic Instructions," and "Bit Operators" sections is that I have lumped instructions together even if they use different addressing modes (which, in some cases, causes an extra instruction cycle). In doing this, I will be able to pare down the 101 op codes (or 255 unique instructions) of the 8051 to 62.

For the 8051's instructions for indirect addressing, I have indicated the index register (either R0 or R1 of the current bank) with an "i" (i = 0 for R0, and i = 1 for R1). For register addressing, "rrr" is used to specify which register is being accessed within the bank.

Data Movement Instructions

Before you can begin to process data, you have to understand how to move it within the microcontroller (and in the 8051's case especially, how to access external memory). The 8051 has a number of instructions for transferring data within (and without) the device.

One of the most important aspects of the 8051 that you will have to understand is the instruction format and conventions. The typical data movement (and arithmetic) instructions are in the format:

```
ins  Parm1, Parm2
```

`Parm1` is the destination of the operation's result and the optional first parameter of the operation. `Parm2` is generally a value brought into the instruction. This means:

```
mov  A, R0
```

executes as A = R0, and

```
add  A, R0
```

executes as A = A + R0.

The Atmel AVR uses this instruction format whereas the Motorola 68HC05 never has two parameters in an instruction, and the Microchip PICMicro operates somewhat differently with the result of an operation is optionally stored in the PICMicro's accumulator (or "w" register). For the AVR, the same right-to-left data-movement convention is used.

When you see "A" in the 8051's instruction set, this means the accumulator (ACC at address 0E0h). The A register is actually part of the instruction and cannot be replaced with another register. If there is only one direct address parameter in an instruction, "ACC" will have to be explicitly specified rather than just "A" (just putting in "A" for a parameter will result in a syntax error when the code is assembled).

With this background, you should have a good idea how the `mov A, operand` instruction works (Fig. 3-1). For each of the four addressing modes (immediate, register, indirect, and direct), the operand is copied into the accumulator.

The same goes for the `mov operand, A` instructions (Fig. 3-2). The contents of the accumulator are copied into the operand (addressed as a register as an 8-bit direct address or an 8-bit indirect address).

`mov Operand1, Operand2` (Figs. 3-3 and 3-4) is used to pass data without using the accumulator as an intermediate step (i.e., using the two instructions `mov A, Operand2`/`mov Operand1, A`).

The `mov C, bit` and `mov bit, C` instructions (Figs. 3-5 and 3-6) are really useful instructions for getting and storing individual bits without affecting other bits in a register.

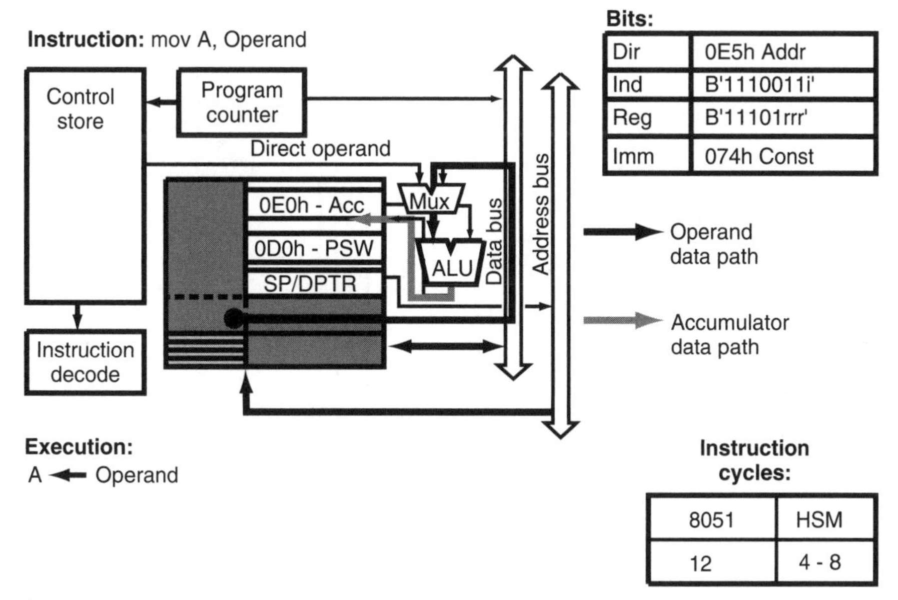

Example: **MCU:** 8051

```
mov A, #123             ;  Put 123 into the Acc
```

FIGURE 3-1 `mov A, operand` **instruction.**

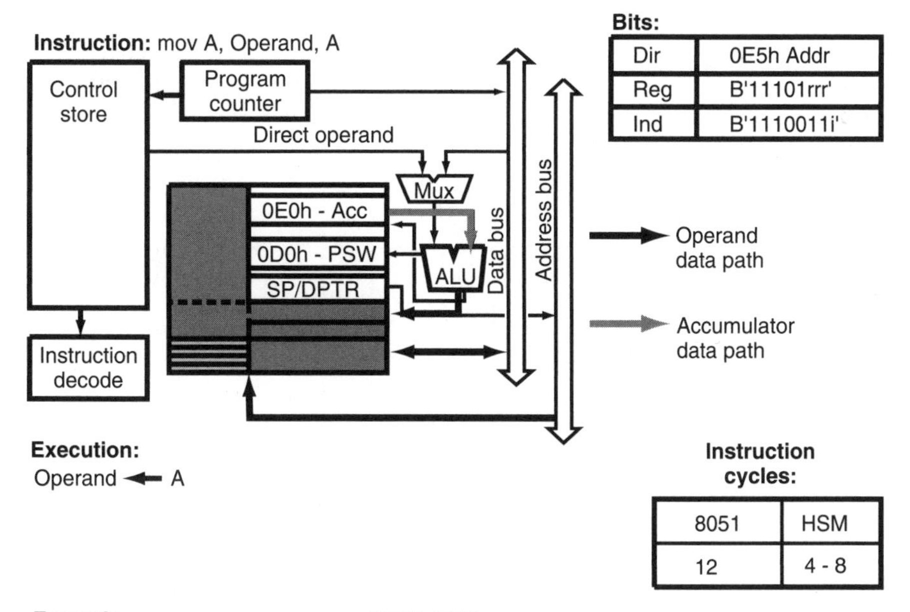

Example: **MCU:** 8051

```
mov A, #123             ;  Put 123 into the "R2"
mov R2, A
```

FIGURE 3-2 `mov operand, A` **instruction.**

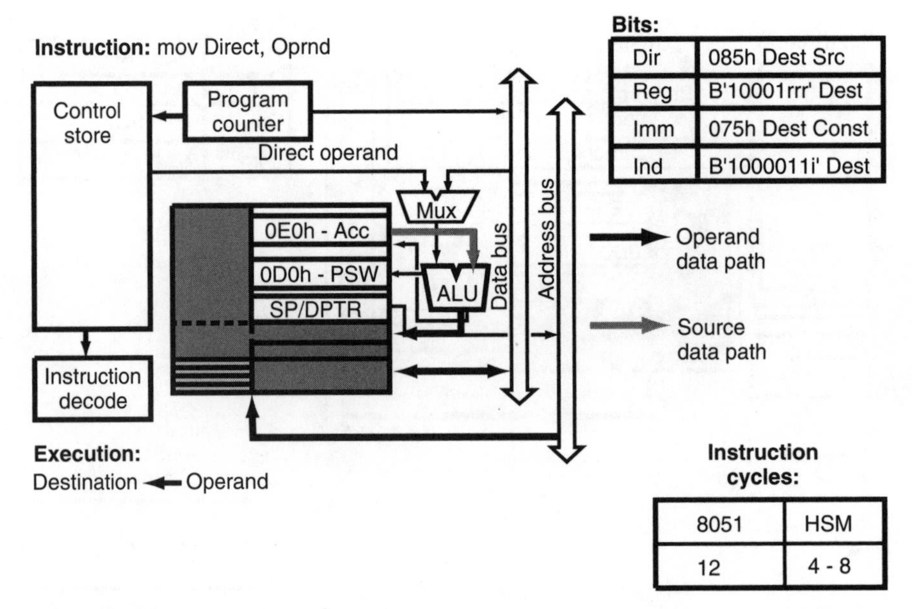

Example: **MCU:** 8051

```
mov B, #123                ;  Put 123 into "B"
```

FIGURE 3-3 `mov direct, operand` **instruction.**

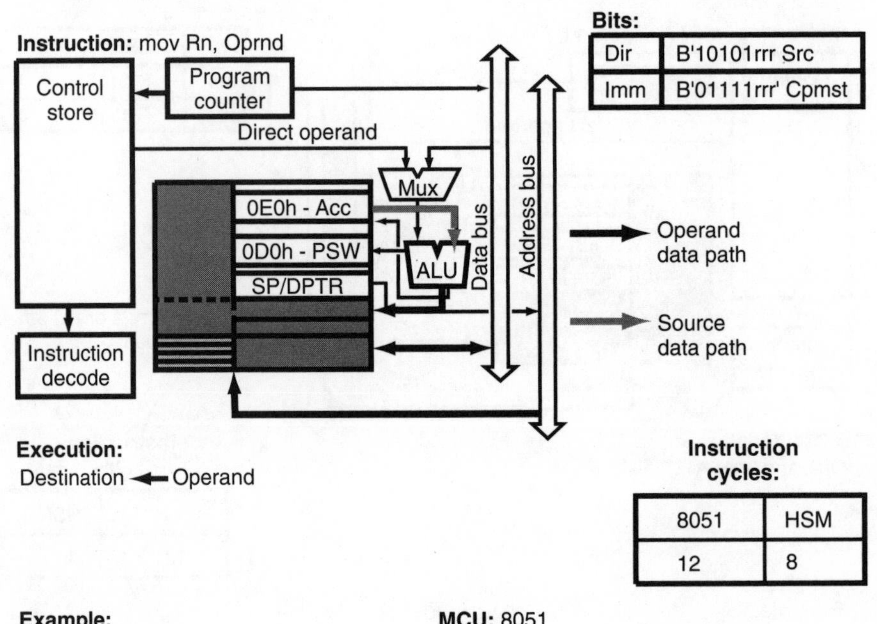

Example: **MCU:** 8051

```
mov R2, #123                ;  Put 123 into R2
```

FIGURE 3-4 `mov rn, operand` **instruction.**

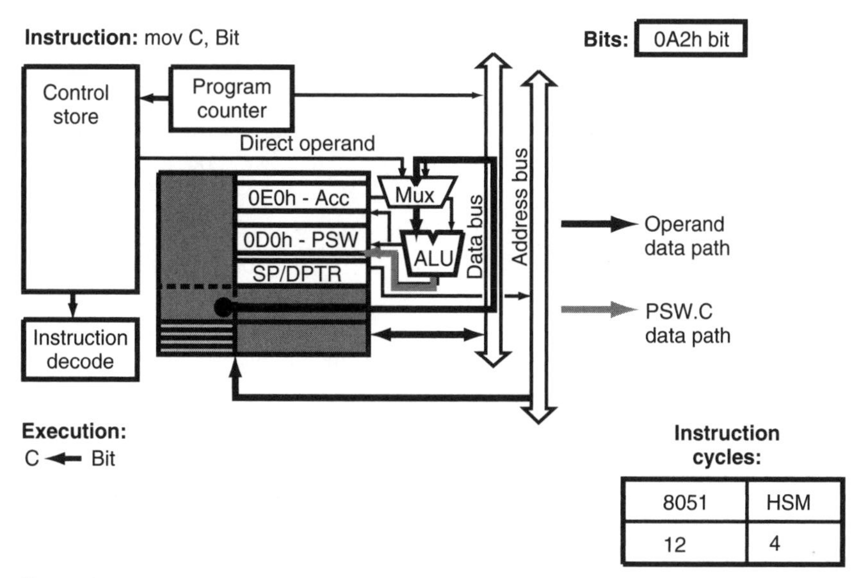

Example: **MCU:** 8051

```
mov C, P0.7          ; Move P0.7 to P2.6
mov P2.6, C
```

FIGURE 3-5 `mov C, bit` **instruction.**

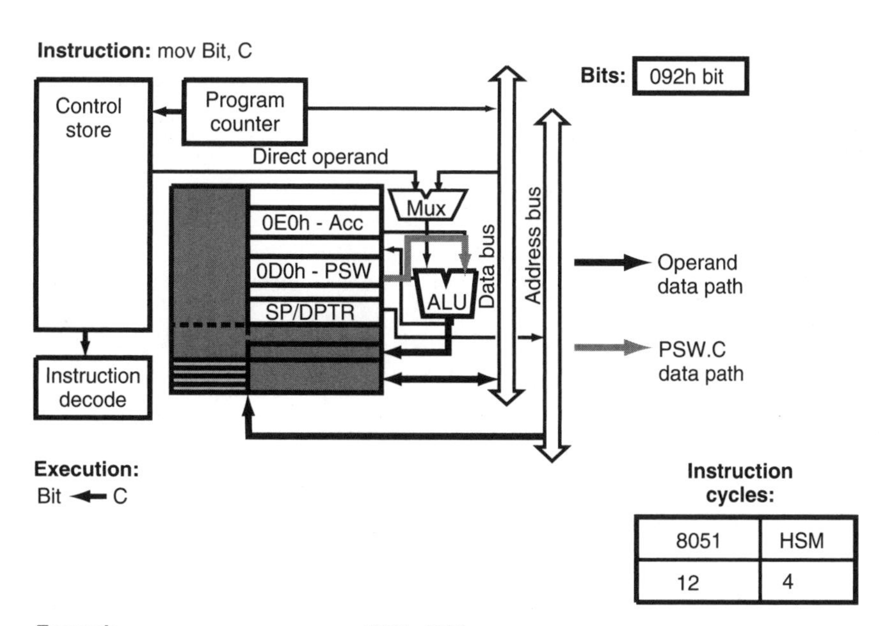

Example: **MCU:** 8051

```
mov C, P0.7          ; Move P0.7 to P2.6
mov P2.6, C
```

FIGURE 3-6 `mov` *`bit, C`* **instruction.**

If you look at the bit pattern, you'll notice that 8 bits are available for addressing. However, when I described the 8051's architecture, I pointed out that there were 128 general-purpose bits located at address 020h to 02Fh (which only requires 7 bits for addressing). When bit 7 of the bit address is set (in these and the bit instructions), bits in the special function register space are accessed. These bits are in the ACC, PSW, B, P0, P1, P2, and P3 registers.

So, to retrieve the least significant bit from P1, the instruction:

```
mov  C, P1.0
```

is used where the digit after the period after `P1` is the bit number.

Indexed addressing into the 64K data space can use either a bank index register (R0 or R1 in the current 8-byte register bank) or the DPTR register using the `mov` and `movx` instructions (Figs. 3-7 through 3-9). As discussed elsewhere, some versions of the 8051s have 256 internal RAM bytes that can be accessed internally only using index registers. When the Ri register is used, only the first 256 addresses can be accessed. When DPTR is used for the index, then the entire 64K data space can be accessed.

Loading the two 8-bit index registers in the current bank is as simple as `mov Rn, #Const`, and loading the 16-bit DPTR register can be done in a similar manner with the `mov DPTR, #Const` instruction (Fig. 3-10).

There are two other forms of the indexed address instruction. The first is the stack operations `push` and `pop` (Figs. 3-11 and 3-12). These two instructions work as you would probably expect except that the instructions `push A` and `pop A` are invalid and will return an assembly error. These are the single parameter instructions I was referencing earlier.

These instructions operate on a direct address (rather than an implied one such as the accumulator), so the correct format for pushing and popping the accumulator is `push ACC` and `pop ACC`.

Tables can be implemented easily using the `movc A, @A+Index` instructions (Fig. 3-13). If the program counter is used as the index register, the accumulator will be loaded from control store. When DPTR is specified as the index register, the data is loaded from the 64K data space.

To implement a table read from control store (i.e., reading a message), the following code could be used:

```
ReadTable:                     ;  Return the offset in ACC in "Table"

  add       A, #(Table-GetTable)
  movc      A, @A+PC     ;   The program counter
GetTable:                ;  Get the offset at "A" plus
  ret                    ;  Return to the caller

Table:                   ;  The table elements are located here
  db        'H'
  db        'e'
  db        'l'
  db        'l'
  db        'o'
  db        0            ;  ASCIIZ string end
```

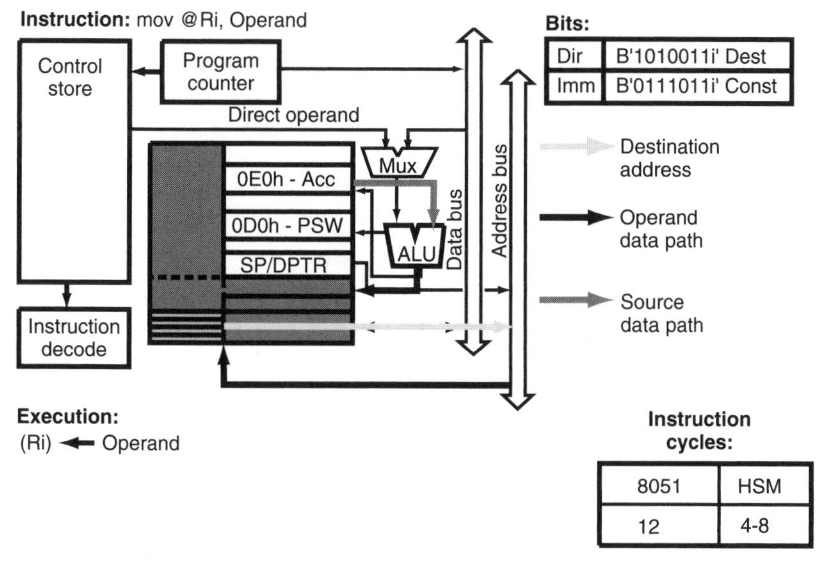

8051	HSM
12	4-8

Example: **MCU:** 8051

```
mov @R0, #123            ;  Put 123 into byte pointed
                         ;   to by R0
```

FIGURE 3-7 `mov @Ri,` *`operand`* **instruction.**

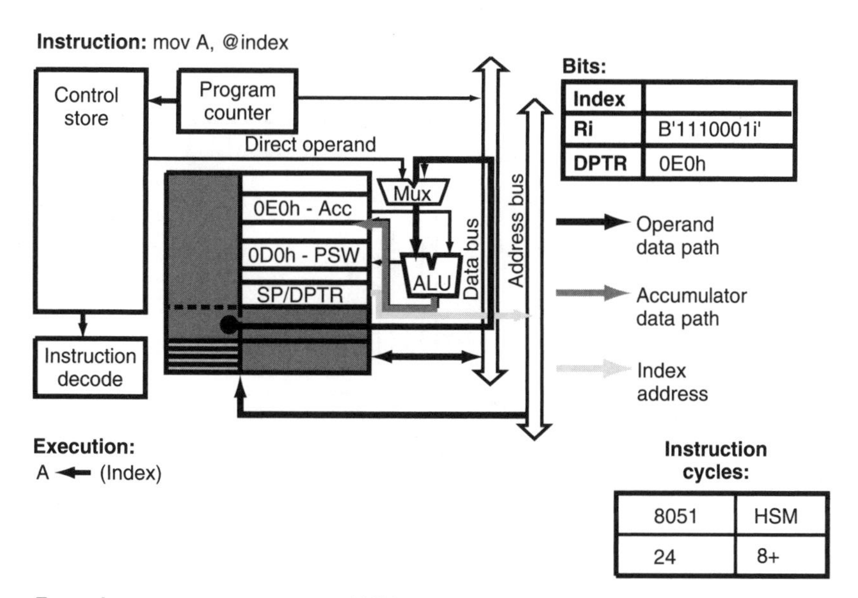

8051	HSM
24	8+

Example: **MCU:** 8051

```
mov DPTR, Buffer         ;  Read from the start
mov A, @DPTR             ;   of the buffer
```

Note: When DPTR is used as the index, the data read can be altered by wait states.

FIGURE 3-8 `mov A, @Index` **instruction.**

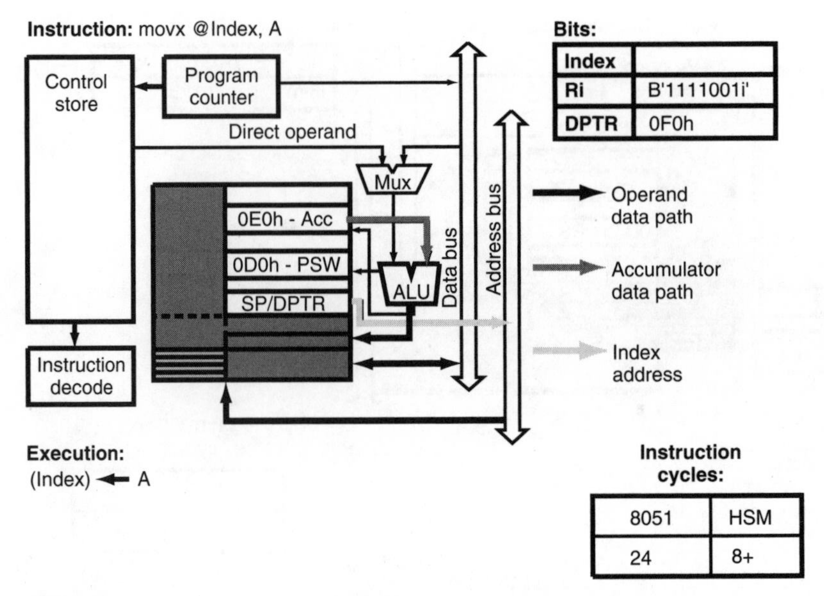

Example: **MCU:** 8051

```
mov DPTR, Buffer          ;  Write to the start
mov @DPTR, A              ;   of the buffer
```

Note: When DPTR is used as the index, the data read can be altered by wait states.

FIGURE 3-9 `movx @Index, A` **instruction.**

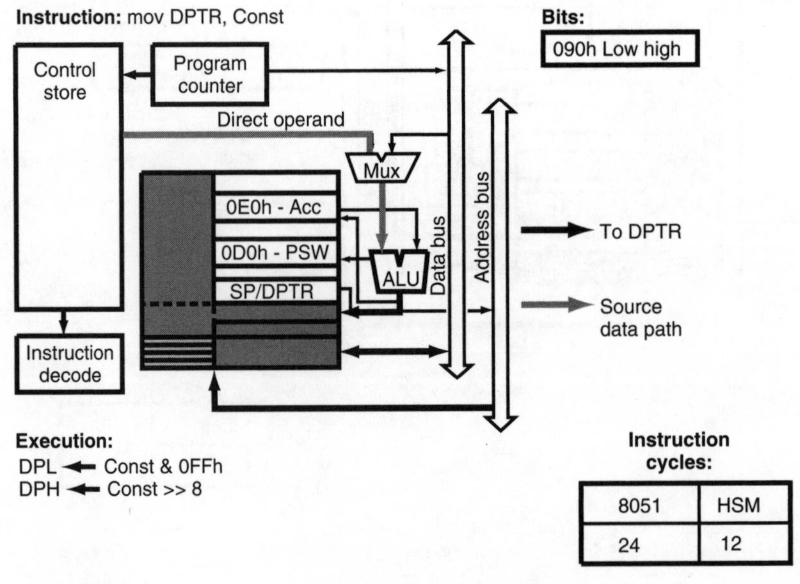

Example: **MCU:** 8051

```
mov DPTR, ButStart        ;  Set DPTR to start of
                          ;   the buffer
```

FIGURE 3-10 `mov DPTR, #Const` **instruction.**

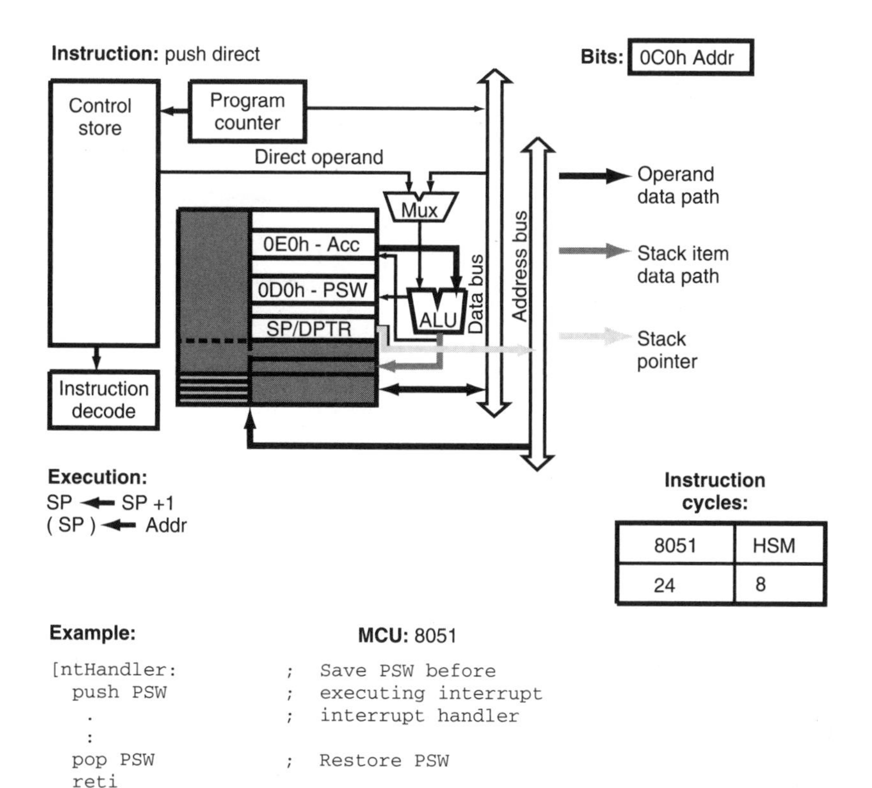

FIGURE 3-11 `push direct` **instruction.**

The last two data movement instructions are the exchange instructions `xch` and `xchd` (Figs. 3-14 and 3-15). The `xch A, `*`Operand`* instruction exchanges the contents of the accumulator with the specified register. It's operation can be modeled as:

```
Temp = ACC
ACC = Operand
Operand = Temp
```

`xchd A, @Ri` executes in a similar manner to `xch` except that only a bank index register can be used to specify the operand register's address and only the lower 4 bits of the two registers are exchanged.

Arithmetic Instructions

The 8051 really has a very complete set of arithmetic instructions when compared to other microcontrollers. This might be surprising to you because the 8051 has fewer arithmetic instructions than some of the others, but a very complete set of functions can be implemented in the 8051.

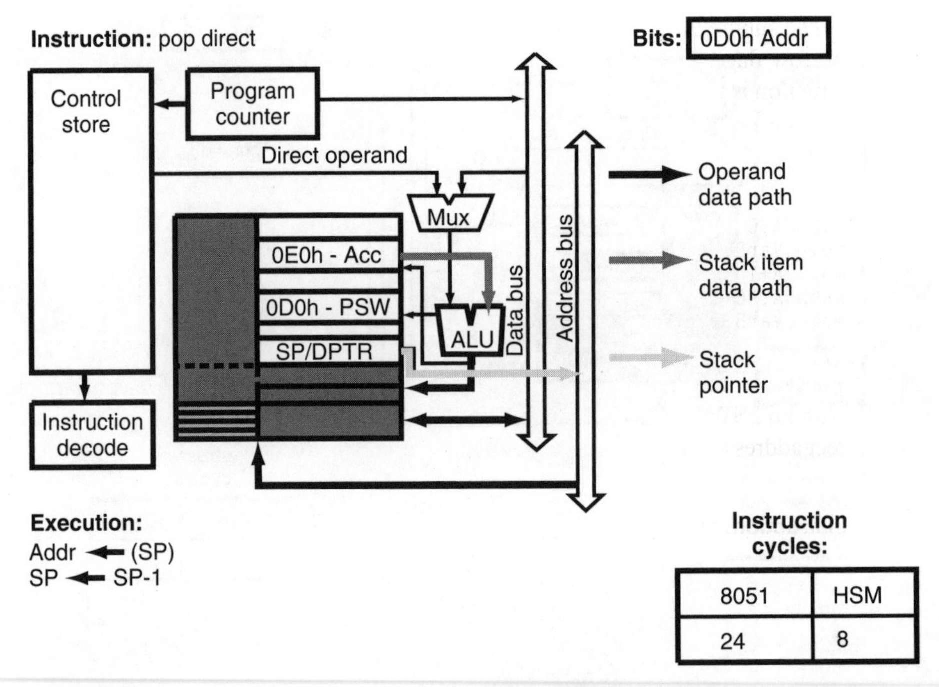

Example: **MCU:** 8051

```
[ntHandler:                 ;  Save PSW before
  push PSW                  ;  executing interrupt
   .                        ;  interrupt handler
   :
  pop PSW                   ;  Restore PSW
  reti
```

FIGURE 3-12 `pop direct` **instruction.**

Addition (Fig. 3-16) is very straightforward in the 8051 with the operand added to the contents of the accumulator. As I have pointed out elsewhere in this section, there is no zero flag available in the 8051's status register (the PSW). So, if a `jz` or `jnz` instruction is going to execute based on the result of an operation, that result has to be in the accumulator before the conditional jumps are executed.

The `addc` instruction (Fig. 3-17) is an addition instruction in which the carry flag is added to the result. This instruction is used to allow addition of 16 bit and larger numbers. For example, adding two 16-bit values in the current bank (or even the first 256 addresses) would be:

```
mov  A, VarB              ; VarB = VarB + C
add  A, C                 ; Add the low 8 bits first
mov  VarB, A              ; Store the low 8 bit result
mov  A, VarB+1            ; Add the higher 8 bits
addc A, C+1
mov  VarB+1, A
```

The subtraction instruction, `subb` (Fig. 3-18) is a subtract with carry (which is used as a borrow flag). This means that you must be aware of the carry flag at all times the `subb` instruction is used. This can be shown in a 16 bit subtract with the carry flag reset before the lower 8 bit operation:

```
mov   A, VarB              ; VarB = VarB - C
clr   C                    ; Clear carry before subtraction
subb  A,C                  ; Subtract the low 8 bits
mov   VarB, A
mov   A+1                  ; Subtract the high 8 bits
subb  A, C+1
mov   VarB+1, A
```

Incrementing and decrementing (Figs. 3-19 and 3-20) in the 8051 is very straightforward with no PSW bits being changed by the result. As well as the register and direct and indirect addressing modes, the accumulator can be operated upon directly.

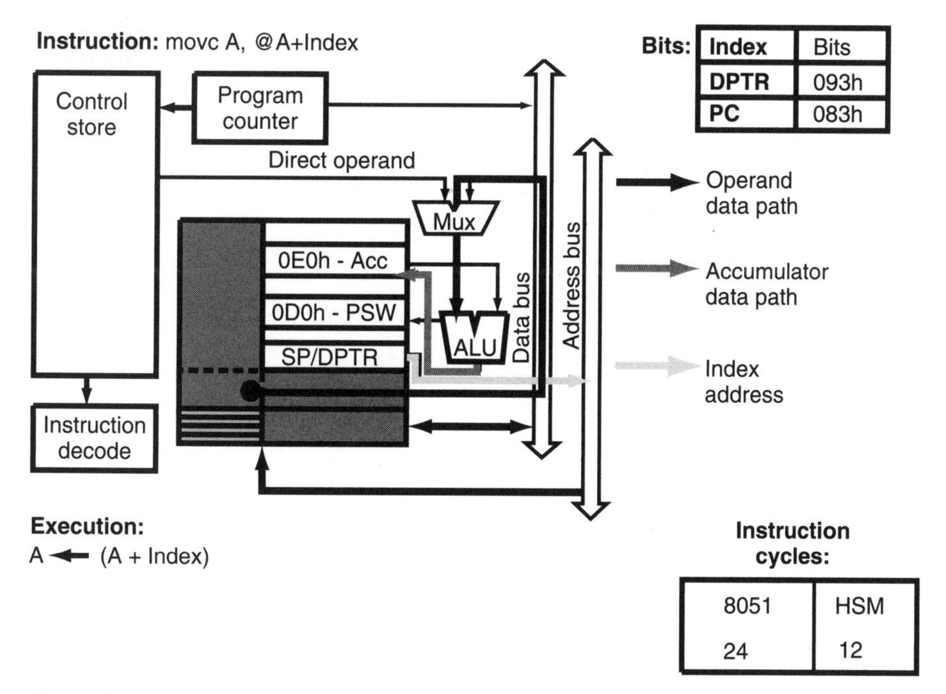

Example:

MCU: 8051

```
mov   DPTR, Buffer          ;  Read from the third
mov   A, #2                 ;  byte in the buffer
movc  A, @A+DPTR            ;  located in 64K
                            ;  memory Space
```

Note: When DPTR is used as the index, data memory is read. When PC is used, then control store is read.

FIGURE 3-13 `movc A, @A+Index` **instruction.**

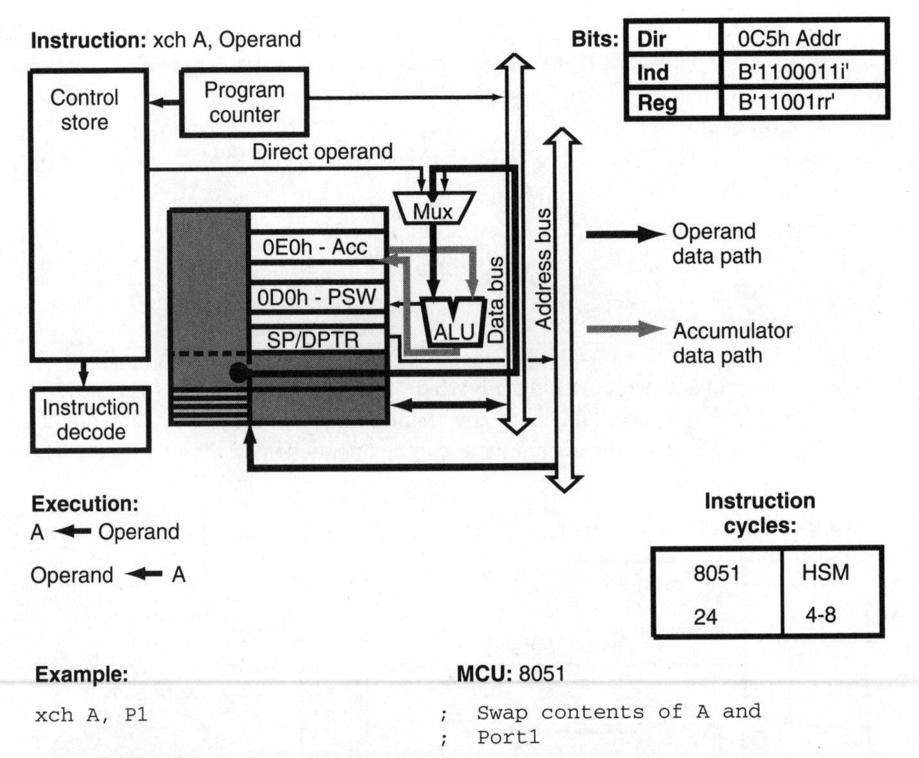

FIGURE 3-14 `xch` **instruction.**

If you go through data sheets of other microcontrollers, you'll see that the zero flag of the status register is integral for performing basic operations like a 16-bit increment. In the 8051, this can be carried out in a similar manner, but it's important to remember that it is the *result* for testing zero on (when the lower 8 bits of a number are equal to zero after an increment, the upper 8 bits have to be incremented as if the carry flag was set from the lower 8-bit increment).

```
  inc  Var              ;   Increment the lower 8 bits
  xch  A,Var
  cjne A,#0,incSkip     ;   The lower 8 bits are not equal to zero
   inc Var+1            ;    Increment the high 8 bits
incSkip:
  xch  Var,A            ;  Save the result of the lower 8-bit increment
```

A 16-bit decrement requires using the `subb` instruction instead of `dec` to set the carry/borrow flag when the lower 8 bits have gone from 0 to 0FFh (which requires a decrement of the upper 8 bits):

```
  mov  A, Var           ;   Decrement the lower 8 bits
  clr  C
  subb A, #1
  mov  Var, A
```

```
  jnc  decSkip             ;  If carry set, then decrement the upper 8 bits
   dec Var+1
decSkip:
```

I've separated the `inc DPTR` instruction (Fig. 3-21) from the increment/decrement instructions because it operates on the 16-bit DPTR register (which is made up of the DPL and DPH special function registers). Unfortunately, there isn't a `dec DPTR` instruction that goes along with this one.

However, the `dec DPTR` can be simulated with the following code:

```
  dec  DPL                  ;  Decrement the low 8 bits
  xch  A,DPL                ;  Put the value in ACC to test it
  cjne A,#0FFh,Skip         ;  Do we have 0FFh after decrement?
   dec DPH                  ;  Yes, decrement the high 8 bits
Skip:
  xch  A,DPL                ;  Replace DPL
```

`DA A` (Fig. 3-22) is executed after two BCD (Binary Coded Decimal) values have been added/subtracted together. A BCD number is defined as "0" to "9" in each of the two nybbles of a byte. After executing `DA A`, the contents of the accumulator are valid BCD, and the carry flag is set appropriately for the next BCD operation (i.e., if two 16-bit [4 digit] numbers are added together).

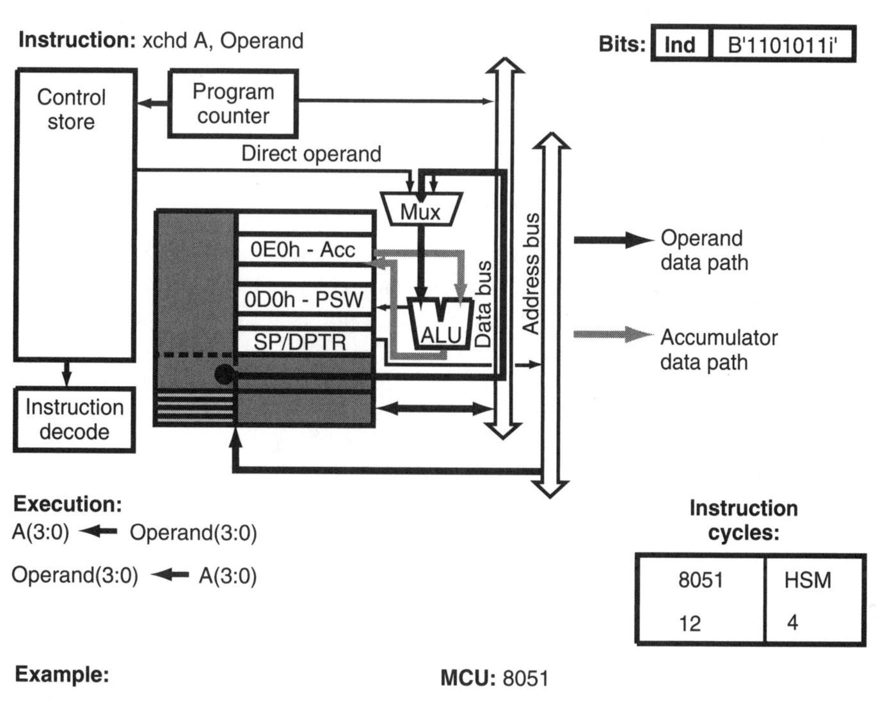

FIGURE 3-15 `xchd` **instruction.**

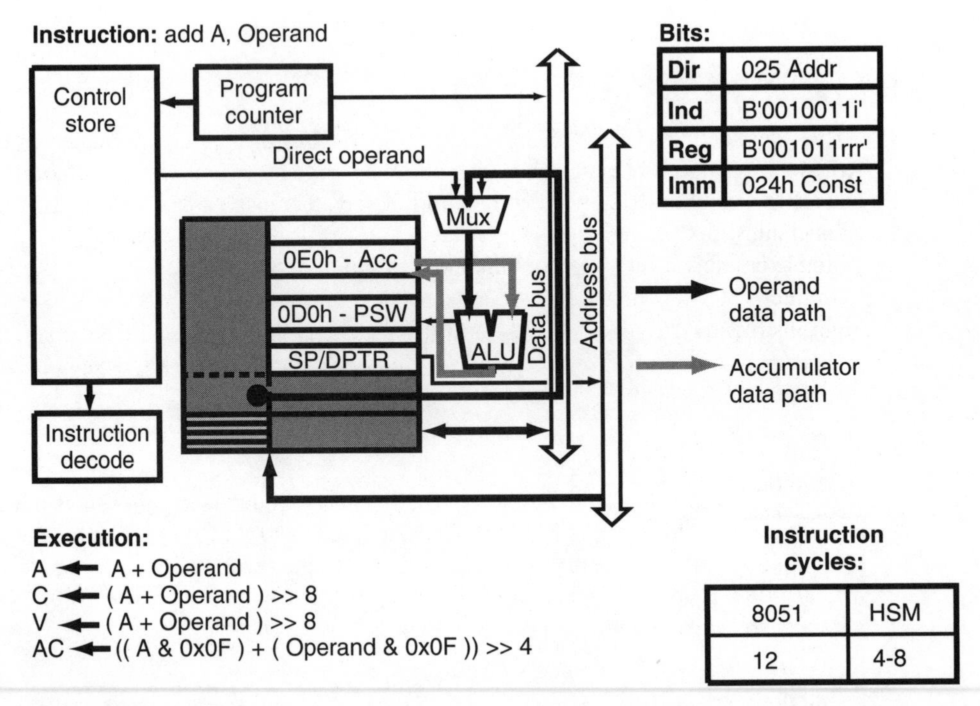

Example: **MCU:** 8051

```
mov A,   #77                    ;  Add 100 to 77
add, A, #100
```

FIGURE 3-16 `add` **instruction.**

A few notes on `DA A`. The first is that `DA A` does not work correctly after subtraction. For BCD in the 8051, there is no way to indicate negative values (so there is no easy way to subtract BCD values). Finally, `DA A` is not used to convert a hex byte to a BCD value.

The `mul AB` instruction (Fig. 3-23) will multiply two 8-bit numbers together (which were stored in the accumulator and B register). The result is 16-bits long and is stored back into the accumulator (low byte) and B (high byte). This hardware multiplication is quite fast and can be used for implementing audio frequency signal DSP functions on the 8051.

The `div AB` instruction (Fig. 3-24) divides the contents of the accumulator by the contents of the B register. The quotient is put into A and the remainder is put into B.

This instruction is probably most useful for data conversion. For example, converting a byte into a 3-digit decimal number could be accomplished by:

```
mov   A, Number           ; Store the number
mov   B, #100             ; want to get the hundreds and remainders
div   AB
mov   Hundreds, A         ; Store the hundreds of the number
```

```
mov   A, B                ; Repeat for tens and ones
mov   B, #10
div   AB
mov   Tens, A             ; Store the tens of the number
mov   Ones, B             ; Store the ones of the number
```

This can be compared to the code used in other microcontrollers to do hex-to-decimal conversions. This code runs much faster, is simpler, and takes up less space than what is possible in many other microcontrollers.

In Intel's 8051 documentation, it is suggested that `div AB` be used as a fast multiple shift right instruction. Personally, I wouldn't use it for that purpose because it cannot be used easily for 16-bit numbers (which is what I find I'm often shifting), but it could be very effective carrying out a virtual shift on single bytes.

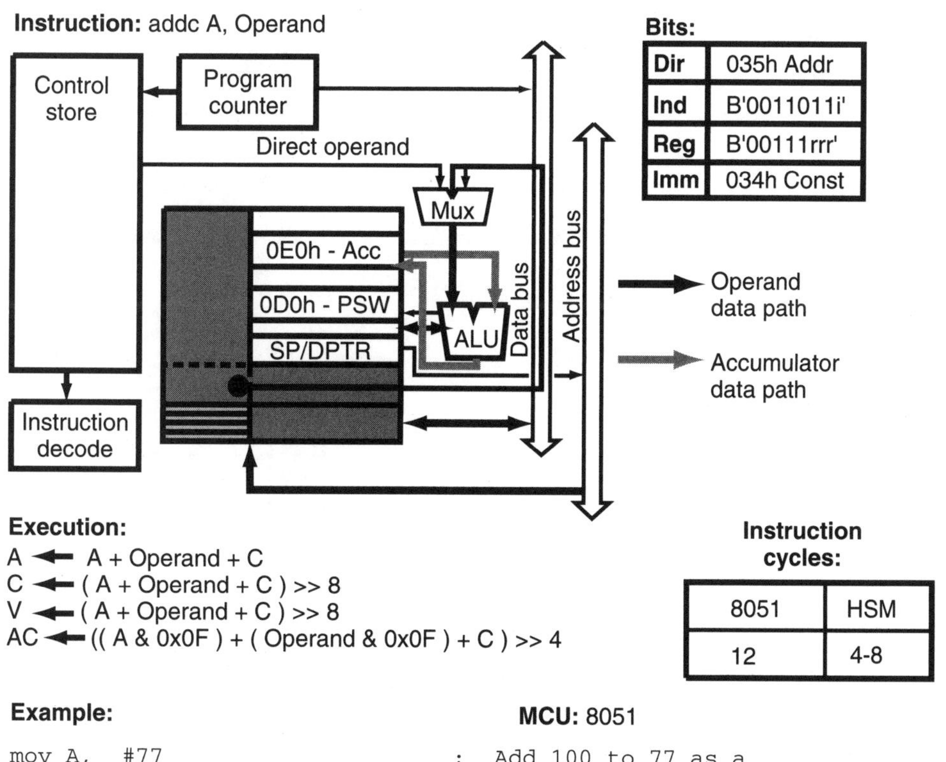

Example: **MCU:** 8051

```
mov A,   #77              ;  Add 100 to 77 as a
add A,   #100             ;  16 bit addition
mov Temp, A
clr A
adc A, #0
mov Temp+1, A             ;  Store result's high
                          ;  8 Bits
```

FIGURE 3-17 `addc` **instruction.**

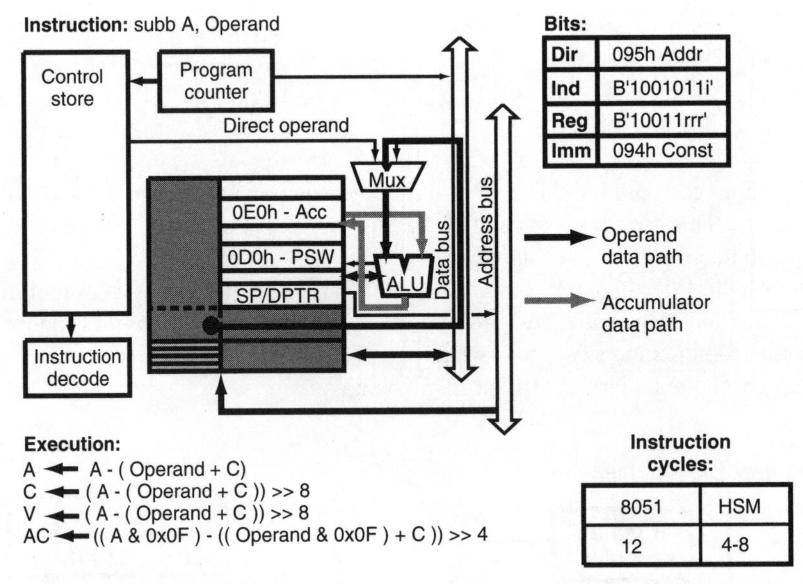

Bits:

Dir	095h Addr
Ind	B'1001011i'
Reg	B'10011rrr'
Imm	094h Const

Execution:

A ← A - (Operand + C)
C ← (A - (Operand + C)) >> 8
V ← (A - (Operand + C)) >> 8
AC ← ((A & 0x0F) - ((Operand & 0x0F) + C)) >> 4

Instruction cycles:

8051	HSM
12	4-8

Example: **MCU:** 8051

```
mov A,  #100                    ;  Subtract 77 form 100
add A,  #77
```

Note: Subb is a subtract with borrow operation. Before subtracting, make sure Carry is in the correct state

FIGURE 3-18 `subb` **instruction.**

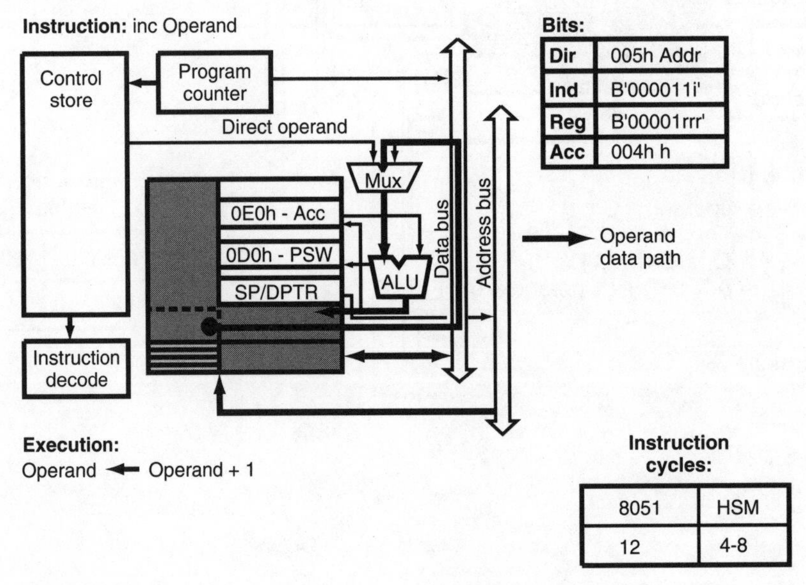

Bits:

Dir	005h Addr
Ind	B'000011i'
Reg	B'00001rrr'
Acc	004h h

Execution:

Operand ← Operand + 1

Instruction cycles:

8051	HSM
12	4-8

Example: **MCU:** 8051

```
inc R0                          ;  Increment R0 of the
                                ;  current bank
```

FIGURE 3-19 `inc` **instruction.**

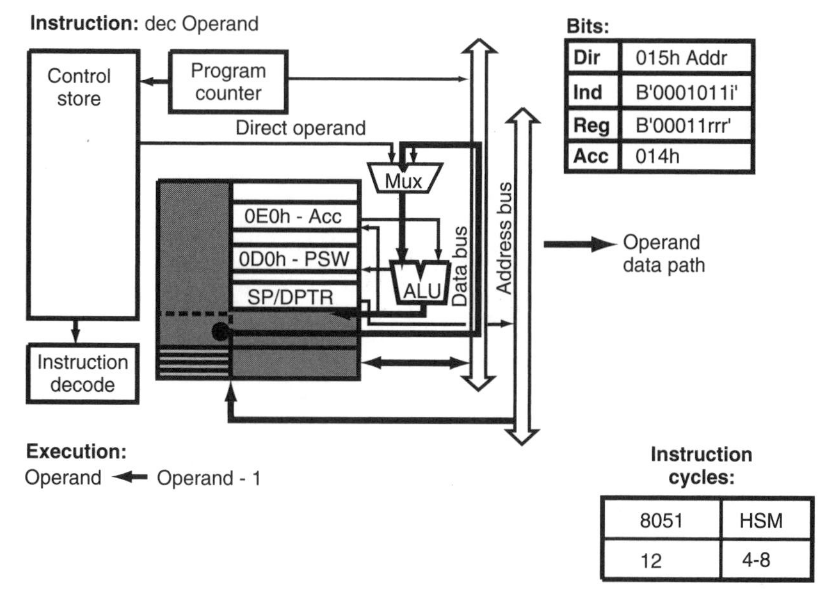

Example:

MCU: 8051

```
dec @R0                    ; Decrement the byte
                           ; pointed to by R0
```

FIGURE 3-20 `dec` **instruction.**

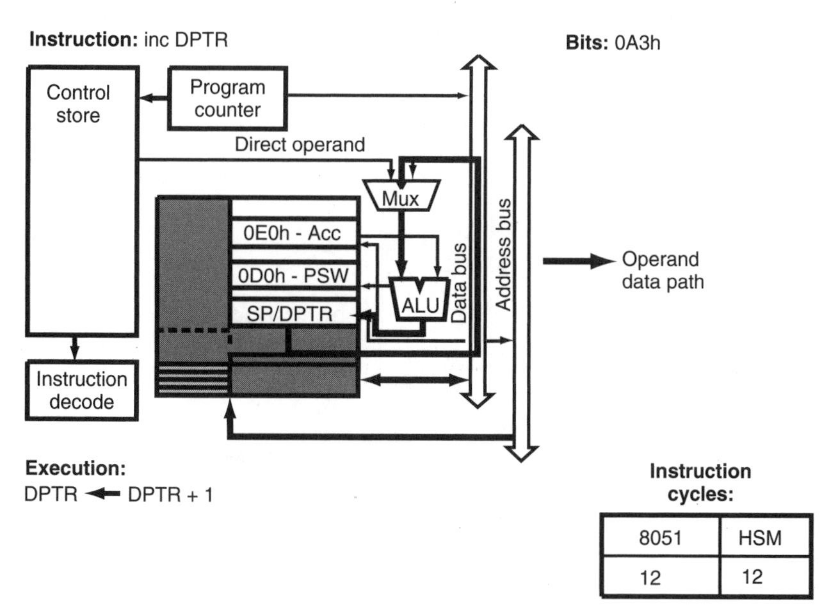

Example:

MCU: 8051

```
inc DPTR                   ; Increment the pointer to
                           ; external memory
```

FIGURE 3-21 `inc DPTR` **instruction.**

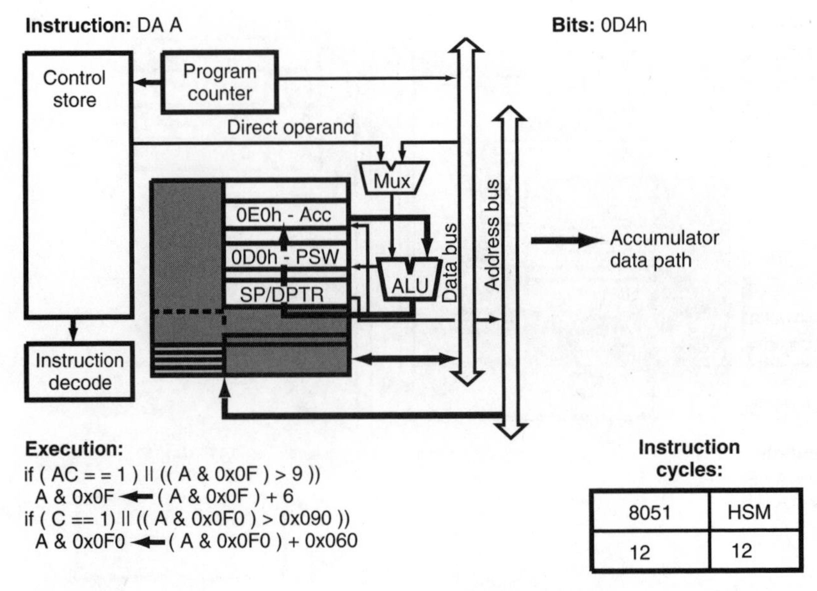

Execution:

if (AC = = 1) || ((A & 0x0F) > 9))
 A & 0x0F ← (A & 0x0F) + 6
if (C == 1) || ((A & 0x0F0) > 0x090))
 A & 0x0F0 ← (A & 0x0F0) + 0x060

Instruction cycles:

8051	HSM
12	12

Example: **MCU:** 8051

```
mov A, #077h                    ;  Add two BCD numbers
add A, #0442h                   ;  Note that the high
da  A                           ;  4 bits will be "1"
```

FIGURE 3-22 `DA A` **instruction.**

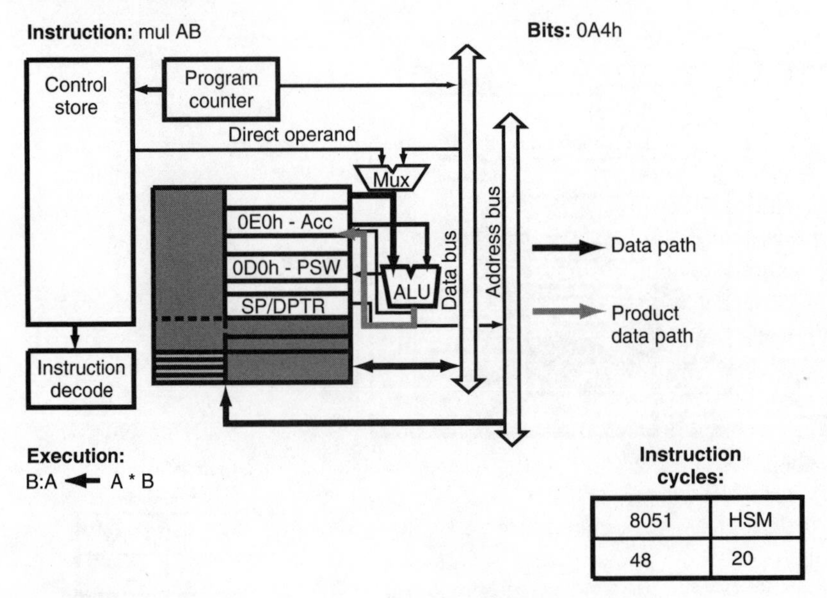

Execution:

B:A ← A * B

Instruction cycles:

8051	HSM
48	20

Example: **MCU:** 8051

```
mov A,  #077h                   ;  Find the product of
mov B,  #042h                   ;  two numbers
mul AB
```

FIGURE 3-23 `mul AB` **instruction.**

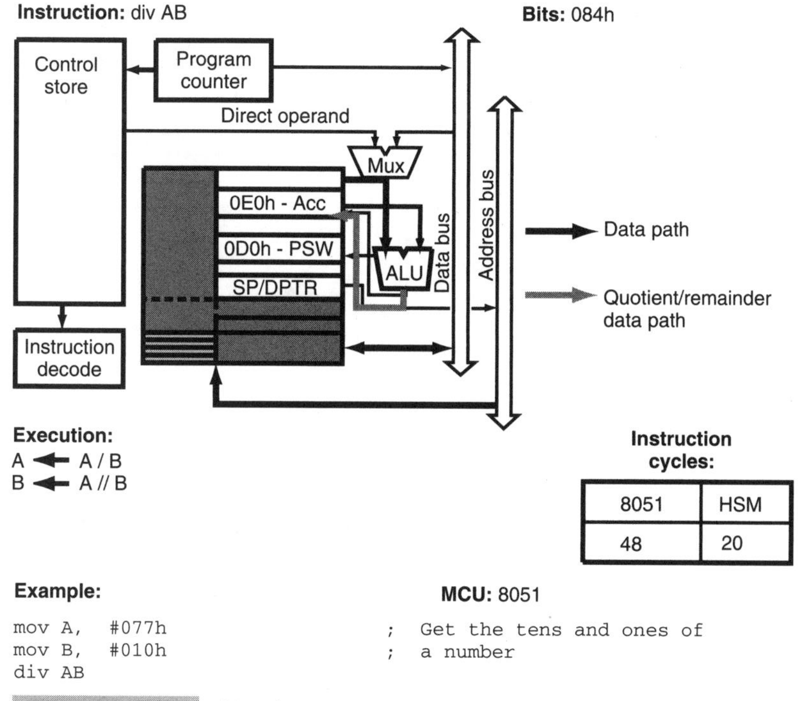

FIGURE 3-24 `div AB` **instruction.**

Bit Operators

Manipulating bits is important in any computer processor; however, in a microcontroller, these functions take on very significant importance. This is due to the requirement of many applications to handle bit-sized I/O. The 8051's bit instructions (which are known as *logical instructions* and *boolean variable instructions*) are well-defined and will make your applications easier to develop.

The "standard" bitwise operations are available in the 8051. These include ANDing an 8-bit value with another value (Figs. 3-25 through 3-27), ORing the bits of value with another (Figs. 3-28 through 3-30) and XORing a value with another value (Figs. 3-31 through 3-33). These three types of instructions do not affect any of the PSW bits (although a conditional jump on zero can be implemented with the `jnz` and `jz` instructions).

One of the nice features of the 8051 is that individual bits can be manipulated as if they were full bytes. The `anl C, Bit` and `orl C, Bit` (Figs. 3-34 and 3-37) instructions perform logical operations on the carry flag and another bit with the result being stored in the carry flag. The results of these operations can be stored or used in `jc` and `jnc` (jump on carry flag state) instructions. The bit being operated on can be inverted as well.

This feature is useful if you wanted to implement a bit XOR. Remembering that XOR is defined as:

$$A \wedge B = (A \,\&\, !B) \mid (!A \,\&\, B)$$

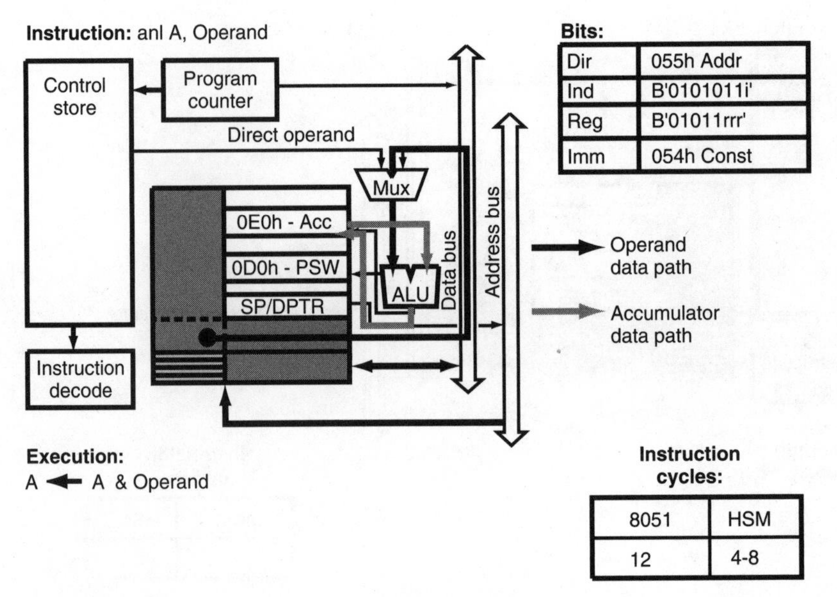

Example: **MCU:** 8051

```
mov A,   #077                  ;  Add 100 to 77
anl A,   #100                  ;  to get 68 decimal
```

FIGURE 3-25 `anl` **instruction.**

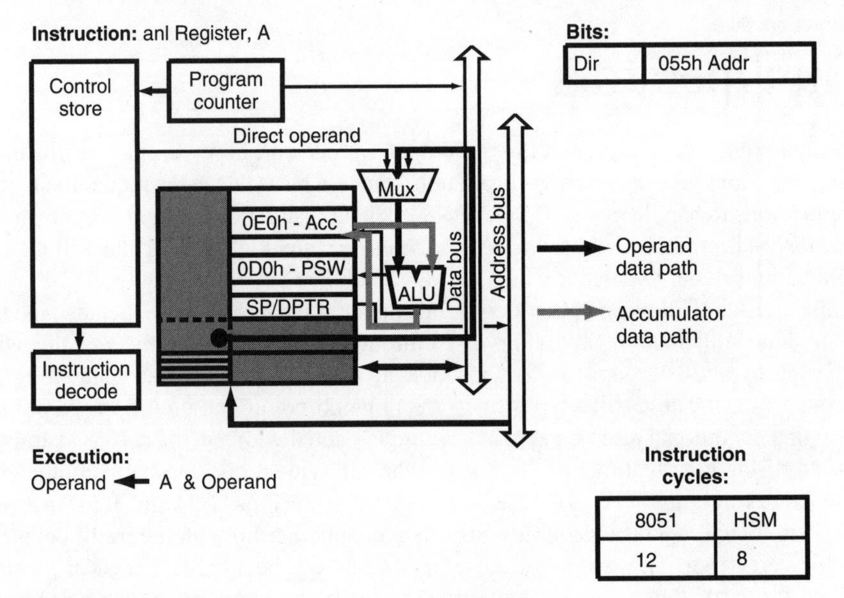

Example: **MCU:** 8051

```
mov A,   #77                   ;  And the register with
anl Register, A                ;  77
```

FIGURE 3-26 `anl direct, Acc` **instruction.**

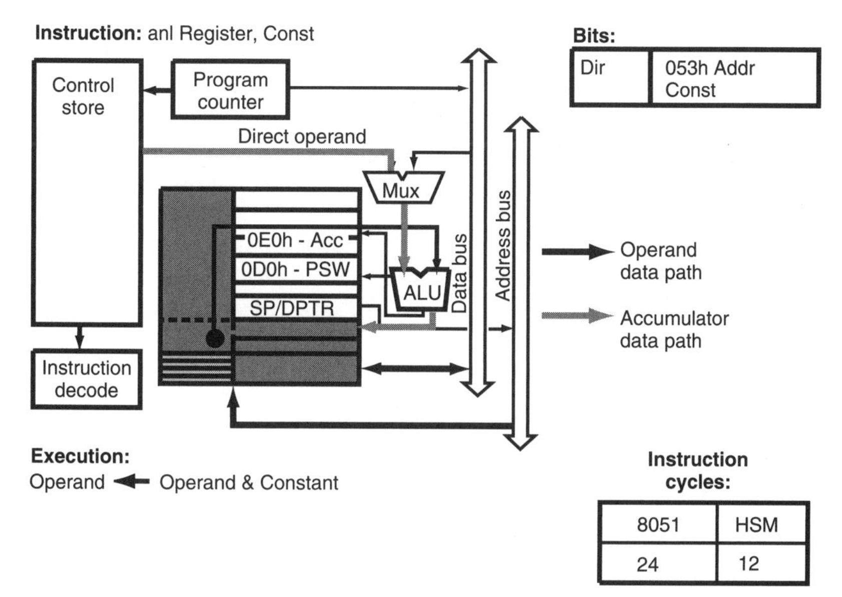

Example: **MCU:** 8051

```
anl P0, #1                  ;  Clear everything but
                            ;  P0.0
```

FIGURE 3-27 `anl direct,` *`Const`* **instruction.**

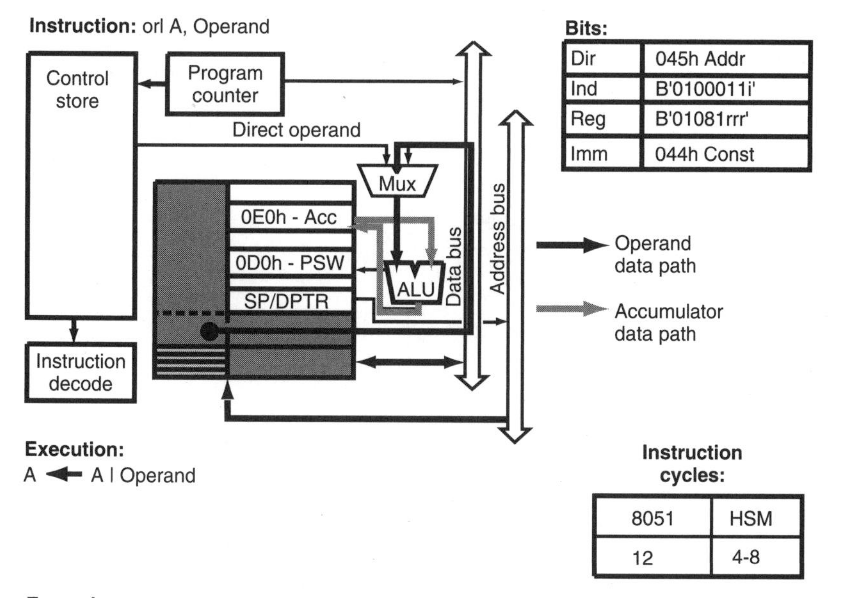

Example: **MCU:** 8051

```
mov A, #77                  ;  OR 100 to 77
orl A, #100                 ;  to get 109 decimal
```

FIGURE 3-28 `orl` **instruction.**

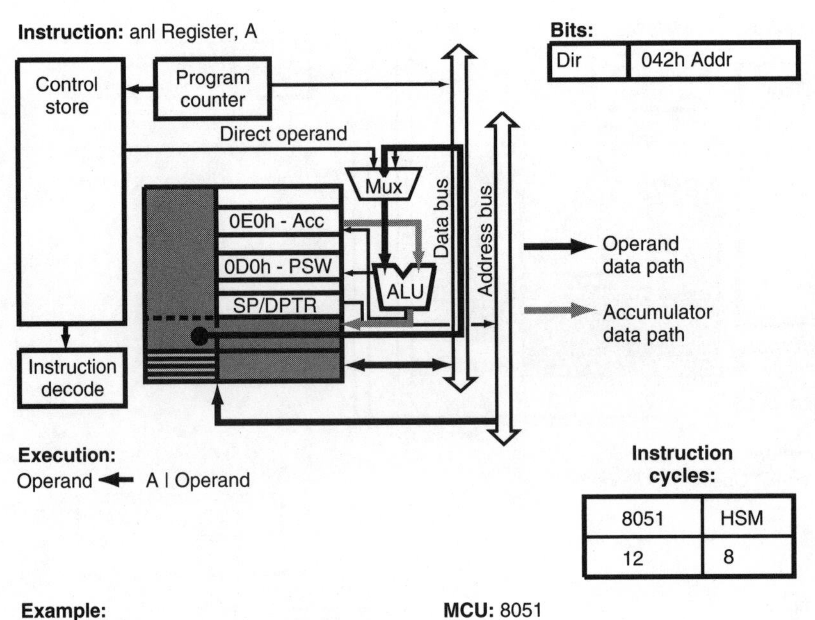

Example: **MCU:** 8051

```
mov A, #77                  ;  OR the register with
orl Register, A             ;  77
```

FIGURE 3-29 `orl direct, Acc` **instruction.**

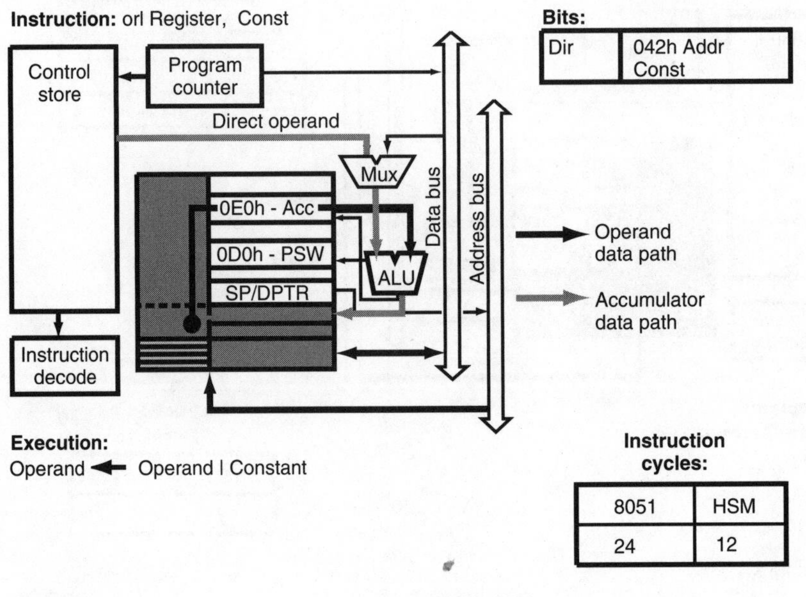

Example: **MCU:** 8051

```
orl P0, #1                  ;  Set P0.0
```

FIGURE 3-30 `orl direct, Const` **instruction.**

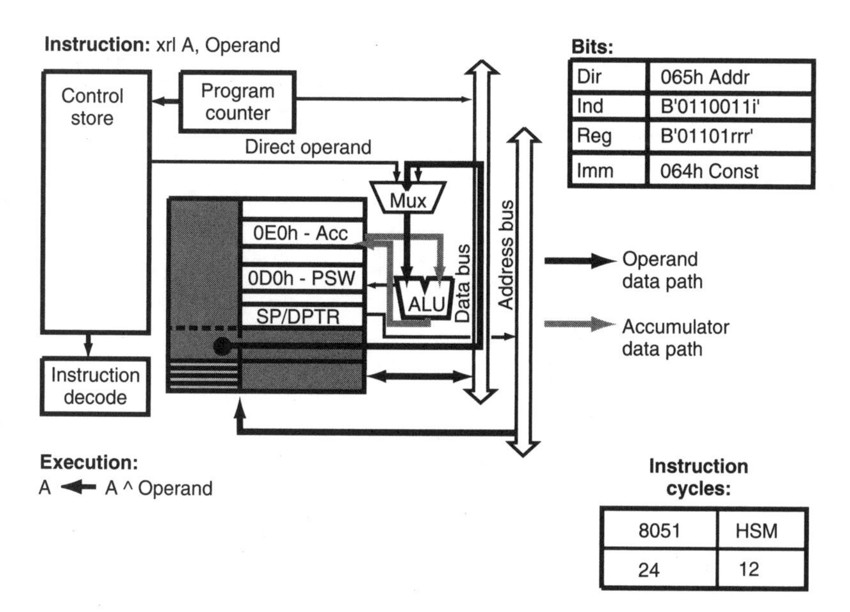

Example:

MCU: 8051

```
mov A, #77           ; XOR 100 to 77
xrl A, #100          ; to get 41 decimal
```

FIGURE 3-31 `xrl` **instruction.**

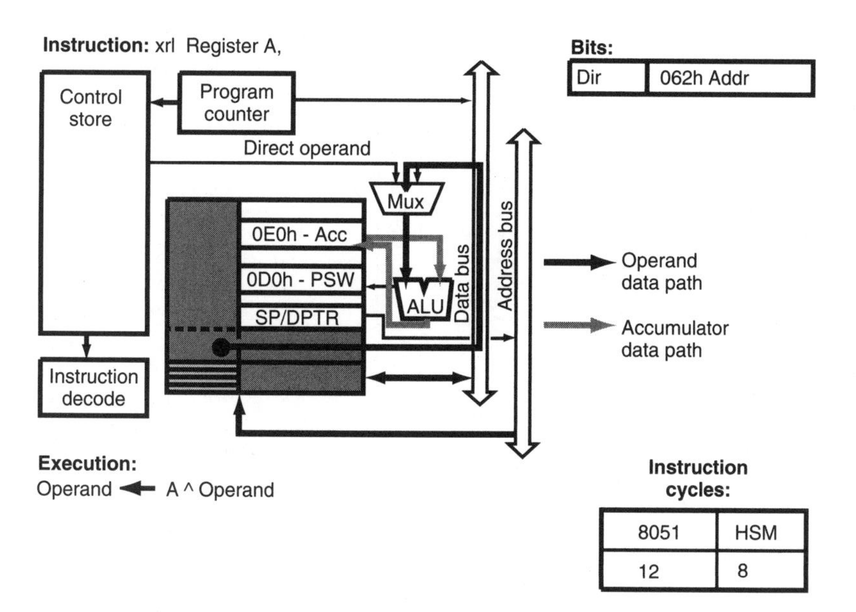

Example:

MCU: 8051

```
mov A, #77           ; XOR the register with
xrl Register, A      ; 77
```

FIGURE 3-32 `xrl direct, Acc` **instruction.**

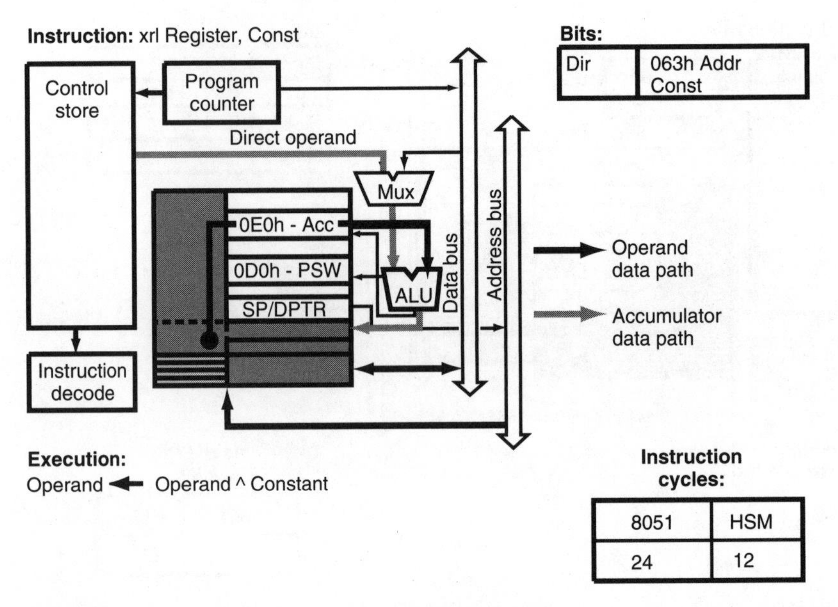

Example: **MCU:** 8051

```
xrl P0, #1                    ; Toggle P0.0
```

FIGURE 3-33 `xrl direct,` *`Const`* **instruction.**

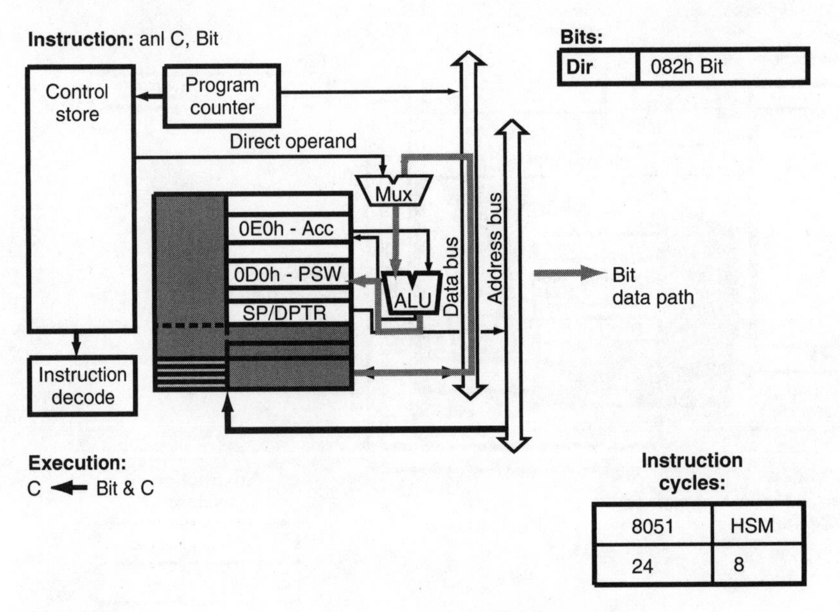

Example: **MCU:** 8051

```
anl C,P1.4                    ; AND the bit with carry
```

Note: Bits greater than 0x07F are in the special-function register block

FIGURE 3-34 `anl C, Bit` **instruction.**

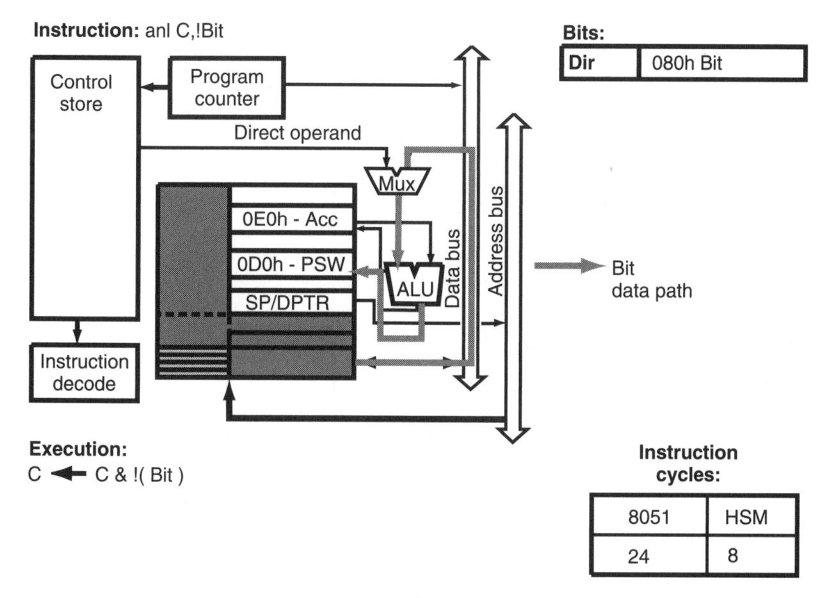

8051	HSM
24	8

Example: **MCU:** 8051

```
anl C, ! (P1.4)            ; AND the inverted bit with
                           ; carry
```

Note: Bits greater than 0x07F are in the special-function register block

FIGURE 3-35 `anl C, !Bit` **instruction.**

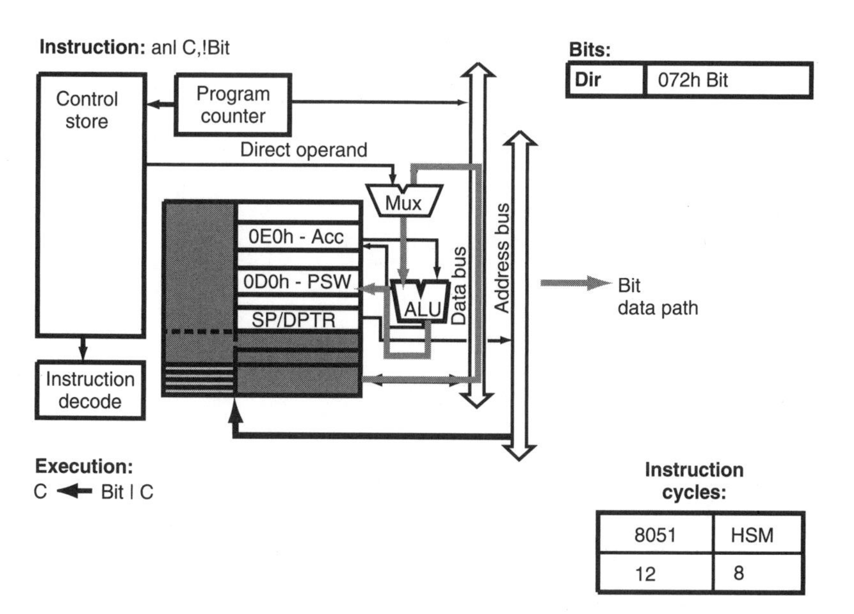

8051	HSM
12	8

Example: **MCU:** 8051

```
orl C,P1.4                 ; OR the bit with carry
```

Note: Bits greater than 0x07F are in the special-function register block

FIGURE 3-36 `orl C, Bit` **instruction.**

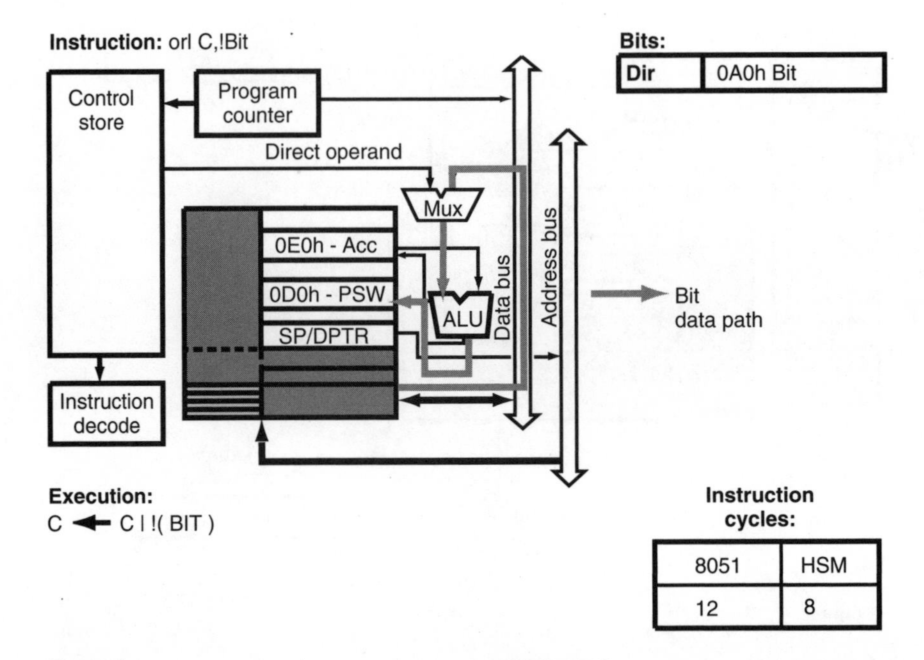

Example: **MCU:** 8051

```
orl C, ! (P1.4)                ;  OR the inverted bit with
                               ;  carry
```

Note: Bits greater than 0x07F are in the special-function register block

FIGURE 3-37 `orl C, !Bit` **instruction.**

you could code a bit XOR operation as the following macro:

```
MACRO xorlbit Parml, Parm2      ; XOR Parml and Parm2 and store result in
  mov  C, Parml                 ;  "Carry"
  anl  C, !Parm2                ; Do "A & !B"
  mov  ACC.5, C                 ; Store the result
  mov  C, Parm2                 ; Do "!A & B"
  anl  C, !Parml
  orl  C, ACC.5                 ; Combine the results
ENDMACRO
```

Note that I store the temporary result (the value of `Parml & !Parm2`) in one of the temporary flags in the PSW. This operation might be halted halfway through by an interrupt. If this is to be used as an autonomous instruction, you might want to mask interrupts that change this flag before executing the XOR function.

Along with logical operations on bits and bytes, full bytes and bits can be cleared (Figs. 3-38 through 3-40), bits can be set (Figs. 3-41 and 3-42), and bits and bytes can be complemented (Figs. 3-43 through 3-45). The complement instructions invert the state of the bits (either as a single bit or XORing the byte with 0FFh). Note that, for each operation, when a bit is specified, it can either be the carry flag or a specific bit in the register space or a bit in ACC, B, P1, P2, P3, PSW or IE.

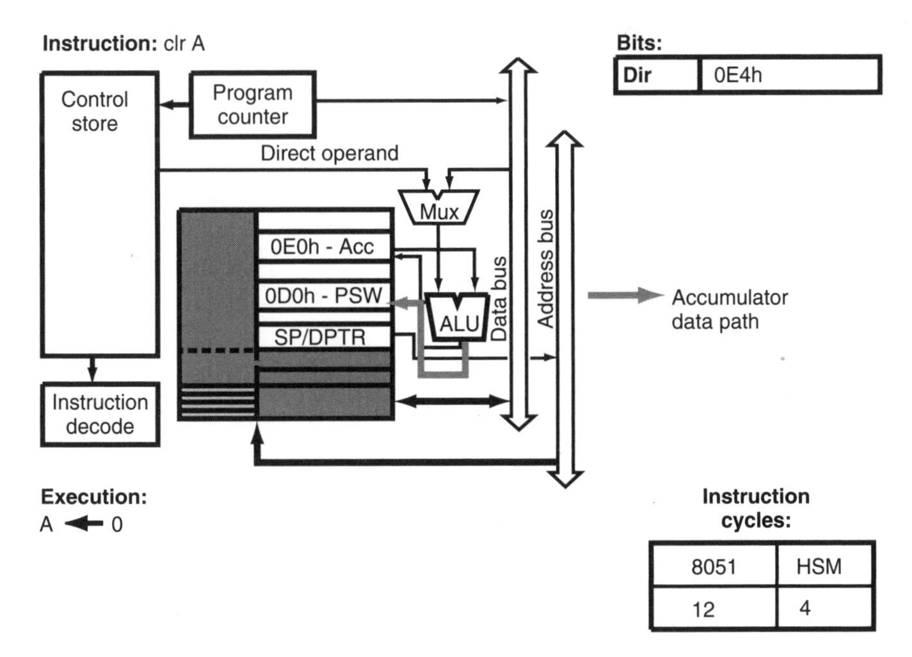

Example: **MCU:** 8051

```
clr A                              ; Clear the accumulator
```

FIGURE 3-38 `clr A` **instruction.**

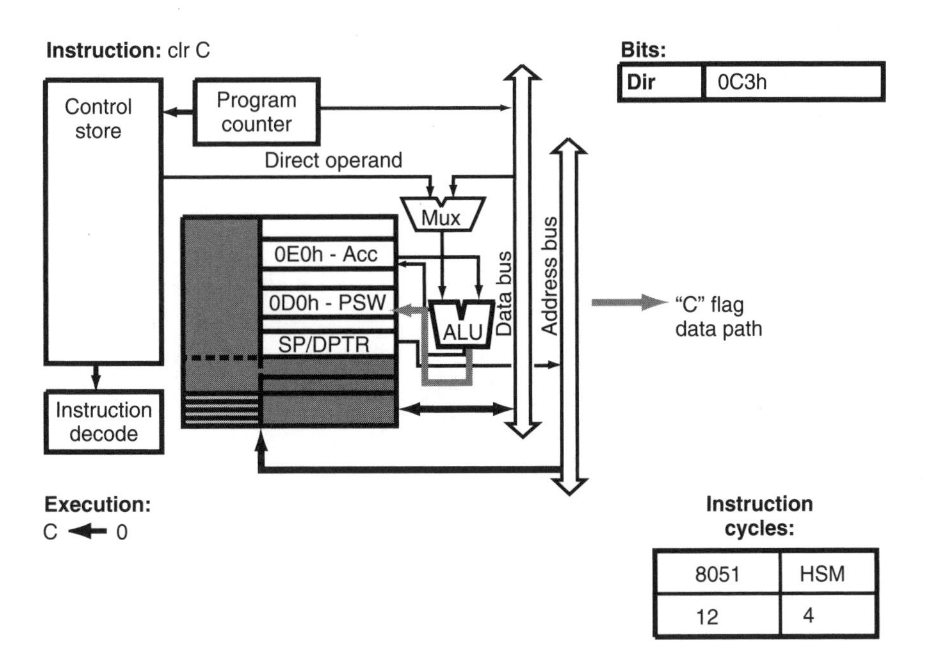

Example: **MCU:** 8051

```
clr C                              ; Clear the carry flag
```

FIGURE 3-39 `clr C` **instruction.**

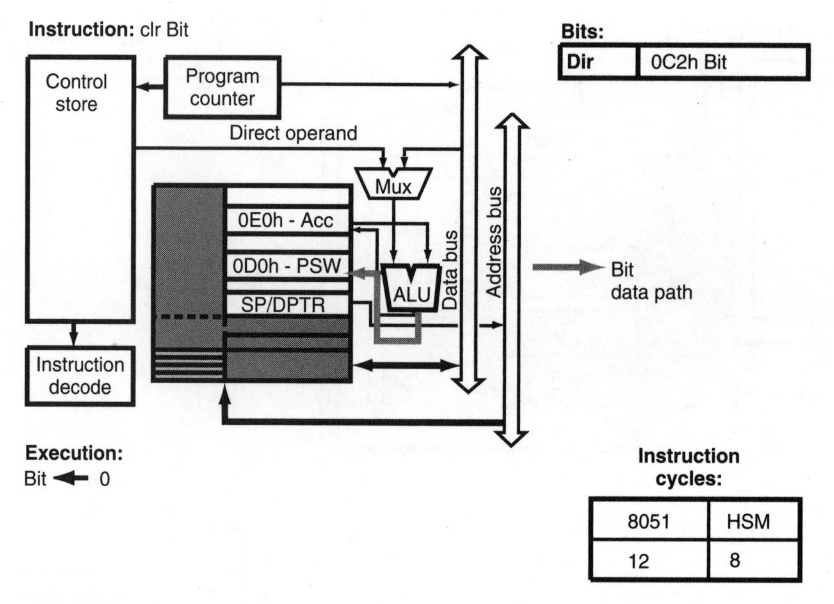

Example: **MCU:** 8051

```
clr P2.3                          ;  Clear the I/O bit
```

Note: Bits greater than 0x07F are in the special function-register block

FIGURE 3-40 `clr Bit` **instruction.**

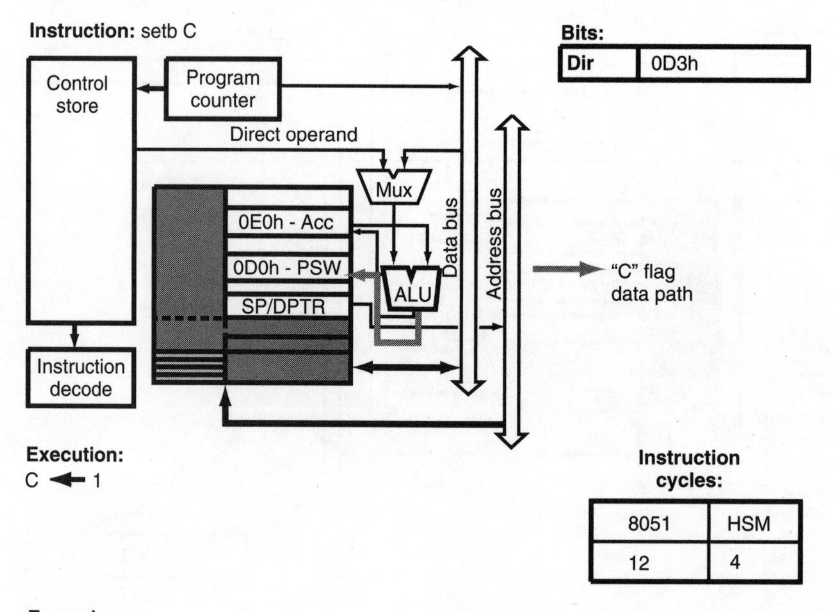

Example: **MCU:** 8051

```
setb C                            ;  Set the carry flag
```

FIGURE 3-41 `setb C` **instruction.**

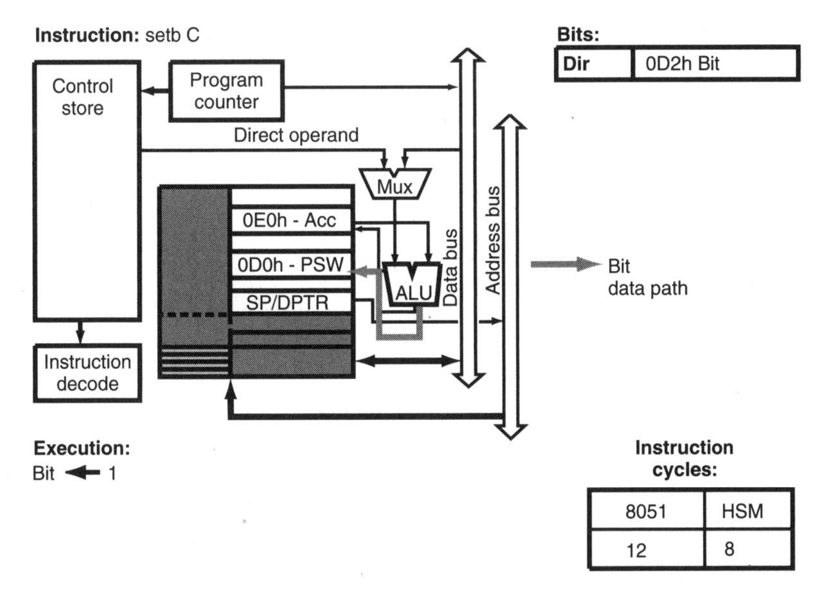

Example: **MCU:** 8051

```
setb PSW.RS0                  ; Change the current
                              ; register bank
```

Note: Bits greater than 0x07F are in the special-function register block

FIGURE 3-42 `setb Bit` **instruction.**

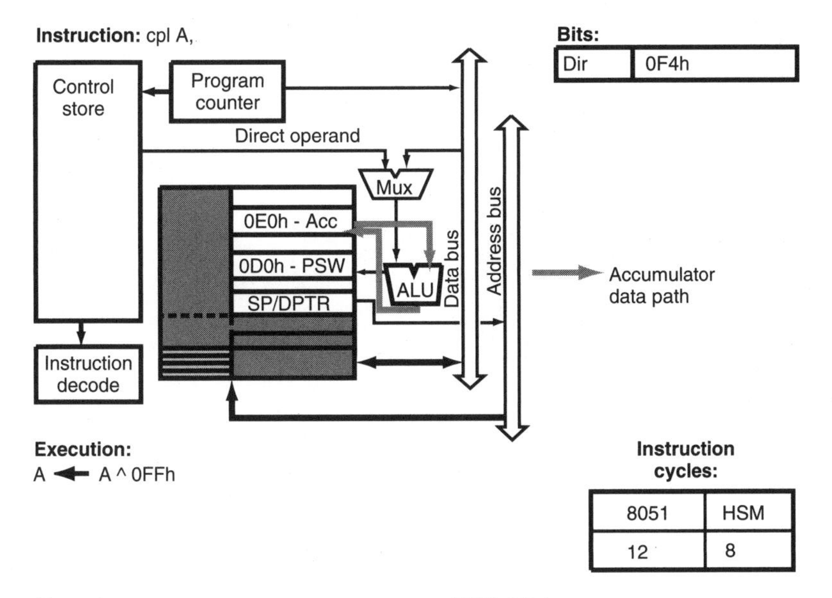

Example: **MCU:** 8051

```
cpl A                         ; Negate the value in the
inc A                         ; accumulator
```

FIGURE 3-43 `clr` **instruction.**

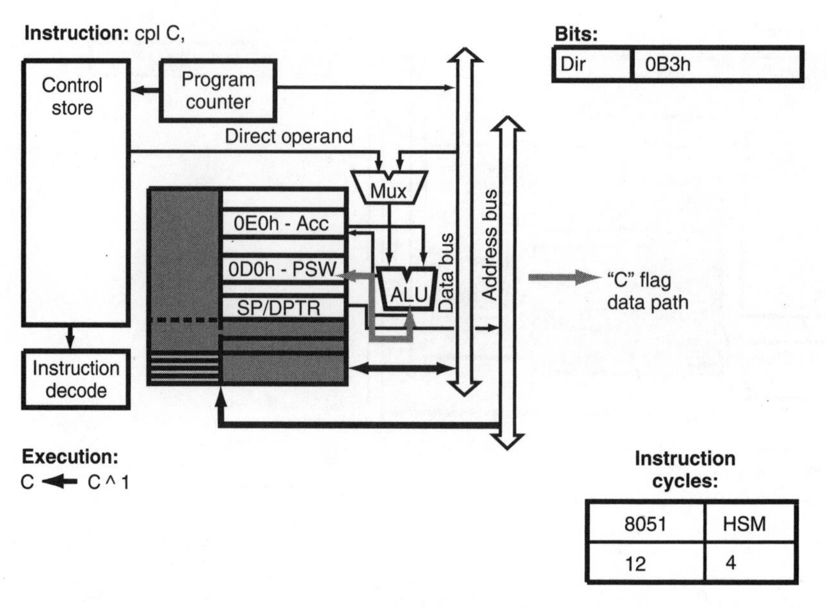

Example: **MCU:** 8051

```
cpl C                          ;  Complement the carry flag
```

FIGURE 3-44 `cpl C` **instruction.**

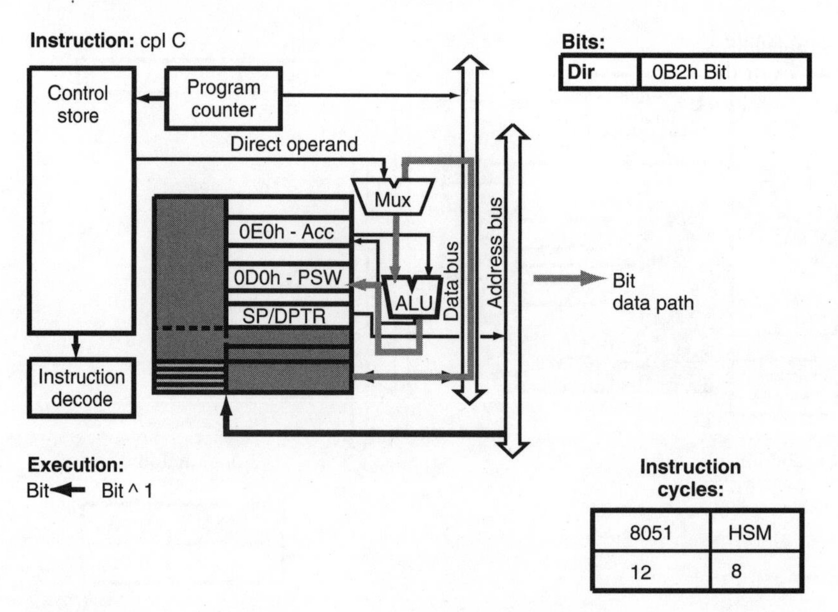

Example: **MCU:** 8051

```
cpl P1.4                       ;  Complement the I/O bit
```

Note: Bits greater than 0x07F are in the special-function register block

FIGURE 3-45 `cpl Bit` **instruction.**

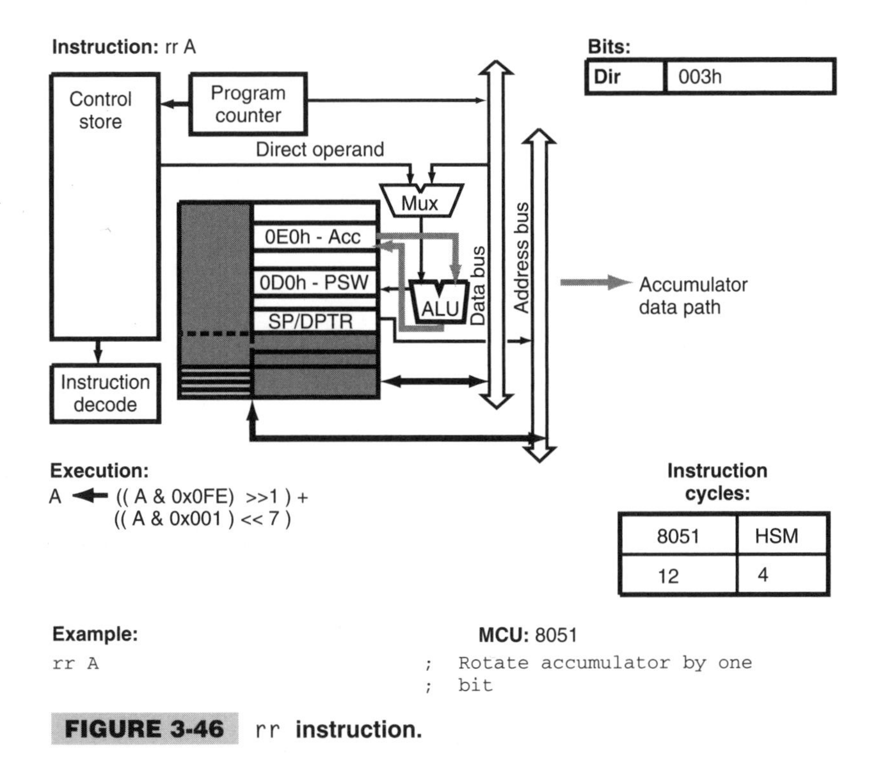

FIGURE 3-46 `rr` **instruction.**

The rotate instructions are useful for moving data bits into specific places for testing (i.e., from the carry flag) or shifting in and out (Figs. 3-46 through 3-49). Personally, I mostly only use the `rlc` and `rrc` instructions. Simply rotating a value within itself is not that useful and potentially disastrous when used as a power of two multiply or divide.

The last bit instruction doesn't modify bits at all. The `swap` instruction (Fig. 3-50) exchanges one nybble for another. This instruction is useful when you want to display the two digits in a byte (as two nybbles).

Execution Change Operators

Jumping around in the 8051 is very straightforward, with the only real wrinkle being a result of how the PSW register works. Without a zero flag, the `jz` and `jnz` instructions work differently than you might expect and definitely differently than how the other microcontrollers execute this instruction.

For jumping to new addresses, there are three different modes of operation. The *small jump* (`sjmp`) is a relative jump from –128 to +127 byte addresses from the start of the next instruction. Despite its limitations, the small jump is probably the instruction that will be used the most in your 8051 applications.

The *page jump* (`ajmp` and `acall`) instruction (Fig. 3-51) jumps to an offset within the current 11-bit (2K) address page. This means that care must be taken *not* to jump outside the current page.

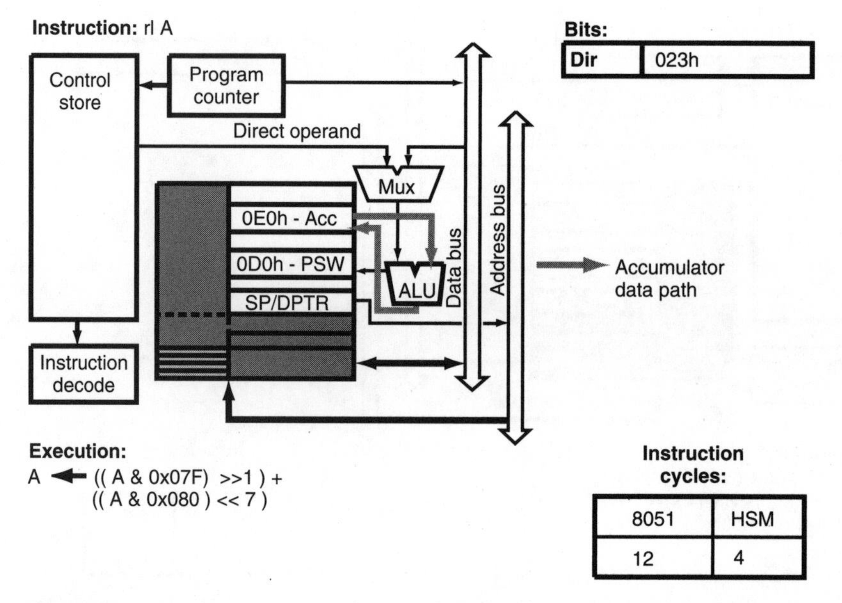

Example: **MCU:** 8051

```
rl A                    ; Rotate accumulator by one
                        ; bit
```

FIGURE 3-47 `rl` **instruction.**

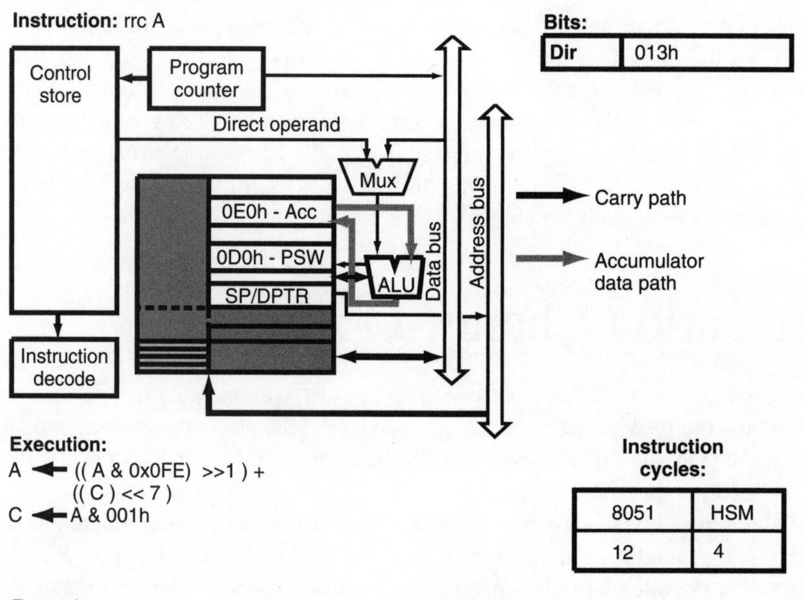

Example: **MCU:** 8051

```
rrc A                   ; Rotate accumulator by one
                        ; bit through the carry flag
```

FIGURE 3-48 `rrc` **instruction.**

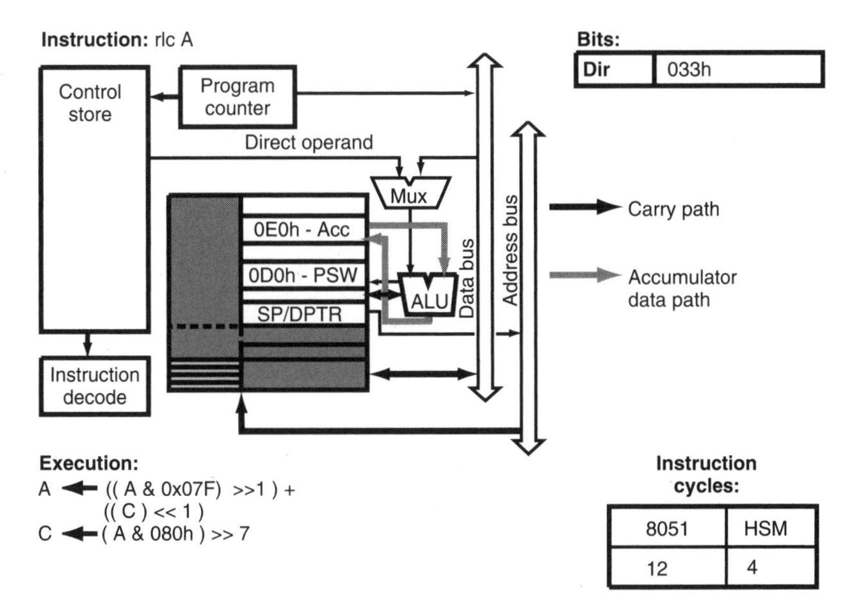

8051	HSM
12	4

Example: **MCU:** 8051

```
rlc A                              ;  Rotate accumulator by one
                                   ;  bit through the carry flag
```

FIGURE 3-49 `rlc` **instruction.**

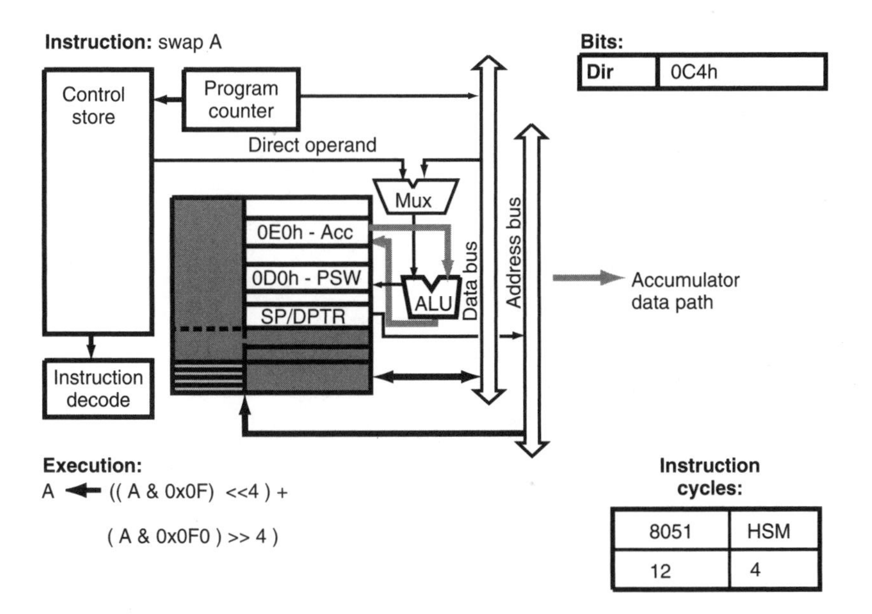

8051	HSM
12	4

Example: **MCU:** 8051

```
swap A                             ;  Swap the nybbles in the
                                   ;  accumulator
```

FIGURE 3-50 `swap` **instruction.**

In Fig. 3-51, you can see how you can get into trouble with the `ajmp` instruction. The second time it is used, the jump actually goes to the "LabelB" offset within the current page, but "LabelB" is not in the same page. This means that the jump will stay in the current page and go to the wrong absolute address.

The last type of jump is the *long jump* (`ljmp`), which will allow jumps anywhere in the 64K control store address space. This jump would probably be preferable in all cases, but it requires an extra byte of storage as well as an extra instruction cycle in some 8051s (namely the HSMs).

The different types of jumps are shown in the `jmp` instruction explanation (Fig. 3-52).

Subroutines in the 8051 work largely as expected with the `call` instruction (Fig. 3-53) return address pushed onto the stack, and at the end of the subroutine, a `ret` instruction (Fig. 3-54) is executed to pop the return address from the stack. The `call` instruction can use either page or long addressing.

In some 8051 assemblers, simply specifying `jmp` or `call` as the instruction will cause the assembler to select the "best" jump addressing mode. This type of operation is preferable because it cuts down on the amount of thinking the programmer has to do and automatically compensates when the program grows in size.

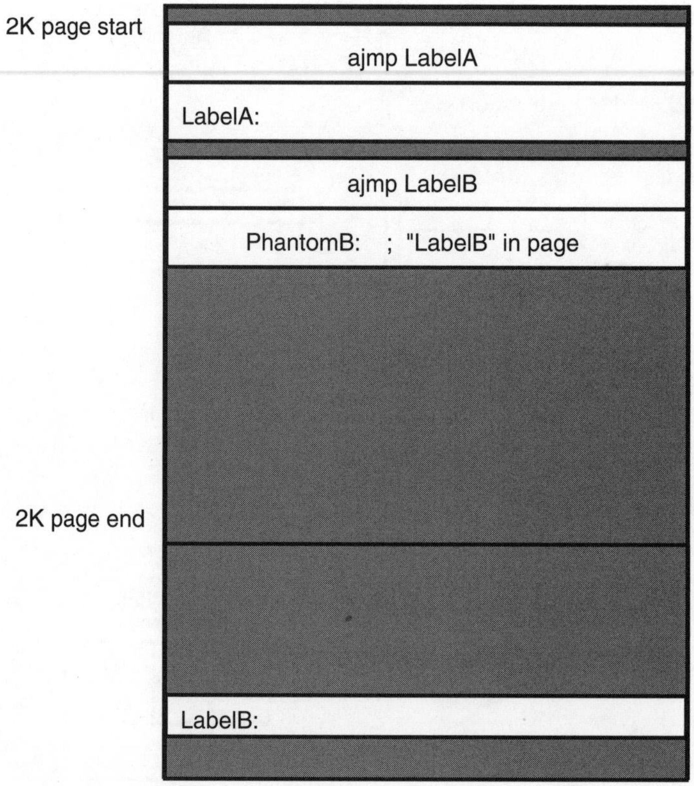

FIGURE 3-51 Jumps within the current 2K page.

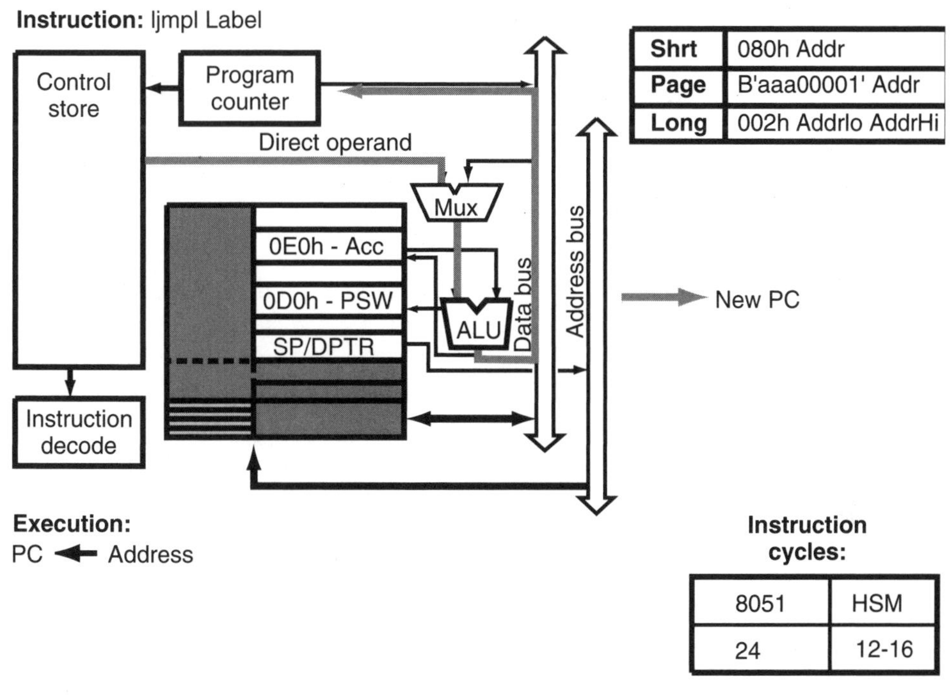

Example: **MCU:** 8051

```
sjmp Label                     ;  jump to "Label"
  :
Label:
```

FIGURE 3-52 `sjmp, ajmp, ljmp` **instructions.**

If you are familiar with the operation of other processors, you might be surprised to discover that the stack pointer *increments* as part of the `push` operation. This is opposite to many other devices where a `push` causes the stack pointer to be decremented. The issue with this is in setting up stack areas. The stack pointer will have to be initialized to the *start* of the stack area (rather than at the *end* of the stack area, as is done in most other processors).

`reti` (Fig. 3-54) works just like the `ret` instruction except, if it is executed within an interrupt handler, the 8051s interrupt hardware will be reset to allow waiting for the next interrupt.

The operation of the `reti` instruction really precludes the use of nested interrupts (which is possible in some of the other devices). This is another reason why the interrupt handlers should be as short (in terms of instructions and cycles) as possible.

Table jumps are accomplished using the `jmp @A + DPTR` instruction (Fig. 3-55). In this case, DPTR is set up with the start of a table and the accumulator is set up with the offset within the table.

```
mov  DPTR, #Table          ;  DPTR = Start of table
mov  A, Index              ;  Get the index address
clr  C                     ;  Multiply by 2 for the actual address
```

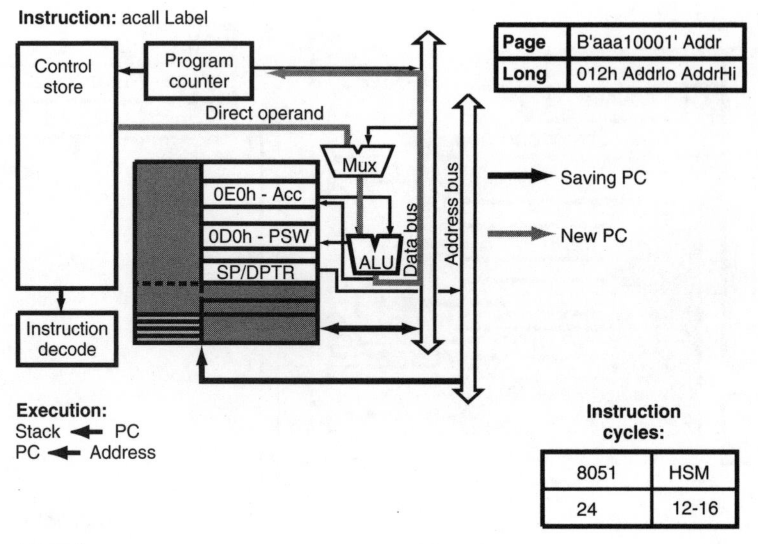

Example: **MCU:** 8051

```
lcall Label                  ;  jump to subroutine "Label"
  :
Label:
```

FIGURE 3-53 `acall` **and** `lcall` **instructions.**

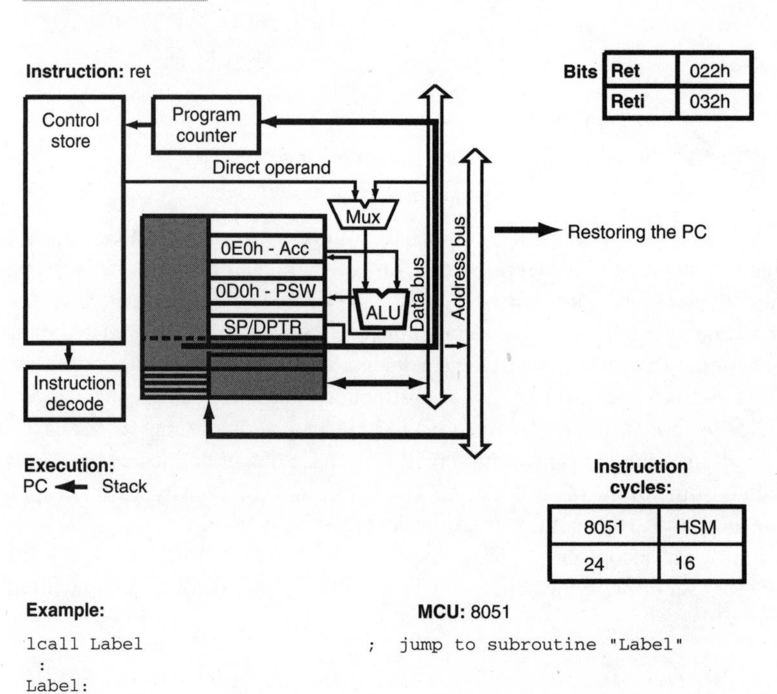

Example: **MCU:** 8051

```
lcall Label                  ;  jump to subroutine "Label"
  :
Label:
  :
ret                          ; Return from subroutine
```

Note: "reti" works similarly to "ret" but resets the 8051's internal interrupt hardware.

FIGURE 3-54 `ret` **and** `reti` **instructions.**

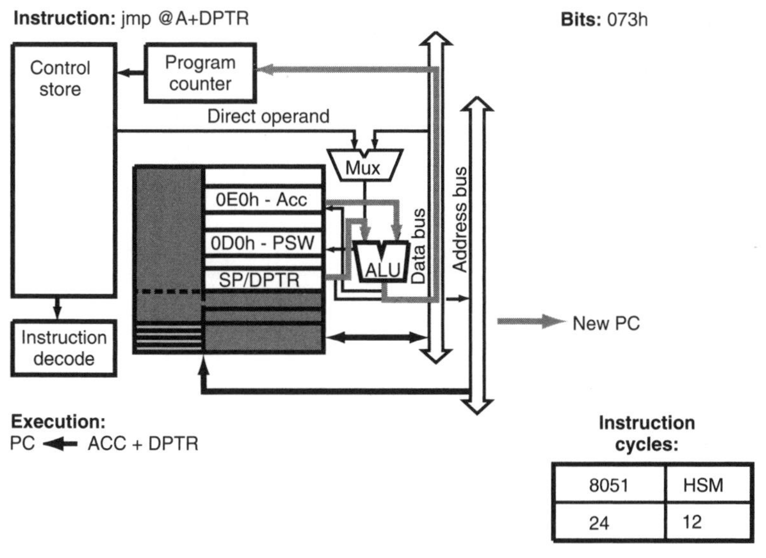

Example: **MCU:** 8051

```
mov DPTR, #Table_Start
mov A, #Table_Element
jmp @A+DPTR                    ;  jump to element in table
```

FIGURE 3-55 `jmp @A + DPTR` **instruction.**

```
  rlc  A
  jmp  @A + DPTR                 ;  Jump to the table

Table:
  ajmp       Element0            ;  Jump to the different elements
  ajmp       Element1
   :
```

In this example, the index address is multiplied by two (shifted left by one) to get the correct table offset (each "element" is 2 bytes long).

The basic conditional jumps use bit states (Fig. 3-56 through 3-58). These relative jumps are taken if the carry flag or the specified bit are either set or reset.

The `jbc Bit, Label` is an interesting instruction because the bit tested for the jump is cleared if it is set. This instruction is useful as a *semaphore* instruction in an RTOS.

As use for a semaphore test-and-set instruction, the jump would be taken if the semaphore flag is available. The code the operation jumps to would notify the task that it now has control of the bit.

The jump on zero/not zero instructions (Fig. 3-59) test the contents of the accumulator for zero as the condition for the jump. This means that, if these instructions are to be used, then after an operation affecting a value, the accumulator cannot be changed.

The last conditional jump instruction is `cjne` (Figs. 3-60 and 3-61). `cjne` does a comparison and jumps if the result is not equal to zero. These instructions can be used with a test register (the accumulator or another register) not set to zero or set to a value that is

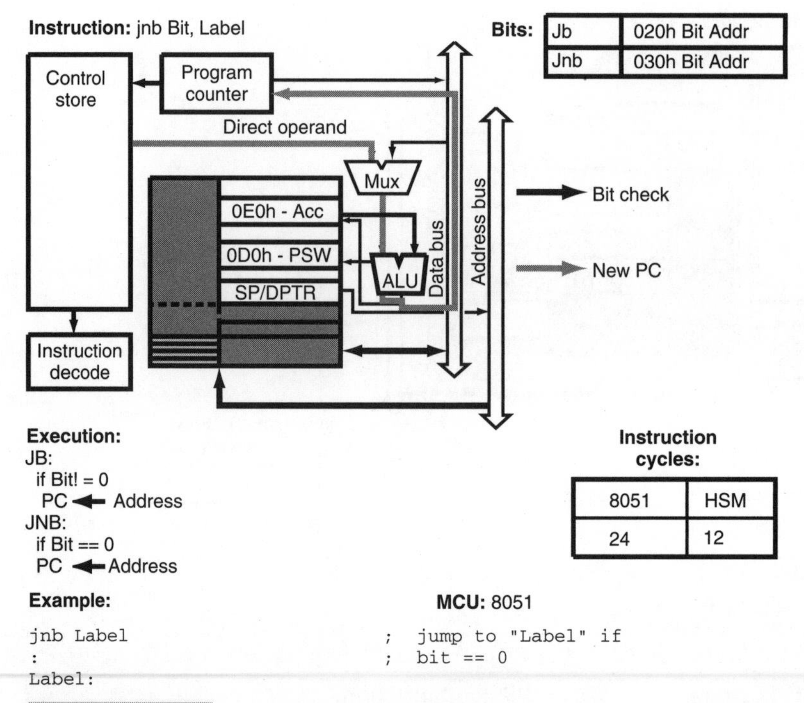

FIGURE 3-56 `jb` **and** `jnb` **instructions.**

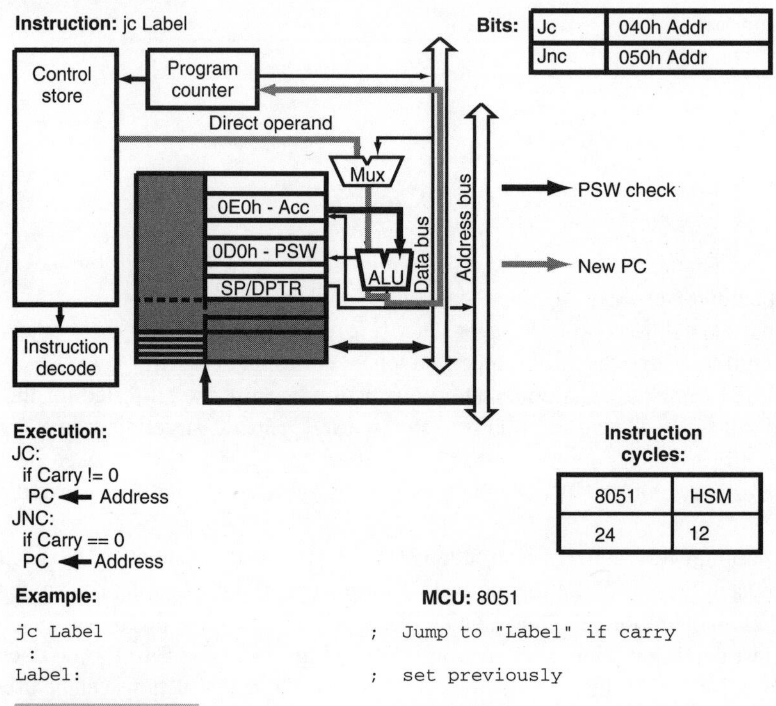

FIGURE 3-57 `jc` **and** `jnc` **instructions.**

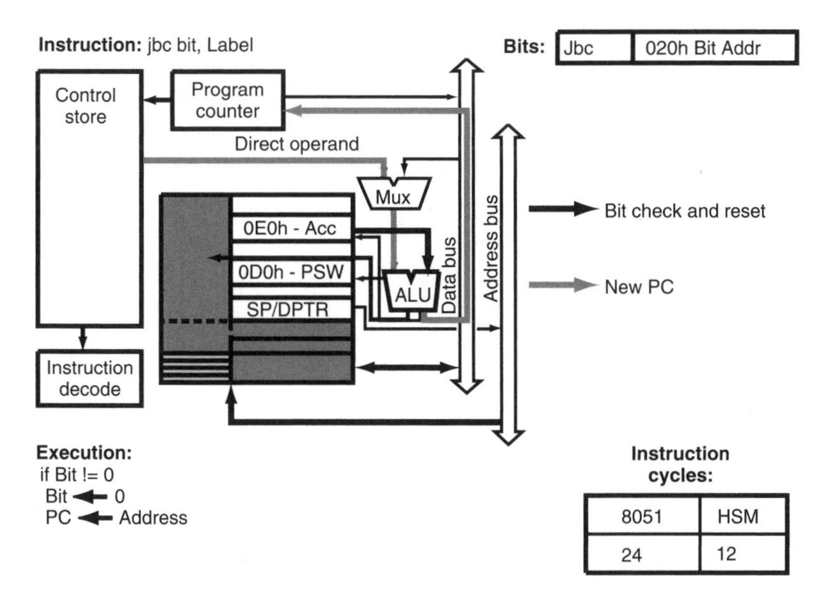

FIGURE 3-58 `jbc` **instruction.**

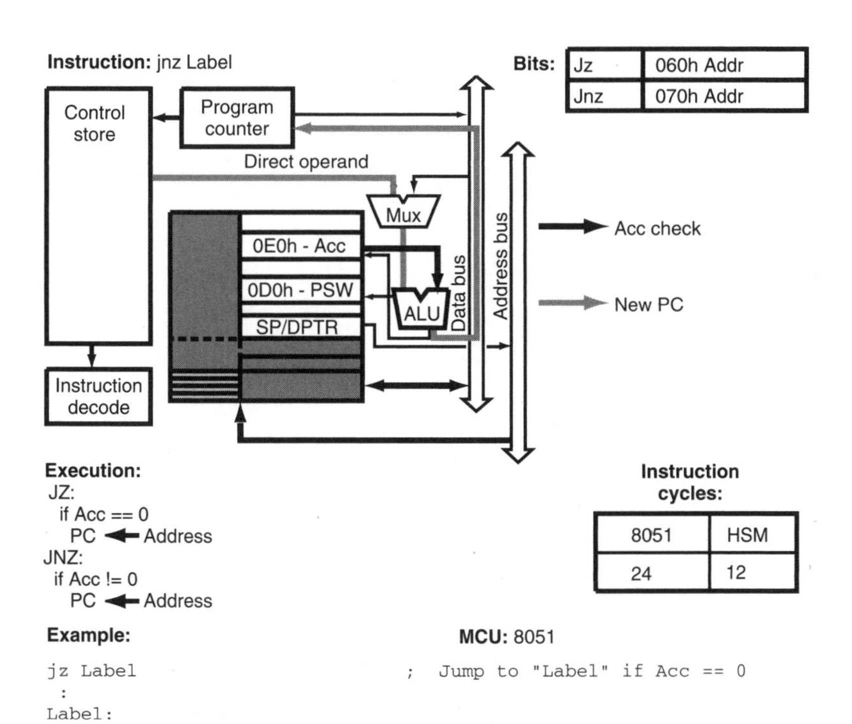

FIGURE 3-59 `jz` **and** `jnz` **instructions.**

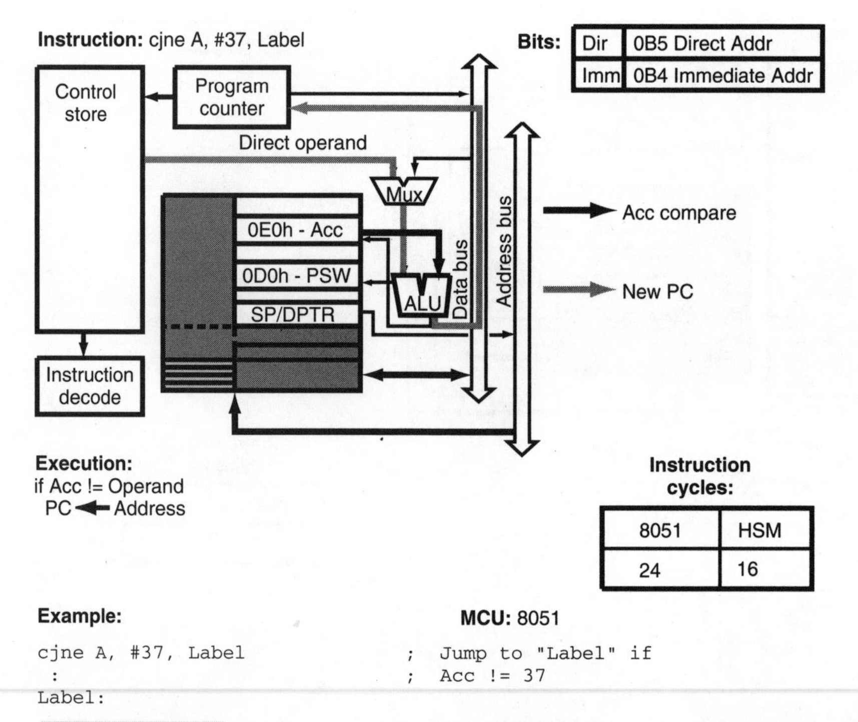

FIGURE 3-60 `cjne A, Operand` **instruction.**

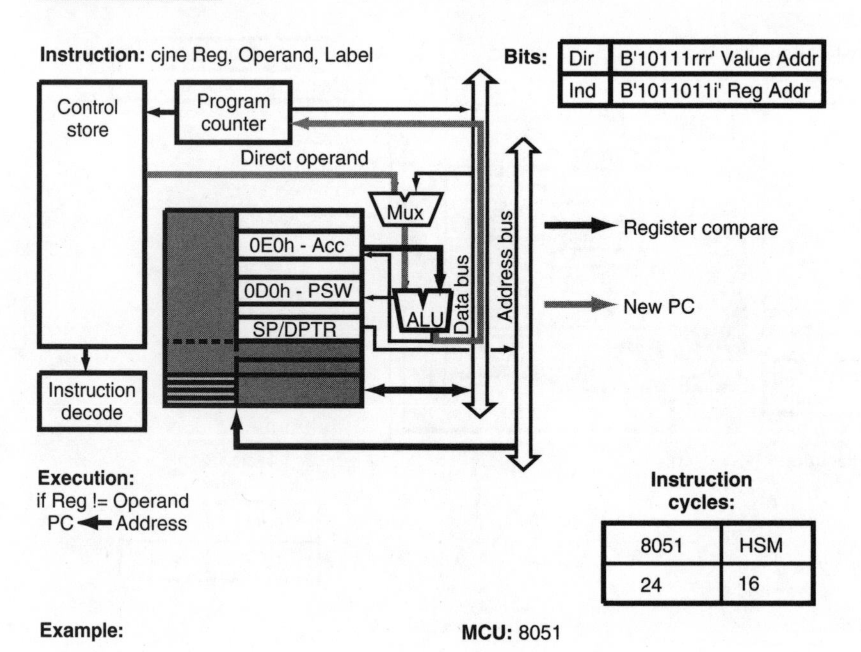

FIGURE 3-61 `cjne Register, #Data` **instruction.**

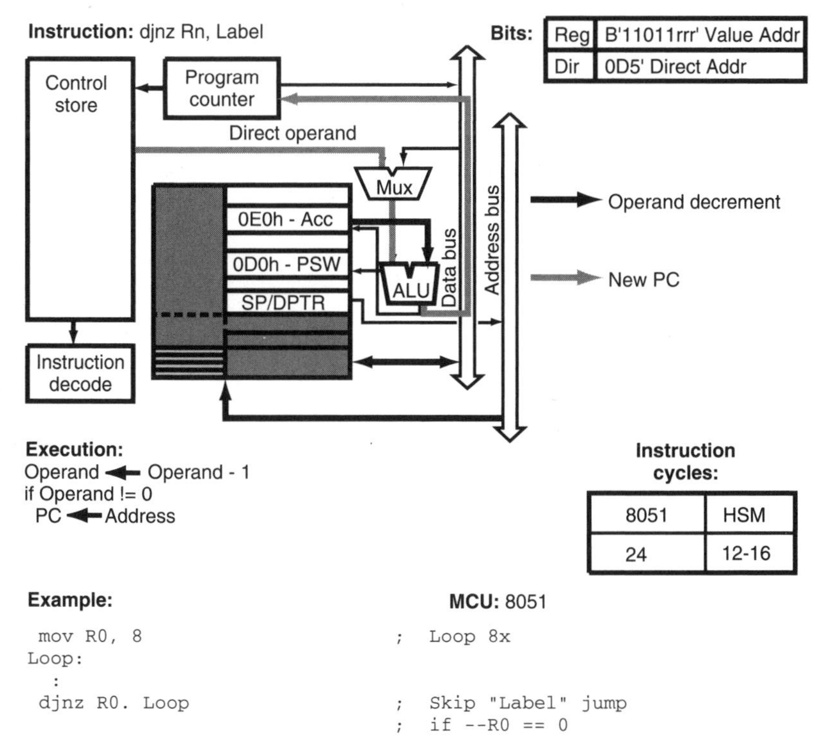

FIGURE 3-62 `djnz` **instruction.**

used for the jump instruction. Instead of just checking for zero, the compare value can be any arbitrary byte as well.

`djnz` instruction (Fig. 3-62), which decrements and jumps if not zero, is used for counting loops.

```
  mov  R2, #7
Loop:                             ; Loop 7
  :
  djnz      R2, Loop              ; Decrement counter and jump if not zero
```

This example code will loop seven times before continuing.

The last instruction (and in many ways the least) is the `nop` (Fig. 3-63). The `nop` will simply take up a byte and one instruction cycle to operate. The primary purpose of the `nop` is to help ensure timing is correct. Although as it is equal to 00h, it can be used to “take out” erroneous instructions when debugging an application, rather than erasing and reprogramming the device.

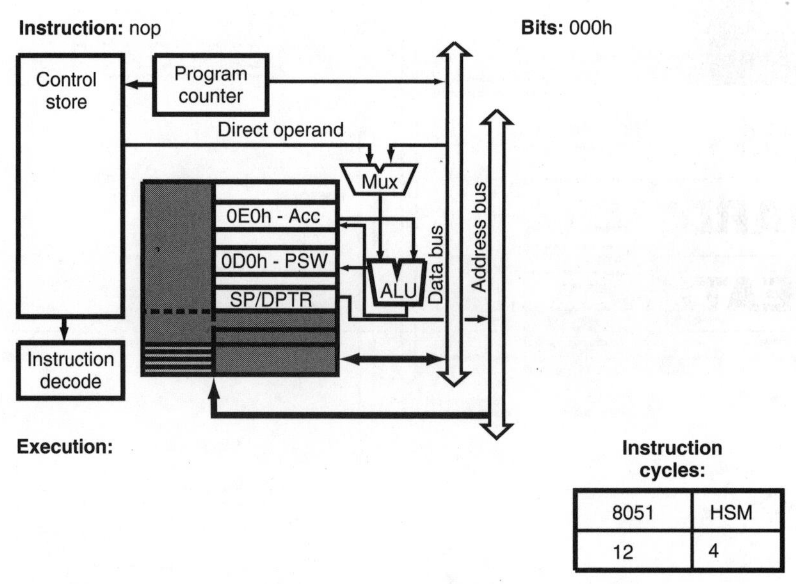

Execution:

Instruction cycles:

8051	HSM
12	4

Example: **MCU:** 8051

```
nop                               ;  Delay one instruction cycle
```

FIGURE 3-63 `nop` **instruction.**

4

8051 HARDWARE FEATURES

CONTENTS AT A GLANCE

As I indicated earlier, the "basic" 8051 is a fairly complete device with timers, interrupts, and 32 parallel I/O pins capable of interfacing to external devices via microprocessor busses and serial I/O. Understanding how these features work is critical in being able to create your own applications. In the next two chapters, I will introduce you to the 8051 hardware features and their various characteristics and then follow up by discussing the practicalities of designing 8051 applications in the most efficient manner possible.

FIGURE 4-1 Various 8051 chips.

When I talk about the 8051, I am really referencing a number of different devices (Fig. 4-1). The devices ending in a "1" (i.e., 87C51) are the "basic" 8051, with two timers and a single serial port. The devices ending in a "2" (i.e., 80C52) are "enhanced" 8051s, with three timers and dual serial ports (which will be discussed in the next chapter). It is also important to make the distinction between the devices with no built-in control store (the 803*x* parts) and the 805*x* parts that do.

The special function registers (the SFRs from 080h to 0FFh) are shown for the three different devices in Table 4-1. You might be surprised to see that I have marked the 8031 as not having P0 and P2 ports; the pins are definitely there when you look at a data sheet. These two ports are not available in the 8031 because they are used for interfacing to external memory. (This is also true for the 8051 and 8052 when they have external memory).

Device Packaging

When I use the term *device packaging*, I am describing the material (known as *encapsulant*) that is used to protect the chip and the interconnect technology used to connect the chip electrically to the printed circuit card (which I call the *raw card*). There are quite a few options in this area, and selecting the appropriate ones to use can have a significant impact on the final application's cost, size, and quality.

There are two primary types of encapsulation used to protect chips: plastic and ceramic. Plastic encapsulants are the most prevalent and use an epoxy "potting compound" that is injected around a chip after it has been wired to a lead frame.

The lead frame becomes the pins used on the package and is wired to the chip via very thin aluminum wires ultrasonically bonded to both the chip and the lead frame. Some chips are attached to the lead frame using C4 technology, which is described later.

Once the encapsulant has hardened, the chip is protected from light, moisture, and physical damage from the outside world. EPROM microcontrollers in a plastic package are

TABLE 4-1 THE SPECIAL-FUNCTION REGISTERS FOR THE 8031, 8051, AND 8052 CHIPS

ADDRESS	REGISTER IDENTIFIER	8031	8051	8052
080h	P0		X	X
081h	SP	X	X	X
082h	DPL	X	X	X
083h	DPH	X	X	X
087h	PCON	X	X	X
088h	TCON	X	X	X
089h	TMOD	X	X	X
08Ah	TL0	X	X	X
08Bh	TL1	X	X	X
08Ch	TH0	X	X	X
08Dh	TH1	X	X	X
090h	P1	X	X	X
098h	SCON	X	X	X
099h	SBUF	X	X	X
0A0h	P2		X	X
0A8h	IE	X	X	X
0B0	P3	X	X	X
0B8h	IP	X	X	X
0C8H	T2CON			X
0C9H	T2MOD			X
0CAh	RCAP2L		X	
0CBh	RCAP2H		X	
0CCh	TL2			X
0CDh	TH2			X
0D0h	PSW	X	X	X
0E0h	ACC	X	X	X
0F0h	B	X	X	X

generally referred to as *one-time programmable* (OTP) packages (Fig. 4-2). Once the EPROM has been programmed, the device cannot be used for anything else.

The primary purpose of putting a microcontroller into a ceramic package (Fig. 4-3) is that a quartz window can be built into the package for the purpose of erasing the EPROM control store.

When a ceramic package is used, the chip is glued to the bottom half and is wired to the lead frame. Ceramic packaging is usually only available as a PTH device, where plastic devices can include surface mount packages.

Ceramic packaging can drive up the cost of a single chip dramatically (as much as 10 times more than the price of a plastic OTP packaged device). This makes this type of packaging only suitable for such uses as application debugging, where the advantage of the window for erasing outweighs the extra cost of the package.

The technology used to attach the chip to the board has changed dramatically over the past 10 years. In the 1980s, pretty much all devices were only available in *pin-through hole* (PTH) *technology* (Fig. 4-4) in which the lead frame pins are soldered into holes in the raw card.

This type of attach technology has the advantage that it is very easy to work with (very little specialized knowledge or equipment is required to manufacture or rework boards

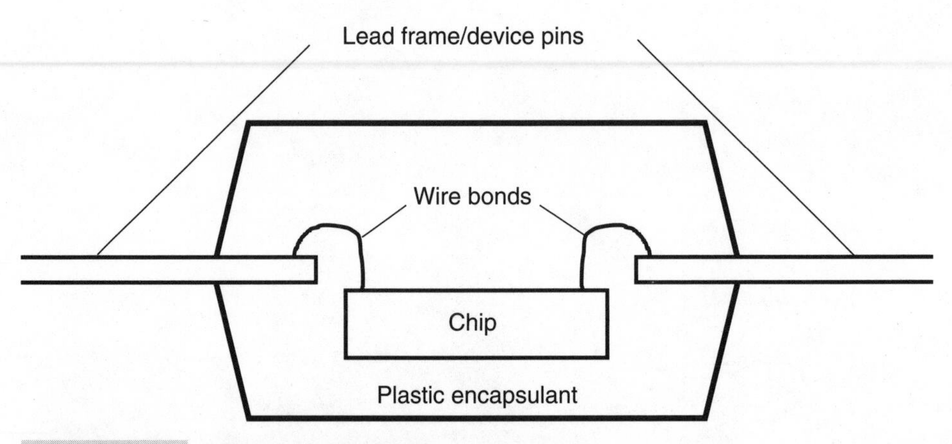

FIGURE 4-2 **OTP plastic package.**

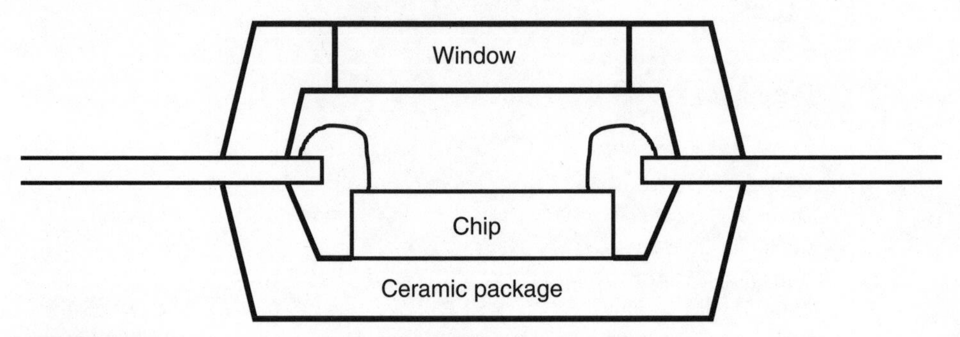

FIGURE 4-3 **Windowed ceramic package.**

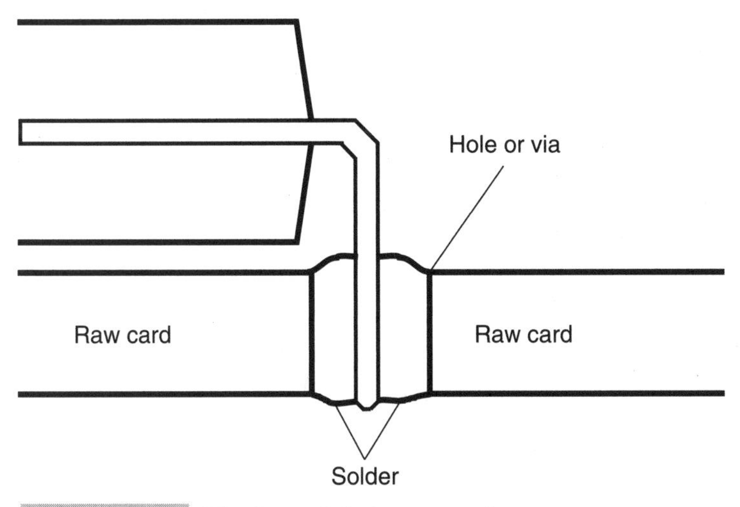

FIGURE 4-4 **Pin-through hole connection.**

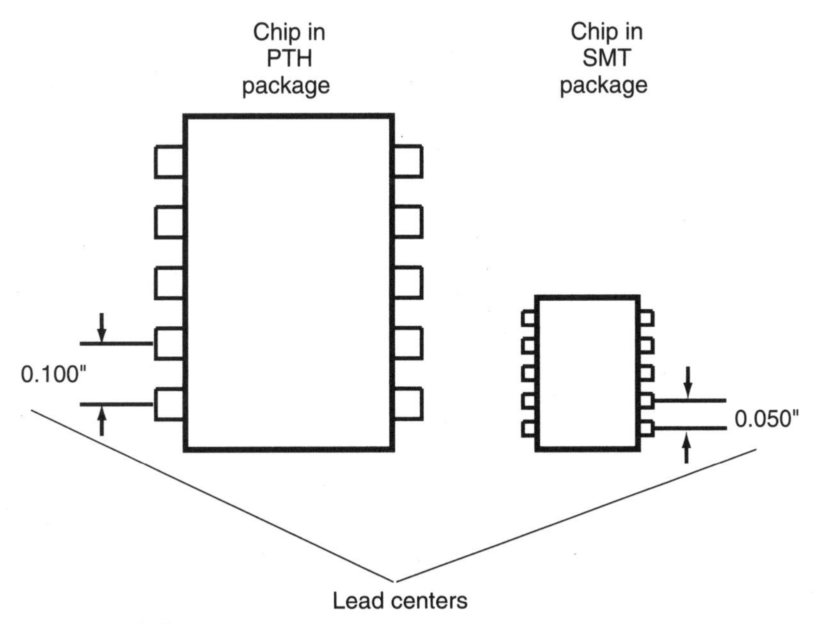

FIGURE 4-5 **PTH package versus SMT.**

built with PTH chips). The disadvantages of PTH is the amount of space required to put the hole in the card and requirements for space around each hole makes the spacing between lead centers quite large in comparison to *surface-mount technology* (SMT) in which the pins are soldered to the surface of the card. (See Fig. 4-5.)

There are two primary types of SMT leads used (Fig. 4-6). The two different types of packages offer advantages in certain situations. The *gull wing* package allows for the hand

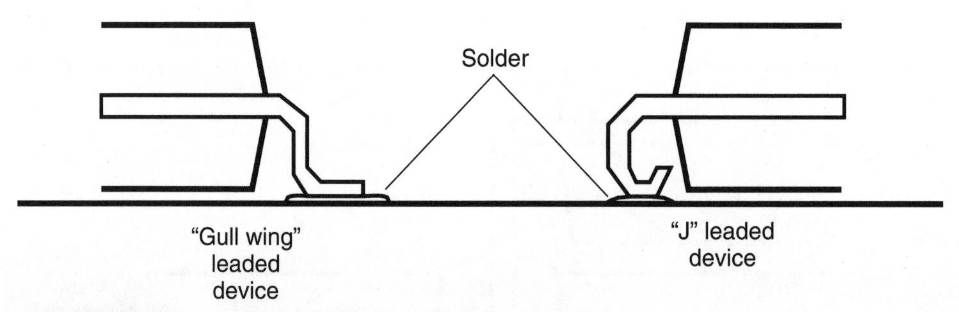

FIGURE 4-6 **Surface-mount technology packages.**

assembly of parts and easier inspection of the solder joints. The *J leaded* parts reduces the size of the overall footprint. Right now, gull wing parts are significantly more popular because this style of pin allows easier manufacturing and rework of very small leads (with lead centers down to 0.016").

The smaller size and lead centers of the SMT devices have resulted in significantly higher board densities (measured in chips per square inch) than PTH. Typical PTH lead centers are 0.100" (with the minimum for interstitial pins being 0.071"), while SMT starts at 0.050" and can go as low as 0.16". The SMT parts with small lead centers are known as *fine pitch parts*.

To give you an idea of what this means in terms of board density, let's look at a PTH package that has pins at 0.100" lead centers and an SMT package with pins at 0.050" lead centers. With the smaller lead sizes, the SMT package can be about half the size of the PTH part in each dimension (which means that four SMT parts can be placed in approximately the same space as one PTH part). As well, without holes through the card, components can be put on both sides of the card meaning that, in the raw card space required for one PTH part, up to eight SMT parts can be placed on the card.

To both increase the board density and support chips with more pins, SMT pin lead centers have shrunk to the point where 0.020" lead center dimensions are not unusual. However, as these pin counts go up, new packaging technologies have been invented to make board assembly easier.

Assembly and rework of SMT parts is actually easier in a manufacturing setting than PTH. Raw cards have a solder/flux mixture (called *solder paste*) put on the SMT pads and then run through an oven to melt the solder paste, soldering the parts to the board. To rework a component, hot air (or nitrogen gas) is flowed over the solder joints to melt the solder allowing the part to be pulled off. While SMT is easier to work with in a manufacturing setting, it is a lot more difficult for the hobbyist or developers to work with (especially if parts have to be pulled off a board to be reprogrammed).

For very high pin counts (300+), PTH technology is usually impractical because of its size, and SMT parts will have problems because of the difficulty in making sure all the pins stay *co-planar* (which is a fancy way of saying "undamaged" and all still bent the same way). For very high pin count devices, which really excludes the 8051, there are other packaging technologies available.

Chip-on-board (COB) packaging (Fig. 4-7) is very descriptive because, in this type of packaging, a chip is literally placed on the raw card. Chip-on-board is useful in microcontroller applications that require a very small form factor for the final product. Because the chip is used directly, there is no overhead of a package for the application. Typical applications for COB include telephone smart cards and satellite or cable TV descramblers.

There are two methods of COB attachment that are currently in use. The first method is to place the chip on the card and wire the pads of the chip to the pads on the card using the same technology as wiring a chip inside a package (using the same aluminum wires ultrasonically welded to the chip and raw card as is used with a lead frame).

The chip itself can either be glued or soldered to the raw card. Soldering the chip to the raw card is used in applications where the raw card is to be used as a heat sink for the chip (which reduces the overall cost of the assembly).

The other method of COB is known as *C4* (Fig. 4-8) and is actually very similar to the SMT process described previously.

Solder balls are used in this process and are called *bumps* (because they are so small). This technology was originally developed by IBM for attaching chips to ceramic packages (without having to go through a wire-bonding step).

C4 attachment requires a very significant investment in tools for placement and a very specialized process (because of the small distance between the chip and the card, water used for washing the card can be trapped with flux residue, causing reliability problems later). C4 attachment is really in the experimental stage at this point, both because of the difficulty in reliability in putting the chip down onto a raw card as well as the opportunity for fatigue failure in the bumps, caused by the chip and raw card expanding and contracting at a different due to heating and cooling.

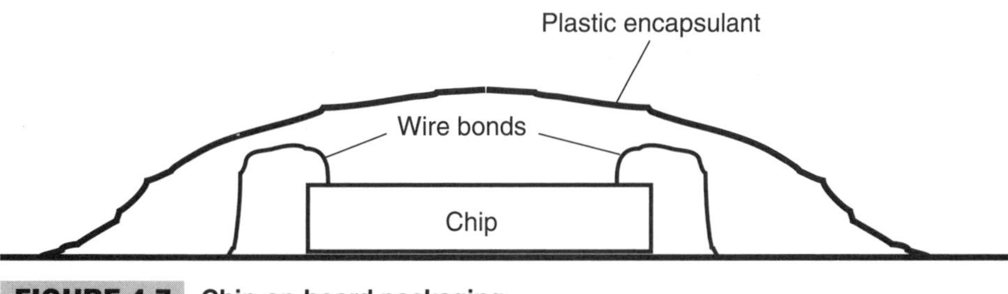

FIGURE 4-7 **Chip-on-board packaging.**

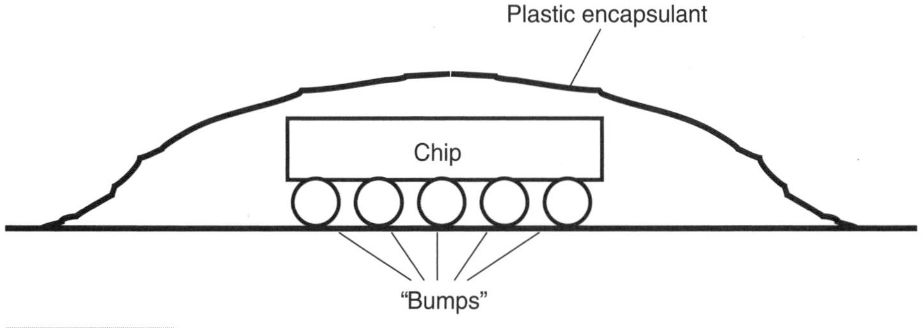

FIGURE 4-8 **C4 chip-on-board packaging.**

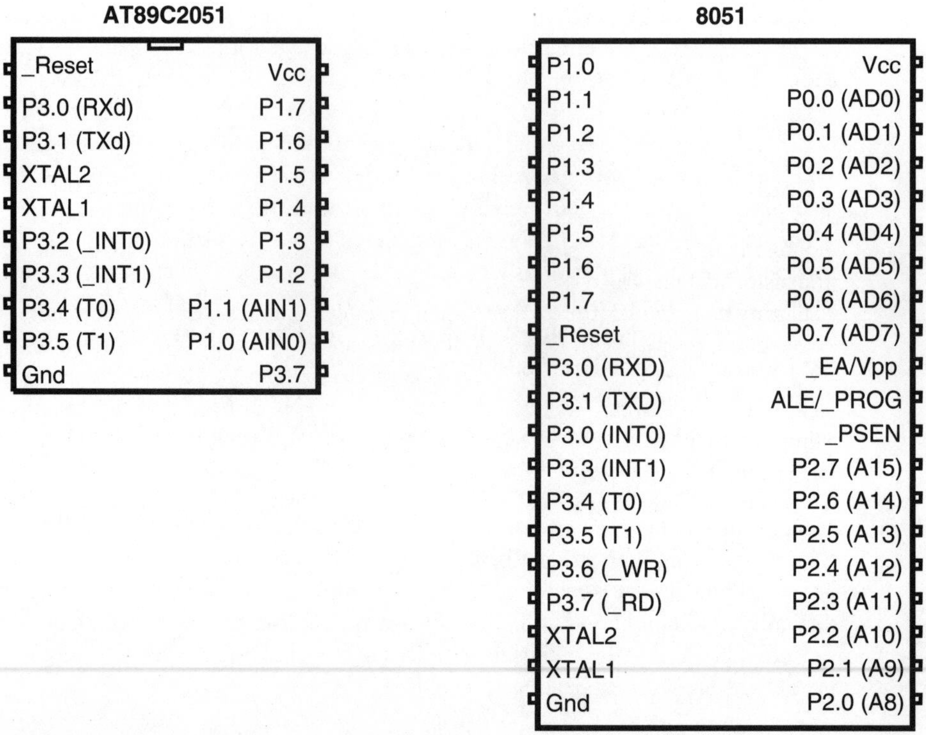

AT89C2051

_Reset	Vcc
P3.0 (RXd)	P1.7
P3.1 (TXd)	P1.6
XTAL2	P1.5
XTAL1	P1.4
P3.2 (_INT0)	P1.3
P3.3 (_INT1)	P1.2
P3.4 (T0)	P1.1 (AIN1)
P3.5 (T1)	P1.0 (AIN0)
Gnd	P3.7

8051

P1.0	Vcc
P1.1	P0.0 (AD0)
P1.2	P0.1 (AD1)
P1.3	P0.2 (AD2)
P1.4	P0.3 (AD3)
P1.5	P0.4 (AD4)
P1.6	P0.5 (AD5)
P1.7	P0.6 (AD6)
_Reset	P0.7 (AD7)
P3.0 (RXD)	_EA/Vpp
P3.1 (TXD)	ALE/_PROG
P3.0 (INT0)	_PSEN
P3.3 (INT1)	P2.7 (A15)
P3.4 (T0)	P2.6 (A14)
P3.5 (T1)	P2.5 (A13)
P3.6 (_WR)	P2.4 (A12)
P3.7 (_RD)	P2.3 (A11)
XTAL2	P2.2 (A10)
XTAL1	P2.1 (A9)
Gnd	P2.0 (A8)

FIGURE 4-9 **Atmel AT89C2051 and Intel 8051 DIP packages.**

The 8051's packaging is very straightforward with the primary packaging being a 0.600" DIP. Most 8051 derivatives follow this pin-out with enhanced features being available on P1 I/O port (this leaves P0 and P2 available for external memory).

In Fig. 4-9, I have shown the standard 8051 along with an ATMEL 20-pin (0.300" wide) version of the 8051. In the experiments and applications presented in this book, these are the two device pin-outs that I will be working with.

The 8051 is also available in a variety of different SMT packages. The SMT packaging pin-out does not correlate directly to the PTH pin-out, which means that, before specifying an SMT part when you go from a prototype (which used a PTH part) to production, you should understand the differences between the two pin-outs and make allowances for it in any raw card designs.

Chip Technologies

Microcontrollers, like all other electronic products, are growing smaller, running faster, requiring less power, and are cheaper. This is primarily due to improvements in the manufacturing processes and technologies used (and not the adoption of different computer architectures).

Virtually all microcontrollers built today use *CMOS* (*complementary metal oxide semiconductor*) logic technology to provide the computing functions and electronic interfaces. CMOS is a *push-pull* technology in which a PMOS and NMOS transistor are paired together.

Figure 4-10 is of a CMOS *inverter* or *NOT gate*. When the input signal is low, the PMOS transistor will be conducting (or "on") and the NMOS transistor will be "off". This means that the switch (or transistor) at Vcc will be on, providing Vcc at the signal out. If a high voltage is input to the gate, then the PMOS transistor will be turned off and the NMOS transistor will be turned on, pulling the output line to ground.

During a state transition, a very small amount of current will flow through the transistors. As the frequency of operation increases, current will flow more often in a given period of time (put another way, the charge transferred per unit time, which is defined as "current", will increase). This increased current flow will result in increased power consumption by the device. Therefore, a CMOS device should be driven at the slowest possible speed, to minimize power consumption.

"Sleep" or "idle" mode, if available in the 8051 device you're planning on using, can dramatically reduce a microcontroller's power consumption during inactive periods because, if no gates are switching, there is no current flow in the device.

An important point with all logic families is understanding the switching point of the input signal. For CMOS devices, this is typically 1.4 Volts to one half of Vcc. However, it can be at different levels for different devices. Before using any device, it is important to understand what the input threshold level is (this will be discussed in greater detail later in the section "Level Conversion").

CMOS can interface directly with most positive logic technologies, although you must be careful of low-voltage logic, to make sure that a high can be differentiated from a low in all circumstances (i.e., a high input is always above the voltage switching threshold level).

There are many more issues to be considered with different logic families, but I think I have brought up the major considerations. Each device is built with different technologies, so these issues are different for different devices.

Just to give you an idea of how 8051 technology has improved over the years, Table 4-2 is a comparison of two devices. The first is a mid-1980s vintage 8751, which is similar to

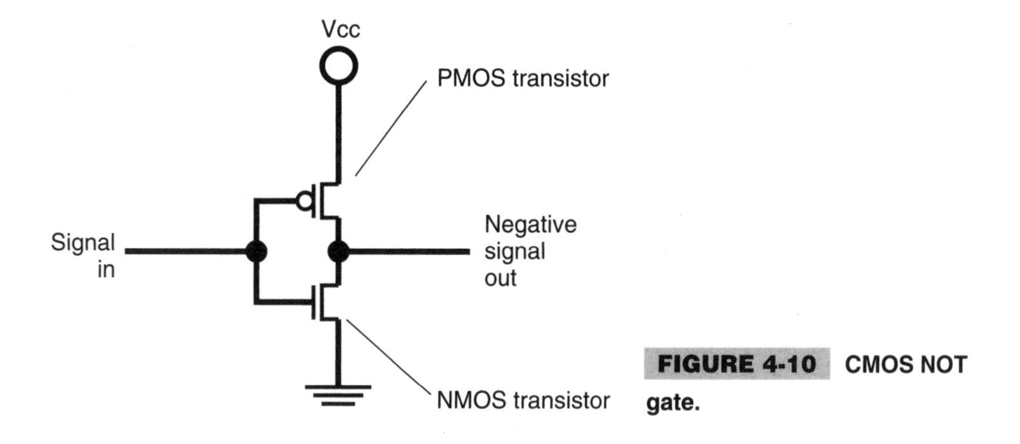

FIGURE 4-10 CMOS NOT gate.

TABLE 4-2 COMPARISON OF THE 8751 AND 87C520 CHIPS

TECHNOLOGY	8751	87C520
EPROM/logic technology	HMOS	CMOS
EPROM memory	4K	16K
RAM registers	128 bytes	256 bytes
Max clock speed	12 MHz	33 MHz
Clock/ins. cycle	12	4
Current required	20 mA	12 mA (in power mgmt mode)
Idle current req'd	5 mA	50–100 μA
Built-in functions	32 I/O pins	32 I/O pins
	2x 8/16-bit timer	3x 8/16-bit enhanced timers
	Async/synch I/O	2x asynch/synch I/O

a Dallas Semiconductor 87C520. Both are 40-pin 0.600" devices, and both cost approximately the same to buy.

To help show the magnitude of the improvement over the past 15+ years, the 87C520 HSM architecture actually executes instructions at roughly 2.5 times faster than the "stock" 8051 while using a bit more than half the power. This means that at 33 MHz, the 87C520 is processing data at an 8751 equivalent speed of 82.5 MHz.

Power Considerations

The 8051, like most microcontrollers, is pretty simple from the perspective of connecting power. Typically, all that's required is a positive voltage and ground connection with a decoupling capacitor in between. (See Fig. 4-11.)

As I said in *The Microcontroller Handbook*, if the microcontroller you're planning on using requires more complex power than this, find another microcontroller. There are very few reasons why anything more complex should be required for a device.

Going along with the previous statement, you should also be looking at devices that can take a fairly wide range of voltages. Some applications might be best suited to run from radio batteries, and being able to use two 1.5-V alkaline batteries in series (which produces 3 V) might be the best method of providing power to the 8051.

The advantages of using batteries extends beyond just simplifying the application, but to power consumption requirements as well. Most 87C51s (which use CMOS logic as opposed to the "original" 8051's NMOS logic, which requires considerably more power to operate) require 10 to 20 mAmps of current to run at 5 V. Reducing the Vcc voltage to 3.0 V can reduce the power required by the 8051 by two-thirds or more of what is required to run at 5.0 V.

Note that, in the previous sentence, I quoted the power figure for just *running* the 8051 microcontroller. I call this *intrinsic power dissipation*. If the microcontroller is sourcing or sinking current, the actual current required by the microcontroller can be a lot more.

This is an important point before releasing the specifications for an application. The *total* power required by the application should be measured instead of using only the various devices' data sheets for determining how much power the application will require. As the different devices interact, the total power can be different.

A good example of this is the digital clock/thermometer project that I present in chapter 11. The 87C520 (the 8051 microcontroller device that I am using) is specified to require about 5 mAmps. The LCD that I used, when it is only connected to Vcc and ground, requires 6 mAmps. When both devices are connected together for the application, the total current required by the application is 7 mAmps.

Occasionally you might require less power than what the sum of the specifications will lead you to expect when you connect multiple devices. However, as a gentleman by the name of Murphy would have it, often you will have to source more power than you will expect from the specifications. The only way for sure to understand what the total required power is would be to wire the circuit and check it with an ammeter.

When planning your application's power requirements, always use the worst case for current consumption, and you'll never be disappointed.

To reduce the total power consumption, operate as much as possible in "power down" or "sleep" modes (if they are available), lower the operating clock frequency and I/O operation, or lower the input voltage. You can also reduce the 8051's power consumption if it does not source or sink any current for loads during operation.

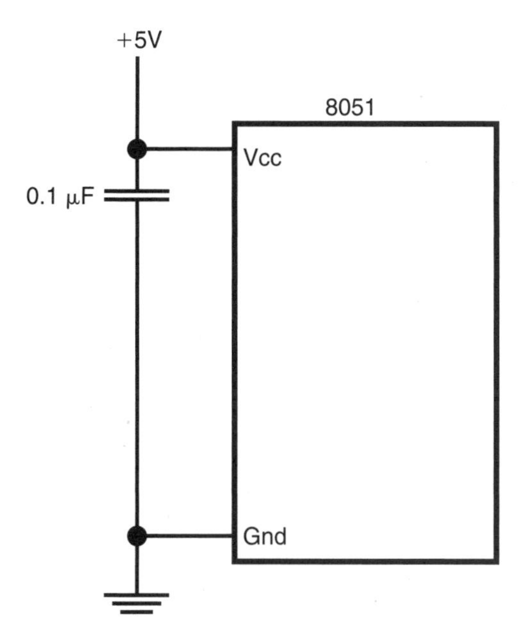

FIGURE 4-11 **8051 power connections.**

This last point might seem impossible; however, with the 8051 driving only high-impedance loads and not changing its I/O pins, you will see the 8051 operate at close to the theoretical minimum (intrinsic) current consumption given in the data sheets.

Reset

Reset in the 8051 is *positive active*, which means that the processor runs when the Reset pin is held low. This is in contrast to the other devices that all have a negatively active reset (i.e., the microcontroller runs when the line is high).

Many of the 8051s available have an internal pull-down and RC delay circuit built in to delay the processor's start-up until the built-in oscillator's operation has stabilized. This built-in circuitry eliminates the need for connecting the reset pin up to anything in many applications.

However, I've found that you can't count on it. I've found that, while many devices will work at modest speeds (up to 4 MHz) without anything connected to the Reset pin, I've found that, for higher speeds and for all speeds in some devices, connecting a capacitor between Reset and Vcc is required to delay the power up sequence long enough for the built in oscillator to stabilize.

To be on the safe side, in all my applications for all 8051 devices, I make sure a 0.1 µF capacitor (which is what I use for power decoupling, so I always have a lot of them around) is connected as shown in Fig. 4-12.

System Clock/Oscillators

The 8051's oscillator circuit is quite simple and reliable. The clocks can run with either a built-in oscillator or an external clock circuit.

When the built-in oscillator is used, the typical circuit is as shown in Fig. 4-13. For power-sensitive applications, a fairly low-speed crystal (such as a 32.756 KHz clock crys-

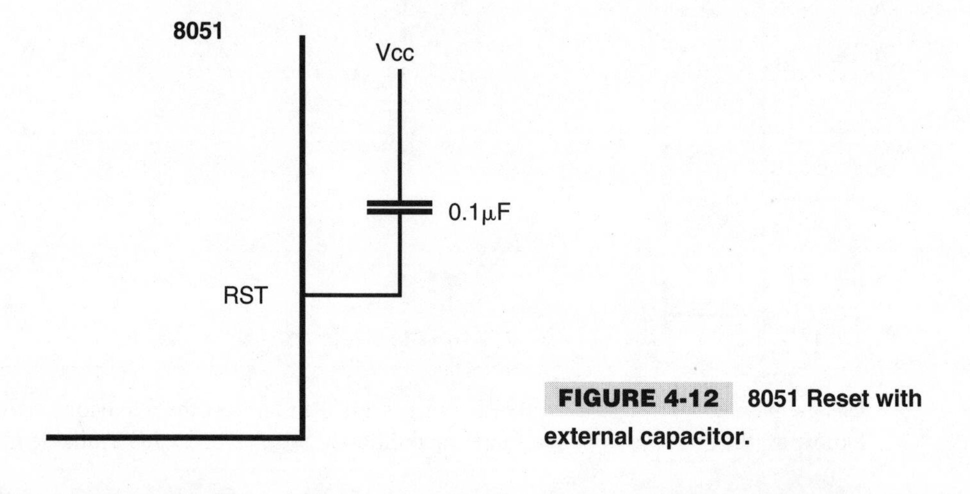

FIGURE 4-12 **8051 Reset with external capacitor.**

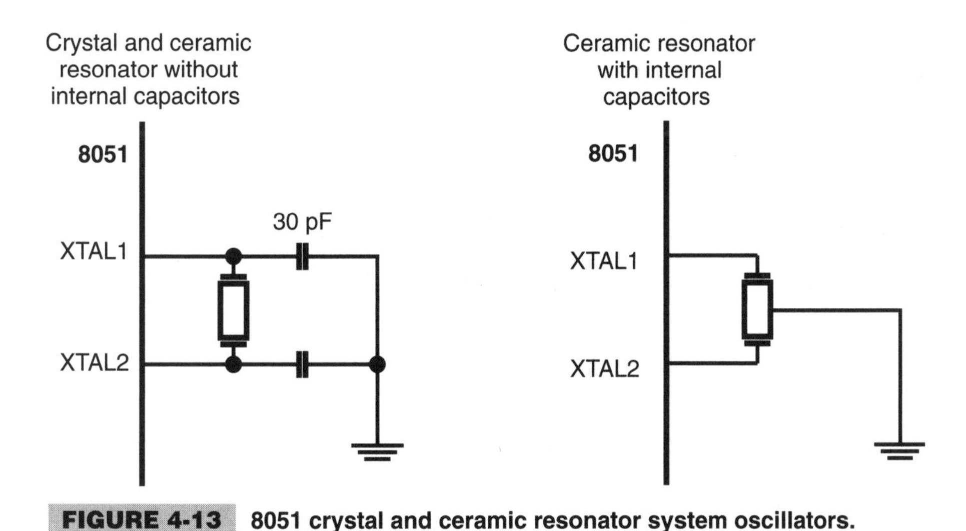

FIGURE 4-13 **8051 crystal and ceramic resonator system oscillators.**

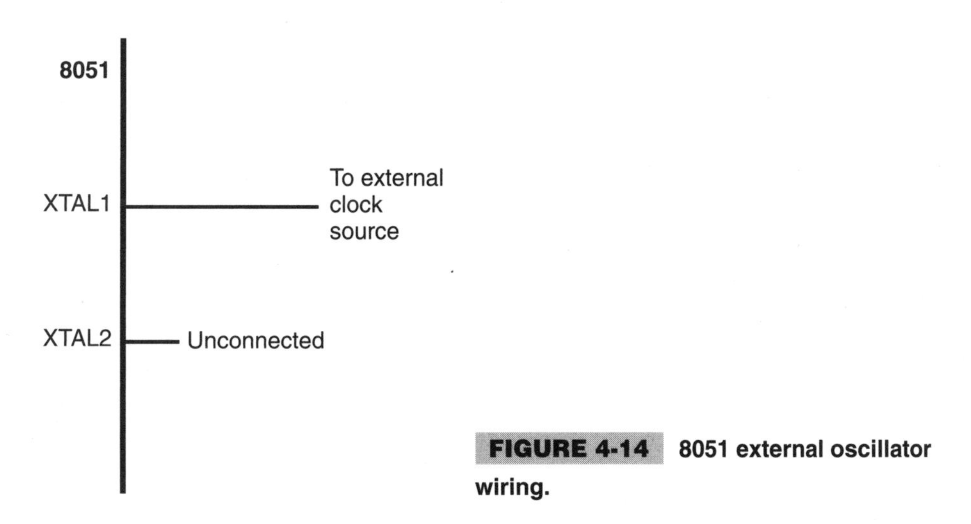

FIGURE 4-14 **8051 external oscillator wiring.**

tal) should be used. The two 30-pF capacitors are generally not specified in the data sheets, but they're good values to use for general-purpose oscillator circuits.

If a clock is already available in the application, or if you want to use an oscillator "can," the external clock signal is input into the 8051 as is shown in Fig. 4-14. The XTAL2 pin should not be used to output the clock signal to other devices. Instead, the clock circuit should be redriven to allow more "fan out" of the clock signal.

Parallel Input/Output

Despite having the simplest parallel I/O pins available in any of the microcontrollers that I know of, the 8051s provide the same capabilities as more complex I/O available in other

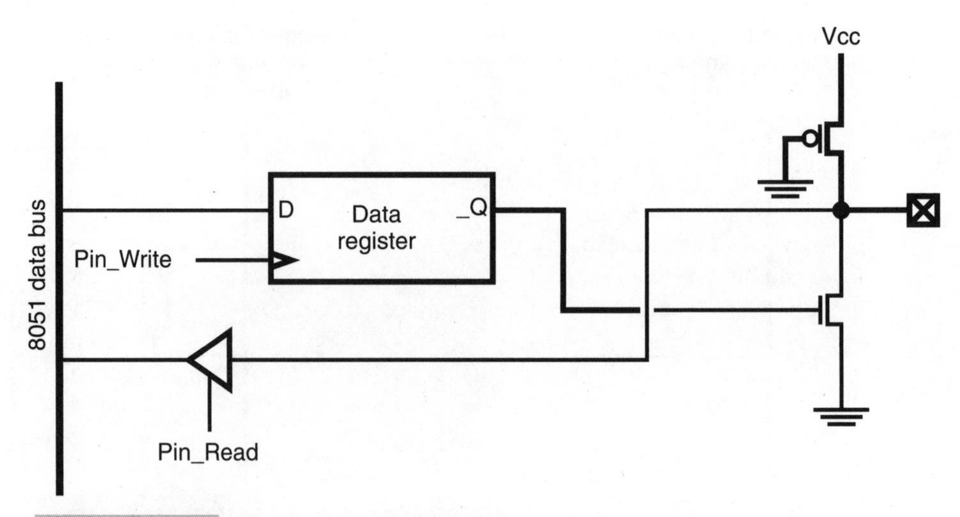

FIGURE 4-15 **8051 parallel I/O pins.**

devices. For the most part, output is carried out exactly the same way as in any other device. To use the input capabilities, you will have to understand both how the pin works as well as how the logic of drivers works.

The pull-up (and associated transistor) is not available on all ports (the "true" 8051 has port 0 without pull-ups on the pins, while ports 1, 2, and 3 all have the integral pull-ups). The best way to describe the 8051's parallel I/O ports (Fig. 4-15) is as being *open-collector* (actually *open-drain*) with pull-ups on most of the pins.

The pull-up is usually capable of sourcing 1 to 2 mAmps, while the pin can sink 20 to 30 mAmps (which is enough current to allow an LED to be lit).

Looking Fig. 4-15, you're probably wondering how the 8051s I/O pins actually work as bidirectional logic pins. When you first look at Fig. 4-15, you can see the "totem pole" output of the pin quite clearly. When the bit is loaded with a "1," the bottom transistor is turned off and current only flows through the pull-up. A "0" will turn on the lower transistor, which will pull the pin to ground.

If the pins is always being driven, how is a logic value read from the pin?

When a "1" is loaded into the output flip-flop, the (weak) pull-up can be easily overpowered by other logic drivers. In many ways, this I/O pin design will make wiring some circuits easier (e.g., dotted AND applications that require an integral pull-up) and some harder—ones that the pin has to source current). If the 8051 has to source anything more than tens of microamps, the output will have to use a buffer to provide the current source.

Even if a pin is used that does not have a pull-up (just the open-drain driver), it will still have to have a "1" written to it. In this case, the transistor will be off (not pulling the pin to ground), and the pin will float and can be driven by the external circuit.

With just the weak pull-ups, the pin might take a relatively long time (on the order of microseconds) to change from a low to a high voltage output. To avoid this problem, Dallas Semiconductor and some other 8051 manufacturers provide high-current drivers on the I/O pins that drive the pin high (without the pull-up) for two clock cycles. As well, when using peripheral I/O devices (such as a serial port), you might find that the pins are actually being driven high, rather than just pulled up. This is a function of the peripheral, and the I/O pin cannot take advantage of this.

Level Conversion

Often when working with microcontrollers, you will have to interface devices of different logic families together. For standard positive logic families (e.g., TTL to CMOS), this is not a problem. The devices can be connected directly. However, interfacing a negative logic to a positive logic family (e.g., ECL to CMOS) can cause some problems.

While there are usually chips available for providing this interface function (for both input and output), typically they work in only one direction (which precludes bidirectional busses, even if the logic family allows it). The chips can add a significant cost to the application.

The most typical method of providing level conversion is to match the switching threshold voltage levels of the two logic families.

As is shown in Fig. 4-16, the "Ground" level for the COMS microcontroller has been shifted below ground (the microcontroller's "Ground" is actually the CMOS "0" level) so that the point where the microcontroller's input logic switches between "0" and "1" (known as the *input logic threshold voltage*) is the same as the ECL logic. The resistor

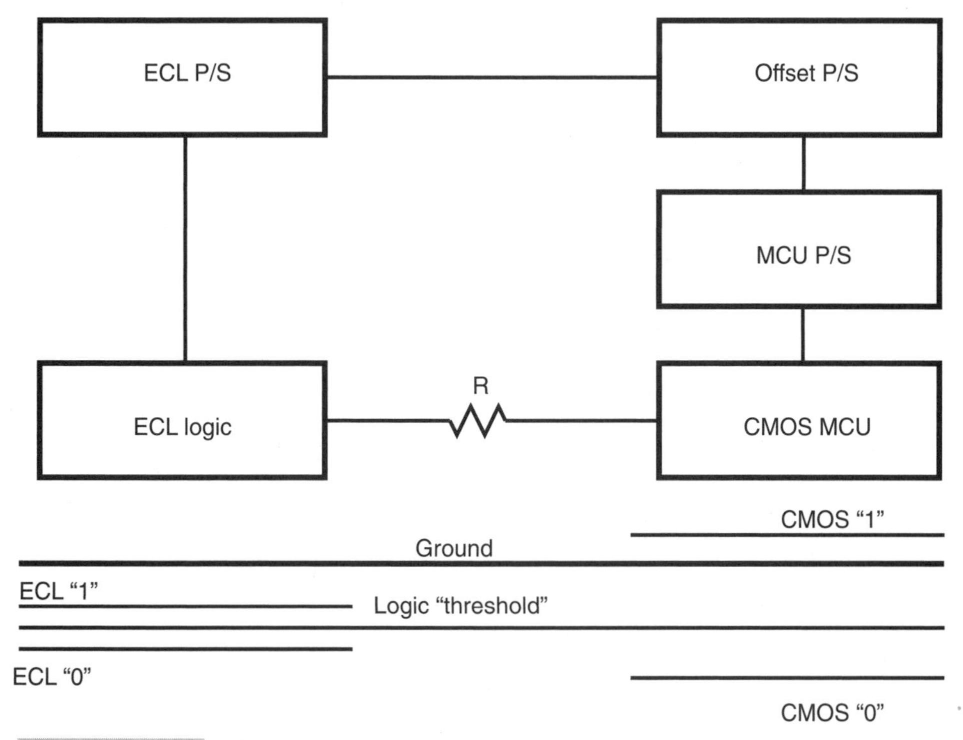

FIGURE 4-16 **ECL to CMOS logic-level conversion.**

(which is between 1K and 10K) is used to limit the current flow due to the different logic swings of the two different families.

Looking at Fig. 4-16, you're probably thinking that the cost of shifting the microcontroller power supply is much greater than just a few interface chips.

Actually, this isn't a big concern because of the low power requirements of modern CMOS microcontrollers. In the previous example, the ECL logic's 0–5-V reference can be produced by placing a silicon diode (which has a 0.7-V drop across it) from the microcontroller's ground to the ECL's –2-V power supply negative output (and the –5-V supply is referenced from "Ground" of the –2-V supply) to balance the logic thresholds. This example might seem simplistic, but it would provide the ability to connect a CMOS 0- to +5-V microcontroller to ECL logic (and allow signals to be sent bidirectionally) at a very low cost.

Timers

The 8051's timers have been well designed to provide general-purpose functions for application software. One of the timer ports can also be used as a baud rate generator for the on-board serial port. At first glance, the timer ports might seem quite limited; however, with the appropriate software, they are capable of providing a wide range of functions needed by applications.

The 8051 has two 8/16-bit timers that can run in four different modes (Fig. 4-17). The clock source can either be the instruction clock (the external clock divided by 12) or an ex-

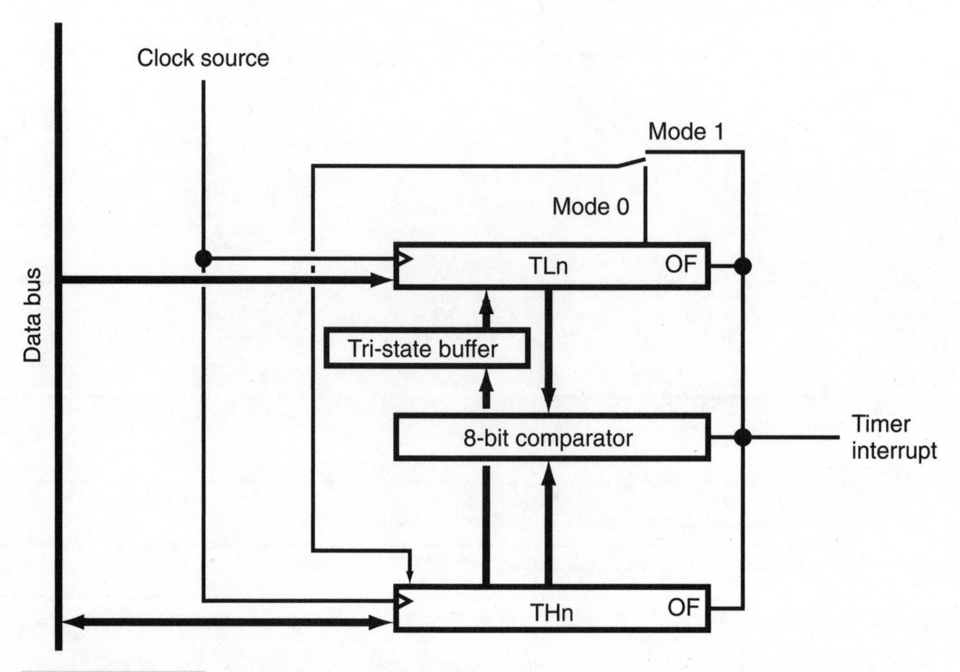

FIGURE 4-17 **8051 timer block diagram.**

ternal source. The timer registers can be read from or written to at any time, including when the timer is running.

An interrupt request can be generated by the timer hardware when the timer overflows.

Each of the two timers' operations is controlled by 4 bits devoted to it in the timer mode (TMOD) register at address 089h (Table 4-3).

The gate bit is used as a secondary execution-control bit. If the gate bit is reset, the timer is enabled to run at any time. If the bit is set, then only when the appropriate _INT*n* bit is high will the timer run.

The timer run control (TR*n*) bits are located in the TCON register (at address 088h) along with the overflow interrupt request pins (and the interrupt pin control bits). (See Table 4-4.)

The timer overflow bits are set when the timer rolls over and are reset either by the execution of an `reti` instruction or by software manually clearing the bits. I recommend

TABLE 4-3 DESCRIPTION OF THE TMOD REGISTER

BITS	DESCRIPTION
1–0	Timer0 mode select bits 00—Mode 0, 13-bit timer 01—Mode 1, 16-bit timer 10—Mode 2, 8-bit reloadable timer 11—Mode 3, operate Timer0 as two 8-bit timers
2	Timer0 "C/_T" clock select bit 0—Internal instruction clock (clock/12) 1—External clock
3	Timer0 gate—Reset to allow the clock to operate
5–4	Timer1 mode select bits 00—Mode 0, 13-bit timer 01—Mode 1, 16-bit timer 10—Mode 2, 8-bit reloadable timer 11—Timer1 stop
6	Timer1 "C/_T" clock select bit 0—Internal instruction clock (clock/12) 1—External clock
7	Timer1 gate—Reset to allow the clock to operate

TABLE 4-4 DESCRIPTION OF THE TCON REGISTER (HIGH-ORDER NYBBLE)

BIT	DESCRIPTION
4	TR0—Timer0 Run control bit
5	TF0—Timer0 overflow bit
6	TR1—Timer0 run control bit
7	TF1—Timer0 overflow bit

clearing the overflow bits before enabling timer interrupts in case they have already been set and you end up jumping immediately to the timer interrupt handlers before you expect to.

Timer mode 0 and mode 1 are similar with the timer being configured as either a 13-bit (mode 0) or 16-bit (mode 1) counter. When the timer reaches the limit of the count (8192 for mode 0 and 65,536 for mode 1), the appropriate overflow flag is set, and the counter is reset back to zero.

In chapter 11, I will show how these modes can be used for creating a real-time clock.

Modes 0 and 1 can also be used to time external events. This is done by setting the gate bit of the TMOD register so that the _INT*n* pin enables and disables the timer running. The timer will run when the _INT*n* pin is high.

To access the counter values, the TL*n* and TH*n* registers are read from and written to. For most delay applications, they are first written to before the timer is allowed to start running. During execution, the values can be read from them.

Mode 0 and 1 can be used as specific time delays by loading them with an initial value before allowing them to execute and overflow (which indicates the time delay interval has passed).

For mode 1 (the 16-bit counter), a specific time delay can be created by loading TL*n* and TH*n* with values derived from the formula:

$$TimeDelay = \frac{(12 * (65{,}536 - InitValue))}{Freq}$$

where

$$InitValue = \mathrm{TL}n + (256 * \mathrm{TH}n)$$

It is important to note that, after the overflow flag is set, the timer is loaded with zero (except in mode 2) and continues counting. If you don't reload the timer with the desired delay values, you will end up waiting for the full number of cycles instead of the delay that you want.

If you reload a timer's TL*n* and TH*n* registers after noting an overflow (either by polling the TF*n* flag or waiting for the timer interrupt), you will be overwriting some initial number of cycles rather than zero.

This is especially true if you are reloading the timer in an interrupt handler. The number of cycles executed after the overflow to the start of the interrupt handler is not necessarily the same value each time the overflow has occurred. The value could be different by one or two cycles if no other interrupts are enabled in the 8051. If other interrupts are enabled, then it could be randomly several more cycles (assuming it's a short interrupt handler) than the one or two cycles you'd normally expect because, before the timer interrupt handler can execute, a currently executing interrupt handler would have to complete.

Timer1 can be used to generate a stable overflow frequency suitable for driving the serial port on the 8051. Using mode 0 or 1, as I've just explained, would not give the stable clock that would be required for serial data communications. To allow asynchronous serial communications to run stably, a programmable interval mode has been provided in the 8051's timer. This is known as *mode 2* (Fig. 4-18).

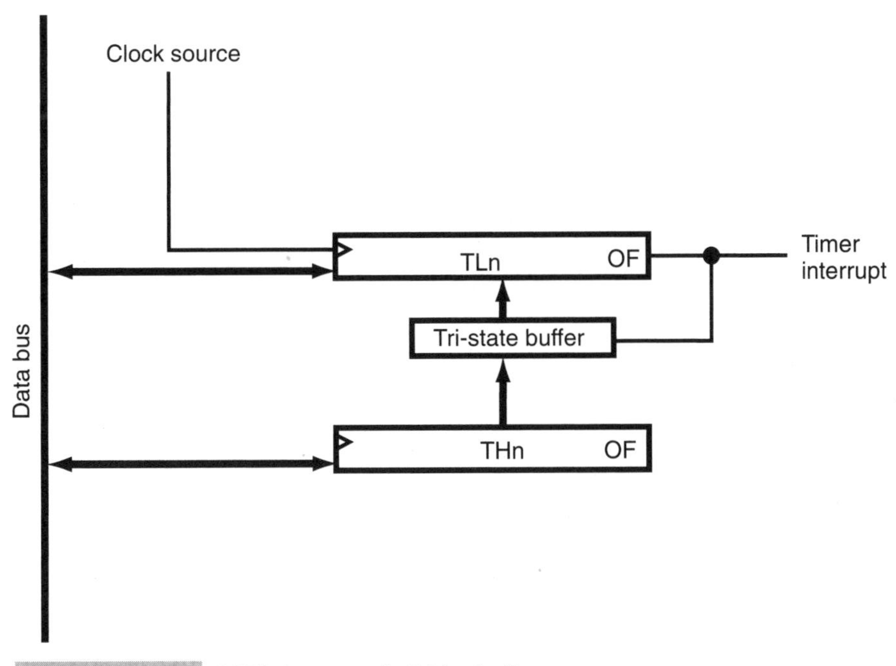

FIGURE 4-18 **8051 timer mode 2 block diagram.**

In this mode, when TL*n* overflows (goes from 0FFh to 0), it is reloaded with the value set in TH*n*. The delays between overflows can be calculated from:

$$TimeDelay = \frac{(12 * (256 - \mathrm{TH}n))}{Freq}$$

Mode 2 works like modes 0 and 1 with software InitValue reloads, but provides a hardware-based reload that does not have any of the cycle uncertainty of implementing the reloads in software. The overflow of Timer1, when in mode 2, will continue to be passed to the serial port even if the TF1 bit is not cleared in software. This helps make the serial ports "set and forget".

Mode 3 (Fig. 4-19) works differently for the two timers built into the 8051. If Timer0 is set in mode 3, the two 8-bit timer registers (TL0 and TH0) are used as two different 8-bit timers.

In mode 3, the TH0 timer uses the Timer1 enable (TR1) and overflow flag (TF1). Because of this translation of timer resources, if Timer0 is in mode 3, then Timer1 is enabled to run all the time (although it cannot request interrupts) except when it is put into mode 3. When Timer1 is put into mode 3, it will not run (regardless of what the state of TR1 or Timer0). The purpose of implementing mode 3 this way in the two timers is to allow Timer1 to be used as the data-rate generator for the serial port without requiring any additional resources.

In this section, I have given you an introduction to the basic 8051 timer subsystem. Different versions and enhancements of the 8051 might have additional changes to Timer0

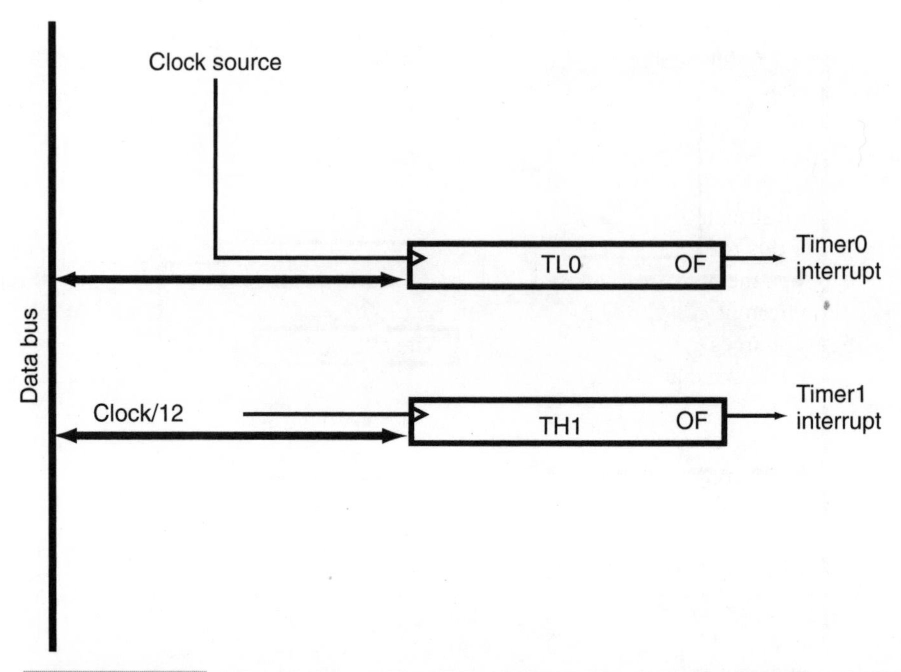

FIGURE 4-19 **8051 timer mode 3 block diagram.**

and Timer1 or provide additional timer features. I will present some of these enhancements in chapter 5. To understand and use many of them, you will have to read through the device's data sheets.

Interrupts

Elsewhere in the book, I have explained what interrupts are and how they work and given you some examples in how they are used. In this chapter, I want to introduce you to the hardware used to implement the interrupt controller and explain how interrupt requests are generated in the 8051.

As I have explained in chapter 1, each interrupt is given an 8-byte *vector* in control store for implementing the interrupt handler or a jump to the interrupt handler. Later in the book, I will go into more detail about how these interrupt handlers should be coded.

Each device capable of requesting an interrupt has a mask bit in the interrupt enable (IE) register at address 0A8h (Table 4-5).

If you had written an application in which the _INT0 pin interrupt was used to enable the interrupt, the instruction:

```
mov  IE, #%10000001     ; Enable _INT0 pin interrupt
```

would be used. By setting these two bits, the conditions that will cause the pin interrupt are unmasked as well as the global interrupts.

To set the desired pin interrupt functions, the appropriate TCON register (at address 088h) bits must be set (Table 4-6).

Before using the pin interrupts, the correct state for the interrupt control register must be chosen. The only time level-triggered interrupts should be used in the 8051 is when the interrupt source is reset from within the interrupt handler. Not resetting the interrupt source within the handler will result in the interrupt handler being invoked again, each time the `reti` instruction is executed.

For this reason, I typically use only edge-triggered interrupts.

When the `reti` instruction is executed at the end of the pin and timer interrupt handlers the interrupt control hardware is reset and prepared for the next interrupt request from these sources.

This means that pulses coming in on an _INT*n* pin could be counted using the code:

```
_INTn:                          ;  _INTn pin interrupt handler
  inc  PulseCount       ; Increment the pulse counter
  reti
```

This will allow you to create some remarkably complex interrupt handlers using only a few simple instructions.

TABLE 4-5 DESCRIPTION OF THE IE REGISTER

BIT	DESCRIPTION
0	EX0—_INT0 pin interrupt enable
1	ET0—Timer0 overflow interrupt enable
2	EX1—INT1 pin interrupt enable
3	ET1—Timer1 overflow interrupt enable
4	ES—Serial port interrupt enable
6–5	Reserved for additional interrupt hardware
7	IE—Global interrupt enable

TABLE 4-6 DESCRIPTION OF THE TCON REGISTER (LOW-ORDER NYBBLE)

BIT	DESCRIPTION
0	IT0—_INT0 control 0—Low-level triggered 1—High-to-low edge triggered
1	IE0—Set when interrupt 0 requested
2	IT1—_INT0 control 0—Low-level triggered 1—High-to-low edge triggered
3	IE1—Set when interrupt 0 requested

TABLE 4-7 INTERRUPT PRIORITY SEQUENCE

INTERRUPT SOURCE	PRIORITY
_INTO pin	Highest
Tmr0 overflow	
_INT1 pin	
Tmr1 overflow	
Serial port	Lowest

TABLE 4-8 DESCRIPTION OF THE IP REGISTER

BIT	DESCRIPTION
0	PX0—_INT0 pin
1	PT0—Tmr0 pin
2	PX1—_INT1 pin
3	PT1—Tmr1 pin
4	PS—Serial port
7–5	Reserved

The serial port interrupt flags (RI and TI) have to be reset manually. They are not reset automatically by the `reti` instruction or reading from or writing to the serial buffer register address. Because the interrupt vector is shared, the software in the interrupt handler will have to determine whether or not the interrupt was caused by a character being received or the transmit shift buffer being empty.

If multiple interrupt requests are pending at the same time, then the priority sequence shown in Table 4-7 is used.

In your application, this priority sequence might not be appropriate. To give different interrupt sources different priorities, the interrupt source bit of the interrupt priority (IP) register at address 0B8h can be set (Table 4-8).

By setting a bit, the interrupt is put into a higher priority level. If multiple bits are required to be at the higher priority level, then they will have the same relative priorities that they had if they were in the original, lower priority level.

If the 8051 device you're using has additional interrupt requesting hardware (such as additional timers and serial ports), you will find that, in the device's data sheet, the extra interrupt control bits will be placed in the "Reserved" locations listed in the registers described earlier. In some cases, additional interrupt control registers might be provided by the device.

Serial I/O

The 8051 serial port (Fig. 4-20) is a very complex peripheral, able to send data synchronously and asynchronously in a variety of different transmission modes.

The SCON register (at address 098h) is used to control the operation of the serial port (Table 4-9).

Figure 4-20 will probably appear to be different than the other serial ports shown for other microcontrollers (and peripherals for microprocessors). A big reason for this is the ability of this serial port to run in both synchronous and asynchronous modes and in asynchronous mode, it is a bit simpler than other devices.

For synchronous mode (mode 0), the instruction clock is used. Data is transmitted and received by the RXd pin, and the clock is provided by the TXd pin. (See Fig. 4-21.) In

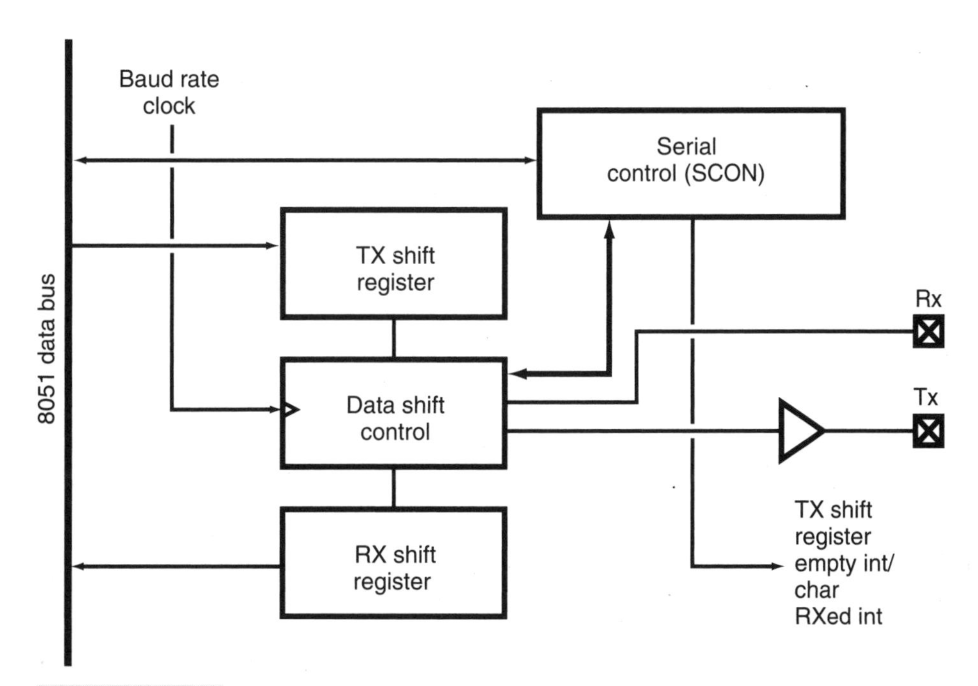

FIGURE 4-20 8051 serial Rx/Tx block diagram.

TABLE 4-9 DESCRIPTION OF THE SCON REGISTER

BIT	DESCRIPTION
0	RI—Set by hardware when a character has been received
1	TI—Set by hardware when the stop bit is being transmitted
2	RB8—Ninth data bit received in modes 2 and 3
3	TB8—Ninth data bit sent by software in modes 2 and 3
4	REN—Set to enable serial data reception
5	SM2—Enable multi-processor communications mode
7–6	SM1/SM0—Serial port mode bits 00—Mode 0, synchronous serial communications 01—Mode 1, 8-bit UART with timer data rate 10—Mode 2, 9-bit UART with set data rate 11—Mode 3, 9-bit UART with timer data rate

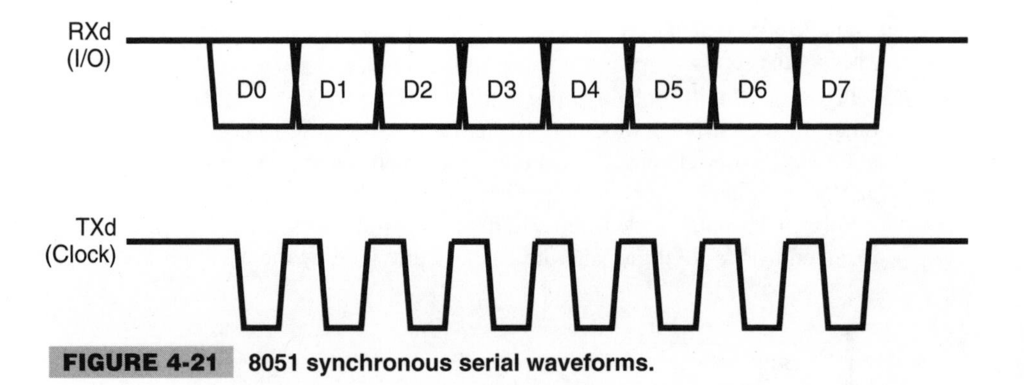

FIGURE 4-21 8051 synchronous serial waveforms.

synchronous mode, the 8051 can only master (i.e., provide the clock to move data in or out).

The synchronous data transfer is initiated by writing to the serial data port address (SBUF at address 099h). Note that the TXd pin is used for the clock output, while the RXd pin is for data transfer (both in and out). Data is transferred or, more accurately, is valid on the rising edge of the clock pulse.

When a character is received, a hardware interrupt can be requested or the status of the data transfer can be monitored by polling the RI_n bit in the serial control register (SCON). For synchronous transmission, the data rate is the oscillator frequency divided by 4 or 12.

In mode 0, the data rate is the oscillator frequency divided by 12 (or simply the instruction clock frequency).

For asynchronous data transmission, the remaining three modes provide two different ways of clocking as well as 8- and 9-bit data transfers. Mode 1, which is probably the mode that you will most commonly use, will transmit 8 bits (for "8-N-1" serial communications) and is driven by Timer1. Modes 2 and 3 both consist of 9 data bit packet communications with the ninth (parity) bit being calculated in software and loaded into the TB8 bit of the SCON register.

Sending data is initiated by writing a byte into the TX shift register. The 8051 does not have the TX holding register that many other devices have in their asynchronous serial communications hardware. This means that loading the holding register while a byte is being shifted out is not possible. To get the maximum string transmit rate, you will have to poll the TI bit of the SCON register and load the TX shift register after the previous byte has been shifted out.

When a byte (which can be 8 or 9 bits long) has been received, the RI bit of SCON is set.

Hardware interrupt requests can be made when the TX shift register is empty or the RX shift register is full. The serial port interrupt-enable bit is located in the interrupt enable (IE) register. When in the interrupt handler, this bit must be specifically reset (it is not reset automatically when the interrupt handler begins to execute, as the timer interrupt hardware does). As I have pointed out elsewhere, the RI and TI bits share an interrupt vector, so in the handler, you will have to poll these two bits to determine which one caused the interrupt request.

In the 8051, the Timer1 overflow rate (in Timer1's mode 2 of operation) can be used for driving the serial ports data rate (in the serial port's mode 1 and mode 3 of operation).

$$DataRate = \frac{(1 + \text{SMOD}) * Timer1OverflowRate}{32}$$

The SMOD bit is bit 7 of the PCON register.
The previous formula can be rearranged to calculate the reload value:

$$Timer1OverflowRate = \frac{(32 * DataRate)}{(1 + \text{SMOD})}$$

which is in the units of cycles/second (Hertz). Inverting the formula to get the actual overflow interval, we get:

$$Timer1OverflowPeriod = \frac{(1 + \text{SMOD})}{(32 * DataRate)}$$

which is in seconds. However, we want to get a time value in terms of clock cycles (which can be used to determine the value to load in the timer to get the overflow value):

$$Timer1OverflowCycles = \frac{((1 + \text{SMOD}) * ClockFrequency)}{(32 * DataRate)}$$

So, if we wanted 9600 bps in a 80C320 running at 12 MHz (and assuming SMOD is zero):

$$Timer1OverflowCycles = \frac{12 \text{ MHz}}{(32 * 9600 \text{ bps})} = 39.063$$

A value of 39 will be used in Timer1 to determine the reload interval. To get the actual value, this value has to be taken away from 256 (the value in the Timer during an overflow). This means that 217 will be loaded into Timer1 to get 9600 bps in the 80C320.

With a cycle count of 39 being used rather than 39.063, there will be an error rate of 0.16%. This error rate is quite minuscule and, over a 10-bit packet (assuming "8-N-1" data transmission) will have a total error of 1.6% from the computed center of the first bit.

For the serial port's mode 2, the data rate is defined as:

$$\frac{DataRate = ((1 + \text{SMOD}) * ClockFrequency)}{64}$$

Standard RS-232 data rates can be achieved by selecting a clock frequency that is a multiple of the data rate that you want. For example, a clock frequency of 614,400 KHz will provide a 9600 bps data rate. Unless very high data rates are desired, this mode (mode 2 in the serial port) might not be very useful for most applications.

RS-232 Level Conversion

The most common application for asynchronous serial communications is known as *RS-232*. This is a communications standard first created in the 1950s. Because of its usefulness, it has remained active for many years. The 8051's asynchronous serial port can support RS-232 communication, although some additional circuitry is required to interface RS-232 to an 8051.

RS-232 voltage and logic levels are a bit unusual (Fig. 4-22). The first aspect that might cause some problems is the high voltage swings. However, as we saw earlier, any unusual voltages and current flows can be attenuated by placing a current limiting resistor in line between the receiving pin on the microcontroller and the RS-232 connector. This resistor will prevent any clamping diodes in the microcontroller from causing any large current flows and also keep the voltage levels at the microcontroller within specified levels and not drive the microcontroller into an invalid hardware state.

This works well because the negative voltages (RS-232 logical 1) will be sensed as 0 Volts in the microcontroller. Positive voltages (RS-232 logical 0), even when clamped to 5 Volts, will be above the *switching region* (which is -3 Volts to +3 Volts) and a good read will be made.

This solution *only* works for receiving RS-232 signals with a *bit-banging* serial algorithm. If data is to be sent to an RS-232 host, the negative voltage levels required for a 1 cannot be output with this circuit (only 0 Volts will be available, which is within the switching region of an RS-232 receiver). Often, this will be ignored (just a resistor will be put in to make sure current limits are not exceeded), the microcontroller will be connected to the true RS-232 receiver, and it will work. Transmitting to an RS-232 device just using a resistor is *very* application- and device-specific, and I really don't recommend it.

In the previous paragraph, I added an important warning that receiving RS-232 could only be carried out with a bit-banging algorithm. What I mean by a "bit-banging algorithm" is one in which the 8051 will wait a set number of cycles for incoming data and

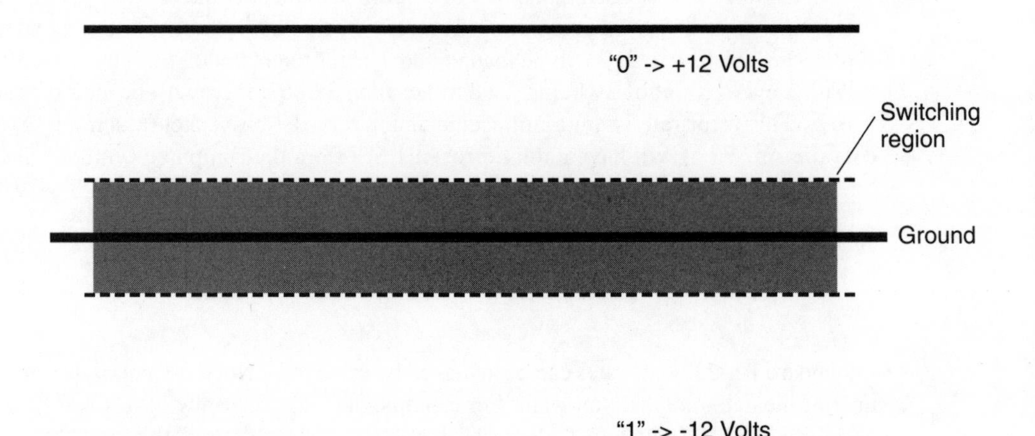

FIGURE 4-22 RS-232 voltage levels.

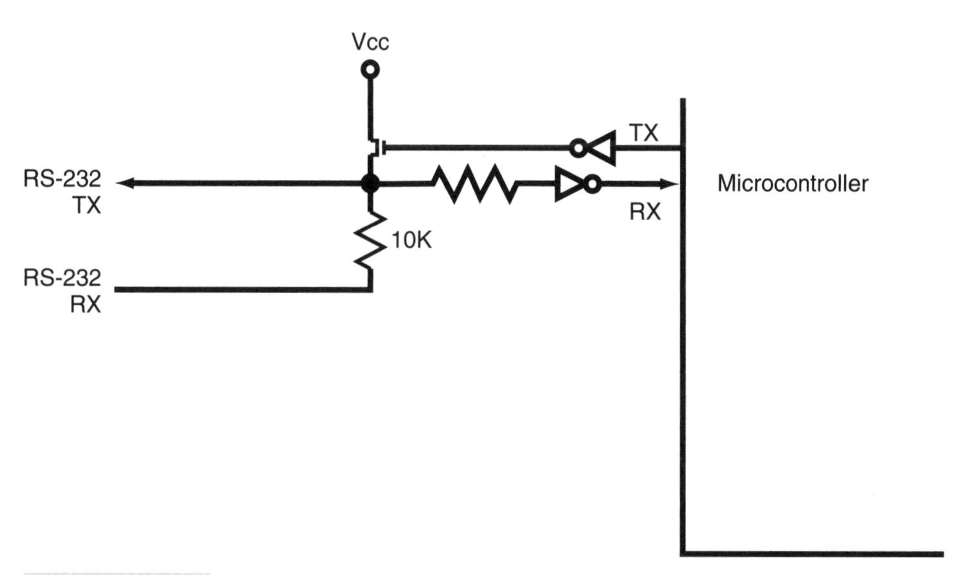

FIGURE 4-23 RS-232 to CMOS microcontroller.

process it in software rather than using hardware to receive the data. I will show and give a transmit and receive algorithm later in the book.

Instead, a popular method of creating negative voltages for transmission is to "steal" the negative voltage from the receiver (which is negative most of the time anyway). (See Fig. 4-23.)

In this circuit, received data will be passed to the microcontroller through the 10K resistor (as was discussed earlier). When transmitting, the line normally will be negative (sending a 1) from the received line. When transmitting a 0 (a positive voltage), the output voltage will be at Vcc. This is a very cheap and elegant method of creating an RS-232 transmit-and-receive interface (basically a 3-wire RS-232 interface with the third wire being Ground).

This method, while it does work, has some issues to be considered before specifying it as part of an application. The receive and transmit data will have to be inverted from positive logic levels as I showed in Fig. 4-23 to allow the built-in serial ports to be used.

A popular chip that provides this function, along with built-in inverters, is the Dallas Semiconductor DS1275. I will show how it's used in chapter 12.

The interface circuit I have shown is only acceptable for 3-wire RS-232. If RS-232 handshaking is required by the host, then another interface will have to be used or the handshaking lines from the host will have to be shorted together (i.e., DTR to DSR and CTS to RTS).

Another concern of using this method is that data transmitted to the microcontroller will be sent right back to the transmitter. When the receive voltage is positive (i.e., when a 0 is sent), the transmitted voltage will also be positive and the data being received will be sent right back out. This might not be a problem and might eliminate the need for you to write code to echo back what has been received, but it might cause some problems with some applications that aren't expecting data to be echoed back. The echoing means that data cannot be transmitted in a duplex manner, in which data is be sent simultaneously in two directions.

If any of these issues are a problem for the application, you will have to use a solution like the Maxim MAX232 chip, which has an onboard negative voltage generator and converts TTL/CMOS voltage levels to RS-232 voltage levels.

Control Store

The control store of the 8051 is quite straightforward with the application software (and table information) stored as bytes. This allows external memory to be added to the device using standard byte-wide ROM, PROM, and EPROM parts.

The "classic" 8051 has 4096 bytes of built-in control store. Other 8051 derivations might have more or less control store on the chip.

After an 8051 has been programmed and released as a product, you might want to protect your code from somebody removing the 8051 and placing it in a programmer and reading out the application code. To help prevent this from happening, when the data is output, it is XORed with the value in a 64-byte portion of EPROM that is used to scramble what is read back after programming (which is known as the "encryption array").

The algorithm used to read back a byte is:

$$Output = \text{Control}(Address) \wedge !\text{Encrypt}(Address \ \& \ 3\text{Fh})$$

where Control is the EPROM control store array of instructions and Encrypt is the 64-byte array (which only uses the lower 6 bits of the address). Normally, after device erasure, these 64 bytes are cleared and all will be read back as 0FFh, which means, if it is left unprogrammed, then the data can be read back by the programmer without being modified.

Another thing to notice with this scheme is, if areas of the EPROM are left unprogrammed (0FFh, the same as the unprogrammed encryption array), then the contents of the encryption EPROM can be read out directly. To avoid this, a pattern other than 0FFh should be stored in the remaining EPROM. Ideally, this code should not be repeating. I have done things like append source code, e-mail notes, or whatever to the end of the program source to provide random data. You could also repeat the program to the end of memory or put in an pseudo-random value (which can be surprisingly easily, as is shown in appendix D).

I recommend using a different algorithm for placing a different pseudo-random value in the unused control store and encryption array.

Programming the encryption array should be done after the application code has been debugged and qualified and is ready to be shipped with the product. Doing it before the code has been fully debugged will result in more work for you. If you want to check the level of code in the device, you'll have to write a program that will take the data read from the 8051 and XOR it with what you think you loaded into the encryption array. To avoid all this hassle, just add the encryption array contents *after* you are happy with the way the application runs.

External Memory Devices

One of the features I really like about the 8051's architecture and hardware is its ability to simply interface with external devices and memory. This gives a lot of flexibility to the 8051 that isn't available in a lot of other microcontrollers (or only in special versions of other microcontrollers).

During execution of the 8051, data is fetched continuously. Most of this data is executed out of the 8051's built-in control store. However, when an address is outside the internal control store, an external memory access is generated.

This memory request causes the 8051 to make a memory read to an external device every 6 clock cycles (one half of an instruction cycle). (See Fig. 4-24.) This timing is important because it allows the 8051 to keep constant instruction timings whether it is executing out of internal or external control store.

The typical circuit used to provide external control store in ROM to an 8051 is shown in Fig. 4-25. In this scheme, the address is output first, and the lower 8 bits (output from P0) are latched into the '373 (using ALE). After the address is valid, the data is read out of the external ROM. The _PSEN actually takes the place of a _RD line. When it is negative, the external ROM should be driving the 8 port P0 bits.

The 8 port P2 bits output the upper 8 bits of the address. Chip select decoding (not shown in Fig. 4-25) should use only these bits and not the least significant 8 bits. The reason for not using the lower 8 bits is that they are valid for a much shorter period of time than the upper bits. Adding additional delays to these bits will reduce the amount of time available

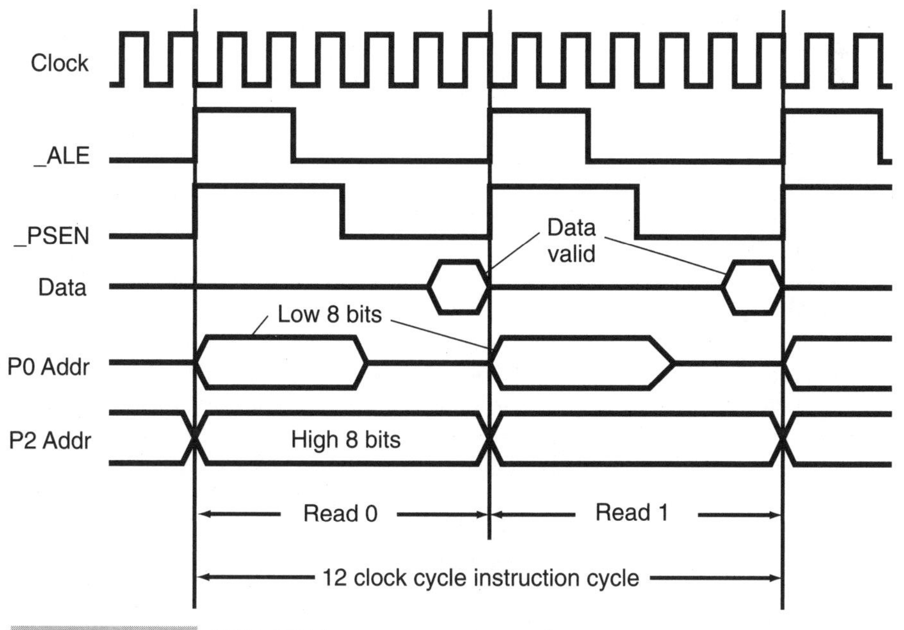

FIGURE 4-24 8051 ROM memory read waveform.

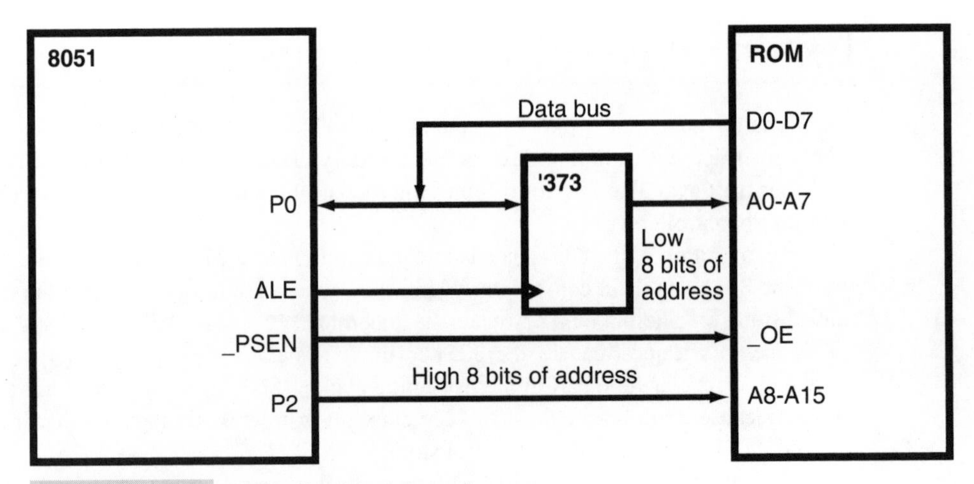

FIGURE 4-25 **8051 wired to external memory.**

to the ROM for outputting the valid data. (At 30 MHz, each clock cycle is approximately 30 nsec in length, and reading out from ROM has 1.5 to 2 clock cycles available for the ROM to output the correct data, which is 45 to 60 nsec in total.)

When I first saw the interconnect scheme, I thought that it would be very easy to add the memory. What I really didn't fully understand was the timing constraints (which I've just introduced you to in the previous paragraph). When you are developing the specifications for an application using external memory, this timing has to be fully understood and appropriate parts chosen before building the circuit.

To try to keep the timing problems down, you might want to keep the clock as slow as possible or work with a device like the Dallas Semiconductor HSM microcontrollers, which allow you to insert "wait states" before reading the memory back.

The _EA pin of the 8051 is used to select whether the 8051 will boot out of internal or external control store. The _EA line is especially useful to parts scroungers who might want to work with 8051's without having to buy a device programmer (they might already have an EPROM programmer) or want to buy surplus mask programmed 8051s and run their own applications.

RAM can be added to the 8051 for some applications in a similar manner to control store ROM.

The _RD and _WR pins are used to control the static RAM (SRAM) connected to the 8051 (Fig. 4-26). Each RAM read or write requires one full instruction cycle.

The timings in this case (Fig. 4-27) are a lot less stringent than in the external control store case. In the control store case, the tight timings are required to maintain the instruction cycle counts of the instructions. For data reads and writes, the execution cycle counts can be maintained much more easily because the read or write take place in 8 to 9 clock cycles (240 to 270 nsec in the previous example).

External reads and writes are invoked by the `movx` instruction, which carries out indexed reads and writes using the DPTR register.

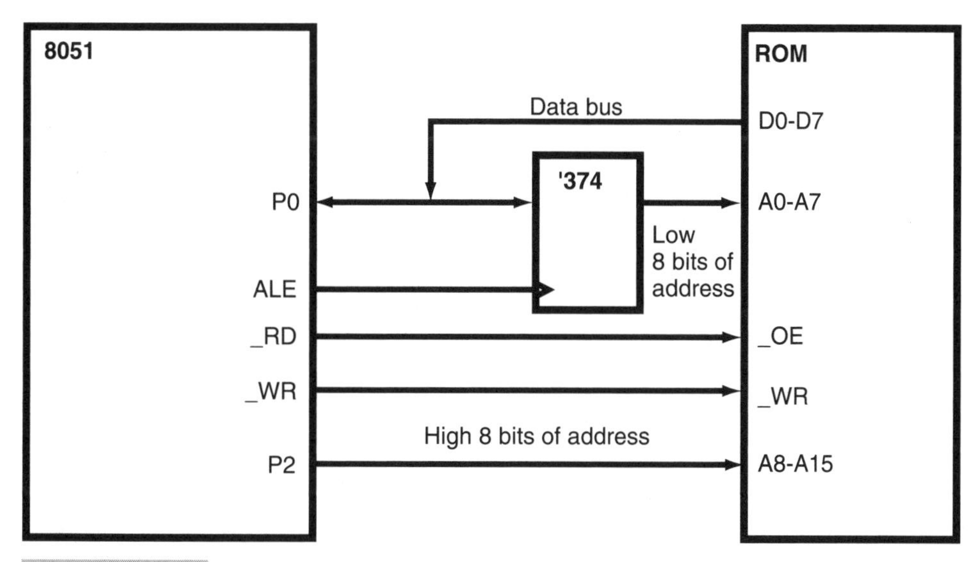

FIGURE 4-26 8051 wired to external RAM memory.

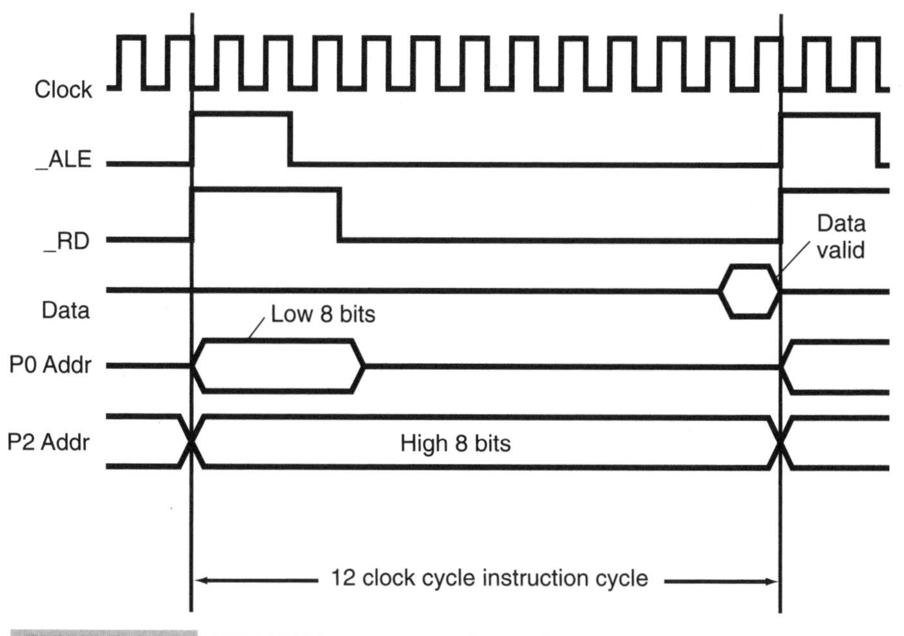

FIGURE 4-27 8051 RAM memory read waveform.

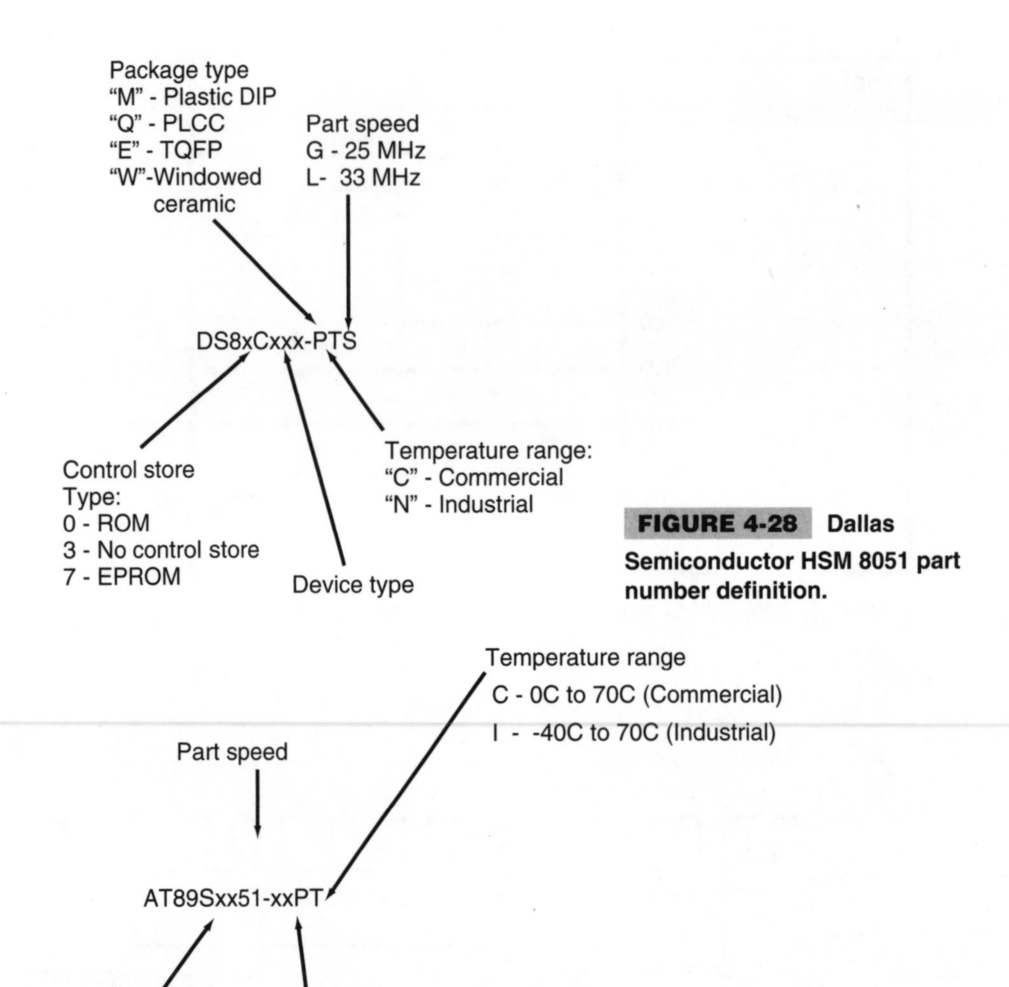

FIGURE 4-28 **Dallas Semiconductor HSM 8051 part number definition.**

FIGURE 4-29 **Atmel 8051 part number definition.**

Ordering Information

I find it incredibly confusing to try to figure out how to order a specific part that I want for the first time. The way devices are specified is generally manufacturer-specific. Few, if any, conventions are followed between the manufacturers.

To help you avoid problems when you need to order parts for working through the experiments, I have shown the 8051 part number definitions for Dallas Semiconductor and Atmel parts in Figs. 4-28 and 4-29.

5

ENHANCED 8051 FEATURES

CONTENTS AT A GLANCE

A lot of my early experiences with microcontrollers was with the Intel 8748 and 8031 in the mid-1980s. The devices I worked with used NMOS technology and did work at a certain basic level with a set number of features that provided microcontrollers with some additional frills, such as a relatively sophisticated interrupt capability, external memory, and a built-in serial port. For the test equipment and products I was working with at the time, these devices were more than adequate. If any additional features were required, I could add them via bit-banging routines or external memory busses.

As I've started writing books about microcontrollers, I've become more sophisticated about what they are required to do and what features are available. In the previous chapter, I presented how the standard 8051 operates. This does not reflect on the many companies that are producing 8051-compatible devices that have enhanced features. In this chapter, I would like to go beyond what the original Intel 8051 offered and look at some features and interfaces that are available in 8051- compatible devices.

As I pointed out in the previous chapter, many of these enhancements are available in the 80*x*2 series of parts. The 80*x*2 executes the same instruction set as the 8051 but has a number of additional features that are not available in the 8051.

8051 Architecture Enhancements

The 8051 processor architecture is often regarded as being old-fashioned or not very efficient because it is of microcoded design (Fig. 5-1). The original 8051 was designed with a processor with 12 clock cycles instruction cycle. Its modern competitors typically have 4 clock cycles or even 1 clock cycle per instruction cycle, and the manufacturer's of these devices make the most of this fact in comparison with the 8051.

To help speed up the 8051, there are several new processor cores on the market that significantly reduce this 12 clock cycles per instruction cycle "handicap" and provide devices

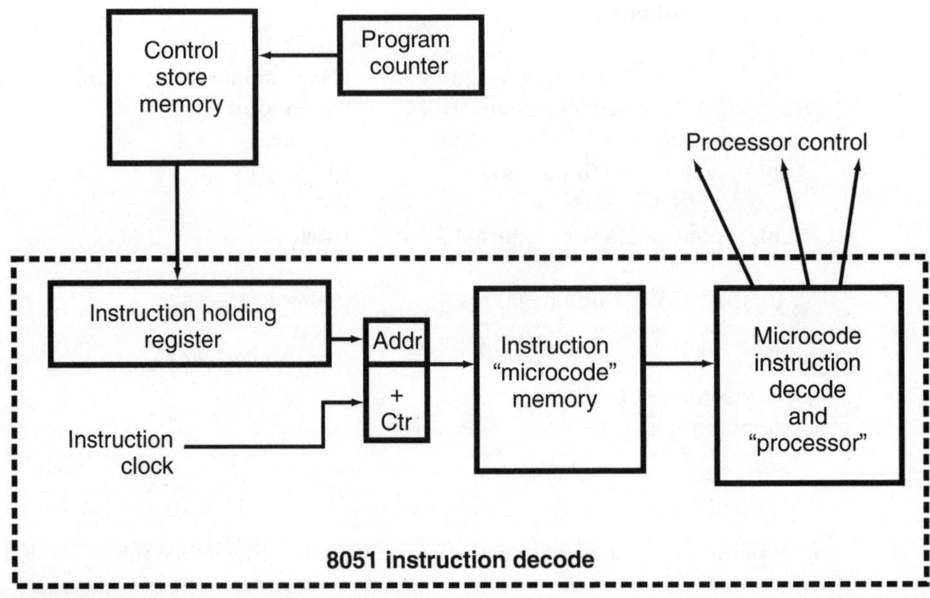

FIGURE 5-1 **8051's microcoded processor.**

that offer significantly more performance than the original 8051 while running at the same clock speed.

INTEL MCS-151/251

Intel's MCS-151 and MCS-251 microcontrollers are an enhancement to the original 8051 with a number of changes and improvements. These improvements include enhancing the instruction set for 16-/32-bit data transfers, up to 16MB of SRAM addressibility (and, going with this, a stack capable of being 64K bytes, rather than the original 256 bytes) and most importantly (to this discussion anyway) the instructions execute in as few as two or four clock cycles.

The MCS-151 and MCS-251 are available versions that are pin and software (binary code) compatible with the 8051. This is interesting to me because I usually associate a change like this to being able to get *more* speed out of the device, not use it as a method to stay at the same speed. I guess I've been brainwashed by the continual improvements in PC and workstation technology into thinking that technological improvements only mean faster computers.

The MCS-151 is basically a replacement for the 8051 with the same instruction set and features. The MCS-251 is an enhancement to the 8051/MCS-151. The MCS-251 can run up to 15 times faster than a stock 8051. Both the MCS-151 and MCS-251 use a hardwired processor capable of *pipelining* (reading ahead) instructions, which is a primary reason for the faster program execution.

DALLAS SEMICONDUCTOR HIGH-SPEED MICROCONTROLLERS

For the 8051 applications and experiments presented in this book, I will be focusing on the Dallas Semiconductor High-Speed Microcontrollers (HSMs). Dallas Semiconductor was the first to come up with the idea of changing the 8051's microcoded processor core with a hardwired one. This change has resulted in instructions taking 4, 8, 12, or 16 clock cycles per instruction for an improvement of 1.5 to 3 times over a true 8051 running at the same clock speed.

This change to the instruction timing means that the Dallas Semiconductor HSM parts cannot just be dropped into an application and expected to run with just a slower clock. The nonlinear improvement in instruction execution (Dallas Semiconductor's research has shown that a 2.5 times improvement can be expected going to the HSM microcontrollers from the basic 8051s) might cause some problems when porting a working application. However, if you work exclusively with the timers providing critical delays and make sure that a divisor of 12 is used with the devices (which is shown later in this chapter), you shouldn't have any problems with moving the code between a true 8051 and a Dallas Semiconductor HSM.

In the diagrams showing how the 8051's instructions execute, I have included both the original 8051's as well as the HSM's instruction clock cycle counts.

Control Store and External Memory

Dallas Semiconductor has made a few enhancements to the 8051 standard with regards to how control store, with and without external memory, works. These enhancements should give you some options in adding external memory (both control store and RAM) that will make your applications easier and give you some flexibility not present in other 8051 microcontrollers.

RESIZING DALLAS SEMICONDUCTOR HSM CONTROL STORE

One interesting feature in the Dallas Semiconductor HSM parts is the ability to resize the control store ROM. This feature will allow you to test code destined for other devices or mix-and-match the amount of EPROM/RAM that is available for control store.

The built-in control store that can be accessed is controlled by the ROMSIZE register at address 0C2h. (See Table 5-1.) On power-up, ROMSIZE is initialized to the maximum size of the HSM microcontroller's EPROM control store. For the DS87C520, these bits will be set to 101 because the device has 16K of on-board EPROM available. The three RS*x* bits are written to by a timed-access instruction sequence (which is explained later in this chapter).

ADDING WAIT STATES TO DALLAS SEMICONDUCTOR HSM MICROCONTROLLERS WITH EXTERNAL RAM

With the sped-up instruction cycle of the Dallas Semiconductor HSM microcontrollers, there are a number of potential problems that might preclude the ability to drop the HSM right into an application that used to use an 8051 and have it run without any problems. One of the most obvious concerns is the interface to external memory. In both the 8051 and HSM, the external memory access is timed as part of the instruction cycle. By going to the HSM with the shorter instruction cycle, external memory accesses that used to work might not work in the HSM because of the faster memory access.

TABLE 5-1 CONFIGURING THE ROMSIZE REGISTER

ROMSIZE BIT	FUNCTION			
7–3	Not used			
2–0	Maximum on-chip ROM to be accessed			
	RS2	*RS1*	*RS0*	*ROM*
	0	0	0	0K
	0	0	1	1K
	0	1	0	2K
	0	1	1	4K
	1	0	0	8K
	1	0	1	16K
	1	1	0	32K
	1	1	1	64K

TABLE 5-2 CONFIGURING THE CKCON REGISTER

MD2	MD1	MD0	MEMORY CYCLES PER ACCESS	CLOCK CYCLES PER ACCESS
0	0	0	2	2
0	0	1	3*	4
0	1	0	4	8
0	1	1	5	12
1	0	0	6	16
1	0	1	7	20
1	1	0	8	24
1	1	1	9	28

In the 8051, two control store reads take place in a 12 clock cycle period, and one external memory access takes place in a single 12 clock cycle period. In the HSM, one control store byte read requires 4 clock cycles, and one external memory access nominally requires 8 clock cycles. This means that, in an HSM running at the same clock speed as an 8051, both the control store reads and the external RAM accesses run 33% faster than in a regular 8051.

The most obvious solution to the problem of the faster external memory accesses is to either run the HSM at a slower speed than the 8051 or replace the memory with faster devices. If the external control store reads are unreliable, then one of these two solutions will have to be implemented.

For external data memory, the actual access time can be varied by adding or taking away instruction cycles during external memory accesses. This is accomplished by changing the values of the MD*x*" bits of the CKCON register (located at address 08Eh). (See Table 5-2.) I have marked the "3" Memory Cycles per Access with an asterisk because this is the default value (the access time can be halved by resetting all three bits). In this case, two actual instruction cycles (using eight clock cycles) will be used for the external memory access.

To time the actual access time, the number of clock cycles used is multiplied by the HSM's clock period. For a 33-MHz HSM clock, accessing external memory with six memory cycles per access, the actual access time will be 485 nsec (30 nsec clock period times 16).

Scratchpad RAM Enhancements

If you've looked at the data sheets of the Dallas Semiconductor HSM and other enhanced 8051 microcontrollers (such as the 8052), you'll probably notice that they are listed as having 256 bytes of scratchpad RAM. This might seem confusing because the first 256 addresses of the 8051 are occupied by 128 bytes of scratchpad RAM and 128 addresses of special-function registers.

One of the aspects of the 8051 that I really like is the strict adherence to a philosophy of keeping everything simple and, if enhancements are going to be made, they are done with an eye toward not complicating or changing the basic architecture. This is especially true in this case.

With the special-function registers, the assumption was made that they would never be accessed indirectly (i.e., using an index register) and would only be accessed directly (with the register address explicitly specified in the instruction). With this assumption, it meant that additional scratchpad RAM could be accessed indirectly and not affect the special-function registers.

By doing this, the first 256 addresses of the 8051's data space could be represented as shown in Fig. 5-2. You can see that, for directly accessing addresses in the first 256 addresses of the enhanced 8051's data space, if the address is less than 080h, then a scratchpad RAM register will be accessed. If the address is greater than or equal to 080h, then a special-function register will be accessed.

Using an index register (R0 or R1) to access anywhere in the first 256 addresses of the enhanced 8051's data space will only access RAM, regardless of what address is loaded into the index register. I typically use the upper 128 bytes of indirectly addressed scratchpad RAM for the processor's stack and (relatively) large arrays.

With 256 bytes of scratchpad RAM, you can actually address 384 addresses in the first 256 addresses of the enhanced 8051's data space. The first time I saw this in a data

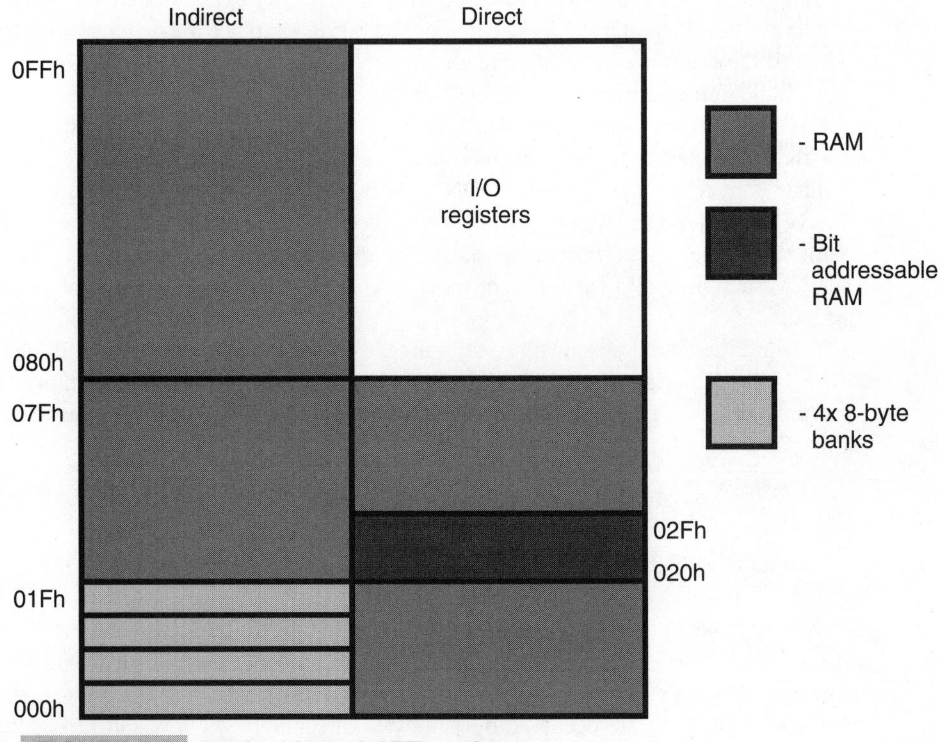

FIGURE 5-2 8051 address 0-0FFh registers.

sheet, I really puzzled over it. Rather than get hung up on this point, it's important to remember that, for these devices, the upper 128 addresses are special-function registers when direct addressing is used and additional scratchpad RAM when indirect addressing is used.

Timers

The basic 8051 timer design is quite easy to understand and use. Unfortunately, it is also quite limited in its capabilities to simplify interfacing with external devices and providing advanced functions. For many enhanced interfaces, specialized timers are available. In the 8052 versions of the 8051, a third 16-bit-only timer has been added to provide some additional software timing options in the 8051.

DALLAS SEMICONDUCTOR HSM EXTENSIONS

Earlier in this chapter, I introduced you to the Dallas Semiconductor HSM microcontrollers, which run with a 4 clock cycle instruction cycle instead of the 12 clock cycle instruction of the stock 8051. Throughout the book, I discuss the importance of keeping the clock constant for different devices to execute applications without any changes. Understanding how clock delays work for different applications will help you use different microcontrollers without requiring major code rewrites.

In the normal 8051, a timer tick takes one instruction cycle (12 clock cycles). This left the Dallas Semiconductor engineers with a problem because of the fewer clock cycles the HSM executes at. The question was, how many clock cycles should result in a timer tick? Following the 8051 precedent and using one timer tick for each instruction cycle, the timers would be incremented once every four clock cycles. This meant that the HSM part would increment three times faster than the base 8051 running with the same clock.

Another solution could have been to just divide the clocks by 12 for each timer tick, but this causes a problem with using the HSM at a reduced speed (one of its major selling points); the timer tick rate would also be reduced.

The final solution was to provide a switch between a 4- and 12-cycle delay for the timers (Fig. 5-3). The timer count switch is controlled by the CKCON Register at address 08Eh. (See Table 5-3).

When the TxM bit for a timer is reset (which is the power-up default condition), the input clock is divided by 12. Setting the bit will result in a divide by 4. This solution satisfies both concerns with the default value of divide by 12, which means that the HSM part can be dropped into the original application's 8051 socket without changing any of the code.

This option of dividing the clock into the timers by four means that an HSM should really work at one-third the speed of an 8051, not divided by 2.5. If an application is currently running at 12 MHZ with a straight 8051, an HSM should be run at 4 MHz to ensure that the timers will work exactly the same between the two applications.

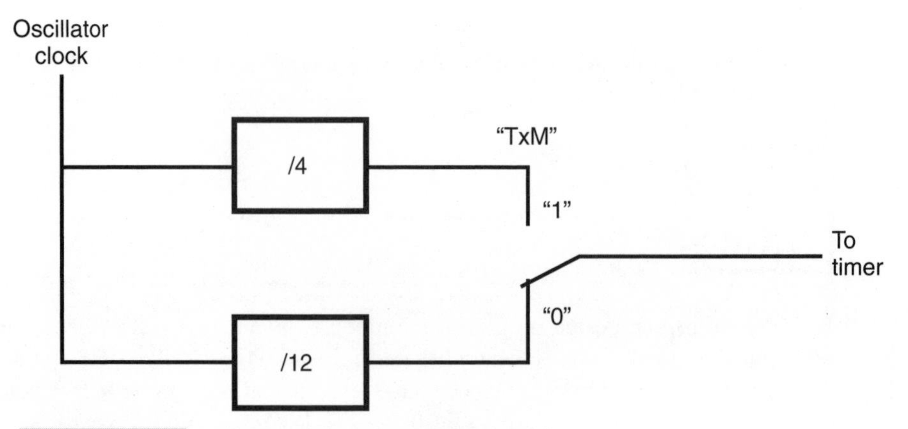

FIGURE 5-3 HSM timer source extensions.

TABLE 5-3 CKCON REGISTER TIMER-COUNT SWITCH SETTINGS

CKCON BIT	FUNCTION
7–6	WDT mode selects
5	T2M—Timer2 clock delay select
4	T1M—Timer1 clock delay select
3	TOM—Timer0 clock delay select
2–0	`movx` instruction delay specifier

TIMER2

As part of the 8052 set of enhancements to the base 8051 is the inclusion of a third 16-bit timer. This timer also provides some additional I/O capabilities when implemented in the Dallas Semiconductor HSM parts. Timer2 can also be used as a secondary baud rate generator source for the first serial port (leaving Timer1 for use with the second serial port).

Timer2 can be run in two different modes. *Capture mode* (Fig. 5-4) uses Timer2 as a free running clock, which saves the timer value on each high to low transition. This mode can be used for recording bit lengths when receiving Manchester- encoded data.

The second Timer2 mode is known as *auto-reload mode* (Fig. 5-5). When the timer has overflowed, a value is written into the TH2/TL2 registers from the RCAP2H/RCAP2L registers. In this mode, the T2 pin on the microcontroller can also be used to initiate a reload of the timer. This feature can be used to implement a system watchdog timer. If the reset signal is not received in a timely manner, the Tmr2 interrupt can be used to initiate a system reset by the 8052.

The T2CON register (at address 0C8h) is used to control the operation of Tmr2 (Table 5-4).

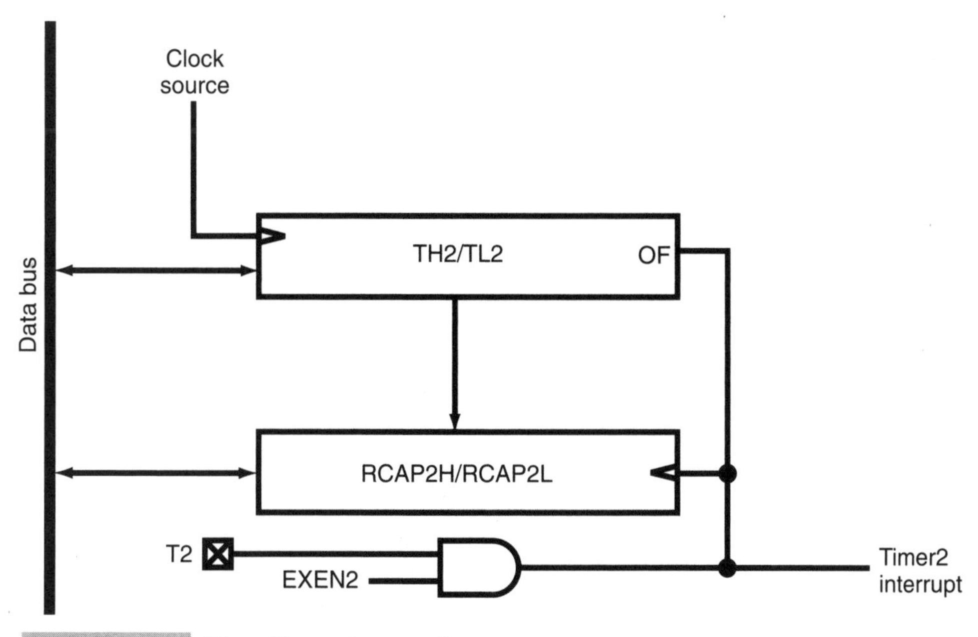

FIGURE 5-4 Timer2's capture mode.

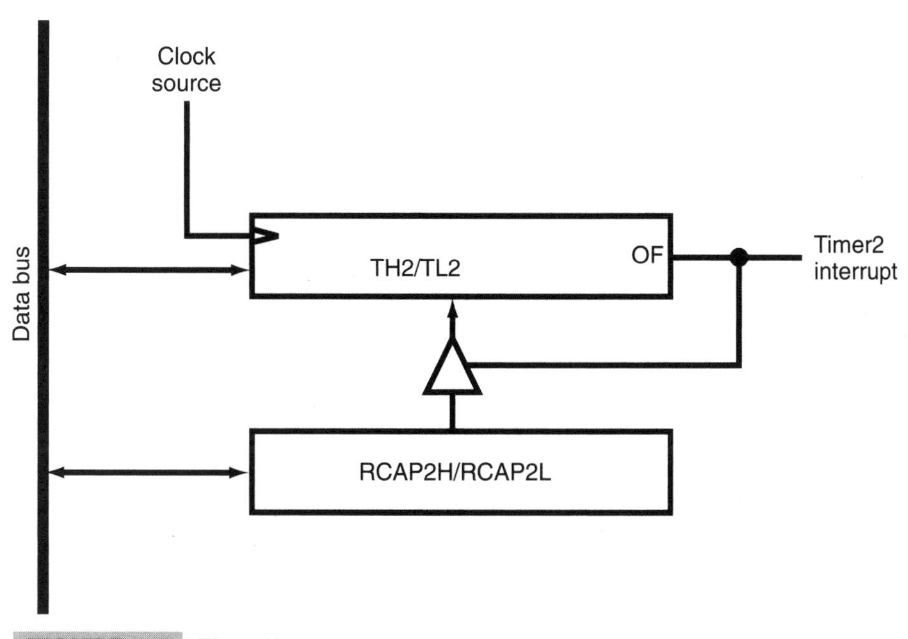

FIGURE 5-5 Timer2's auto-reload mode.

TABLE 5-4 DESCRIPTION OF THE T2CON REGISTER

T2CON BIT	FUNCTION
7	TF2—Tmr2 overflow flag—Must be cleared in software
6	EXF2—Set upon reload or capture—Must be cleared in software
5	RCLK—When set, Tmr2 overflow is used for serial Rx port clock
4	TCLK—When set, Tmr2 overlow is used for serial Tx port clock
3	EXEN2—When set, capture/reload from external source
2	TR2—Set to start Tmr2 running
1	C/_T2 select—Reset, internal instruction clock for Tmr2 source
0	CP/_RL2—Select between capture or reload modes

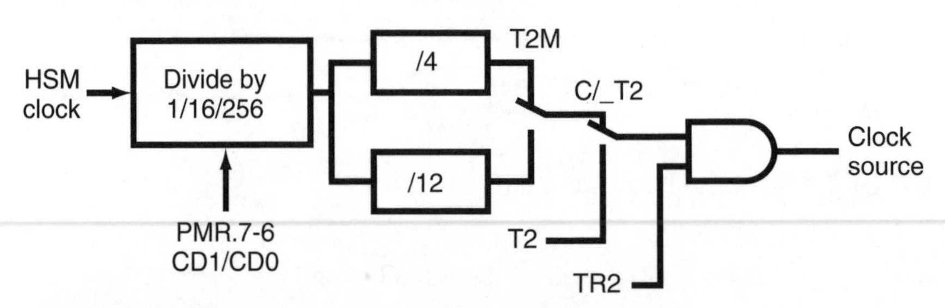

FIGURE 5-6 **Dallas Semiconductor HSM TMR2 clock source.**

Like TMR0 and TMR1, TMR2 can use either the internal clock or an external one for the clock source. Unlike TMR0 and TMR1, when an interrupt request is acknowledged and the TMR2 interrupt handler is executing, the overflow flag (TF2) or reload/capture flag (EXF2) must be reset in software and not rely on `reti` to reset them.

For a typical 8052, the clock source is identical to TMR0 and TMR1. However, in the Dallas Semiconductor HSM, this input is a bit more complex. (See Fig. 5-6.) In the Dallas Semiconductor HSMs, the clock cannot only be divided by 4 or 12 (as was discussed in the previous section), but there also is an additional divider, controlled by the CD*x* bits of the PMR register (at address 0C4h).

When TMR2 runs in auto-reload mode, the overflow can be used as a data rate generator for the serial port (or the first serial port in the Dallas Semiconductor HSM parts). By setting the RCLK and TCLK bits, the serial port will be driven from TMR2 and not TMR1. In the Dallas Semiconductor HSM parts, this means the two serial ports can be driven by different clocks and run at different speeds.

DALLAS SEMICONDUCTOR HSM WATCHDOG TIMERS

A *watchdog timer* (usually referred to as WDT) is used to protect an application in case the controlling microcontroller begins to run amok and execute randomly rather than the preprogrammed instructions written for the application. Causes for this behavior include large electrical and magnetic fields and disturbances (such as found under a car's hood or in a video monitor) that can change the electrical state of a silicon gate. The purpose of the watchdog timer is to reset the microcontroller if orderly execution is lost.

The watchdog timer detects a problem with application execution if it is not reset within a given period of time. The usual way watchdog timers are used is to start them running and let the application reset the timer before the watchdog timer overflows and resets the application. If the application does not reset the watchdog timer in the appropriate time, it is assumed that the microcontroller has made a "branch to the boonies" and has to be reigned in.

The watchdog timer provided in the Dallas Semiconductor HSM microcontrollers (Fig. 5-7) is similar to that provided in other microcontrollers. The microcontroller clock (not the instruction clock) is passed through a series of counters, and the overflow of each is passed either to another stage or used to trigger the watchdog timer reset. The first divisor selects the first delay in clock cycles for the incoming signal. This delay is specified by the CD0 and CD1 bits of the PMR special- function register. If the PMR register is not present in the HSM, then the clock isn't divided at this stage (or can be thought of as a divisor of 1). For most applications, I would recommend leaving the divisor at 1 because of the large range of time-outs possible using the other delay selections.

The watchdog timer divisor circuit can time out after two to the power of 17, 20, 23, or 26. These values are selected from the WDCON register at address 0D8h. I always find that describing numbers as powers to be meaningless to me. A better way of describing the watchdog timer time-out intervals is after 131,072, 1,048,576, 8,388,608 or 67,108,864 clock cycles. If the HSM was running at 12 MHz, the time-out is selectable

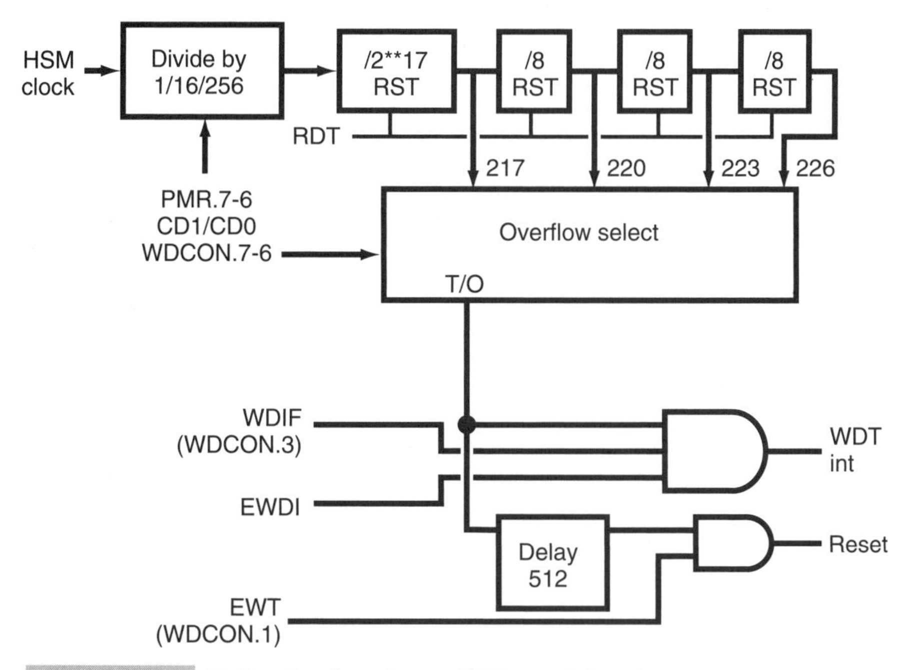

FIGURE 5-7 Dallas Semiconductor HSM watchdog timer.

TABLE 5-5 DESCRIPTION OF THE WDCON REGISTER

WDCON BIT	FUNCTION
7	SMOD—Used to select the speed range of serial port 1 (the second serial port in the HSM).
6	*PowerOnReset—Set if the reset was caused by the WDT.
5	Set to enable an interrupt if input power drops below 4.5 V (brown-out).
4	Power Fail Interrupt Flag—Set if power is below 4.5 V and requests an interrupt.
3	*Watchdog Interrupt Flag.
2	Watchdog Timer Reset Flag—Set if watchdog reset has occurred.
1	*Enable Watchdog Timer—When set, the WDT is running.
0	*RDT—Reset WDT.

between 11 msec, 87 msec, 699 msec, or 5.6 seconds. With a divide by 256, this can be increased to almost 24 minutes.

To control watchdog timer operations, the WDCON register (at address 0D8h) is used (Table 5-5). While I don't really discuss them in this book, the power-failure (brown-out) functions should be self-explanatory.

To start the watchdog timer, bit 1 of WDCON is set to start the watchdog timer running. Before the watchdog reset occurs, an interrupt can be generated by the watchdog timer hardware. Using the watchdog timer interrupt request circuitry, the application will be warned 512 clock instruction cycles before the reset, to allow the application time to either power down or reset the watchdog timer.

In the list of WDCON bits in Table 5-5, you'll notice that I've marked four bits with an asterisk. These bits all have to be modified using a *timed-access write* in the HSM. A timed-access write is defined as setting or resetting a bit using a specific write sequence to allow the bits to be changed. This sequence consists of writing 0AAh and then 055h to the TA register (at address 0C7h) before modifying any of the marked bits in WDCON.

This timed access consists of the instruction sequence:

```
  clr  EA                 ; Disable interrupts during T/A sequence
  mov  0C7h,#0AAh         ; Write the T/A sequence
  mov  0C7h,#055h
;  #### - Change timed-access bit using "clr" or "setb" instructions
  setb EA                 ; Interrupts can now be enabled
```

Changing a timed-access bit could be a `mov`, `setb`, or `clr` instruction (although I recommend just using the bit set or clear instructions). If there are any delays (such as `nops` or interrupts) after the first `mov 0C7h,#0AAh` instruction, the bit change operation will fail and the bit won't be updated.

If multiple bits are to be changed, then multiple timed-access instruction sequences are required.

Along with the four WDCON bits, several others in the Dallas Semiconductor HSM require timed-access updates as well. These include the ROMSIZE bits (used to control how much built in ROM can be accessed).

Serial I/O

Serial interfacing is one area that the original 8051 specification could be improved upon significantly. This is not because the basic modes provided within the 8051 don't work very well, but because over the years, new standards in interfacing to memory and other devices have emerged that make wiring simpler and require much fewer pins and glue logic than interfacing to external devices using a parallel, memory-mapped I/O interface.

These interfaces might not be implemented in exactly the same way across different manufacturer's devices. For this reason, I will introduce you to the interface standards to familiarize you with them, but I won't present any 8051 manufacturer's registers and programming information. This information can be found in the appropriate device's data sheets.

DALLAS SEMICONDUCTOR HSM'S SECOND SERIAL PORT

There is one major difference between the Dallas Semiconductor HSM's peripheral equipment and that of a standard 8052: The HSM has two serial ports whereas all true versions of the 8051 only have one.

This additional serial port works in almost exactly the same way as the first one (including having its own interrupt handler). The second serial port is known as serial port 1, and the first is serial port 0.

The only physical difference between the two serial ports is that the second serial port can only be driven from Timer1's overflow, whereas Port 0 can be driven from either Timer1's or Timer2's overflow. The operation of the two serial ports working together will be shown in the "Electronic RS-232 Breakout Box" example application in chapter 12.

MICROWIRE

The Microwire synchronous serial protocol is capable of transferring data at up to 1 megabit per second. Sixteen bits are transferred at a time.

To read 16 bits of data, the waveform would look like Fig. 5-8. After selecting a chip and sending a start bit, the clock strobes out an 8-bit command byte (labelled OP1, OP2, and A5 to A0 in Fig. 5-8), followed by (optionally) a 16-bit address word transmitted and then another 16-bit word either written or read by the microcontroller.

With a 1 megabit-per-second maximum speed, the clock is both high and low for 500 nsec. Transmitted bits should be sent 100 nsec before the rising edge of the clock. When reading a bit, it should be checked 100 nsec before the falling edge of the clock is expected. While these timings will work for most devices, you should make sure you understand the requirements of the device being interfaced to.

SPI

The SPI protocol is similar to Microwire, but with a few differences:

- SPI is capable of up to 3 megabits-per-second data transfer rate.
- The SPI data word size is 8 bits.

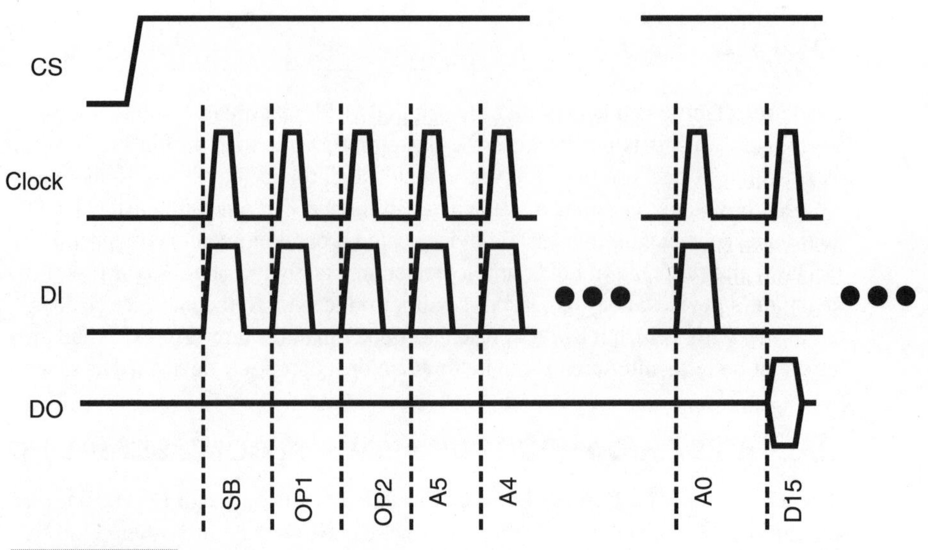

FIGURE 5-8 Microwire data read.

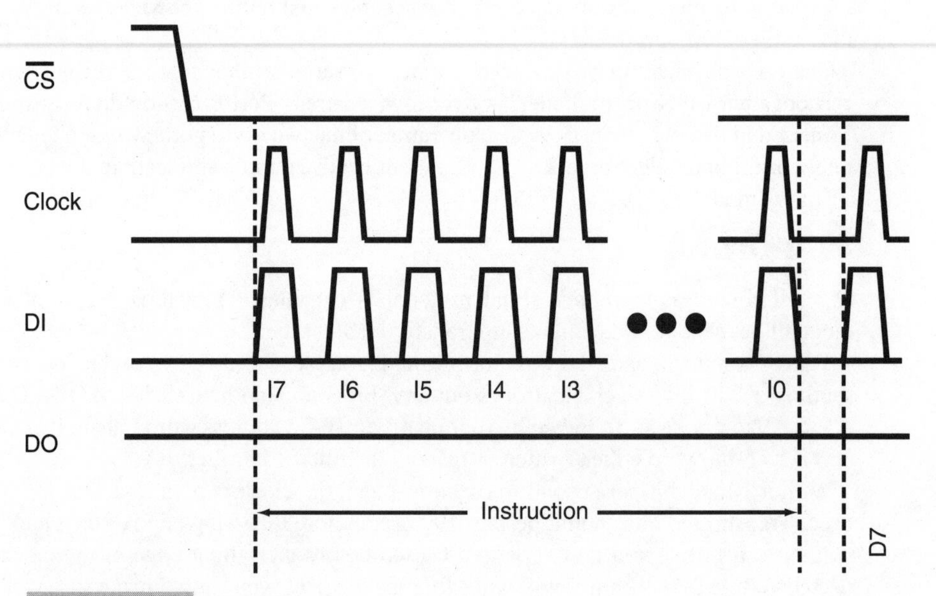

FIGURE 5-9 SPI data write.

- SPI has a hold that allows the transmitter to suspend data transfer.
- Data in SPI can be transferred as multiple bytes known as *blocks* or *pages*.

Like Microwire, SPI first sends a byte instruction to the receiving device. After the byte is sent, a 16-bit address is optionally sent, followed by 8 bits of I/O (Fig. 5-9). As noted earlier, SPI does allow for multiple byte transfers.

The SPI clock is symmetrical (an equal low and high time). Output data should be available at least 30 nsec before the clock line goes high and read 30 nsec before the falling edge of the clock.

When wiring up a Microwire or SPI device, there is one trick that you can do to simplify the microcontroller connection is to combine the DI and DO lines into one pin Fig. 5-10. In this method of connecting the two devices, when the Data pin on the microcontroller has completed sending the serial data, the output driver can be turned off and the microcontroller can read the data coming from the device. The current limiting resistor between the Data pin and DI/DO limits any current flows when both the microcontroller and device are driving the line.

This trick can be used in a variety of different situations, but it must be done so carefully. The resistor will prevent any possible I/O pin damage due to *bus contention* (two devices driving the same line to different levels at the same time). However, if the resistor is not properly wired into the circuit or if the 8051 attempts to read while it is driving, then invalid data will be stored or read.

I2C

When I discuss networking with regards to microcontrollers, I'm generally talking about buses used to connect additional hardware devices to the microcontrollers and allowing communications between microcontrollers. (This is opposed to local area networks, or LANs, like the ethernet that is used to connect your PC to other computers in your office.)

There are a variety of standards (which can be broadened to include Microwire and SPI discussed in the previous sections) that lead to some confusion over what is and what isn't a network. I personally define a microcontroller network as one that has a single communications medium (i.e., wire) and multiple devices connected to this medium that can initiate message transfers and respond to messages directed towards them.

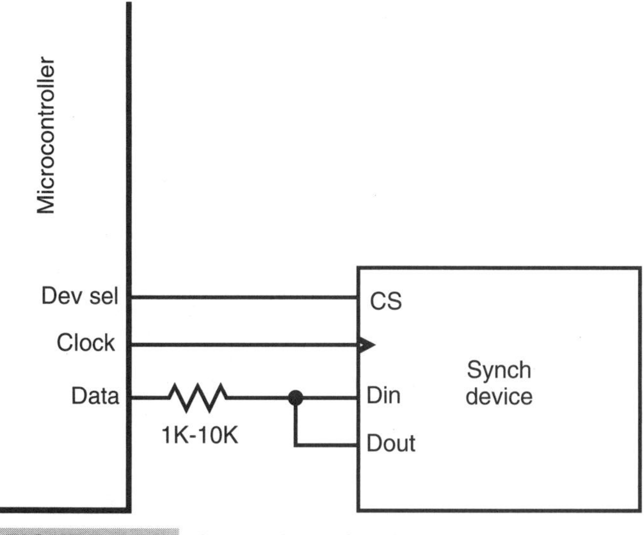

FIGURE 5-10 Combining DO and DI.

In these types of networks, a *master* is an intelligent device that can initiate data transfers. Devices that respond to requests (and can't initiate them) are known as *slaves*. Microcontroller networks can have multiple masters, which means that the network protocol requires an arbitration scheme that allows multiple masters to transmit without ruining each other's messages.

Typically, microcontroller networks transmit very little data, so the bandwidth required is quite modest when compared to something like an ethernet connected PC that has an owner who is a Web surfer. It's not unusual to have a microcontroller network that is transferring bytes per second (compared to a PC network that is transferring megabytes per second).

I will introduce you to two of the most popular microcontroller networks. While what I have written is quite complete, it's not sufficient to use as a reference for developing network applications. If you are planning to use these network protocols, you will have to get more information, in the form of device data sheets or standards. What I have done is give you enough information to understand the basics of the protocols and evaluate them for an application, to decide which is the most appropriate.

The most popular form of microcontroller network is *I2C*, which stands for Inter-Intercomputer Communications. This standard was originally developed by Philips in the late 1970s as a method to provide an interface between microprocessors and peripheral devices without wiring full address, data, and control buses between devices. I2C also allows sharing of network resources between processors (which is known as *multi-mastering*).

The I2C bus consists of two lines: A clock line (SCL) is used to strobe data from the SDA line from or to the master that currently has control over the bus. Both these bus lines are pulled up (to allow multiple devices to drive them).

A I2C Controlled Stereo system might be wired as shown in Fig. 5-11. The two bus lines are used to indicate that a data transmission is about to begin as well as pass the data on the bus.

To begin a data transfer, a Start Condition is put on the bus. Normally, when the bus is in the idle state, both the clock and data lines are not being driven (and are pulled high). To initiate a data transfer, the master requesting the bus pulls down the SDA bus line followed by the SCL bus line. During data transmission, this is an invalid condition (because the data line is changing while the clock line is active/high).

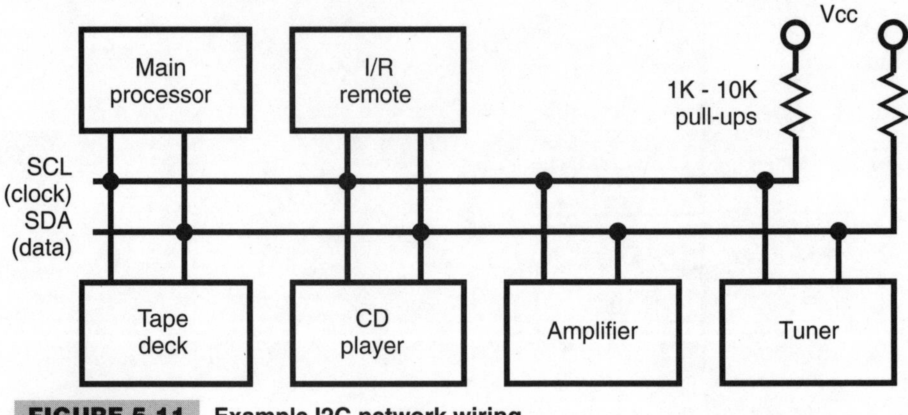

FIGURE 5-11 Example I2C network wiring.

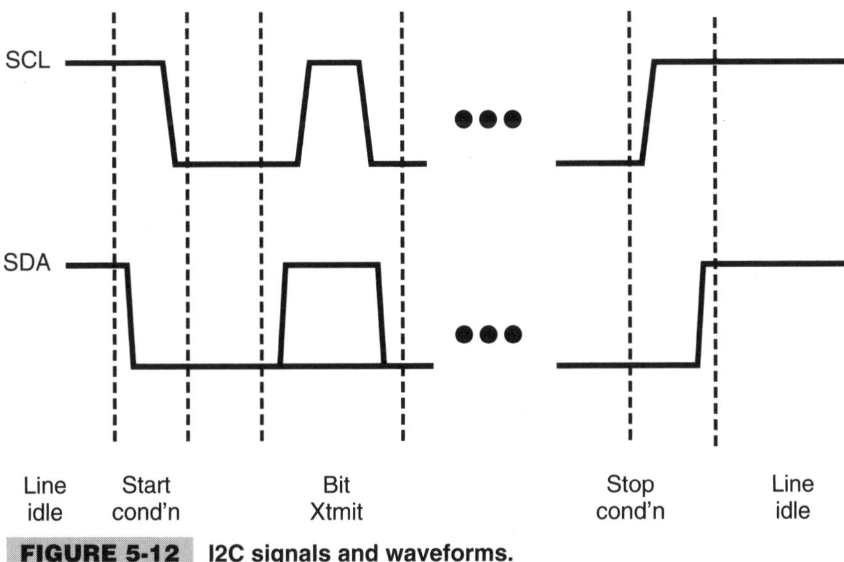

FIGURE 5-12 I2C signals and waveforms.

To end data transmission, the reverse is executed: The clock line is allowed to go high, which is followed by the data line (Fig. 5-12).

Data is transmitted in a synchronous fashion, with the most significant bit sent first. The byte transmitter (which might or might not be the master) allows the data line to float (it doesn't drive it low) while strobing the clock to allow the receiving device to pull the data line low as an acknowledgment that the data was received. After the acknowledge bit, both the clock and data lines are pulled low in preparation for the next byte to be transmitted or a Stop/Start Condition is put on the bus. (See Fig. 5-13.)

Sometimes, the acknowledge bit will be allowed to float high, even though the data transfer has completed successfully. This is done to indicate that the data transfer has completed

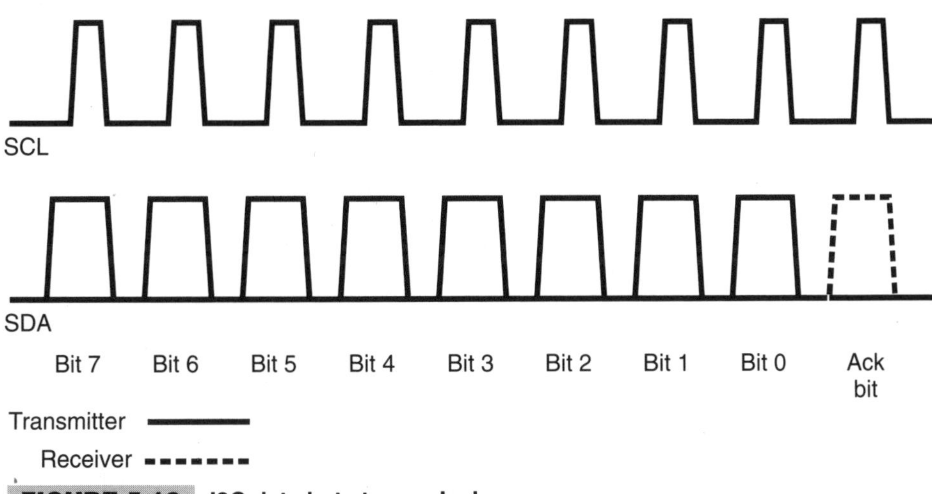

FIGURE 5-13 I2C data byte transmission.

and the receiver (which is usually a slave device or a master that is unable to initiate data transfer) can prepare for the next data request.

There are two maximum speeds for I2C (because the clock is produced by a master, there really is no minimum speed): "Standard mode" runs at up to 100 kbps, and fast mode can transfer data at up to 400 kbps. (See Fig. 5-14.)

A command is sent from the master to the receiver in the following format shown in Fig. 5-15.

The receiver address is 7 bits long and is the bus address of the receiver. There is a loose standard to use the most significant 4 bits to identify the type of device, while the

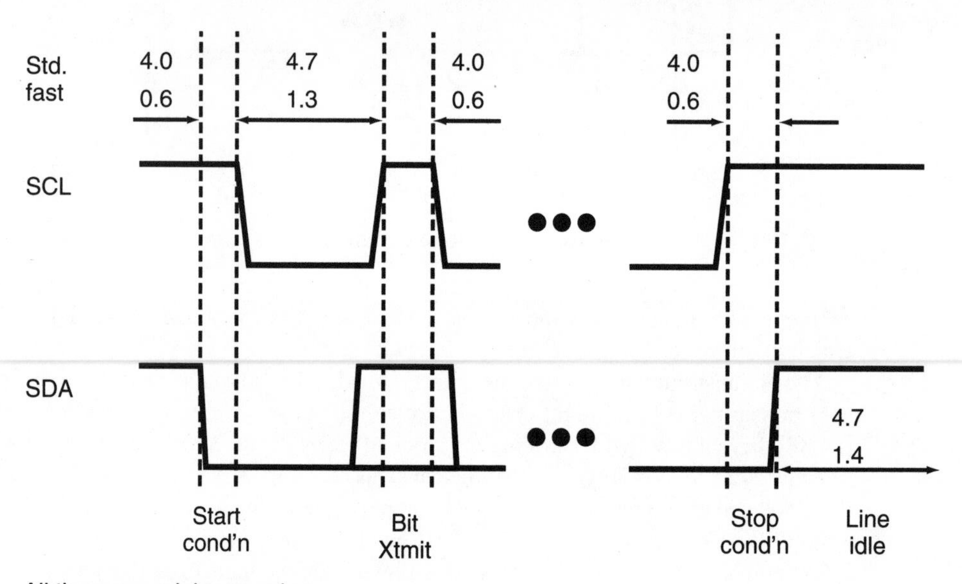

FIGURE 5-14 I2C signal timing.

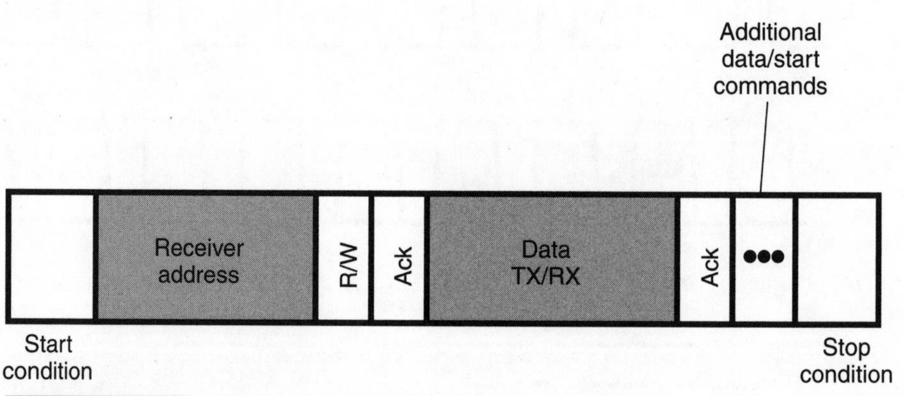

FIGURE 5-15 I2C data transmission.

TABLE 5-6 CONVENTION FOR THE NEXT 4-BIT PATTERNS

BIT PATTERN	PURPOSE
0000	Reserved address
0010	Voice synthesizer
0011	PCM audio interface
0100	Audible tone generation
0111	LCD/LED displays
1000	Video interface
1001	A/D and D/A interfaces
1010	Serial memory
1100	RF tuning/control
1101	Clock/Calendar
1111	Reserved/10-bit address

next 3 bits are used to specify one of 8 devices of this type (or further specify the device type).

As I said earlier, this is loose standard. Some devices require certain patterns for the second 3 bits, while others (such as some large serial EEPROMs) use these bits to specify an address inside the device. As well, there is a 10-bit address standard in which the first 4 bits are all set, the next bit reset, the last two are the most significant 2 bits of the address, with the final 8 bits being sent in a following byte. All this means is that it is very important to map out the devices to be put on the bus and all their addresses.

The next 4-bit patterns generally follow the convention shown in Table 5-6 for different devices.

This is really all there is to I2C communication, except for a few points. In some devices, a start bit has to be resent to reset the receiving device for the next command (e.g., in a serial EEPROM read, the first command sends the address to read from, the second reads the data at that address).

The last point to note about I2C is that it's a multi-mastering protocol, which is to say that multiple microcontrollers can initiate data transfers on the bus. This obviously results in possible collisions on the bus (which is when two devices attempt to drive the bus at the same time). Obviously, if one microcontroller takes the bus (sends a Start Condition) before another one attempts to do so, there is no problem. The problem arises when multiple devices initiate the Start Condition at the same time.

Actually, arbitration in this case is really quite simple. During the data transmission, hardware (or software) in both transmitters synchronize the clock pulses so that they match each other exactly. During the address transmission, if a bit is expected to be a 1 by a master is actually a 0, then it drops off the bus because another master is on the bus. The master that drops off will wait until the Stop Condition, then it will re-initiate the message.

I realize that this is hard to understand with just a written description. In the next section, "CAN," I will show how this is done with an asynchronous bus that works in a very similar way to this situation.

I2C can be implemented in software quite easily. However, due to software overhead, the fast mode probably cannot be implemented. Even the standard mode's 100 kbps will be a stretch for most 8051s. I find that implementing I2C in software to be best as the single master in a network. That way it doesn't have to be synchronized to any other devices or accept messages from any other devices that are masters and are running a hardware implementation of I2C that might be too fast for the software slave.

CAN

The *Controller Area Network protocol* was originally developed by Bosch a number of years ago as a networking scheme that could be used to interconnect the computing systems used within automobiles. At the time, there was no single standard for linking digital devices in automobiles. I read an interesting statistic when researching about CAN: Before the advent of CAN (and J1850, which is a similar North American standard), cars could have had up to three miles of wiring weighing 200 pounds interconnecting the various computing systems within the car. With a CAN network installed, the wiring is often just a few hundred feet long and weighs about 10 pounds.

CAN was designed to be:

- Fast (1 Mbit/second)
- Insensitive to electromagnetic interference
- Simple with few pins in connectors for mechanical reliability

Devices could be added or deleted from the network easily (and during manufacturing).

While CAN is similar to J1850 and does rely on the same first two layers of the OSI 7-layer communications model, the two standards are electrically incompatible. CAN was the first standard and is thoroughly entrenched in European and Japanese cars and is rapidly becoming the standard of choice for North American automotive manufacturers.

CAN is built from a dotted-AND bus that is similar to that used in I2C. Electrically, RS-485 drivers are used to provide a differential voltage network that will work even if one of the two conductors is shorted or disconnected (giving the network high reliability inside the very extreme environment of the automobile). This dotted-AND bus allows arbitration between different devices (when the device's drivers are active, the bus is pulled down, like in I2C).

An example of how this method of arbitration works is shown in Fig. 5-16. In this example, when a driver has a miscompare with what it is transmitting (e.g., when it is sending a 1 and a 0 shows up on the bus), then that driver stops sending data until the current message (which is known as a *frame*) has completed. This is a very simple and effective way to arbitrate multiple signals without having to retransmit all of the colliding messages over again.

The frame is transmitted as an asynchronous serial stream (which means there is no clocking data transmitted). This means that both the transmitter and receiver must be working at the same speed (typically data rates are in the range of 200 kbps to 1 Mbps).

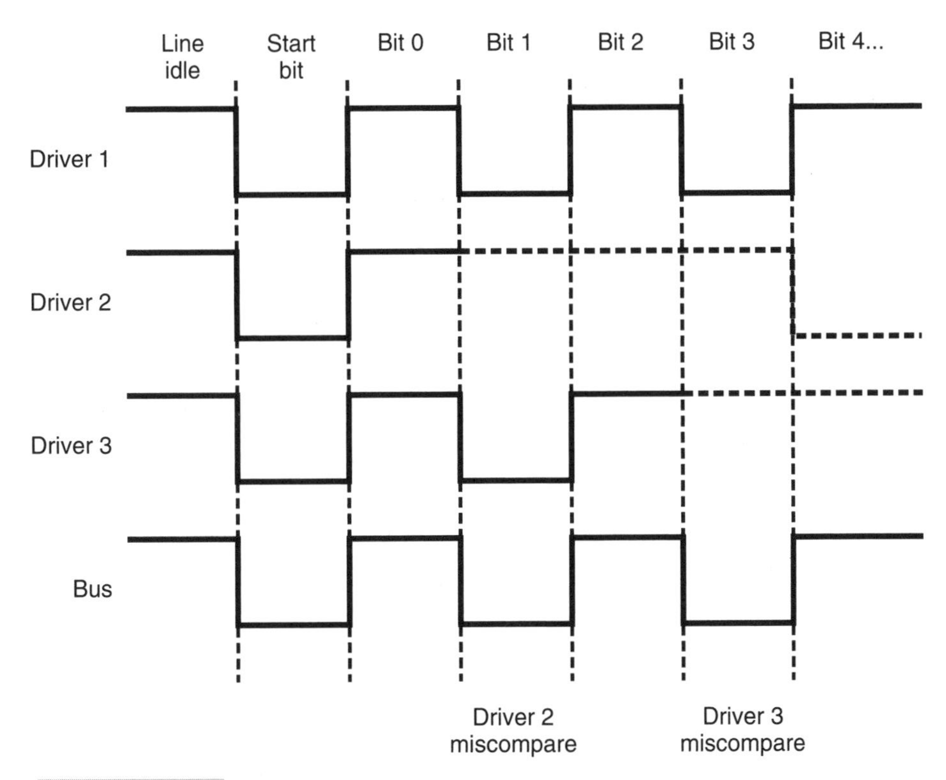

FIGURE 5-16 **CAN transmission arbitration.**

A frame looks like Fig. 5-17. In CAN, a 0 is known as a *dominant bit*, and a 1 is known as a *recessive bit*.

The different fields of the frame are defined in Table 5-7.

The last important note about CAN is that devices are not given specific names or addresses. Instead, the message is identified (using the 11- or 19-bit message identifier). Devices on the network either wait for message identifier's meant for them or, if an event occurs that they have to send a message to other devices about, they send a message identifier appropriate to the receiving devices. This method of addressing can provide you with very flexible messaging (which is what CAN is all about).

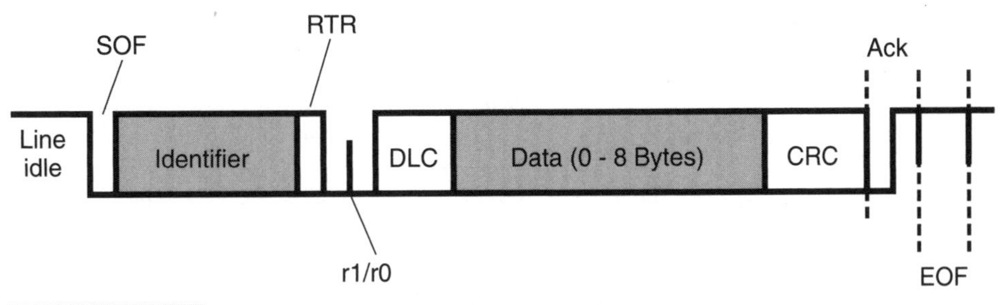

FIGURE 5-17 **CAN 11 bit identifier frame.**

TABLE 5-7 DESCRIPTION OF THE FRAME FIELDS

FIELD	DESCRIPTION
SOF	Start of frame, a single dominant bit
Identifier	11- or 19-bit *message identifier*
RTR	This bit is set if the transmitter is also transferring data
R1/R0	Reserved bits, should always be dominant
DLC	4 bits indicating the number of bytes that follow
Data	0 to 8 bytes of data, sent MSB first
CRC	15 bits of CRC data followed by a recessive bit
Ack	2-bit field, dominant/recessive bits
EOF	End-of-frame, at least 7 recessive bits

The CAN frame is very complex, as is the checking that has to be done, both for receiving a message and transmitting a message. While it can be done using a microcontroller and programming the functions in software, I would recommend only implementing CAN using hardware interfaces. There are several microcontroller vendors that provide CAN interfaces as part of the devices. There are quite a few different standard chips (the Intel 82527 is a very popular device) available that will carry out CAN interface functions for you effectively and cheaply.

Analog I/O

The world outside the microcontroller is not all black and white (or ones and zeros); it is really varying shades of gray (or values between zero and one). Often a microcontroller will have to interface with analog signals (between Vcc and ground), both receiving them (and interpreting the value) as well as outputting them. In many models of the different 8051 microcontroller families, there are analog-to-digital convertors (ADCs) and digital-to-analog convertors (DACs). In this section, I will introduce you to analog voltage I/O.

There are three types of analog voltage input for microcontrollers. The first is a physical positional control sensor in which the microcontroller determines the position of a potentiometer. Next, an analog voltage comparator can be used to determine whether or not an input voltage is greater than or less than a reference voltage. The last type of ADC used in microcontrollers is known as an *integrating ADC*. It can determine a voltage level independently. Each of these different types of ADCs are best suited for different applications.

The first type of ADC isn't really an ADC at all, but a method of reading a potentiometer's current resistance using the 8051's digital I/O pins (Fig. 5-18). To read the resistance, a simple RC network is attached to a microcontroller I/O pin. To ensure damaging currents are not passed when the potentiometer is at the 0 Ohm position, a 100–200 Ohm resistor is placed between the "sensor" pin and the potentiometer.

The potentiometer resistance is read by measuring how long the voltage of the current flowing from the Charge pin to the capacitor through the potentiometer takes to get above the digital threshold level. The higher the resistance, the longer the circuit takes to be read as a 1. While the Sensor pin is being polled for a 1, a counter is incremented to record the value of the resistor in the network.

For what should be obvious reasons, the Sensor pin should not have any internal pull-ups. This pin simply monitors the voltage level of the R/C network. The typical threshold voltage for the 8051 is 1.4 V.

To determine the values for R and C, you can use the approximation:

$$t = 2.2 * R * C$$

where t is the time for the time required to discharge the capacitor.

When I say that this will give you the approximate time, I mean to say that it will really only get you in the ballpark. This is generally *not* a precision circuit because of the tolerances of the resistor and capacitor and the nonlinear behavior of the CMOS input bit. You can get more exact values by specifying a precision capacitor (which is often hard to find and is expensive), but for the circuit, the application shouldn't really require it.

Now, you're probably asking yourself, if this is a very imprecise method of reading a potentiometer, where and why is it used?

It's used in circuits that simply provide a relative value for the potentiometer's position. For example, the IBM PC uses this type of circuit to read the current joystick position. This imprecision is often reflected in the games where the user is asked to move the joystick to extremes to train the program how the capacitor and joystick resistors interact.

Typically, I use a 0.1-µF tantalum cap and a 10-K potentiometer for this circuit. This gives me an approximate maximum time delay of 22 msec, which is long enough for a microcontroller to accurately read the time delay and is fast enough that the user won't be aware that the resistance measurement is taking place. I use 0.1-µF tantalum caps because

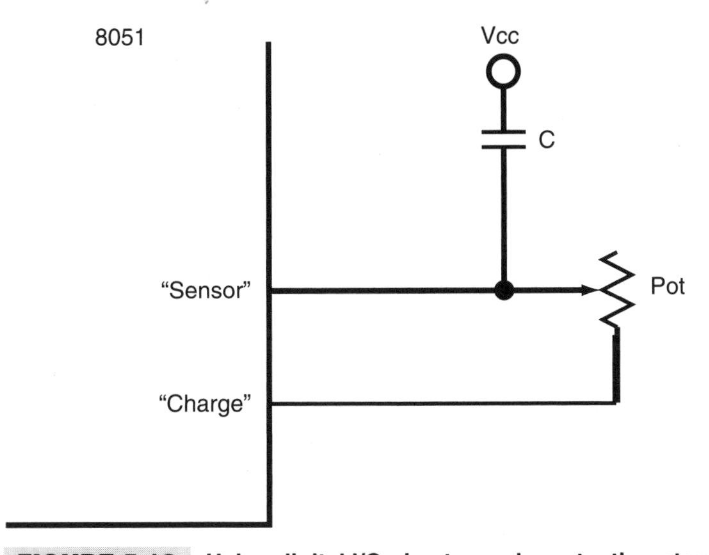

FIGURE 5-18 **Using digital I/O pins to read a potentiometer.**

I use them for decoupling Vcc and ground in digital chips and I always have a bunch of them around. The problem with using tantalum caps is that they can often be out by as much as 100% from their rated value. This means that I usually have the microcontroller calibrate the extreme values of the potentiometer before the application is usable.

In chapter 12, I will show you an application that uses this circuit to determine a joystick's position.

The next type of ADC commonly available in microcontrollers (such as the Atmel AT89C*x*051 devices) is the analog voltage comparator. A comparator is a simple circuit that compares two voltages and returns a 1 when the input signal voltage is greater than the reference voltage (Vref). (See Fig. 5-19.)

This circuit is best used for applications like thermostats, where something happens at specific input voltage points.

Often, in microcontrollers that use comparators for ADCs, the reference voltage (Vref) is generated internally using a resistor ladder and an analog multiplexer to select the desired voltage to output (Fig. 5-20).

This circuit can actually give some modicum of voltage specification. A simple algorithm could be used to run through each of the different voltages until the comparator output changes. The point of change is the actual input voltage. The reason why I say this gives a modicum of voltage determination is because typically the Vref circuit has a fairly wide range between each voltage step (this is also known as a large *granularity*). If the re-

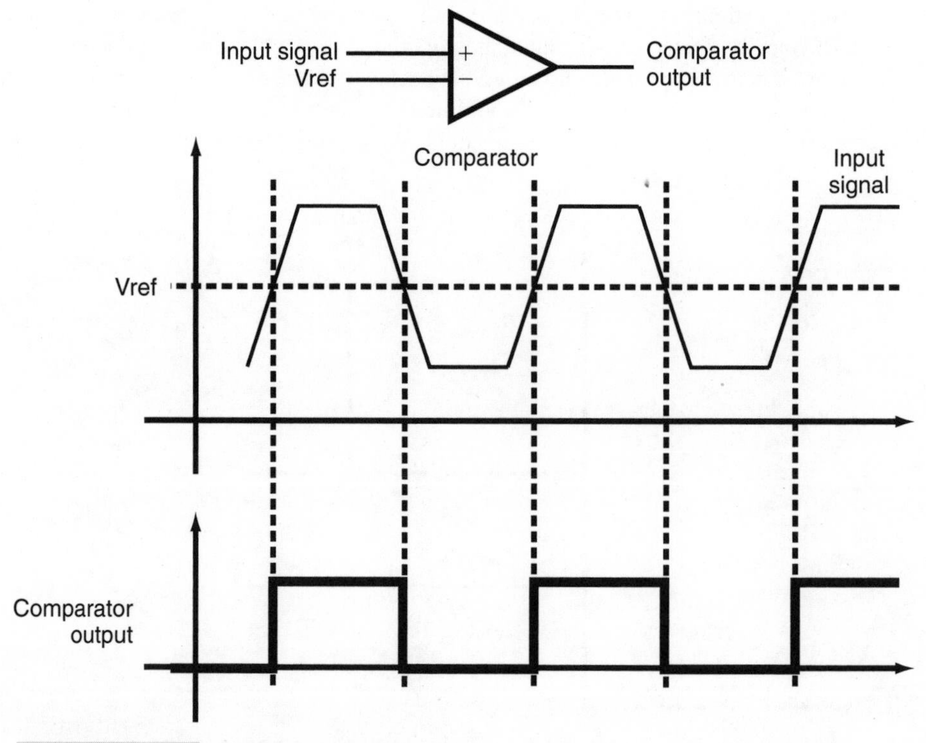

FIGURE 5-19 Comparator response.

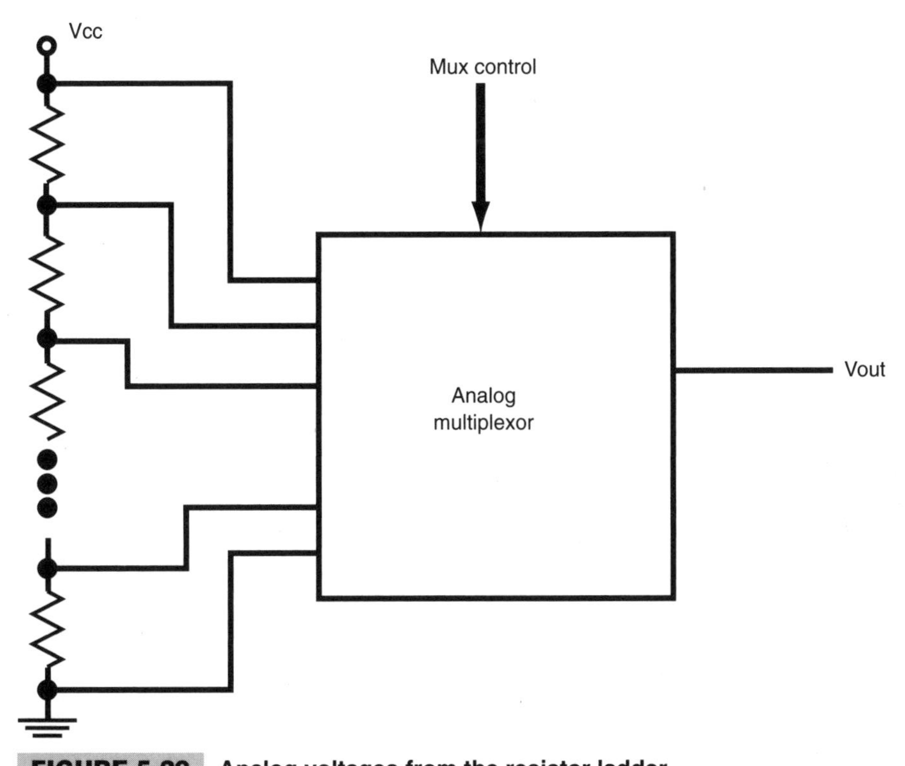

FIGURE 5-20 **Analog voltages from the resistor ladder.**

sistor ladder/analog multiplexer had 8 voltage levels, in a 5-V system, the granularity between voltage steps is over 700 mV.

A method of creating an ADC using comparators is to arrange them in a flash configuration (Fig. 5-21).

The reason why this is known as a flash ADC is because it is very fast when compared with other types of ADCs. The time delay to get an analog voltage measurement is limited by the delay through the comparators and priority encoder. This type of ADC tends to be very expensive because of the chip real estate required for multiple ADCs (for example, to get 8-bit precision, 256 comparators would be required, which would require special buffers to deal with internal signal fan out).

The last type of ADC also uses a comparator, but it uses an analog voltage source that starts at zero volts and runs linearly up to Vcc (this is known as a *sweep generator*). (See Fig. 5-22.) In this circuit (known as an *integrating ADC*), when the conversion is about to start, the timer is reset and the sweep generator is set to 0 V. Then the timer clock and sweep generator are enabled. When the sweep generator output is greater than Vin, the clock is turned off to the timer and an ADC Stop signal is asserted (which can cause an interrupt). At the end of the comparison, the timer will contain a value proportional to Vin.

This method, while being quite accurate, has a few issues associated with it. The first is the time required to do the conversion. The higher Vin is, the longer it will take to

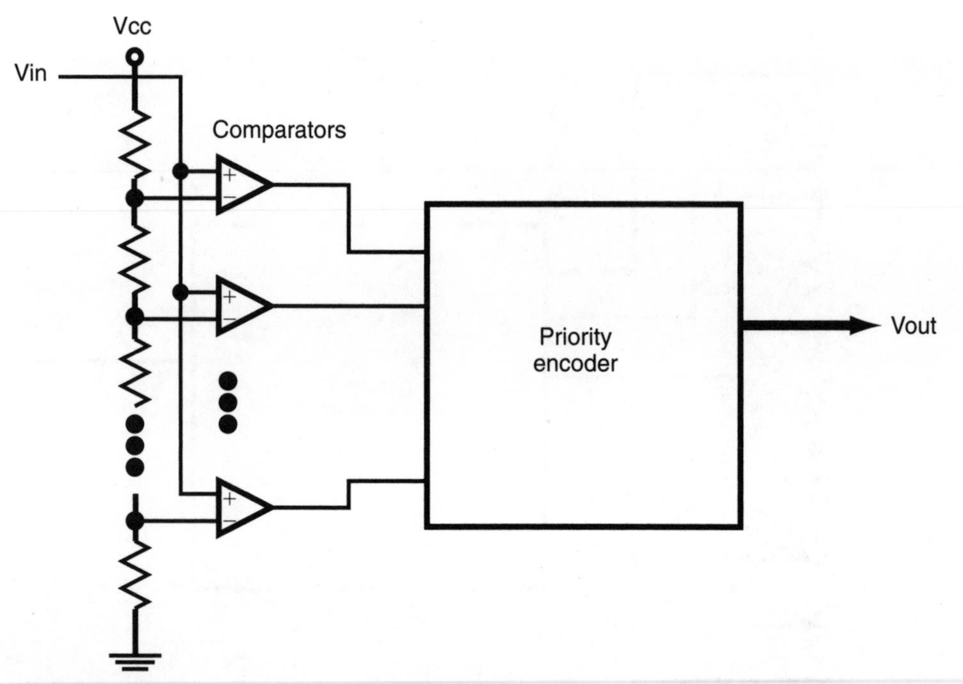

FIGURE 5-21 Flash analog-to-digital comparator.

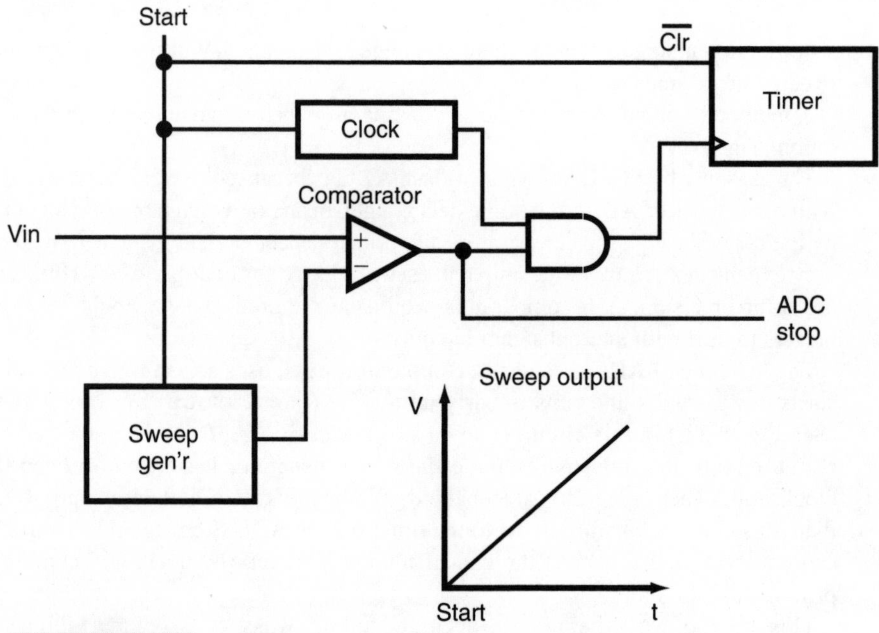

FIGURE 5-22 Integrating analog-to-digital converter.

complete the conversion. Some microcontrollers have sweep generators built in their ADCs that can run at different speeds (although running at a faster speed will decrease the accuracy of the measurement).

Another problem is what happens if the input signal changes during the sample. This problem is avoided by allowing a capacitor to charge very quickly and then sampling the voltage level on the capacitor.

The last issue to be considered with integrating ADCs is the time required with regards to the sampling of very fast input waveforms (Fig. 5-23). For the previous example, a completely different waveform will be read than what is actually present at the microcontroller's ADC pin. Understanding what is the correct sample frequency goes beyond this book, but explanations can be found in DSP text books.

When you look at microcontroller specifications, you will find very few devices with analog voltage output. Of the ones that do, the output is typically generated from the comparator Vref resistor ladder circuit. The reason for this is the many different circuits an analog voltage that would be required to drive. If these circuits are not properly interfaced to the microcontroller output, the device drivers might be overloaded (which will cause an incorrect voltage output) or rapidly changing outputs might cause reflections, which will appear as an incorrect output waveform.

So, to generate analog output voltages, external DACs, digital potentiometers (wired as voltage dividers), or PWM signals filtered by an RC network are usually used in applications. These devices offer very good voltage granularity (often in the range of millivolts).

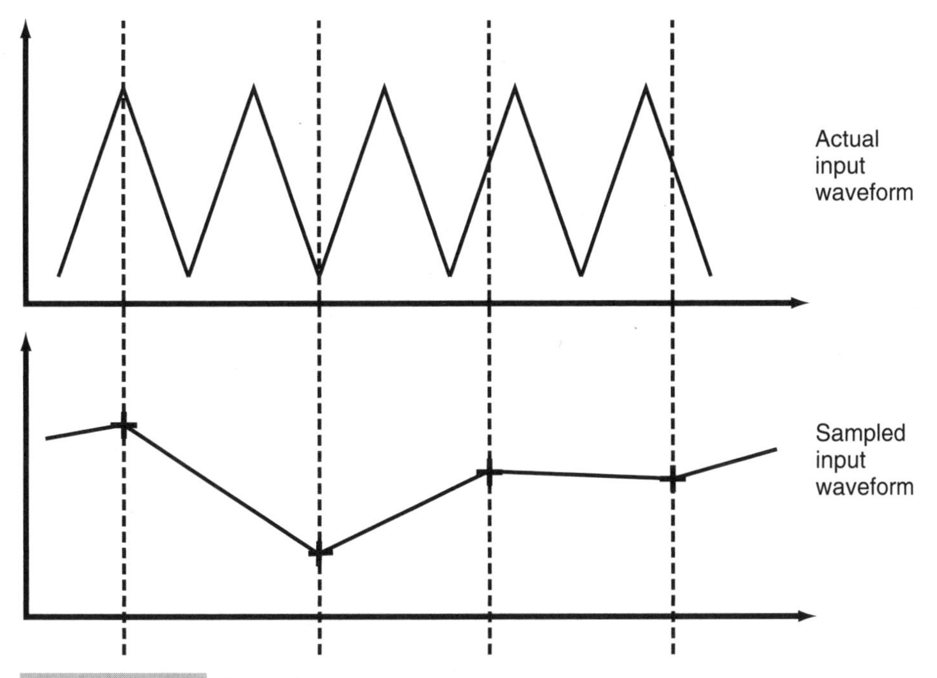

FIGURE 5-23 **Missed sample read waveform.**

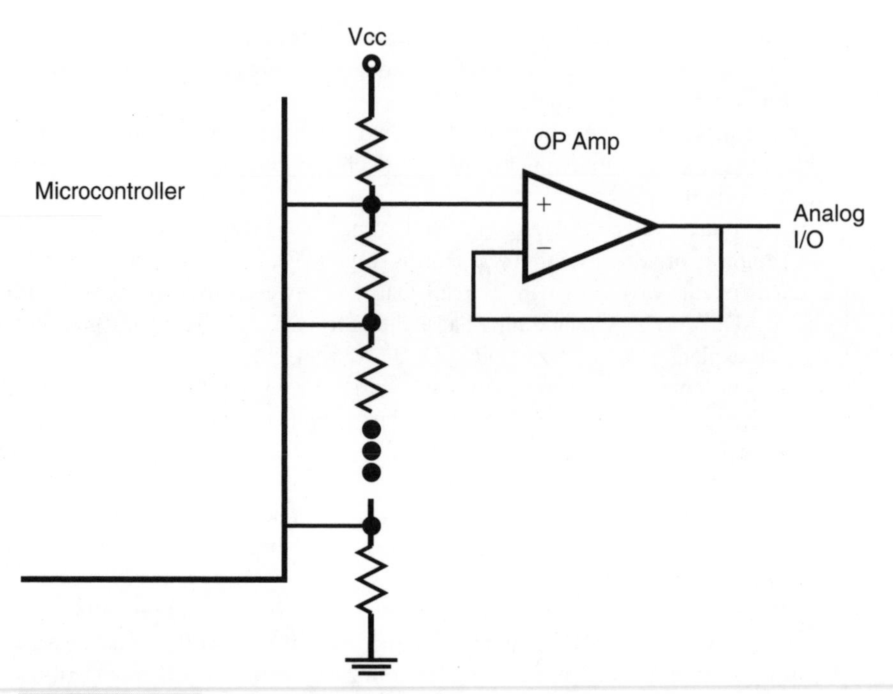

FIGURE 5-24 Analog output circuit.

However, Fig. 5-24 shows a simple circuit that can be used on a microcontroller to provide an analog voltage output. In the circuit in Fig. 5-24, by enabling different pin outputs (and outputting a 0 on the I/O pin), different voltages will be available at the input of the Op Amp. The Op Amp is used to isolate the resistor ladder from the load circuit. The resistor values are dependent on the number of pins used to control the output.

Looking at this circuit, you might be thinking that it can be used to provide greater than Vcc voltages (by providing higher than Vcc at the top of the resistor ladder). This will probably not work because of clamping diodes placed on the I/O pins to prevent overvoltage conditions which could damage the device. Instead the Op Amp should be provided with a voltage amplifying resistor network.

Another type of analog voltage output is the *pulse-width modulated signal* (Fig. 5-25). This is a repeating digital signal with a specific pulse width.

The fraction of the cycle the pulse is active is known as the *duty cycle* and is a measure of the fraction of the total possible power being transmitted in each "PWM cycle."

Outputting a PWM signal is usually the function of a timer in the microcontroller, although it can be produced in software easily (and I will show some methods of implementing it in the 8051 in chapter 12).

There's a lot of different options with analog I/O. The good news is that most devices are designed to make analog I/O quite simple (often doing a sample with an integrating ADC is simply setting a start bit in a register and waiting for a complete bit to become set) as long as the analog voltage changes fairly slowly. If a high frequency signal is input, then external flash ADCs coupled to DSPs should be used for the application, rather than an 8051.

Atmel AT89C*x*051 Voltage Comparators

An analog comparator has been included in the Atmel AT89C*x*051 series of devices. In the example applications presented in this book, I use this comparator to provide voltage monitoring functions between two analog voltage sources.

The comparator circuit itself is very simple with the voltages to be compared input to pins P1.0 and P1.1 and the resulting value read from P3.6 (Fig. 5-26).

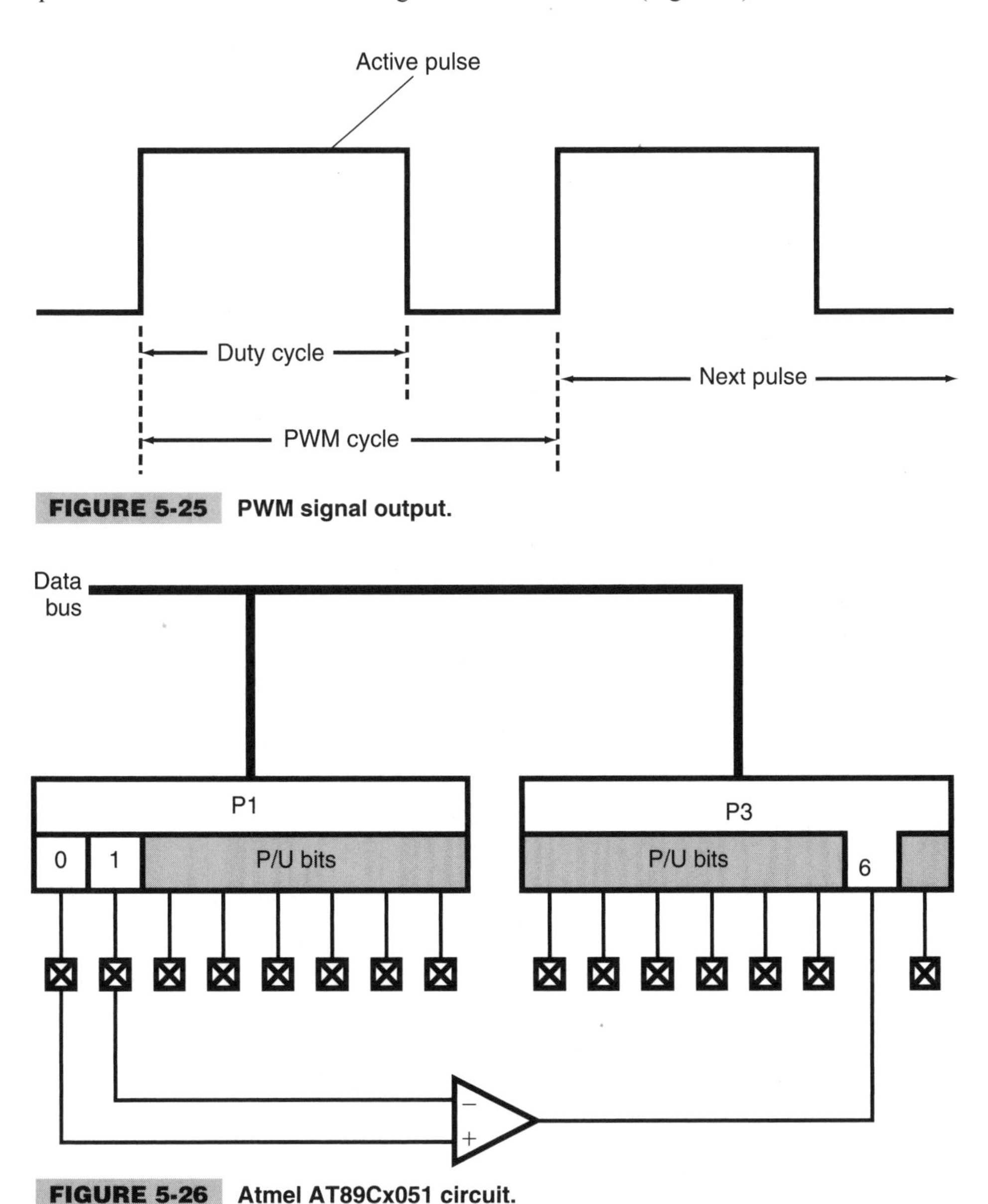

FIGURE 5-25 PWM signal output.

FIGURE 5-26 Atmel AT89Cx051 circuit.

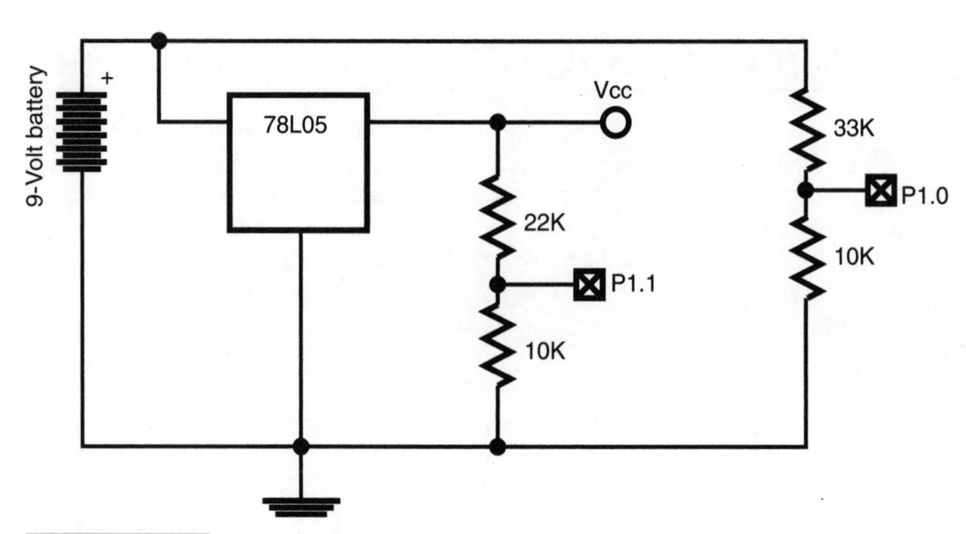

FIGURE 5-27 **Power monitor using a voltage comparator.**

In the AT89C*x*051, the P1.0 and P1.1 pins are not pulled up. This was done to allow the comparator to work without internal pull-ups affecting the circuit driving the pins. However, the pins can be pulled down (when a 0 is loaded into the bits). The output latches must always be loaded with a 1 to turn off the open-drain output transistors.

With both pins' output drivers turned off and analog voltages input, P3.6 can be polled. If it is set, P1.0 is at a higher potential than P1.1.

This circuit could be used to monitor a 9-V battery source that is used to power the application (Fig. 5-27). In this circuit, at a nominal 9 V, P1.0 will have 2.1 V input from the its voltage divider, and P1.1 will have 1.56 V input from its. When P3.6 is polled, a 1 will be returned.

As the voltage output from the battery drops, P3.6 will change from a one to a zero when the battery output is less than 6.7 V (below which point, the 78L05 voltage regulator's output will become unreliable). P3.6 should be polled periodically in software. When the bit changes to a zero, the application should initiate a power down sequence or notify the user.

6

APPLICATIONS DESIGN

CONTENTS AT A GLANCE

Putting an 8051 into a circuit is very straightforward. Most 8051s are very robust and able to take power and clocking from a variety of sources without complaint. This is to be expected for a microcontroller, which can be placed in a variety of different applications, ranging from toy rockets to real spacecraft. Understanding how the different features of the 8051 work is critical to being able to take advantage of this robustness and develop applications that will work the first time and run reliably throughout its life.

Power Input

As I've shown in chapter 4, providing power input to the 8051 is very simple and straightforward. I wanted to show how the 8051 could be used in low-power applications. By following a few simple rules, power required by the 8051 can be reduced to the point where alkaline radio batteries can be used with an application that runs for literally months.

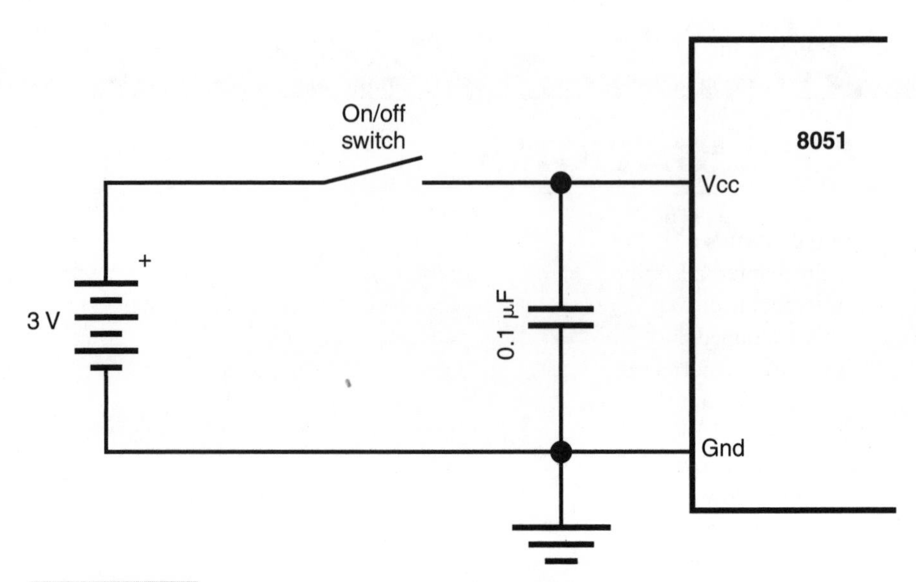

FIGURE 6-1 Battery power for an 8051 application.

Using battery power can reduce the complexity of the final application as well (Fig. 6-1). As you're probably well aware, the power equation is:

$$Power = Voltage * Current$$

Reducing the power used in an application is actually quite easy. The first way to reduce total power required is to reduce the voltage applied to the circuit. By applying 3.0 V to the circuit, rather than 5.0 V, the power required will be reduced by 40%.

If you are going to run the application at 3 V, then all the parts (including the 8051) should be specified to run at something less than 3.0 V (such as 2.0 V or 2.5 V) to allow the application to operate, even as the batteries run down. The logic threshold should remain pretty constant at 1.4 V to 1.5 V.

Reducing the voltage applied to the circuit will reduce the total power required by a reasonably large fraction, but substantial reductions in the 8051's current requirement are possible and will have a much greater impact on the total power required by an application. Reductions in current required can be made by reducing the 8051's clock speed or by using idle and power-down modes of the microcontroller.

As a rule of thumb, I've found that there is a pretty linear relationship between intrinsic current required and operating speed, with 1 mAmp required for each MHz that a CMOS microcontroller runs at. This is true for not only the 8051, but other microcontrollers as well.

With this relationship, reducing the clock speed to the minimum required to run the application will result in the minimum current draw required by the 8051. The improvement can be on the order of 10 times or more.

The use of power-down or idle modes, which turn off the internal oscillator during periods of inactivity, result in a substantial reduction in current requirements as well. For example, in the AT89C2051, if you were to put the microcontroller into power-down mode at 5 V, the total current required by the device is only 12 μAmps, as opposed to 10 mAmps, while running at 10 MHz.

The use of power-down or idle modes can also result in a cheaper, more reliable application (no power switch is required). I should also say that it could also result in a more aesthetically pleasing design if it starts and stops automatically and not with the aid of a toggle switch.

In chapter 4, I introduced the concept of intrinsic power as the power required to run the microcontroller. If the device is sinking or sourcing current, you have to make sure that this is stopped during power down or idle modes. Having any kind of current sourcing will increase the current required by the dormant 8051 significantly.

A good example would be a powered-down 8051 still driving an LED. If current sunk through the LED and 8051 was 16 mA, then the 10 μAmps of current required to maintain the 8051 in power-down mode becomes irrelevant.

Power-down and idle modes are handled differently in different 8051 microcontrollers, so I won't go into a lot of details about them. Some devices, such as the Dallas Semiconductor HSM microcontrollers, have other features (including an internal low-speed oscillator) that provide limited function while requiring much less power than running in full operation.

Reset

In chapter 4, I noted that, for all 8051 applications that I design, I place a 0.1 μF capacitor between the Reset pin on the 8051 and Vcc. By placing this capacitor on the Reset line, power up of the 8051 is guaranteed to be delayed long enough for that built-in oscillator to operate properly. This capacitor might not be required in all applications, but it is cheap insurance to make sure your application will start executing properly each time.

When I'm debugging an application, I will add a momentary on switch across the Reset capacitor (Fig. 6-2). This switch will allow me to reset the 8051 as I am trying to understand problems without having to power down the application.

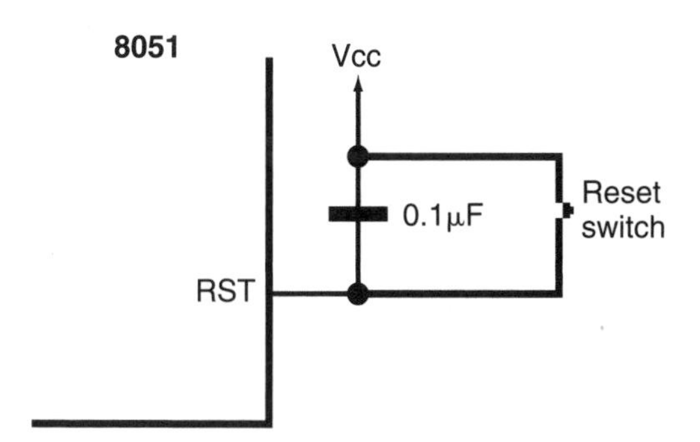

FIGURE 6-2 8051 Reset with capacitor and switch.

System Oscillators/Clocks

I have found the clocking circuit to be very robust in the 8051 when the microcontroller has a proper reset circuit. With proper reset control (the 0.1-μF capacitor between Reset and Vcc), the 8051 will start working reliably over a wide range of operating frequencies.

With a solid power on reset, I found that I could get the 8051's clock working in conditions that should be totally unacceptable. The two 30 pF capacitors between XTAL1 and XTAL2 are not absolutely necessary, but they do serve to make the oscillator waveform symmetrical.

The only things I would draw your attention to are that 12 clock cycles are required for each instruction cycle and each instruction requires one or two instruction cycles. In the Dallas Semiconductor HSM parts, the instruction cycle timing is reduced to one instruction cycle for every four clock cycles.

I find it most useful to think in terms of instruction cycles rather than the actual clock cycles. This is the level of granularity the 8051 works in.

For example, if you had to respond to an input within 20 μsec, you would have to determine the number of instruction cycles required. From this, the clock speed can be easily calculated.

To further flesh out the requirements, you should also assume that you had to output the contents of the accumulator to P0 within 20 μsec of the input bit being set. To do this, you could use the code:

```
Loop:                       ; Loop here for until the input bit is high
  jnb  input_bit,Loop       ; Two instruction cycles required for this
                            ;  instruction
  mov  P0,A                 ; One instruction cycle required for this
                            ;  instruction
```

The maximum timing would be 5 cycles (just missing the compare in `jnb` and having to wait through the next instruction). At 20 μsec maximum for five instruction cycles, one instruction cycle would have to take 4 μsec. This means the instruction clock would have to run at a minimum of 250 KHz.

To find the appropriate clock speed, the minimum instruction clock speed has to be multiplied by 12. So, for the previous example to respond to input within 20 μsec, a 3-MHz clock must be used with the 8051.

I/O Pin Interfacing

To successfully interface the 8051 to external devices, two rules have to be followed:

1. Only use negatively active signals.
2. If a pin isn't actively outputting a signal, set it high.

If you follow these two rules at all times, you will probably never have any electrical interface problems with your 8051 applications.

By keeping the pins high and only having negatively active signals, you will be able to forget about what state a pin is in and whether or not its an input or an output. When the pin is high, it can be used as either.

If you've worked with other microcontrollers, this will probably seem foreign to you, because the 8051 does not have an output driver control that has to be set according to whether or not the pin is to be used for input or output. However, this actually makes many applications quite simple. For example, a simple push button input could be designed as shown in Fig. 6-3.

In this circuit, normally the pin is pulled up internally. However, when the button is pressed, the pin is pulled down to ground, which can be easily polled as a 1 or 0 (or used to trigger an interrupt).

If you wanted an LED indicator that lit when the button was pressed or at other times in the application, you could use two pins (one for the LED and one for the switch) or combine them into one (Fig. 6-4). In this case, to light the LED without pressing the button, the pin is simply reset (loaded with 0). When the period of the LED being on was over, the pin is set (loaded with a 1).

In a typical microcontroller, the pin would have to be changed from input to output: A 0 is written to the output latch, then after the delay, the pin is changed back to an input.

Another very useful thing to note is that the I/O pins are *bit addressable*. This means that you can set, reset, or complement (invert the state of) single I/O pins with a single instruction and not affect any of the pins in the rest of the port.

As I will show in chapter 9 with PROG45, using an `anl Pn,#0xxh` instruction, which ANDs the port with a constant and puts the result back in the port, will not change the pins used for input in some devices. However, this is not something to count on. You should always either write explicitly to an I/O port or set all the input bits of the port explicitly.

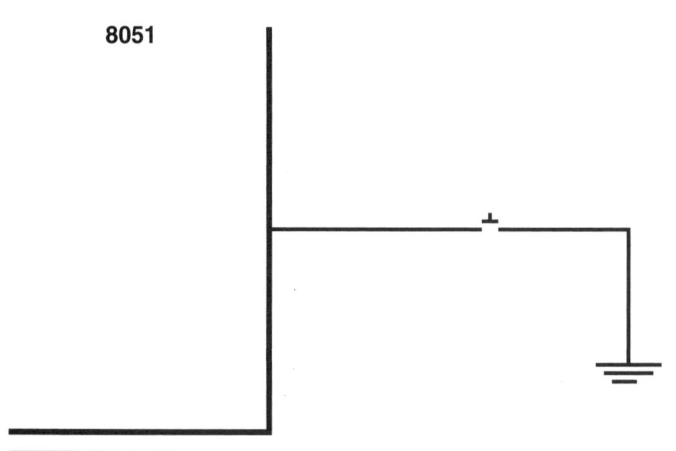

FIGURE 6-3 **Button input using an internal pull-up pin.**

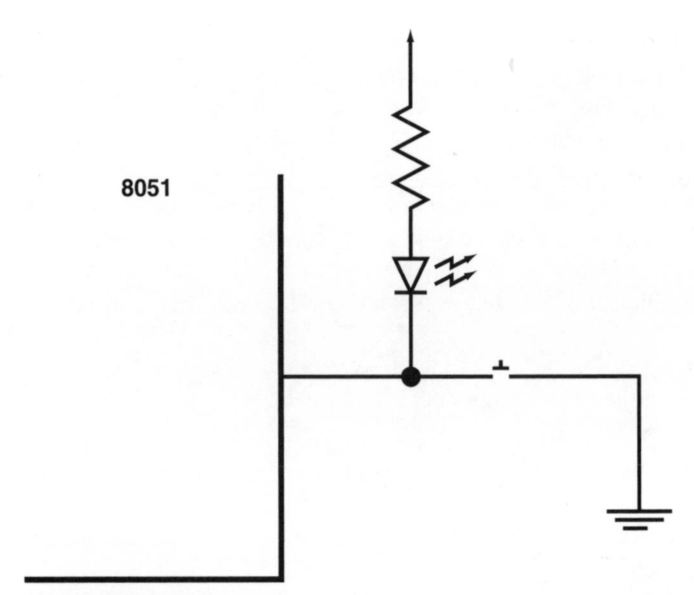

FIGURE 6-4 Button input using an internal pull-up pin.

Interrupts

Earlier in the book, I introduced you to the concept of interrupts and how they are implemented in the 8051. In this section, I would like to discuss a few aspects of creating interrupt handlers for the 8051 and the strategies that should be used to ensure that interrupts will be successful in your application.

As I have indicated elsewhere, interrupts were an important part of the 8051's architecture. When I first began to understand the 8051 and develop applications for it, I realized that the lack of a zero flag in the PSW register was not such a bad thing after all because of the ability to modify variables (setting to a value or incrementing or decrementing) without worrying about how the status (PSW) register was affected.

Many interrupt handlers are used only to keep track of external events. This is done by implementing a counter that the mainline code can poll. A typical interrupt handler in this case could be:

```
Int:

  push ACC          ; Save context registers
  push STATUS       ; Save the status flags

;  #### - Reset the interrupt requesting and control hardware

  mov  ACC,IntCount     ; Increment the interrupt counter
  inc  ACC
  mov  IntCount,ACC

  pop  STATUS           ; Restore the context registers to the state
  pop  ACC              ;  _before_ the interrupt

  reti
```

The `push STATUS/pop STATUS` instructions are required if the `inc` instruction will set or reset the zero flag of the STATUS register based on the result of the increment operation.

The 8051 neatly sidesteps this issue by not having a zero flag that has to be saved before the interrupt. As well, the 8051 has many instructions that execute without affecting the accumulator. These instructions include setting registers and scratchpad RAM bits and bytes along with bitwise updating of scratchpad registers without affecting either the accumulator or the PSW bits.

This means that the interrupt handler shown earlier can be simplified in the 8051 to:

```
Int:
;  #### - Reset the requester and interrupt control hardware
  inc   IntCount          ; Increment the interrupt counter
  reti
```

Further simplifying this routine is the method in which the interrupt hardware works in the 8051. Many interrupts are reset by simply executing the `reti` instruction.

In the example, if the interrupt handler was for one of the two button interrupts or a timer interrupt, the interrupt handler code could simply be:

```
Int:
  inc   IntCount          ; Increment the interrupt counter
  reti
```

which takes up only 3 bytes.

With this capability, you should be looking at making the interrupt handlers as simple as possible and keeping them within the 8 bytes available for each vector.

Part of implementing this strategy is to allocate some of the bank 0 registers to specific interrupt handlers. By doing this, surprisingly complex interrupt handlers can be implemented within the 8 bytes available. For example, if you wanted to implement an ASCIIZ string serial transmit routine using an interrupt handler rather than polling the TX shift register empty flag:

```
TXInt:                    ; ASCIIZ transmit interrupt handler
  clr   TI
  mov   SBUF,@R0          ; Send the current character out
  inc   R0                ; Point to the next character
  reti
```

In the mainline, a loop would simply monitor what R0 is pointing at and when it was pointing to a zero, then the loop would stop:

```
  mov   SCON,#%01010000   ; Run Serial port in 8-bit mode
;  #### - Setup Timer1 for the actual data rate
   :
```

```
  mov  R0,#BufferStart   ; Point R0 to the start of the buffer
  mov  SBUF,@R0          ; Send the first character

;  #### - Enable serial interrupts

Loop:                    ; Loop here to wait for the string to be sent
  cjne @R0,#0,Loop       ; When R0 pointing to zero, code finished

;  #### - Disable serial interrupts
```

This requires 6 bytes of interrupt vector handler code along with a few other bytes in the mainline. In this example, the value at R0 could be polled periodically in the mainline rather than in a separate loop. The only requirement would be for the polling to be done at an interval less than the time required to send a byte to make sure that interrupts could be disabled before the zero was inadvertently sent.

When planning your interrupt handler, mainline code, and register/RAM allocations, you should work at avoiding pushing anything onto the stack during the interrupt handler. This might seem to be sacrilegious if you think back to what teachers and professors have told you; however, in the 8051, there are significant advantages in taking this approach.

Each `push` and `pop` takes up 2 bytes each time they're used. If your interrupt handler just saves the accumulator before processing, then 5 bytes will be used up within the interrupt handler (2 bytes each for the `push` and `pop` and 1 byte for the `reti`) before you start processing the interrupt.

In some cases, you might want to split the interrupt handler for code inside the vector as well as code outside it. In the previous example, the interrupt handler could be modified to stop sending characters when the zero was encountered by placing the character send and pointer update portion of the handler somewhere outside the interrupt handler:

```
TXInt:                    ; ASCIIZ transmit interrupt handler

  clr  TI

  cjne @R0,#0,TXIntSend   ; If the value pointed to by R0 is not zero,
                          ;  send it
  reti

    :                     ; Skip to the end of the interrupt handler

TXIntSend:                ; Can send byte pointed to by R0

  mov  SBUF,@R0           ; Send the current character out

  inc  R0                 ; Point to the next character

  reti
```

What I'm really trying to say here is that the 8051 has some architectural features and instructions that can simplify the space and minimize the cycles required to service an interrupt. This will allow you avoid having to save registers on the stack or carrying out complex functions that will be hard to debug later.

I find that, when I am implementing applications that use the Int0 and Int1 pin interrupt sources, the code between them is just about identical, except for keeping track of which interrupt requested the interrupt. By using the vector space, I can set up parameters to indicate which pin cause the interrupt request:

```
 org   003h                ; Int0 pin
Int0:

  mov   R7,#Int0Indicator ; Indicate that interrupt 0 was active

  ajmp PinInt

    :

 org   013h                ; Int1 pin
Int1:

  mov   R7,#Int1Indicator ; Indicate that interrupt 1 was active

  ajmp PinInt

    :

PinInt:                          ; Handle the Int0/Int1 pin interrupts

;  #### - Put in the interrupt handler that keys off of R7

  reti
```

Note that, in this case, I use a register (which isn't accessed in the mainline) to indicated which pin caused the interrupt.

While I haven't used all of the 8 bytes available for the interrupt handler, I have reduced the interrupt handler code requirements by about 50% and, along with that, reduced the opportunity for errors to creep in when I copy one pin's interrupt handler when I've copied it from another.

External RAM and ROM

The external RAM and ROM capabilities of the 8051 are actually quite straightforward and work as I have described in chapter 4. With this background, I wanted to explore some of the funky things that you can do with the external RAM and ROM interface that will enhance the capabilities of the 8051. What I'm trying to show in this section is that the external RAM and ROM interface can provide some useful functions for enhancing your application while not taking up a lot of extra resources.

As I will show in chapter 10, RAM can be used instead of ROM to give you a simple emulator (Fig. 6-5). In the 8051 itself, a monitor program will write incoming data into the RAM (as well as providing new offsets for jumps). To execute the example program, the monitor can just jump to the starting address in the RAM.

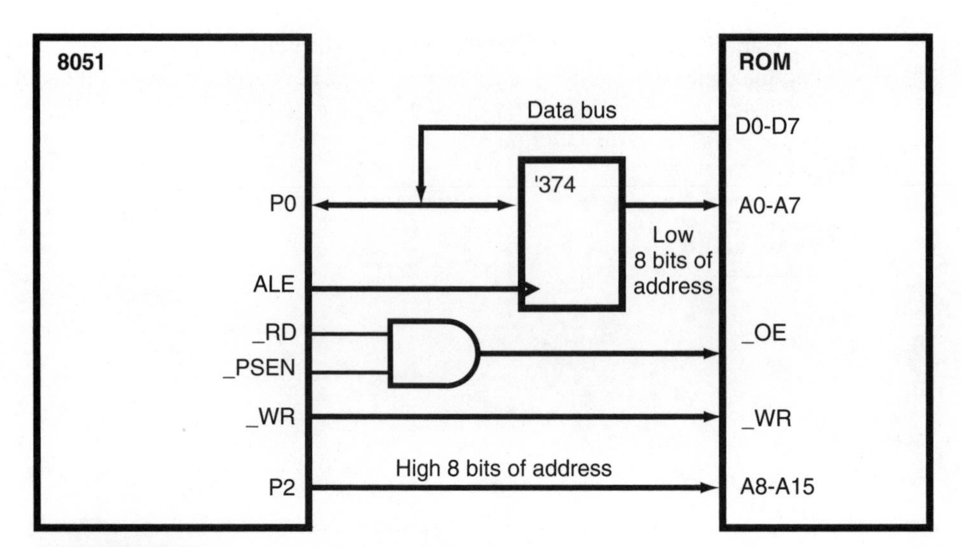

FIGURE 6-5 **8051 wired to RAM as control store.**

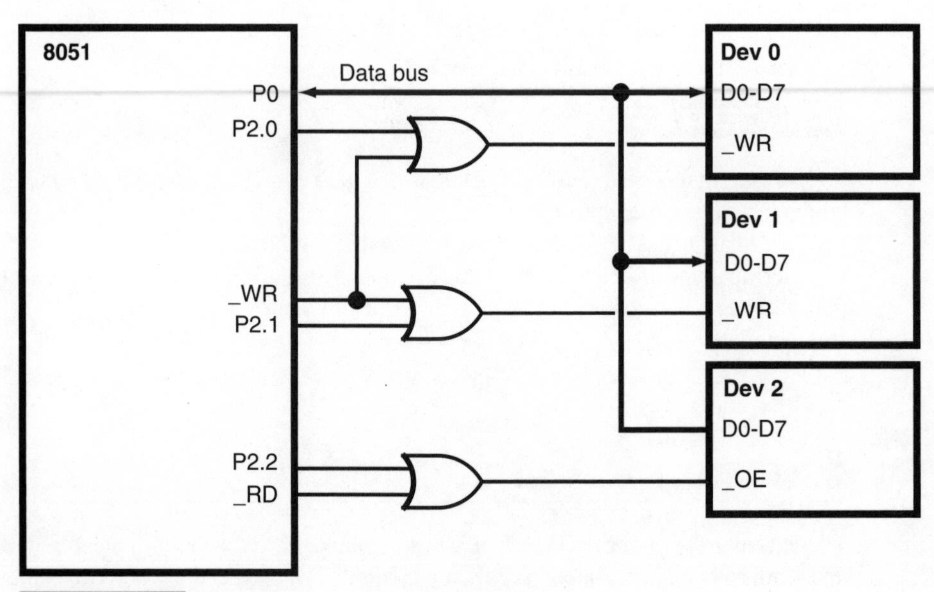

FIGURE 6-6 **8051 wired to memory-mapped I/O.**

Simple memory-mapped I/O can be added to the 8051 as well. Typically with memory-mapped I/O, very few addresses are required (often less than four). In this case, the '373 can be eliminated, and P0 wired directly to the devices as the data bus. This leaves P2 available for the device address decode.

Applying both these concepts, an application with two devices to write to and one to read from could be implemented as shown in Fig. 6-6. For this application, device 0 is at address 0FE00h, device 1 is at 0FD00h, and device 2 is at address 0FB00h.

7

8051 PROGRAMMING

CONTENTS AT A GLANCE

I find that one thing that scares many people off from starting to work with microcontrollers is the perception that the devices are difficult to load with an application and require expensive programmers. This is generally not true for most devices, and programmers capable of working with a variety of parts are available for quite reasonable costs. In this chapter, I will present to you the programming protocol used by 8051 devices presented in this book along with an example circuit for programming the Atmel 20-pin devices.

8051 Programming

The information provided here can be used to program an Intel 8751 and Dallas Semiconductor EPROM based HSMs. For other devices that use other algorithms, you should consult

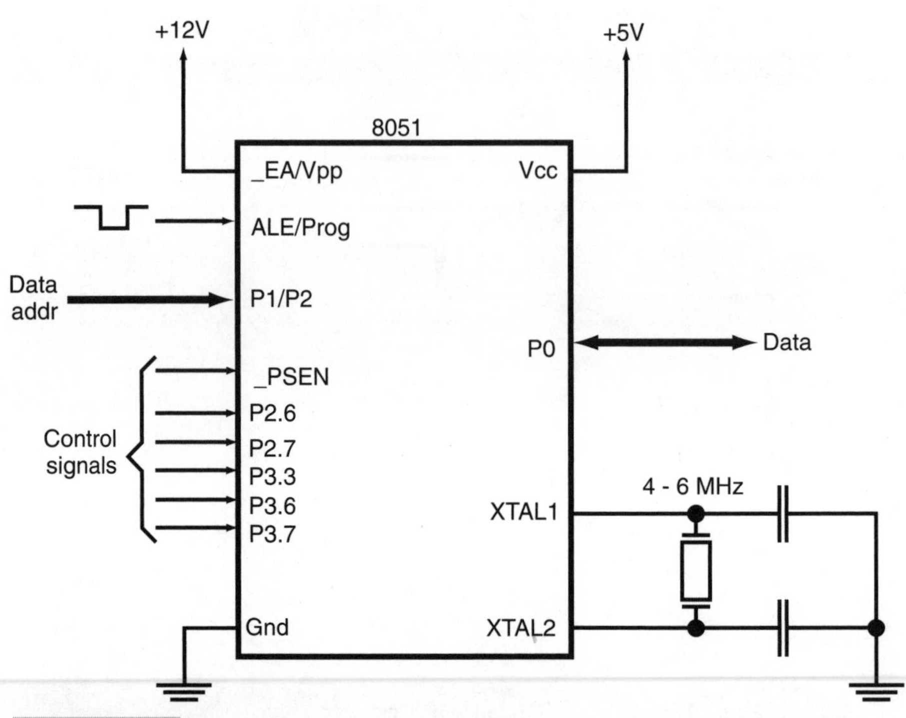

FIGURE 7-1 8051 programming connections.

their data sheets before attempting to burn a program into them. You should also purchase a programmer capable of programming the device you want to use.

When programming the 8051, a circuit has to be created for programming the device. With the 8751 and Dallas Semiconductor HSM parts, the circuit looks like Fig. 7-1. Each byte is set up with an address before it is written into the Control Store EPROM (Fig. 7-2).

The ALE/Prog pin is cycled five times for the EPROM write to take place. The crystal (which is in the range of 4.0 to 6.0 MHz) is used to make sure the bus is operating during the programming cycle to pass data through the microcontroller's buses. A 48-clock-cycle wait is the specified time between programming bytes in the control store EPROM. At 4 MHz, this is 12 μsec.

To enter programming mode, Reset must be high, _PSEN is low, and ALE is driven high waiting for the command to toggle the data in.

From Table 7-1, you can see that there is a number of things I haven't discussed yet about the 8051. The lock bits are three EPROM bits used to specify how external memory is to operate and whether or not the control store contents can be read back. By executing each of these commands, a single lock bit is programmed.

The lock bit operation is defined as shown in Table 7-2. The three signature bytes have address of 30h, 31h, and 60h and are used during programming to read the type of device that the microcontroller is. Address 30h is used for the device's manufacturer (89h is Intel, 0DA is used for Dallas Semiconductor, and 1Eh is used for Atmel). Address 31h returns the model number for the device, and address 60h provides an extension to the model number. For specific 8051-compatible microcontrollers, the device's data sheet must be consulted to get the signature information.

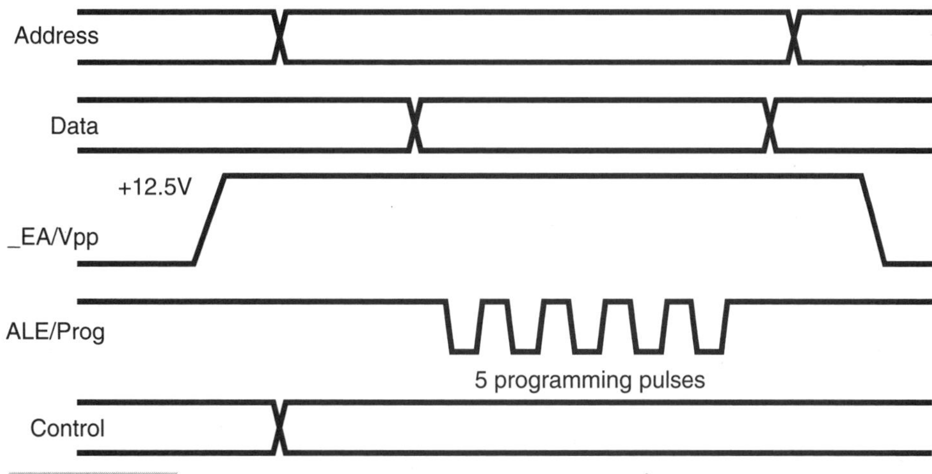

FIGURE 7-2 8051 programming waveform.

TABLE 7-1 THINGS YET TO BE DISCUSSED ABOUT THE 8051

MODE	ALE/PROG	_EA/Vpp	P2.6	P2.7	P3.3	P3.6	P3.7
Pgm Code	Toggle 5x	+12.5V	0	1	1	1	1
Verify	+5V	+5V	0	0	0	1	1
Wrt Encrypt	Toggle 5x	+12.5V	0	1	1	0	1
Pgm Lock	Toggle	+12.5V					
Bit1			1	1	1	1	1
Bit2			1	1	1	0	0
Bit3			1	0	1	1	0
Read Sig	+5V	+5V	0	0	0	0	0

TABLE 7-2 THE LOCK BIT OPERATION

LEVEL	LB1	LB2	LB3	MODE
1	U	U	U	All bits unprogrammed—No features enabled
2	P	U	U	Prevents `movc` from storing internal data to external memory
3	P	P	U	Level 2 plus no verify readback and `movx` can't read internal SRAM
4	P	P	P	Level 3 plus no external program execution

Dallas Semiconductor DS87000 Programmer

To program the Dallas Semiconductor 87C520s used in the example applications and experiments, I used the Dallas Semiconductor DS87000 programmer. This device connects to a PC via a serial port and uses the PC to download object code from ".hex" files and to control the operation of the programmer. This programmer can be used for all true 8051s using the programming algorithm described elsewhere in this section.

When I got the programmer working, it really worked very well although I had two main difficulties with the package. The program software for the PC was located on a 5.25" diskette. I don't know what your PC is like, but I consider 5.25" diskettes to be a relic of the early days of personal computing, and I really just focus on electronically transferring data. In the rare cases where the target PC isn't attached to a network, I rely on using 3.5" diskettes or recordable CD-ROMs for transferring data between PCs.

With this type of attitude, you can imagine that I had problems finding a PC with both a 5.25" drive and a 3.5" drive. Actually, I was never able to do it. However, I did find an old IBM PC/AT with a 5.25" drive that I could connect up to a laptop using a null-modem cable. To save you the same problem if you decide to go with the DS87000, I have included DS87000.EXE, a PC control program, on the CD-ROM.

The second problem was a result of the 25-pin D-Shell RS-232 connector provided for attaching to a PC. This connector attached to an RJ-11 terminated phone cable for connecting the PC to the programmer. Like the 5.25" diskette, my PC uses 9-pin D- Shell connectors, and none of the PCs I have access to just have 25-pin D-Shell serial connectors. My original plan was to use 9-to-25 D-Shell converters and gender changers, but this didn't work out for me and became very complex.

Instead, I got a female D-Shell connector with an integrated RJ-11 connector and wired it to match the 25-pin D-Shell as shown in Fig. 7-3.

With this wiring done, the programmer worked like a charm. The PC's screen allows you to select the port to be used with the programmer as well as downloading the data with a variety of options (including serializing each part programmed). It also provides a script facility for repetitive programming. The programming software also allows you to select how many bytes of the 87C520's EPROM is to be programmed, read, or verified. If for no other reason, the amount of EPROM burned should be the same size as the program to keep the programming time to a minimum (although the devices are programmed very quickly).

Atmel AT89C*x*051 Programming

Programming the Atmel AT89C*x*051 series of 8051 microcontrollers uses somewhat of a different algorithm than what is used for the standard 40-pin devices. The AT89C*x*051 algorithm is actually quite simple to implement (in the next section, I will present the hard-

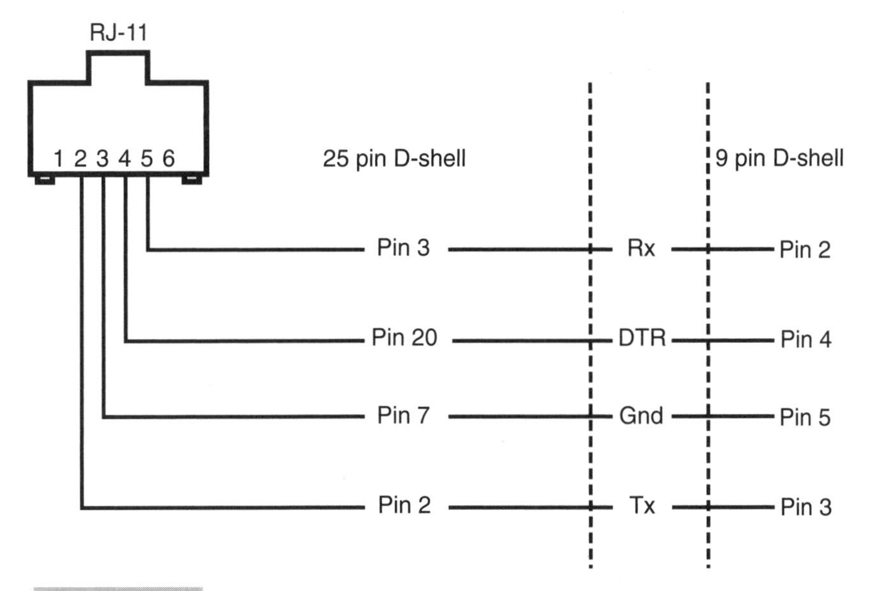

FIGURE 7-3 **Dallas Semiconductor DS8700 cabling.**

ware circuitry and software that you can build). This programmer hardware can also be used to program AVR 20-pin microcontrollers.

The basic connection for programming the AT89C*x*051 is shown in Fig. 7-4. From 30,000 feet, the programming can be described as erasing the control store and then presenting bytes to the microcontroller and latching it in. After the byte is latched in, the programmer waits for the byte to be saved into control store before reading it back and incrementing the AT89C*x*051's program counter to receive the next byte. This process is shown in Fig. 7-5.

To begin the programming cycle, the AT89C*x*51 is powered up with the Reset and XTAL1 pins held low. Then, +5 V is applied to Reset and the PROG pin, P3.2. At this point, the program counter inside the AT89C*x*51 is reset to zero.

After power up, the first thing you should do is a *chip erase*, to prepare the control store for the next program (all the control store bytes are loaded with 0FFh). This is accomplished by setting P3.3 high and P3.4 to P3.7 low (later in this section and in Fig. 7-5, this will be characterized as HLLL to show how the control signals are set) and pulsing PROG low for at least 10 msec.

With the chip erased, the control store can be programmed using the waveforms shown in Fig. 7-5. Note that Reset is cycled between +5 V and +12 V for writes and reads. This means that the Reset driver has to be a circuit that can output 0 V, 5 V, and 12 V to the Reset pin.

The operations that P3.3, P3.4, P3.5 and P3.7 control are listed in Table 7-3. The lock bits are used to limit access to the application in control store of a programmed part. If lock bit 1 is programmed, then the flash control store cannot be updated until it is erased again. If bit 2 is programmed, the verify function (read back) will return invalid data (this is copy protection for the chip, there is no encryption array in the AT89C*x*51) again until

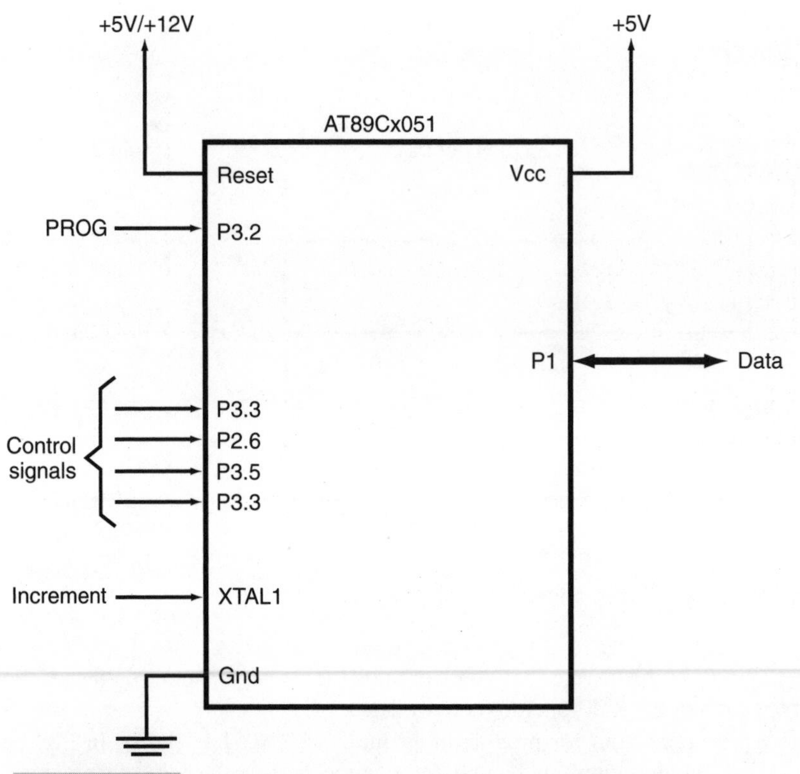

FIGURE 7-4 AT89Cx051 programming connections.

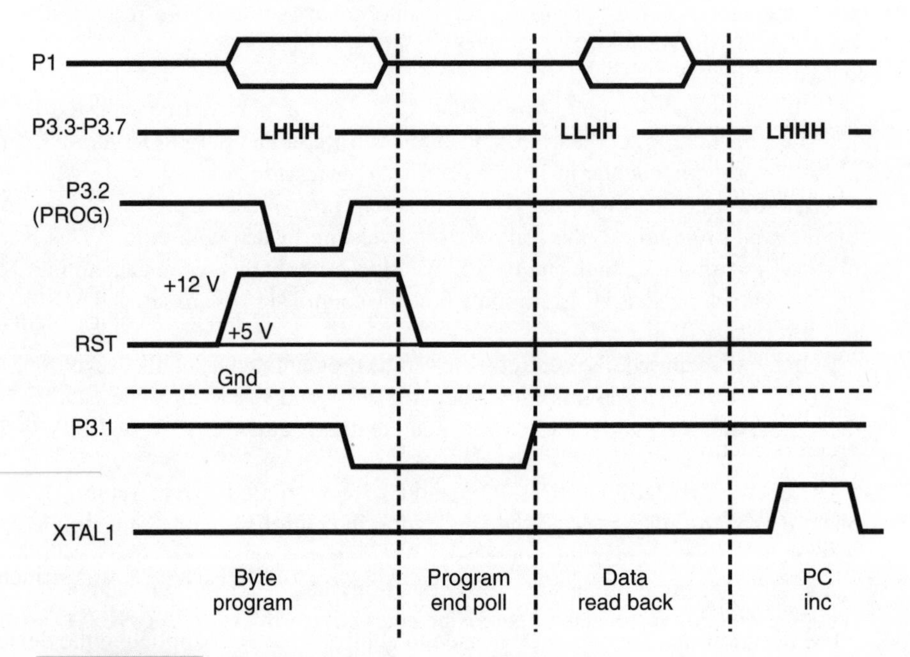

FIGURE 7-5 AT89Cx051 programming waveform.

TABLE 7-3 THE OPERATIONS THAT P3.3, P3.4, P3.5 AND P3.7 CONTROL

MODE	Rst/Vpp	PROG	P3.3	P3.4	P3.5	P3.7
Write code data	+12V	Pulsed low	L	H	H	H
Read code data	+5V	H	L	L	H	H
Write lock bit 1	+12V	Pulsed low	H	H	H	H
Write lock bit 2	+12V	Pulsed low	H	H	L	L
Chip erase	+12V	10 msec low pulse	H	L	L	L
Read sig. byte	+5V	H	L	L	L	L

the control store on the chip is erased. For obvious reasons, these two bits should not be programmed until the application programming is complete.

Often in application programming, there will be gaps in the code, which means there are areas that are not programmed. The AT89C*x*51's program counter can be incremented (by pulsing XTAL1) to skip over these areas. To carry this out, the programmer's control software will have to keep track of the current value of the program counter as it works through programming the device.

PROG35: AT89C*x*051 Programmer Circuit

I'm surprised at the lack of hobbyist programmers available for the AT89C*x*51 series of microcontrollers. For many other devices (including the PICMicro and even the 68HC*xx*), there are actually quite a few simple circuits available for simply programming the microcontroller. While not attempting to fill the gap, I have designed a programmer circuit and used it for all the AT89C*x*51 applications in this book.

One nice feature of the programmer is its ability to be used in-circuit; it can be wired into a prototype circuit and have the AT89C*x*51 run without having to pull the chip in and out of the programmer as circuits are being developed. Another feature is that this circuit could be used for programming a 20-pin Atmel AVR microcontrollers in parallel mode.

The circuit itself, is pretty simple and can be blocked out as shown in Fig. 7-6, with the programmer connected to an IBM-compatible PC via the parallel port. Power is supplied by an ac or dc "wall wart" with at least 16 V peak-to-peak. The power circuit provides switched +5 and +12 V for the 8051's Vcc and Reset (0 V, +5 V, or +12 V). The power circuit is controlled by the programmer control block. If Reset is being driven by something other than 0 V, the programmer drivers are active.

With this circuit, I found that, when going from +12 V to +5 V on Reset, 30 µsec was needed. If you end up writing your own software for this circuit, you might want to make sure you have a long enough delay before attempting to read back what was written. Going from 0 V to +5 V or +12 V (or from +5 V to +12 V) took less than a µsec.

The programmer control block is used to control the power applied to the device being programmed as well as to its Reset (as noted in the previous paragraph) and the

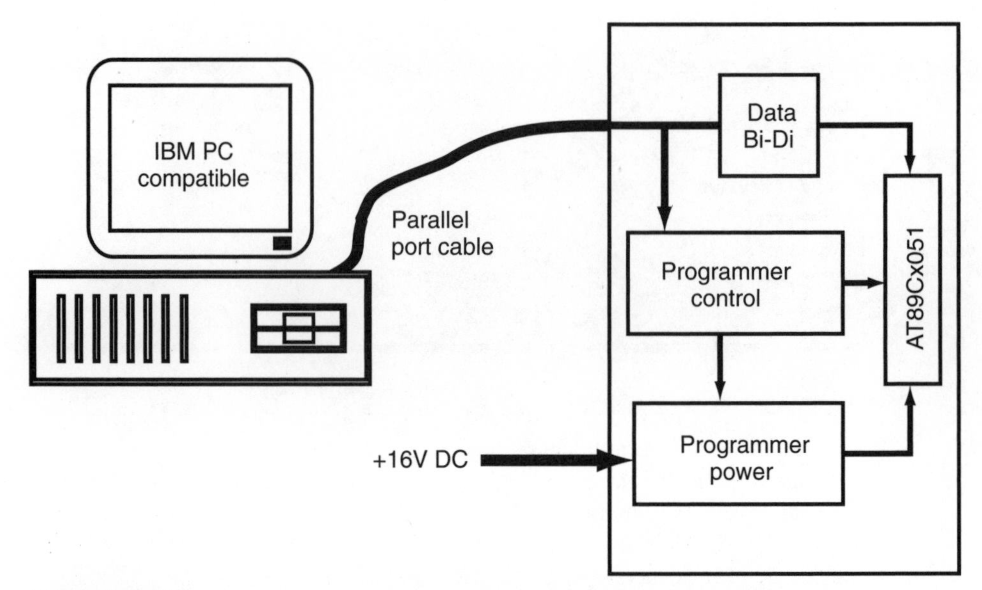

FIGURE 7-6 **AT89Cx051 programmer block diagram.**

programming mode of the AT89C*x*51. A 74LS374 is used with data being latched in from the PC's parallel port. The output of the '374 is always enabled, but all the lines going to the AT89C*x*51 (with the exception of the power and Reset, which are independently controlled) pass through a 74LS244, which allows the AT89C*x*51 to be pulled from the circuit without turning off the power to the programmer.

The '244 is also used to pass the RDY/_BSY signal back to the PC to allow the programmer to poll the RDY/_BSY to determine when the programming operation has finished.

The "Data Bi-Di" in Fig. 7-6 is a 74LS245, which allows a programming byte to be passed to the microcontroller or read from it. I could have eliminated this pin and had the same functionality by simply using the bidirectional features of the PC's parallel Port. However, to ensure that the AT89C*x*51 would run in-circuit, I wanted to make sure I could disable the connection to the PC, to make sure the cable wouldn't affect the operation of the application and, more importantly, make sure that invalid voltages or signals in the application circuit would not damage the PC.

The PC should have a parallel port capable of bidirectional I/O, and I used a switch-box dual male DB-25 connector cable. This cable is used for connecting a PC's parallel port to a printer sharing switch box. On two of the DB-25 connectors, each pin is directly connected (i.e., pin 1 is connected to pin 1, pin 2 to pin 2, and so on), which makes wiring to the application easier.

The final circuit (Fig. 7-7) probably looks pretty complex; however, if you follow the nets, you'll find that it's actually quite simple and easy to understand. What might be surprising is the component reference numbers (they don't go in any order in the schematic). They are not in any kind of logical order because I developed this raw card along with the other three that are presented in the book. As I updated circuits with fixes, the component numbers got somewhat scrambled.

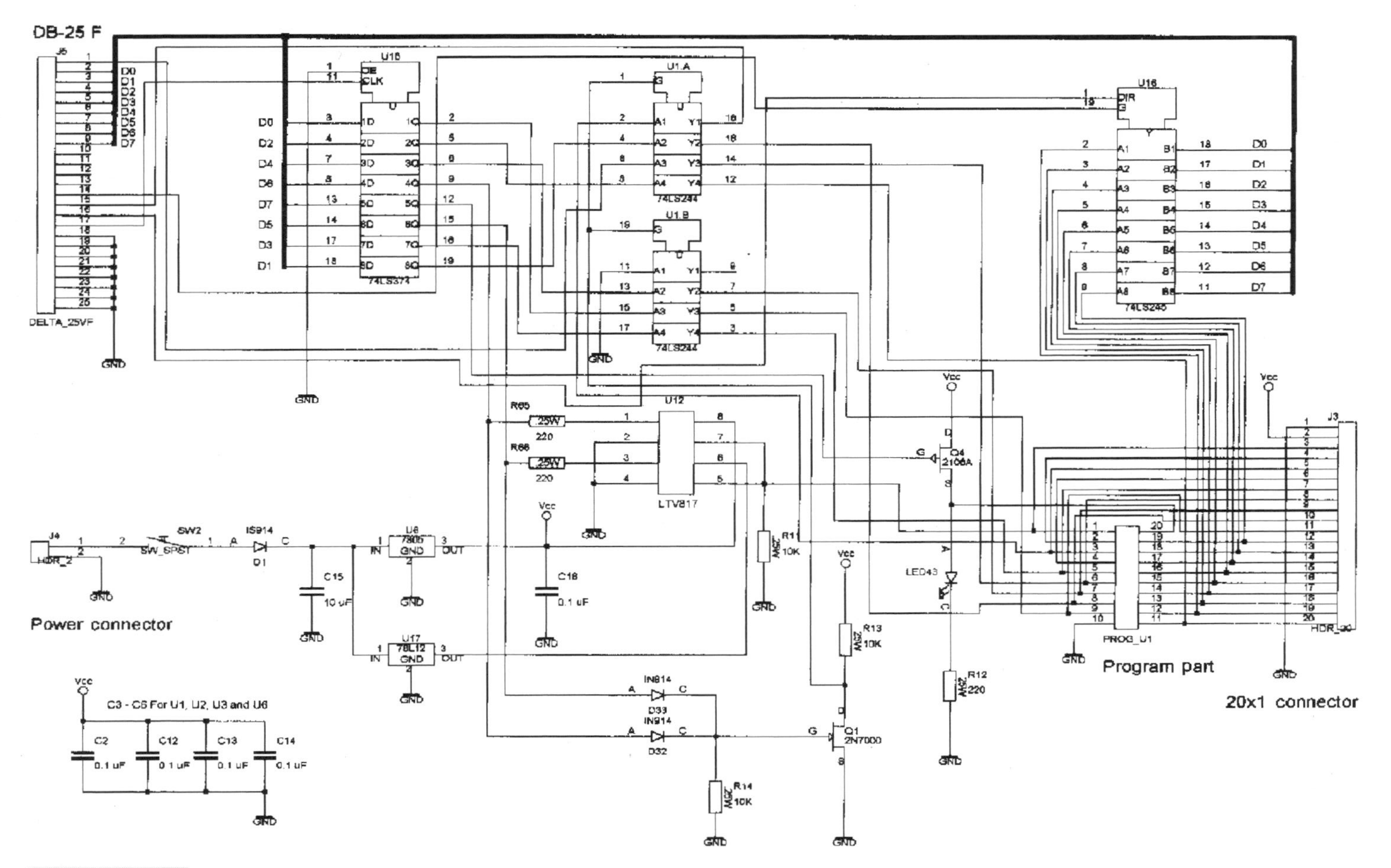

FIGURE 7-7 AT89Cx051 programmer schematic.

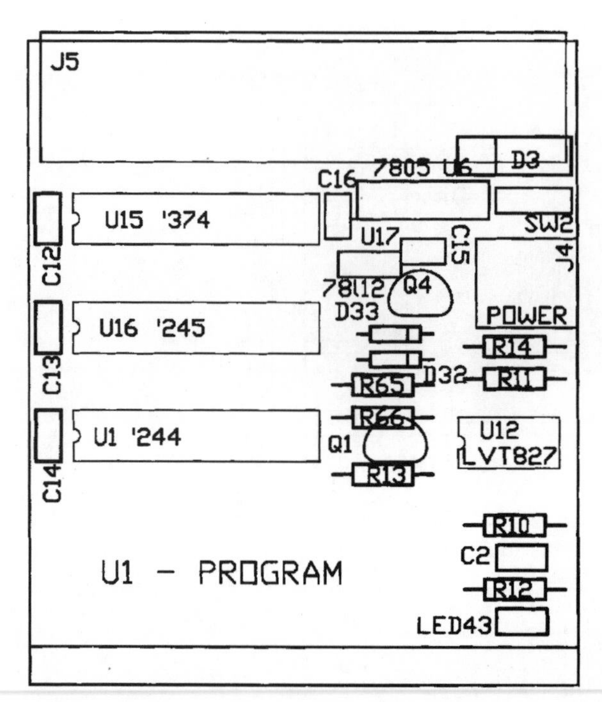

FIGURE 7-8 AT89Cx051 programmer overlay.

The programming software provided on the CD-ROM contains both the source and an executable file for the programmer (PROG35I). The program runs very quickly (taking less than 3 seconds to program more than 1K of code) and does check what was programmed into the device (and stops on the first failure). Power to the AT89C*x*51 is controlled by the PC; therefore, other than turning on the primary power switch, you do not have to play with switches at all.

This code has been tested on a number of different PCs, ranging from 16-MHz 386s to 200-MHz Pentium IIs. I do not like using the parallel port for driving applications and, despite this testing, wouldn't be surprised if some PCs had problems with this code. If yours is one of them, please let me know, and I'll see what I can do about updating the code. Before sending me a note, check on my Web page for updates to the software.

I have created an embedded printed circuit card design for this programmer. This card can either be built from Figs. 7-8 through 7-10 or using the "Gerber" files located on the CD-ROM (see appendix G for more information).

A kit will be made available from Wirz Electronics:

Wirz Electonics
P.O. Box 457
Littleton, MA 01460-0457

You can also call toll-free in the USA and Canada and 1-888-289-9479 (1-888-BUY-WIRZ) or e-mail Wirz at sales@wirz.com

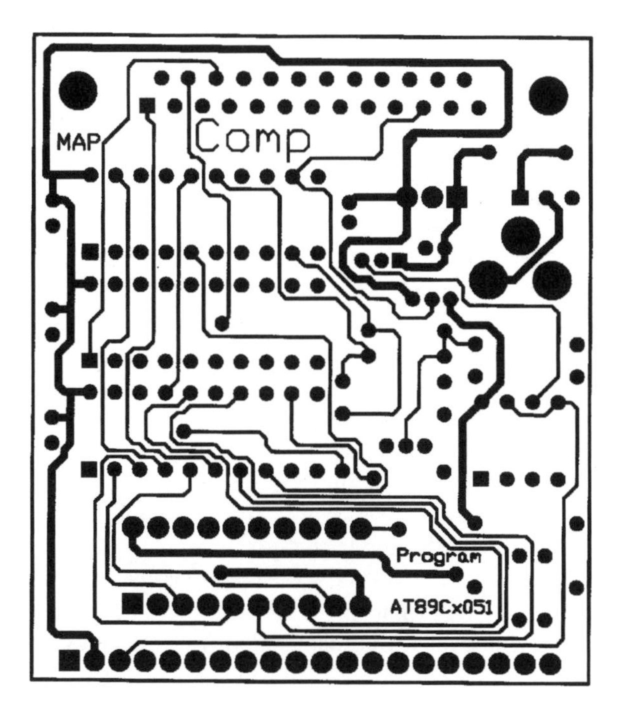

FIGURE 7-9 AT89Cx051 programmer top-side traces.

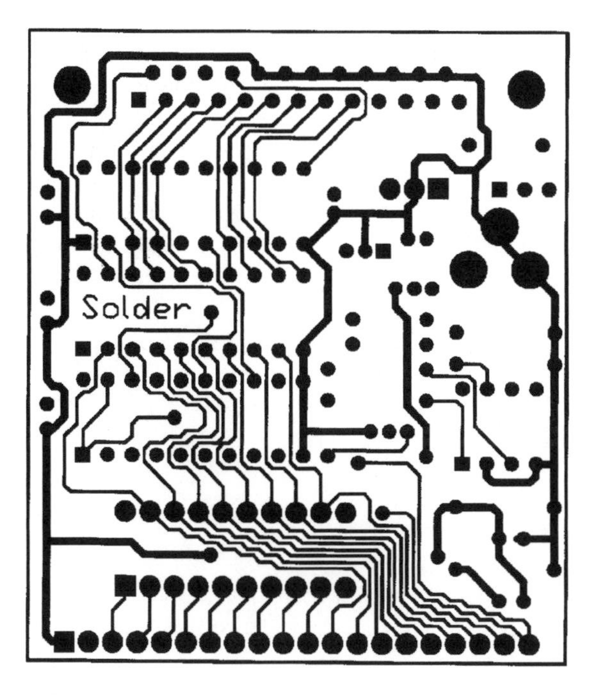

FIGURE 7-10 AT89Cx051 programmer bottom side traces.

Dallas Semiconductor Encrypted Data Microcontroller Programming

The Dallas Semiconductor Secure microcontroller products have their control store programmed using an RS-232 serial cable attached to a PC. The microcontroller can be programmed in-circuit very simply without complex hardware interfaces with varying voltage levels because the Secure devices use RAM instead of some form of ROM. By using RAM and a standard RS-232 interface, some interesting features are available with the Dallas Semiconductor Secure microcontrollers.

What makes the Secure devices interesting is the unexpected capabilities built into the programmer software (which is known as KIT.EXE). The KIT programmer software is run from a PC's command-line interface and performs the standard programming functions including downloading the program object file from the PC into the device and verifying the contents of control store. It also allows loading of an encryption key, which will keep the contents of the control store RAM safe from prying eyes.

Of particular interest is KIT's ability to read and write the I/O ports, control store, and the first 256 addresses of the data space (scratchpad RAM and special-function registers) as well as communicate serially with an executing application. These features are really all the basic functions of a monitor or debugger program. With the Secure microcontroller, all the functions of an emulator could be implemented.

There are three operating modes for KIT. The first is *KIT mode* in which the user can interface directly with the Secure microcontroller and program the device or read and write to the I/O, memory, and registers. *Terminal mode* behaves as a dumb serial terminal, allowing the user to interface with the executing application via the RS-232 serial link. The last operating mode, *batch mode*, runs a script rather than taking commands from the user console. Batch mode is used to program multiple Secure microcontrollers in production or carry out multiple repetitions of an operation.

8

SOFTWARE

CONTENTS AT A GLANCE

Up to now, I have been focusing on the hardware aspects of the 8051. Before going on and experimenting with the 8051, I think that it's appropriate to discuss a few of the software aspects of the microcontroller because, in many ways, it is the most important part of the application. Before starting to develop your own application code, you must have a good understanding of the tools that are available and how they are used.

Having efficient software that requires the least amount of control store and scratchpad RAM is critical to a successful microcontroller application. Part of the software also includes developing interfaces to different types of hardware. When these factors are optimized, the chances for a successful application are just about guaranteed.

Development Tools/Environments

Before starting any applications (or even the experiments in this book), you should load the software tools that you are going use to develop the applications and make sure you have a working programmer before you begin. Having the tools ready before starting will make learning the new device easier and allow you to become productive much faster. Choosing the tools to be used and how they operate are really an individuals choice, but I will make a few comments and recommendations.

If you are just starting out developing applications for the 8051, you'll know that there are no computer systems as pervasive as the IBM PC compatible running the Microsoft Windows operating system. This ubiquity has resulted in a large majority of available 8051 microcontroller development tools being written for this platform. I realize that there are good technical reasons for using non-“Wintel” personal computers and workstations, but these reasons become irrelevant when the choice of a development system for the 8051 is being made.

If you don't have an IBM PC compatible, before attempting to load UMPS (or another assembler and simulator), I strongly urge you to buy a PC and use it as your primary 8051 development station. This machine does not have to be very powerful or modern. As I was writing this book, I was able to find a used '386-SX PC and color VGA monitor for $70. The PC had 4MB of DRAM and a 100MB hard drive. To bring the system totally up to snuff, I had to buy a 3.5" diskette drive, an ISA VGA card (with 256K of VRAM) and a serial mouse. The total cost for the system was approximately $100.

Finding 8051 assemblers, simulators, and programmer interfaces that will work on another system will cost many times this.

After preparing the personal computer, you should determine how your software will work. Modern graphical user interfaces (GUIs), such as Windows, give you advantages in integrating software tools and provide a consistent interface along with built-in features, such as the Clipboard, which provides a much easier method to transfer data and code snippets between applications than manually copying data from a command line.

When I'm setting up the software tools for a new microcontroller, I tend to base it around the editor. The editor is the program that will be used to enter and modify the source code. Choosing the editor that will be used to develop and modify software is not a trivial exercise. There are a large number of options available that you should consider.

For the 8051, I have been using the integrated editor in UMPS, which is a pretty standard Windows editor, with multiple active edit windows and clipboard cut-and-paste being the only features available. For my C programming, I use a programmable editor that I have set up to respond to keywords and put in language prototypes.

For example, after entering:

```
if
```

the “if” prototype is loaded into the file:

```
if (  ) {
} else {
}  //  endif
```

The choice of the editor is very important and should be made by you based on what you want out of the editor, your past experience, and its ability to work with (and invoke) the assembler, simulator, emulator, and programmer. I recommend trying out a number of different editors to discover what works best for you.

The editor and other tools might be part of an *integrated development environment* (IDE), such as UMPS. An IDE combines all of the required different functions into a single program that will allow you to edit, assemble or compile, simulate, emulate, and program your application. The IDE is typically centered around the editor and uses the source code windows to reference back syntax errors for the assembler and compilers and allow you to easily follow the execution of the program.

Using an IDE, while not having all the features you would like, should be considered when you are first starting out developing applications for a device. Most IDEs will protect you from having to keep include and object files straight and having to pass them between applications as well as remembering how to invoke different programs (and what their run time parameters are) which will be a few less things to worry about when you're trying to write a simple program that works.

Assembling and compiling a program is the process of converting applications that you've written into a form that the microcontroller can use. This data type is known as an *object*, or *hex*, *file* and is loaded directly into the microcontroller.

As part of the integration of the development tools, I insist that, at the very least, the assembler/compiler is integrated with the editor. This integration means that the code conversion will be initiated from the editor (without leaving it) and any errors encountered in the source file will be highlighted in the editor when the assembly/compilation has completed.

When you're developing an application, you might be working on multiple source files. These consist of source, include, and library files. *Source files* consist of the actual instructions or code that are unique to the application. Both *include* and *library files* contain code that could be used by other applications.

Include files consist of data that is common across different source files. This data is usually constants definitions, variable declarations, and function prototypes. Libraries are collections of common subroutines that can be used in a variety of applications.

The source and library files can be assembled or compiled individually and then linked together to form the application. This is an excellent method of application development for PCs and workstations that have essentially unlimited memory and disk space, but I have reservations about this method for small 8-bit microcontrollers.

These concerns center about the use of libraries. In the PC world, libraries can have a huge number of subroutines. In the PC, all the subroutines in a library are linked together to form the application.

Ideally, in a microcontroller compiler/linker environment only the subroutines that are required by the source code should be used when the application object files are linked together.

Another concern is the generality of library subroutines and functions. This generality is a result of the subroutine's author not knowing exactly how they are going to be used in the application. The result is that the routines can be larger and take more cycles than would be appropriate for your final application.

An excellent example of this is in a `printf` statement written for an IBM PC C. If you trace through all the instructions used to print a message on the screen (which boils down to writing a number of bytes in video RAM and updating the cursor's position), you'll find that the library routines used are written to allow the operation to work on a variety of different video display adapter types that can run in different modes along with a console I/O. It can take literally hours to figure out how the PC writes an "A" on the screen if you look through all the library code.

If libraries are to be used, ideally the linker should ignore all subroutines that are not used by the application. As I write this, the first compilers/linkers are becoming available with this feature.

If you go through the first 18 experiments with UMPS, you should gain a good insight in how simulators work in an integrated development environment. A *simulator* is a program that simulates the microcontroller processor (and, optionally hardware connected to it) to allow you to watch the execution of your application to make sure it works properly before it is burned into actual hardware.

To simulate input from external hardware, a *stimulus file* can be used. This file is used to provide a series of inputs to the simulator so that you can observe how the program will behave. A typical stimulus file consists of a cycle counter and pin states. At the specific cycle count, microcontroller input pins are set at specific values.

For example, a stimulus file to input a carriage return (00Dh) at 9600 bps to an 8051 running at 12 MHz (with a one million instruction cycle per second execution speed) would be:

```
Clock      P3.0      ;  Input signal into the RxD pin
    1         1      ;  Input line is high
 1000         0      ;  Start bit
 1104         1      ;  Bit 0
 1208         0
 1312         1
 1416         1
 1520         0      ;  Bit 4
 1624         0
 1728         0
 1832         0
 1936         1      ;  Stop bit
 2040         1      ;  00Dh finished
```

In this example stimulus file, each clock cycle is an actually instruction cycle and each bit is 104 μsecs long (which is the bit period of a 9600-bps signal). The 1000-cycle delay is to allow the application code to prepare the simulated hardware for receiving data.

Writing stimulus files is quite boring and can have a number of errors in the process. However, putting the effort in here can save you many hours of debugging later.

An important consideration when choosing a simulator is how fast it can execute the application's instructions. Many microcontroller applications have processes that can take a large fraction of a second to execute. Having a simulator that executes slowly can turn into a hindrance.

A good example of this is an application that I was creating for my book *Programming and Customizing the PIC Microcontroller*. In the application, I had to wait three seconds for hardware to properly initialize. The simulator that I was using could only execute at a rate of about 300 instruction cycles per second. Running this application with the clock set at 250,000 instruction cycle per second meant that, to run through the 3 seconds of simulated time would take approximately 42 minutes.

In the intention of working quickly, when I was simulating the code, I skipped over this delay. After getting the application working on the simulator without this delay, I took out the skip and burned the application code into a part.

As you have probably guessed, the application didn't work. It was pretty bad, the application was supposed to set up a multiple 7-segment LED display, and it would just sit there with nothing displayed.

I spent a great deal of time trying to find a hardware reason why this application wouldn't work. After many frustrating hours, I finally bit the bullet and went back and simulated the code with the full delay. After about one second of simulated time (almost 15 minutes of real time), I discovered that a timer interrupt handler accessed an unitialized variable and caused the application to execute a branch to the boonies. When I had put in my initialization code, I copied previous code and had not correctly changed the variable's name. This variable was accessed *during* the 3-second delay, so this mistake was never picked up in the simulation.

The fix was simple: I changed the initialization to the correct variable label.

The lessons learned were to always run through the application before burning it into a part and to find a faster simulator.

An *emulator* is a piece of hardware that is placed into an application circuit to allow you to run and observe the operation of the application without going through the hassle of developing stimulus files. An emulator is an excellent tool for determining whether or not you have correctly modelled the hardware the microcontroller interfaces to.

Later in the book, I will go through emulators in more detail. At this point, however, I think it's important to note that they should integrate into the editor, for source-code level debugging the same way a simulator does. It is also important to note that an emulator is not a development tool. I feel that it should be only be used for debugging applications. I will expand on this point in chapter 9.

The last tool you should have ready before developing applications is a programmer. While I will go into the mechanics of 8051 device programming later in the book, there are a few points that I should make here.

When choosing a programmer, you have to make sure the programmer will take the assembler/compiler's object file without modification. If it won't, can the object file be changed to a format that will work easily?

This is potentially a very big problem and can cause you a lot of grief. Typically, the Intel hex format (described later in this chapter) is used for the 8051.

A second point is making sure the programmer will work with your PC. Ideally the programmer should be controlled via the PC's serial port, which is insensitive to the timing and architecture quirks of the different PCs that have been sold over the years. I have seen a number of programmer manufacturers support a programmer that requires different control software to work on different PCs.

With all these elements (editor, assembler/compiler, simulator, and programmer) in place and working together, you are ready to begin your application development. As I will point out repeatedly in this book, it's the up-front work that really is an indicator of how well a project will end up.

Choosing these elements is actually more critical in the 8051 than in most other microcontrollers because of the lack of a single IDE provided and supported by a single manufacturer that would "force" you into working with one product.

Assembly Language

When you look at the different types of programming languages, you have to understand the "pay me now, pay me later" rule that exists with programming costs. Assembly language programming is generally the cheapest way to get into microcontroller programming, but it is the most difficult to learn, requires the most effort, and is the least portable to other platforms.

Conversely, using a high-level language (such as BASIC or C) can make it much easier for a beginner to program a microcontroller, but it is the most costly option. Code written for a high-level language is, by definition, portable to other platforms.

Where the "pay me now, pay me later" rule comes into effect is if you are developing 8051 applications professionally. Spending time on assembly language programming is probably costing you money over doing it in a high-level language.

For learning the 8051 or any other microcontroller or computer processor, assembly language is, in my opinion, the best way of doing it. By going through the experiments in this book, you will get a good feeling for how the 8051 processes instructions and how it works.

Assembly language programming is the process of writing code that uses *assembler statements*, which are the actual instructions the 8051's processor executes (the smallest unit of granularity).

Along with assembler statements, *directives* are added to the source file to control the operation of the assembly process. Macros and conditional assembly statements are types of directives that can help you develop code unique to your application. *Macros* are labels that are replaced with code; they're similar to subroutines, except the subroutine code is copied directly into the source before the assembly operation. *Conditional assembly* statements are "if/else/endif" statements that execute during assembly and, depending on the conditions, not allow certain sections of code to be assembled.

A completed assembly language source file is assembled into a *listing file* (showing how the assembler program converted the source into bits for the processor) and an *object*, or *hex, file*, which is the actual bits and bytes to be burned into the 8051. Assembly language programming is the lowest form of "human-readable" source code processing possible. *Interpreters* and *compilers* take high-level language statements and convert them directly into processor instructions.

Now, if you're well heeled and don't want to do the drudgery of assembly language programming, you could buy a compiler, but you will never use the full potential of the 8051. Knowing and being proficient in assembly language programming will allow you to enhance your high-level language applications by allowing you to add code that will reduce the number of cycles required to execute, reduce the number of bytes required for the program, or enhance the operation of the application.

8051 Assembly Language Programming Styles

Assembly language programming is often perceived as the most difficult method of coding applications. This is because of the need to totally understand the operation of the processor and the difficulty in reading through the code and seeing how the code implements the application. This can be particularly onerous if you have to go back and look at code you've written years before or have to read through code that somebody else has written.

This is the primary reason why I have written this section. With poor coding practices and incomplete documentation and comments, being able to read through an application and understand it can be close to impossible.

Assembly language programming should not be approached like high-level language programming because you have to do all the background work. This makes carrying out "what if" experiments difficult to do in assembly language unless you are experimenting with I/O. For example, in a high-level language (C), to experiment with reading and writing a byte in external data memory, you might write the code:

```
main(){                        //  Test writing/reading external memory

int ext Mem[ 65536 ];          //  Declare array in external memory
int Temp;

  Mem[ 01234h ] = 0AAh;        //  Write to it
  Temp = Mem[ 01234h ];        //  Read the value back
}
```

which is quite straightforward to understand and, with few modifications, could be used in a variety of different systems

In 8051 assembly language, this could be:

```
org 0                        ;  Program at the reset vector
 mov  DPTR,#01234h           ;  Set up pointer to the external memory
 mov  A,#0AAh                ;  Write to external memory
 movx @DPTR,A
 movx A,@DPTR                ;  Read from external memory
Loop:                        ;  End - Loop forever
 ajmp Loop
```

At the start of the book, when I introduced the concept of external memory, you probably would have understood the function of the previous C code immediately. If I had introduced the concept of external memory with the previous assembler program, you probably would have closed the book and returned it where you bought it from.

When I created the application code for this book, I tried to make it as readable and simple as possible because it is all written in assembler. Where appropriate, I have used high level code to explain its function, like above. Often, when I am documenting an application, I will include high level source code. This approach can be useful when you are developing your own applications to make it easier for others.

One tool that I use to help with making the code easier to understand is that I apply a template at the start of all my programming. If you look in the \CODE\PROG46 subdirectory of the CD-ROM, you will find the TEMPLATE.ASM file (Listing 8-1). When using this template file, you will have to fill in everything that is inside square brackets ("[" and "]").

This template is designed for the UMPS assembler. This assembler has some features that I like (such as instruction, label and constant highlighting, and color coding) but does not have any list formatting options. These list format options include page size, page titles, and forced page breaks.

LISTING 8-1 **The TEMPLATE.ASM file.**

```
;  [Filename] - [Program description]
;
;  [Detailed description of application]
;
;  [Author]
;
;  Started: [Date]
;  Updated: [Date] - [Description]
;
;  Hardware notes:
;  [Microcontroller part number]
;  [Operating speed/source]
;  [Reset source]
;  [Pin connections]

;  Include files
[Include filename.ext]

;  Variable declarations
;  [Bank register functions]
[Variable_Name] EQU [Address]    ;  [Variable description]

;  Constant declarations
[Constant_Name EQU value]        ;  [Constant description]

;  Macro declarations
[MACRO LABEL parameters]         ;  [Macro comments]


;  Mainline
 org 0                           ;  Execution starts here
  ajmp   MainLine

 org 003h                        ;  _Int0 pin interrupt handler
_Int0:
  [Put in _Int0 handler]

 org 00Bh                        ;  Timer0 pin interrupt handler
Tmr0:
  [Put in Tmr0 handler]

 org 013h                        ;  _Int1 pin interrupt handler
_Int1:
  [Put in _Int1 handler]

 org 01Bh                        ;  Timer1 interrupt handler
Tmr1:
  [Put in Tmr1 handler]

 org 023h                        ;  Serial Port interrupt handler
SP:
  [Put in Serial Port handler]


MainLine:                        ;  Mainline Code

  [Variable initialization]
```

LISTING 8-1 The TEMPLATE.ASM file *Continued.*

```
  [Put in application code]

;  Subroutines

  [Put in subroutine code]
```

I like to place the application's source filename and a few words of description of the application or file on the first line of the source code file. This allows me to quickly look at only the first line of the file when I'm looking for a particular one.

I'm lousy at keeping application code documentation in separate files. Instead, I make sure a detailed description of the application is given at the start of the source file. Along with the description of the application, I also note any special information about the application and any tricks that I've used or come up with. Ideally, this is kept less than 20 or so lines so that it can be displayed on the first screen of an editor. This is where I'll put the high-level language statements explaining how the code works.

When I develop code, I tend to do it in a bottom-up approach, first getting the output user interfaces working, followed by the input user interfaces, device interfaces, internal functions (such as timers) followed by the mainline, which ties together all the functions. In several of the experiments and applications shown in this book, you will see something on the first line like:

```
PROG123A - Light LEDs
PROG123B - Get Button Input
PROG123C - Interface to I2C EEPROM
Application - Finished Application (or may be "PROG123D")
```

By implementing only one new function in each source file and testing and building up the capabilities of the circuit and software, by the time I have to start writing the application, most of the difficult work has been completed. (At this point, all I have to do in the application is intelligently link the different functions together.)

Another advantage of this style of programming is that I can go through the code specific to an interface later if problems with it appear and fix it without having to worry about the other interfaces. Once the problem has been fixed, the changes can be cut and pasted into the later versions of the code fairly simply.

It's not always a good idea to identify yourself as the author of some source code, but you have to do it. I say it's not a good idea to identify yourself as an application's author because you will always live in fear that somebody's going to ask you to fix a problem or enhance it because, as the author, "you know the code best." I have this nightmare of being on my deathbed, the plug has been pulled and somebody from work bursting in to ask me to update some Token Ring functional test code I wrote in 1986. It's a nightmare because I stabilize, get better, and end up having to do the work.

I highly recommend that you keep a running log of when the code was written and the updates that have been made to it. I find that, with a mature application, 80% of the problems found are a result of previous fixes. Having a list of the changes to the application will help to point to where the problem is.

Some people keep track of changes by marking them in the code. This has never worked for me, and adding pointers just seems to make the code more complex and harder to follow. If I haven't said it before, I'll say it now: Keep the code simple and obvious wherever possible.

I find it useful in microcontroller application code to keep a list of the pins and what they are connected to for each application. As part of this effort, I also keep track of the device that the application is written for (which is very important for the 8051, which has many versions), what speed the application runs at (important for creating and calculating instruction delays), and how reset is to be accomplished. Having this information at the top of the application source file lessons your dependency on additional information as you write the application.

With the header information out of the way, I can now begin to define the information that is required within the application code. For information common across files, I start with include files. For most assemblers, this can be anywhere, but I follow the C convention and declare information as early as possible. In C, all variables, functions, and constants have to be declared *before* they are used. In assemblers, this isn't as critical because multi-pass features will allow the information to be retrieved from anywhere in the source.

In most embedded microcontrollers, I try to avoid using include files altogether except for register definition files. I place the entire application in one file, rather than link a series of source files together. The reason for this is that most microcontroller code-level emulators and simulators can only work with one source file. If execution takes place over multiple files, then parts of the application might not be visible or impossible to step through. This also might be true with included files, which are loaded during assembly, but the source can't be accessed during debugging.

Include files might be appropriate for some applications, especially when complex declarations and definitions or standard macros are required, but they should be avoided for passing source code to an application.

Registers should be defined in an include (or be part of the assembler). If the assembler that you want to use doesn't have an include file for the part that you are working with, ask the assembler developer to write one for you or get another assembler. Writing a register definition file can take literally days, and any problems can really ruin a bunch of your days.

The first real information that I define is the variables to be used by the application. In a PC application, when you declare variables, you are setting aside memory from code memory space. How big the variable area required has a bearing on the instruction addressing. In a PC or other Princeton-architected processor, declaring a variable reserves space for just that variable.

In a Harvard architecture, like the 8051s, variable addresses are manually allocated. When variables are declared, you aren't reserving any space. You are just creating a constant that can be used as an address in scratchpad RAM. This constant really only has meaning to you.

As I've discussed elsewhere, I place byte variable declarations at addresses 030h to 07Fh. By doing this, there's no opportunity of overwriting bank registers or bit variables.

For 8051 devices with 256 bytes of scratchpad RAM, I put arrays and the program counter stack into the upper 128 addresses. This frees up the lower 128 addresses for direct addresses and is the most efficient way to use all of the scratchpad RAM available.

Constants are declared the same way as variables, using the format:

```
Const_Label EQU 012h           ; Define the constant
```

When the constant's label is encountered, the constant numeric value is exchanged for the string. I make this distinction to separate constants from defines, which can also be inserted in a program:

```
Define=Define_Label String     ; Define_Label is "String"
```

Defines are usually placed in the constant area. Instead of replacing the label string with a numeric value, the actual string is placed in the code. Defines can be thought of as a simple type of macro.

Defines can also be made without a replacement string. These types of defines are often used for conditional assembly, which will be explained later.

Constants and defines should be used only when they add meaning to the application source code. For example:

```
mov  P1,#DevReset
```

is a lot more meaningful than:

```
mov  P1,#036h                ;  Reset the external device
```

and could eliminate the need for a comment.

Constants and defines should not be used for obvious values such as zero and one or other values that are just as meaningful as straight numerics.

There is one comment that I would like to make about constants. Keep your numbering system (radix) straight. Personally, I think best in base 10, so I keep my default radix set to decimal and mark binary and hex numbers explicitly. I keep this convention for all my programming, regardless of the device or language so that I don't have to switch mental gears.

Macros are small routines that are used to replace specific strings in a program. They can have parameters passed to them, which enhances the functionality of the macro itself. A 16-bit increment, that doesn't affect the contents of the accumulator could be implemented as:

```
MACRO inc16 Variable         ;  Increment two bytes starting at "Variable"
Local inc16End
  inc         Variable       ;  Increment the low 8 bits
  push        ACC
  mov         A,Variable     ;  Are the incremented low 8 bits == 0?
  jnz         inc16End
  inc         Variable+1     ;   Yes - Increment the upper 8 bits
inc16End
  pop         ACC
ENDMAC
```

In this macro, first the lower 8 bits of the 16-bit number are incremented, then they are tested to see if they are equal to zero. To accomplish this test, I have to load the accumulator with the incremented value, which means I have to save the current contents of the accumulator before loading it.

In the `inc16` macro, `Variable` is a parameter, and the corresponding value put in at the macro invocation replaces the parameter within the macro. If `inc16 VarA` was invoked, the code inserted into the source would be:

```
  inc  VarA
  push ACC
  mov  A,VarA
  jnz  inc16End
  inc  VarA+1
inc16End
  pop  ACC
```

This macro example is a good one because it does not affect any other registers. When writing and using macros, you have to make sure you understand what registers and PSW flags are potentially affected by the operation of the macro.

Macros, if used in moderation can be wonderful. I say "in moderation" because some people's code style is to make an application totally out of macros with no assembler statements used in the source code file at all. The problem with this style is that, while the source becomes quite readable, understanding what is actually happening is just about impossible. Macros should be short, concise, and used to enhance the readability of the source code, not replace it.

A potential problem can be encountered when enhancing the 8051's instruction set and using a mnemonic that is already in the instruction set. The following macro implements a clear register or scratchpad RAM instruction (the 8051 can only clear the accumulator in the standard instruction set):

```
MACRO clr Dest                ;  Clear the "Dest" byte
  mov  Dest,#0                ;   By writing zero to it
ENDMAC
```

The problem with this macro is deciding what takes precedence. In some assemblers, the macro will be used every time `clr` is encountered.

This is probably okay, because if `clr A` was encountered, the code:

```
mov        A,#0
```

would be inserted into the application, which is a valid instruction.

The problem comes up if the 8051 instructions take precedence and the instruction `clr R0` is encountered by the assembler. In this case, even though there is a macro that will create valid code, an error is produced because `clr R0` is not a valid instruction in the 8051. Ideally, the assembler should first attempt to assemble each instruction and, if there is a problem, look for macros that meet the format of the problem instruction.

In the first macro example ("inc16"), I had to jump to a label with some data conditions. I used a local label within the macro instead of a global label by declaring it as such within the macro. Local labels are labels that only have meaning within the macro (and usually take precedence over the same labels outside of the macro).

Local labels should only be used for labels within macros to avoid problems if the macro is invoked multiple times. If a label within a macro is not declared local, then at assembly time, there is an opportunity for two types of errors. The first states that there are multiple labels in the source, and the second indicates that jump instructions don't know which one to use.

Conditional code consists of "if/else/endif" as well as other structured language constructs. Conditional code executes only during assembly of the source file and is used to instruct the assembler on which code to use as source. Source for conditional code can be macro labels, constants, or defines.

For example, if you wanted to load the accumulator with the logical inversion of a constant value, you could write the macro:

```
MACRO invert Value
 if (Value == 0)
 mov A,#1
 else
 clr A
 endif
ENDMAC
```

When the conditional code within invert executes, either the `mov A,#1` or `clr A` instruction will be inserted into the source.

There is one interesting thing you can do with conditional code and that is to use it to comment out large blocks of code.

Traditionally, code is commented out by placing a comment character (usually a semicolon) at the start of the line. Using the conditional code "if" statement, this could be accomplished much more easily:

```
   :                    ;  Usable code
 if ( 0 )               ;  This can never be "true"
   :                    ;  Code to be "commented out"
 endif
   :                    ;  Usable code
```

In the 8051, the reset vector is address 0h with the interrupt vectors at low addresses, 8 bytes apart. In my template, you can see that I have set aside space for the different interrupt vectors and jump the reset code around it. The interrupt handler `org` statements and labels can be deleted to make more space for the application (and avoid the confusion of jumping needlessly) if they aren't used.

The `org` statements themselves are assembler directives that specify the address that the code is to be placed at. Except for the reset vector and interrupt vectors, I never use `org` statements; I let the assembler place all the code in line with no space between routines or their mainline.

With UMPS, there is one thing to watch out for with `org` statements. If the `org` statement ends up overwriting code, an error is not produced. You might be used to most assemblers, which flag an error if an `org` statement results in two different values being specified for the same control store address.

Many people, like to start their mainline code somewhere arbitrarily away from the interrupt handler addresses (such as address 0100h). I let mainline be the first address the assembler specifies after the interrupt handlers. I don't like to limit myself or leave unused control store when I create applications. By specifying an arbitrary `org` statement for mainline, I might be setting myself up for an unexpected problem if the mainline code overwrites interrupt handler code that is larger than I expect.

I also don't leave patch areas in the microcontroller's control store, for fixing up problems. Instead, I try to get more than enough windowed (or EEPROM) microcontrollers to allow me to keep a steady supply while I debug failing applications and erase devices with non-working code.

In the mainline, the first thing that you will see is that I have left a space for initializing all the variables. Initializing variables is critical to the proper operation of your application. This probably seems like an overly dramatic statement, but every simulator I've worked with either sets all variable store to initially all zeros or the previous value you were working with. This means that, even if you don't properly initialize the variable, there is a chance that, in the simulator, the application will work due to the value of the variable the simulator starts up with. This means that the code appears to work fine in the simulator; however, when you burn the code into an 8051, it doesn't work at all or portrays strange, erratic symptoms.

Before burning an 8051, especially if you are about to release the application to manufacturing a product, you might want to set all the scratchpad RAM bytes in the simulated 8051 to some nonzero value (011h is fine) after stepping through the `ajmp Mainline` instruction. I have done this in a number of failing applications, and this test is excellent for detecting problems caused by unitialized variables.

The rest of the template is pretty simple and self-explanatory with the application mainline following variable initialization and subroutines put at the end of the code.

The location of the subroutines is largely arbitrary. They could be located after the interrupt handlers or between variable initialization (which could be a subroutine) and the mainline code. I put the subroutines at the end of the application source code file because, if they grow so large that Mainline can no longer be reached by an `ajmp` instruction, I might have a problem due to the application being too large for the device I'm targeting. In any case, I will have to correct the problem and re-assemble the code.

The TEMPLATE.ASM file presented in this book was designed for UMPS with my particular (or is that peculiar) style of 8051 assembly language programming. Other assemblers might require an `end` directive at the end of the file, or directives (including `define` or `MACRO`) might be in different formats than what I have shown here. I recommend for you to develop your own template that works with the assembler you want to use and satisfies your way of doing things.

In your application code, there are a few things I want to bring to your attention regarding how you code your application. Throughout this section, I have tried to use what I consider good coding practices. Before going any further, I want to point out what these are.

In your application, you might have the requirement to increment a 16-bit variable if a condition was met on an input port and then wait for this condition to change before executing any other code. You could code this as shown in Listing 8-2. This code is unnecessarily wordy and hard to read. There are a few things we can do to clean it up.

The first improvement that can be made is with the comments. Most of them parrot what the instruction is doing and do little to explain the big picture.

By assuming that somebody reading the program will understand how the instructions work and just explaining what I want to happen in this code, I can reduce the comments in the code (Listing 8-3). This is a lot easier to read with the comments just explaining what each function of the code does.

LISTING 8-2 **Conditional-increment code.**

```
  mov  DPTR,#DevAddress  ;  Load DPTR with the external device address
  movx A,@DPTR           ;  Get the value at the external device
  xrl  A,#Expected       ;  Does it equal the expected value?
  jnz  Skip              ;  If not zero, skip over increment
  inc  Counter           ;  Increment the low 8 bits
  mov  A,Counter         ;  If the low 8 bits are not equal to
  jnz  Wait              ;   zero, skip over increment high 8 bits
  inc  Counter+1         ;  Increment the high 8 bits
Wait:                    ;  Now, wait for the condition to change
  movx A,@DPTR           ;  Poll the external device
  xrl  A,#Expected       ;  Does it still match the expected?
  jz   Wait              ;  If it does, check again
Skip:
```

LISTING 8-3 **Removing the unnecessary comments.**

```
  mov  DPTR,#DevAddress
  movx A,@DPTR
  xrl  A,#Expected       ;  if DevAddress == Expected
  jnz  Skip
  inc  Counter           ;   Increment "Counter"
  mov  A,Counter
  jnz  Wait
  inc  Counter+1
Wait:                    ;   Wait for DevAddress != Expected
  movx A,@DPTR
  xrl  A,#Expected
  jz   Wait
Skip:
```

LISTING 8-4 **Using whitespace to make the code more readable.**

```
  mov  DPTR,#DevAddress
  movx A,@DPTR
  xrl  A,#Expected       ;  if DevAddress == Expected
  jnz  Skip

  inc  Counter           ;   Increment "Counter"
  mov  A,Counter
  jnz  Wait
  inc  Counter+1

Wait:                    ;   Wait for DevAddress != Expected
  movx A,@DPTR
  xrl  A,#Expected
  jz   Wait

Skip:
```

The code can be made more readable by the use of "whitespace" or blank lines that will help separate blocks of code (Listing 8-4). This example is now quite easy to read, and the operation should be fairly simple to understand.

When writing code, I always try to format it in such a way that I can see what's happening under all circumstances. I have seen lots of code that is formatted to place instructions and parameters on tab stops. This makes the writing of the code easier.

```
Loop:
        mov         A,#1
        anl         A,R0        ;  Isolate bit 0
```

The problem with this type of formatting is that, when you assemble the code and try to look at the listing file with an 80-column editor, you can't see the generated address and code at the same time.

In the previous example, the listing file, when seen from an editor might look like Listing 8-5. This makes debugging your code unnecessarily difficult and hard on your mouse (as you shift back and forth).

To avoid this, I keep my code fairly well squashed over to the left. I normally start the instruction on column 3 and start the instruction's parameters on column 8. This will allow me to see the source code in the listing file in virtually any type of editor.

You'll notice in some of the code that I present in this book that I break away from this convention and put some parameters only one space from the instruction. The reason for doing this is to take advantage of UMPS' highlighting of instructions. Some instructions are set up in such a way that, if anything more than one blank separates the instruction and parameter, the instruction highlighting won't take place. By making sure the highlighting is always correct for all instructions, I never have to worry about syntax errors when I assemble the source code.

As I pointed out at the start of this section, in some assemblers, page breaks or directives indicating the start of a new page can be specified in the source. This function is useful for separating the variables from the constants in the source file as well as the interrupt handler code from the mainline and the subroutines from the mainline. This makes the listing file much easier to go through.

The last tool in programming you'll see is the comment with "####" at the start.

The four pound signs are used to allow me to easily find (either visually or using an editor search) a specific line of code. This line could be where I left off editing or a location that I want to mark for later. If you go through all the source code in this book, you'll probably find five or six different reasons why I've used this convention to mark code. I chose "####" because it is not a valid string in any programming environment that I know of, it is not case sensitive (which could trip up some editors), and it is very easy to spot visually.

You might want to use a marking scheme like this, but don't use too many (one or two is the maximum that I want to use) because you will find that they become confusing.

LISTING 8-5 Aligning code on tab stops can make debugging unnecessarily difficult.

```
Addr        Data

1234                                      Loop:

1234        74 01                                         mov         A,
1236        58                                            anl         A,
```

When you are using a visual flag like this, make sure that you use them consistently, they're appropriately commented, and their purpose is easily understood at first glance.

These three rules are actually what I want to close with. You are going to develop your own style of programming. As long as you code consistently (i.e., use the same format), comment appropriately, and strive to be easily understandable, the applications you code will be appreciated by others and won't cause you to be needlessly pulled from your deathbed.

Interpreters

If you are like me and remember back to the deep, dark days of personal computing, you probably don't have very fond recollections of programming an Apple II or IBM PC in Microsoft BASIC. The source code was executed directly in the PC without an intermediate compilation step. The program that parses each line (to figure out what each line of the source was doing) and then executes it is known as an *interpreter*. The name simply comes from the idea that each line is interpreted and then executed.

Interpreted code is usually the least efficient manner of program execution (although I'm sure I would get an argument from Forth affecionadoes). Before the function of the line can execute, the source has to be fetched from memory and parsed. Even if it took zero time to parse and execute the source code, interpreted code will always take longer to run than compiled code because the source code is always larger than the resulting compiled code. Also, to execute the code, the source has to be read from memory and appropriate subroutines have to be called and executed, rather than running the compiled code directly from the application program.

Many interpreters work as simple compilers to first convert all the instruction statements into tokens as they are entered into the computer system. These tokens are used by the interpreter to execute the desired functions without having to plow through each line. Interpreters that compile the source code first into tokens run the programs much faster than ones that don't.

If you know anything about the Parallax BASIC Stamp, you'd probably think that the device is a PBASIC (the Stamp's BASIC) interpreter. Actually, the Stamp consists of a PC host that compiles code into tokens and downloads them into an on-board EEPROM. Once in the EEPROM, the tokens are used to select subroutines to execute the program.

There are some microcontrollers that contain true interpreters on board. (There is an Intel 8052 that has a version of BASIC built in, and the source is loaded into an integrated RAM. The source code for doing this is available on the Internet). However, these devices do not tokenize the code. This means that they run slower and have less program space available for program statements (because source code takes up more space than tokenized code). However, the source code can be used to debug the application directly inside the microcontroller.

High-Level Languages

The 8051 architecture is interesting because of the extremes it works with. The 8051 probably has one of the most difficult architectures available for which to create a high-level language compiler that produces the simplest, smallest, and fastest assembler code for arithmetic operations. This is despite its being very easy to create one for writing compiler operations that will allow efficient manipulating large blocks of data. These two aspects are not necessarily mutually exclusive, but they will affect how the compiler operates.

Ideally, an 8051 application should run all of its arithmetic operations out of the current 8-byte bank and the accumulator. Keeping all variables within a single bank will result in the smallest and fastest code. If code has to be moved in and out of the banks for being operated upon, the code will become very clumsy very quickly.

For example:

```
VarA = VarB + C              ;  "VarA" and "VarB" variables are NOT
                             ;   the accumulator and "B" register.
```

If all three variables were in the current bank, the code could be:

```
mov  A, VarB                 ;  2 Bytes/1 instruction cycle
add  A, C                    ;  2 Bytes/1 instruction cycle
mov  VarA, A                 ;  2 Bytes/1 instruction cycle
```

Now, if `VarA`, `VarB`, and `C` were all in the first 256 addresses (but not in the same bank), the code would change to using direct addressing on each of the three instructions. This new code would require twice as many cycles to execute and take up twice as many bytes of control store.

If `VarA`, `VarB`, and `C` were located anywhere within the 64K data area, the best code that could be written would be:

```
mov  DPTR, C                 ;  3 Bytes/2 instruction cycles
movx A, @DPTR                ;  1 Byte/2 instruction cycles   - Load "C"
mov  Temp, A                 ;  1 Byte/1 instruction cycle  - Save "C"
mov  DPTR, VarB              ;  3 Bytes/2 instruction cycles
movx A, @DPTR                ;  1 Byte/2 instruction cycles   - Load "B"
add  A, Temp                 ;  1 Byte/1 instruction cycles   - ACC = B + C
mov  DPTR, VarA              ;  3 Bytes/2 instruction cycle
movx @DPTR, A                ;  1 Byte/2 instruction cycles
                             ;     - Save B + C in "VarA"
```

The general case shown here requires more than three times the memory and over four times the cycles to execute over the best case of using three variables in the currently active bank.

With this wide variance in performance, it is definitely worth the compiler writer's time to develop a strategy of how data is to be stored and manipulated. By reducing data movement to and from the current bank, the first 256 addresses, and the 64K data area, the final program can be made much more efficient in terms of space and time.

Some strategies for carrying this out could include keeping arithmetic variables grouped in banks and then switching between banks to operate on the data directly. Another strategy could be grouping all the arithmetic variables in the same block of memory, the start of which is pointed to by the DPTR or one of the two index registers in the current bank. In the latter case, data could be accessed using the index registers much faster and using less space than if it were moved into temporary registers in the current bank.

The ability of the DPTR to access memory across the entire 64K data space and the availability of instructions to update the register quickly means that the large amounts of data can be accessed quickly. While data structures are really beyond the scope of this book, looking back over the instructions should give you some ideas of how to store data in blocks that are pointed to by the DPTR and can be accessed with offsets from the DPTR address.

This capability makes the 8051 ideally suited for implementing data loggers and collection devices, simple terminal data switches, and other applications where large amounts of data must be manipulated within the device quickly.

Intel Hex Format Object Files

The Intel INHX8M is the most common object file format. Pretty well all embedded microcontrollers use this object file format (which is also referred to as a ".hex" file) although the most notable exception is the Motorola microcontrollers, which use the Motorola developed S19 format object file.

If you were to look at a ".hex" (or ".obj") file that was produced by an assembler or compiler, you would see something like the following file, which was taken from a PICMicro microcomputer application. The 8051 information is stored in exactly the same format.

```
:10000000A601A701AB018E288C00030883128D0086
:100010000B11AB0A0D0883008C0E0C0E0900A80111
:10002000A201A3011E081F04031908001F082106CE
  :
:0A04F000A40B782AA50B782A080057
:02400E00F13F80
:00000001FF
```

All the information needed to load the microcontroller's control store is located in this file. You should note that no symbolic information is located in this file (that is usually produced by the compiler or assembler specifically for a specific debugger or emulator).

Each line is broken up into a number of fields that are used to specify to the programmer where the file is going to be used. The fields are defined by columns and work out as Table 8-1. Typically, up to eight 16-bit words can be stored on each line.

TABLE 8-1 DESCRIPTION OF THE INTEL HEX FORMAT

COLUMNS	IDENTIFIER	COMMENTS
1	Line start	Always ":"
2–3	Bytes in line	Number of bytes to be loaded
4-7	Data start address	The address where the data will be stored
8-9	Data type	"00" - Data, "01" - File end
10–11	1st byte in data	Each byte is represented as hex value, in LSByte first
12–13	2nd byte in data	
14 . . .	Program data	Additional bytes of program data
End–2	Checksum	Line total & 0x0FF = 0

If we were to take the third line from the bottom in the previous example, we can break up the data and show what is actually in it.

```
:0A04F000A40B782AA50B782A080057
```

Columns 2 and 3 are "0A," which is two times the number of 8-bit words in the line. In this case, there are five words.

Columns 4 through 7 are "04F0," which is the start of where the line's data is to be loaded into control store. Note that this 16-bit number does *not* use Intel byte ordering, instead the high byte appears before the low byte.

The next two columns are both "0," which indicates that data is being sent in the line.

Starting at column 10, there are ten 8-bit words of program data:

```
A40B782AA50B782A0800
```

The last two columns are the checksum. You've probably noticed that each data byte is broken up into two hex characters. These values are added up and subtracted from 0x0100, and the result is used for the checksum.

For the previous example, the Data bytes are summed as:

```
  0A
  04
  F0
  00
  A4
  0B
  78
  2A
  A5
  0B
  78
  2A
  08
+ 00
----
 3A9
```

This sum is then ANDed with 0x0FF and subtracted from 0x0100 to get the checksum:

```
  100
– A9  (= 0x03A9 & 0x0FF)
-----
   57
```

If you look earlier, you'll see that "57" is the last two columns of this line.

The last line looks pretty boring (just the data type set to 1 and a checksum of FF), but this is all that's required to indicate that the object file is ended.

8051 Debugging Hints

I've said it earlier in this chapter, but I really feel that the most important debugging is done before you burn your application into a part. This is accomplished by understanding how the interfaces to other devices are going to be implemented, making sure the device is properly timed, and simulating the application until you feel comfortable and confident that it will work the first time it's powered up. Even after going through this effort, there's still a good chance the application won't work properly.

There's nothing more frustrating than developing an application, wiring a sample circuit, applying power, and having nothing happen. (Actually there is one thing that's more frustrating, and I'll discuss it later.)

When this happens, you have to start looking at the obvious problems.

First check to see whether the correct power is being supplied to the various components. I've found that some voltage regulators (i.e., versions of the 7805) require substantially more voltage than you would expect (such as 10 V for the regulator to operate properly and produce 5 V output).

If the power is low, check for a hot component or locations on the board. You might have a Vcc-to-ground short.

Next, check the 8051's Reset pin. It should be low when power is applied. If you are running without a capacitor between Reset and Vcc, you should wire one in.

I've found that some power supplies might come up very slowly, and the 8051's internal reset circuit does not work properly. You can discover reset problems by adding a switch that when closed will tie Reset to Vcc. (See Fig. 8-1.)

The 0.1-µF capacitor in this circuit should help keep the reset line high until power has stabilized. The push button that shorts Reset to Vcc is used to reset the device during debugging.

I've caught myself putting a resistor between Vcc and Reset a few times because, in many other microcontrollers, Reset is negatively active. In the 8051, Reset is *positively active*,

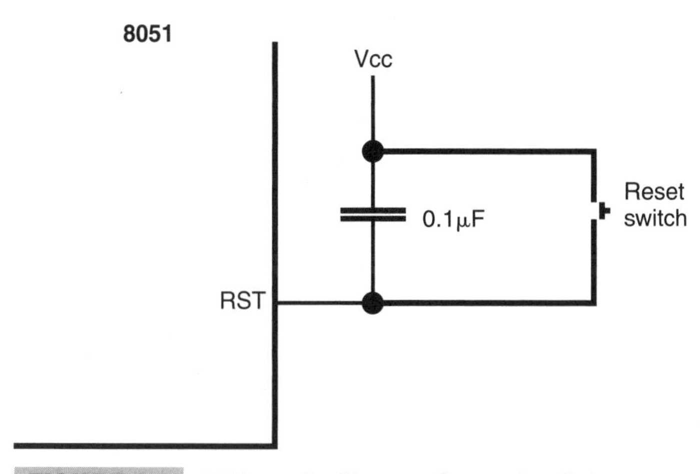

FIGURE 8-1 **8051 reset with capacitor and switch.**

which means that, when a high voltage level is applied to Reset, the device stops the oscillator and sets the special-function registers to their power on default values and resets the program counter back to zero (the reset vector).

Next, the clocks should be checked. This can be done by simply probing the XTAL2 pin with a logic or oscilloscope probe. If you can check with an oscilloscope, make sure that the clock signal is a sine wave of approximately the correct frequency and the amplitude is between 4 V and 5 V.

Another mistake I often make is to forget to tie the EA pin to Vcc. Leaving this pin floating or connected to ground will result in the 8051 attempting to retrieve instructions from external ROM.

If you are using external control store (ROM), then the EA pin should be tied to ground.

Check the address, data, and control lines to make sure that they are correctly wired. As well, make sure the A0 to A7 address buffer is latched by the ALE pin. You might want to probe these lines during program execution to make sure data is coming through (or, at the very least the lines are "wiggling"). You should also check your control store timing to make sure that the access time of the memory chip is not being exceeded by the application.

At this point, you should be pretty confident that the 8051 core hardware *can* run. Now, check the devices that the 8051 interfaces with to make sure that they are wired correctly and have a common ground with the 8051. Not having a common ground could result in the circuits not being read or written to properly (if at all).

With the circuit now completely checked out, check to see whether the 8051 or the external ROMs are programmed correctly. Occasionally, the devices will not program correctly, or you will accidentally put in a device that hasn't been programmed at all in the excitement of seeing the application running. I've done this a few times.

You should also make sure you are loading the correct version of the software as well.

With all these hardware items confirmed, you now have to go through your software. As a gross check, make sure you simulated the same device as you are using. Different 8051s have different amounts of control store or different special-function register sets. You should check to see whether the application is attempting to execute or is accessing any resources that aren't present in the physical device that you are using (e.g., you have simulated a device with 4K of internal control store but are using a device with only 2K).

With the program and device confirmed, you now have to go through the program and look for potential trouble spots. The following suggestions are really for people who don't have access to an emulator. If you have an emulator, then you should be able to find the trouble spots very easily by placing breakpoints where problems could potentially happen and stepping through or examining data values are these points.

One of the most efficient ways of finding problems in a simulator is to set the scratchpad RAM to a random value. This will help to flush out any missing "#" symbols in instructions that should be immediately addressed or variables that are not initialized. In both these cases, the simulator's power up and reset scratchpad RAM values might be correct for the application to run without any problems. By explicitly running the application code with different scratchpad RAM values than normally present at start up, you might be able to identify the problem very quickly.

If all of these processes fail to identify the problem in the application, you will have to start commenting out sections of code or putting the microcontroller into a hard loop to try to figure out what is happening.

For example, if you had a write to an LCD (Listing 8-6), you might want to see if the correct data is getting to the LCD along with the correct control bits. To confirm this, you would mask interrupts and go into a loop after "E" is set high to latch in the data (Listing 8-7).

In the second case, I have commented out the `clr E` instruction so that I could put the hard loop code in the white space between the data output and the polling of the busy flag. Also note that I have used "####" to mark the lines that I have added or changed. The purpose of marking the code is to give me a visual indicator (as well as a search string) of the changes that I have made.

I could have commented out the code as shown in Listing 8-8 to provide the same function, but I find this sub-optimal because it's much harder to see the changes and understand how it works. It is probably not completely obvious that I left the `ajmp Loop` instruction not commented out to provide the hard loop in the code.

LISTING 8-6 **Code to write to an LCD.**

```
   :
  mov   P1,#030h          ;  Initialize the LCD
  clr   RW
  clr   RS
  setb  E                 ;  Toggle out the initialize command
  clr   E

  setb  RW                ;  Now, wait for the operation to complete
Loop:
  setb  E
  jnb   P1.7              ;  Bit 7 of the LCD will be low when the
  clr   E                 ;   operation is complete
  ajmp  Loop
Skip:
  clr   E
   :
```

LISTING 8-7 **Code to mask the interrupts and loop after "e" is active.**

```
   :
  mov   P1,#030h          ;  Initialize the LCD
  clr   RW
  clr   RS
  setb  E                 ;  Toggle out the initialize command
;  clr  E                 ;  ####

  clr   EA                ;  #### - Disable interrupts
CheckLoop:                ;  #### - Loop here forever
  ajmp  CheckLoop         ;  ####

  setb  RW                ;  Now, wait for the operation to complete
Loop:
  setb  E
  jnb   P1.7              ;  Bit 7 of the LCD will be low when the
  clr   E                 ;   operation is complete
  ajmp  Loop
Skip:
  clr   E
   :
```

LISTING 8-8 **Sub-optimal code to mask interrupts and loop after "e" is active.**

```
   :
  mov  P1,#030h           ;  Initialize the LCD
  clr  RW
  clr  RS
  setb E                  ;  Toggle out the initialize command
;  clr E

  clr  EA                 ;  #### - Disable interrupts

;  setb RW                ;  Now, wait for the operation to complete
Loop:
;  setb E
;  jnb  P1.7              ;  Bit 7 of the LCD will be low when the
;  clr  E                 ;   operation is complete
  ajmp Loop
Skip:
  clr   E
   :
```

By placing hard loops in the code, you can figure out where the problems lie with the application. When doing this type of debug, I highly recommend having a number of windowed devices handy or using EEPROM/Flash control store devices. Otherwise, this will become very tedious, slow, and potentially expensive (if you don't have windowed parts).

At the start of this section, I said that having an application that sits there doing nothing is the most frustrating thing there is. This isn't true by any stretch of the imagination.

The most frustrating problem to have to debug is an intermittent problem (constant ones are quite easy in comparison). This is a problem that pops up after some period of time of the application running without any problems.

Rather than looking at your wiring or software, the first thing to do is characterize the problem. By characterizing, I mean to say you should try to understand the parameters of the failure:

1. *Input*—Does a specific bit, pattern, or sequence cause the problem?
2. *Output*—After a specific output value or sequence, does the application fail?
3. *Timing*—Does the problem occur after so many seconds? Does the problem occur a few seconds after an input?

With this information, you should be able to figure out in which part of the code the problem is occurring in. From there, you can look for code that could cause the symptoms that you are seeing.

Some intermittent problems are caused by device specification violations. For example, in the digital clock/thermometer application in chapter 11, I found the DS1820 digital thermometer reading to be somewhat intermittent and erratic with a number of the readings returning 0FFh instead of a valid temperature.

Looking at the command input to the DS1820 from the DS87C520, I found that a "0" pulse was 15 μsec long. Looking at the DS1820 data sheet, the valid timing was 15 to 60 μsec. By increasing the delay to 30 μsec, the application ran rock steady without any more invalid temperature reads.

9

EXPERIMENTS WITH THE 8051

CONTENTS AT A GLANCE

CONTENTS AT A GLANCE *(Continued)*

It's unfortunate, but the first time most engineers work with a new part is when they have to design an application with it. I've always hated being in that situation because it has meant that, when I start to discover better ways of doing things, I'm hitting the "right edge" (end) of my schedule and I don't have any time to make the changes that will improve the application.

To help counter this, when I start working with a new device, I like to take about a month and go through a number of experiments so that I can understand the tools that I am working with. These experiments include setting up and learning the software development tools that I am going to use, going through the addressing modes of the device, working through arithmetic operations, understanding how jumps and subroutine calls are implemented, getting the actual hardware to run in a circuit, and looking at the electrical operation of the device as well.

I highly recommend going through this chapter and running each experiment as I have. I'm sure that how the processor works will be better cemented in your mind, and understanding how the 8051 is wired into a circuit and experimenting with those circuits should be a help for you in developing your own applications.

Tools and Parts Required

My ideal for any microcontroller project is to be able to build and debug the circuit using very simple tools and not require an oscilloscope or logic analyzer. This is especially true for the initial experiments that I do when first learning about a microcontroller. Following this philosophy, I have tried to make the 8051 experiments as simple as possible so that you can implement them with a minimum of problems or require a substantial investment in equipment.

For the experiments, I will use a similar execution core for the Dallas Semiconductor 87C520 and Atmel AT89C2051 devices (Figs. 9-1 and 9-2). These core circuits are probably an excellent place to start for most of your own applications.

Note that I have included a momentary On switch as a reset button to allow you to restart your applications whenever you want to. When you are ready to release your applications, you will probably want to tie Reset down to Ground to make sure there is no possibility of the pins being pulled to Vcc and stopping the microcontroller's operation.

I have tested the experiments on both devices (except where the lack of pins for the Atmel part makes it impossible).

For the experiments that require a simulator, I will use UMPS by Virtual Micro Design. This integrated development environment (I have provided a demonstration version on the CD-ROM that comes with this book) is an excellent tool for developing software as well as seeing how it will work with external hardware. I first discovered this program when I was working on the *Handbook of Microcontrollers*, and I really fell in love with it.

In the next section, I will discuss installing UMPS onto your hard drive. Then I will show how UMPS can be used to develop and debug a program before burning code into a device (and hoping for the best).

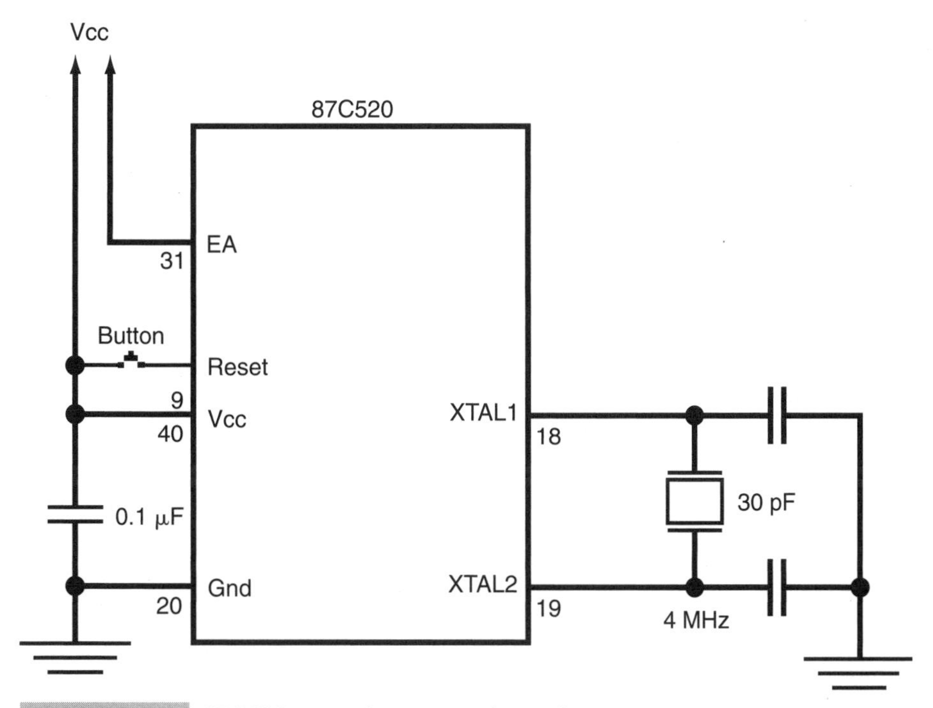

FIGURE 9-1 **87C520 execution core schematic.**

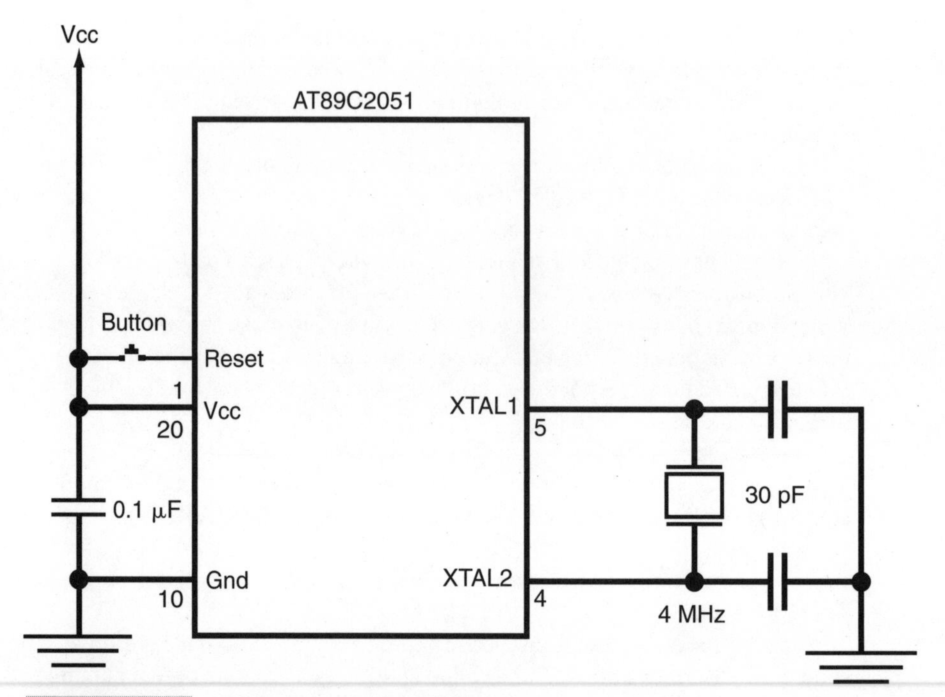

FIGURE 9-2 **AT89C2051 core processing.**

Not including the microcontroller cores discussed earlier, the physical tools and parts required to implement the experiments (and many of the applications) in this book are quite straightforward:

- Device programmer and U-V eraser
- Protoboard and wiring kit
- +5-V 100-mA power supply
- ZIF sockets
- LEDs and 220-Ω resistors
- Momentary-on push buttons
- Digital multi-meter
- Logic probe
- Clippers, pliers, and screwdrivers

The programmer could be either the circuit presented earlier in the book or a commercial device. When choosing a commercial programmer, make sure it can use the Intel IHXM8 object file data format.

I like using the spring-loaded contacts of protoboards. I realize that they aren't everybody's first choice for prototyping, but they are well suited to wiring the experiments presented here. What might be surprising is my use of *zero insertion force* (ZIF) *sockets* for connecting the microcontrollers into the circuits. As great as protoboards are, I find it gets old pretty quick having to insert and pull out (using a screwdriver) the microcontrollers during application debugging (when you pull them out to put them into another circuit). As well, plugging the devices into the holes does leave the pins at risk of being damaged (which can get expensive very quickly with ceramic windowed parts).

Using a ZIF socket pushed into the protoboard eliminates this risk.

For powering your experiments, I recommend the circuit shown in Fig. 9-3. This circuit takes the output from a *Wall Wart* (socket mounted transformer and rectifier with a 6' or so power cord) and outputs a regulated +5-V power supply at up to 1 Amp. This circuit can be used for many of the example applications presented later in the book.

The 7805, if you're not familiar with it, is really a marvelous device; it's a linear voltage regulator capable of outputting very constant current (even with quite a noisy source). The 78*xx* series of regulators will output positive "*xx*" voltages, while the 79*xx* series will output negative "*xx*" voltages. There is a 78L05 available, which is only capable of sourcing 100 mAmps.

The wall-wart input should be greater than 9 V to make sure the 7805 works properly regardless of the load it's under. If currents greater than 100 mA are to be drawn through the 7805, I recommend attaching a heat sink to the 7805.

There are a few things to notice about this circuit that are important for your protection. The first is to use only a UL/CSA approved Wall Wart. Household 110 Vac can burn or kill. Minimize the risk to yourself, and only use a properly operating dc output device (if you are going to use an ac transformer, this circuit will work with a full-wave rectifier, although you might want to add a bit of capacitance as input to the 7805). Note that the switch is upstream from the 10 µF capacitor. This is to eliminate the chance that the capacitor is left charged and could shock somebody.

The best part of this power supply is that, even using new parts, it can be built for around $5.

I hate book and magazine projects that require hard-to-find parts (which, by some strange coincidence, are only available from the author). For this reason, in this chapter (as well as the others that provide you with circuits), I have specified only parts that are easily obtainable.

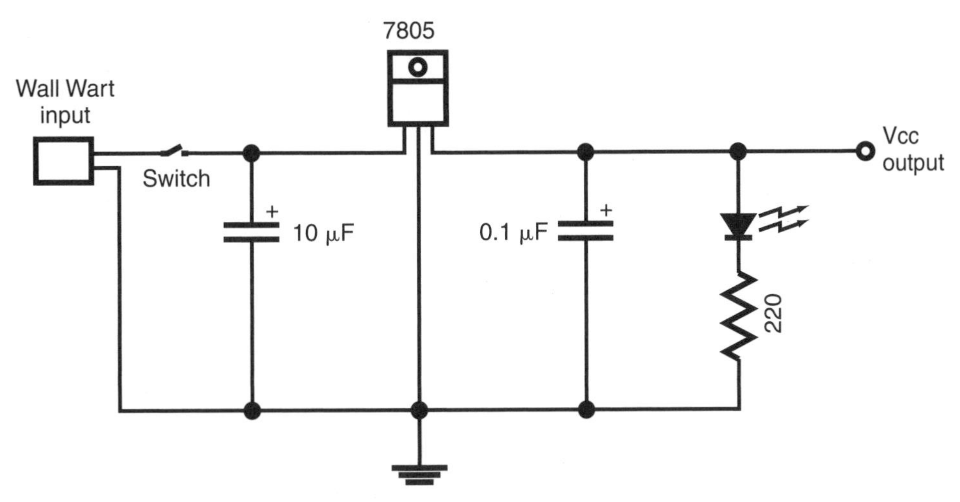

FIGURE 9-3 8051 typical power circuit.

In terms of test equipment, as you can see, I've really kept the list of required parts to a minimum (both in terms of quantity and cost). In the book, I have put in example waveforms to show you what should be happening at the various pins to eliminate the need for you to get an oscilloscope or logic analyzer.

PROG1: Setting up UMPS and Getting a Program to Assemble

If you look at the program shown in Listing 9-1, you're probably going to think that this is an unreasonably simple first example. All this program does is pull an I/O bit low and then go into an endless loop. It sounds pretty simple, doesn't it?

It probably *should* seem pretty simple because I've gone through the instruction set and I/O pins, but the most difficult aspect of learning any new device is how to set up your tools and actually correctly assembling your program. Having a very simple program (like PROG1) eliminates one of the variables needed to get everything running (the variable of whether or not the program works). Once the program is running, then you can start writing more complex applications using the tools.

For the rest of this section, I will go through the steps in setting up UMPS and running PROG1 in UMPS to debug it.

After starting up UMPS, click on `Configure`, `Load CPU`, and the `DS80C320`. This will specify that UMPS is to run the Dallas Semiconductor 80C320 simulator code (and, when assembling source code, use the 8051 mnemonics).

With this completed, we now have to load the source file, click on `Files` and then `Load`. When this is done, the screen will look like Fig. 9-4.

LISTING 9-1 PROG1.

```
;  PROG1 - First Program, Turn on an LED at at P1.0
;
;  This application turns on an LED and loops forever afterward.
;
;  Myke Predko
;  98.01.30
;
;  Hardware notes:
;  80C520 running at 4 MHz
;  P1.0 is the LED (to light, it is pulled down)

;  Variable declarations

;  Mainline
 org 0                          ;  Execution starts here

  clr    P1.0                   ;  Turn on the LED

Loop:                           ;  Loop here forever
  ajmp   Loop
```

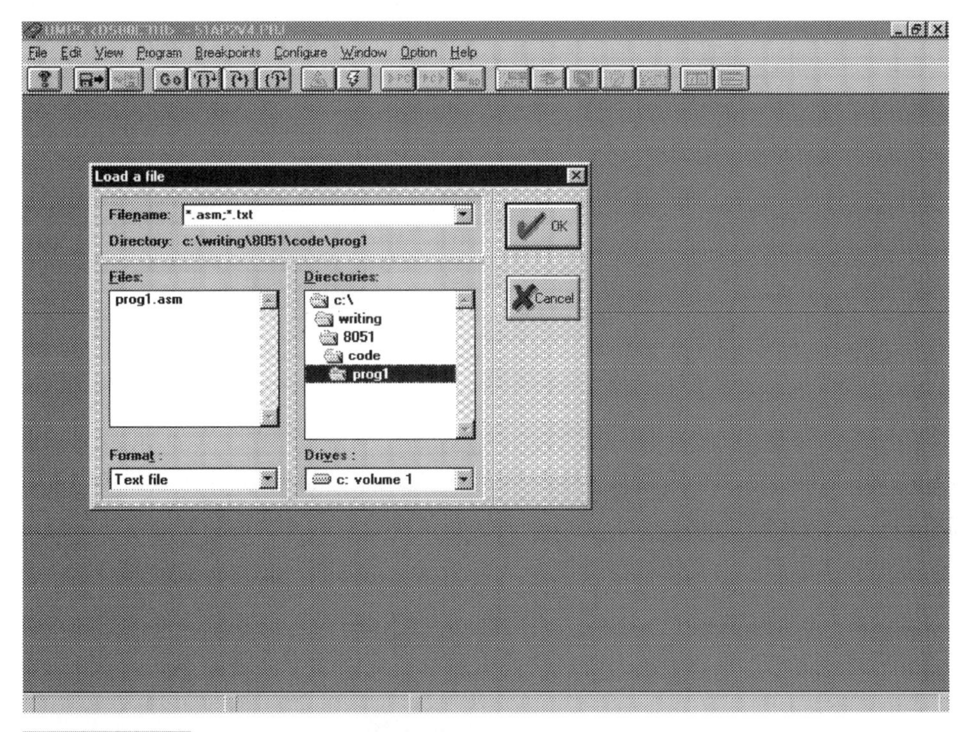

FIGURE 9-4 Loading a source file in UMPS.

Go to the *d*:\8051\code\prog1 subdirectory (the subdirectory shown in the diagram in Fig. 9-4 was taken from the PC that I wrote this book on and is where the program resides). The *d*: of the path is the letter of the drive where the subdirectory and file prog1.asm are located.

When I have selected the source file and its editing window is displayed on the screen, I usually move it at the top of the UMPS window and stretch it so that it goes across the width of the UMPS window (Fig. 9-5). This means that I can put in code/comments that are longer than 80 columns.

With the program loaded into the UMPS window, now it is time to define registers to monitor and hardware to interface the microcontroller to. Click on `Configure` and then `CPU Registers` to get the window shown in Fig. 9-6.

The window that comes up is used to display the simulated microcontroller registers. The registers are selected by clicking on the `Add` button (or the `Add` pulldown on the top line of the UMPS window) and then selecting the register you want to monitor (Fig. 9-7).

The configuration of each register can be specified by clicking on the configuration button (the button with the small CRT icon) and selecting the options that you want for the register (Fig. 9-8). The register can be moved by the mouse, and I tend to leave the colors at their default values (i.e., black on white), which means the only option that I really am concerned with is what the radix (the number base of the values) to be displayed.

With all the registers specified and located in the window, I usually put them together in meaningful (to me) locations and put the window together. As shown in Fig. 9-9, I display

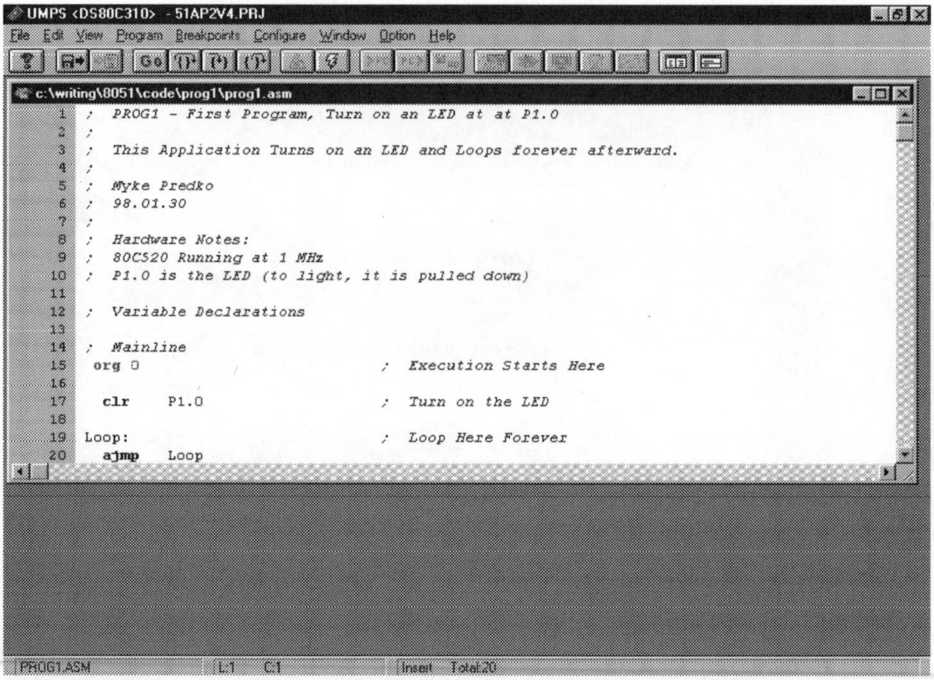

```
UMPS <DS80C310> - 51AP2V4.PRJ
File Edit View Program Breakpoints Configure Window Option Help

c:\writing\8051\code\prog1\prog1.asm
 1  ;  PROG1 - First Program, Turn on an LED at at P1.0
 2  ;
 3  ;  This Application Turns on an LED and Loops forever afterward.
 4  ;
 5  ;  Myke Predko
 6  ;  98.01.30
 7  ;
 8  ;  Hardware Notes:
 9  ;  80C520 Running at 1 MHz
10  ;  P1.0 is the LED (to light, it is pulled down)
11
12  ;  Variable Declarations
13
14  ;  Mainline
15   org 0                          ;  Execution Starts Here
16
17    clr    P1.0                   ;  Turn on the LED
18
19  Loop:                           ;  Loop Here Forever
20    ajmp   Loop

PROG1.ASM        L:1   C:1        Insert   Total:20
```

FIGURE 9-5 Source file loaded into UMPS.

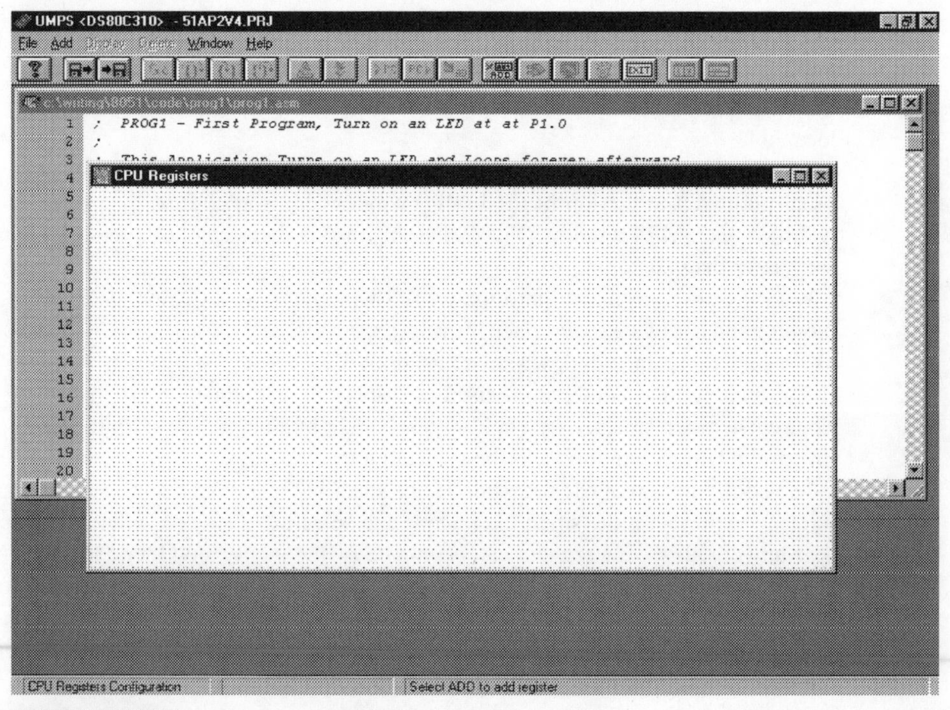

FIGURE 9-6 Opening the register monitor window.

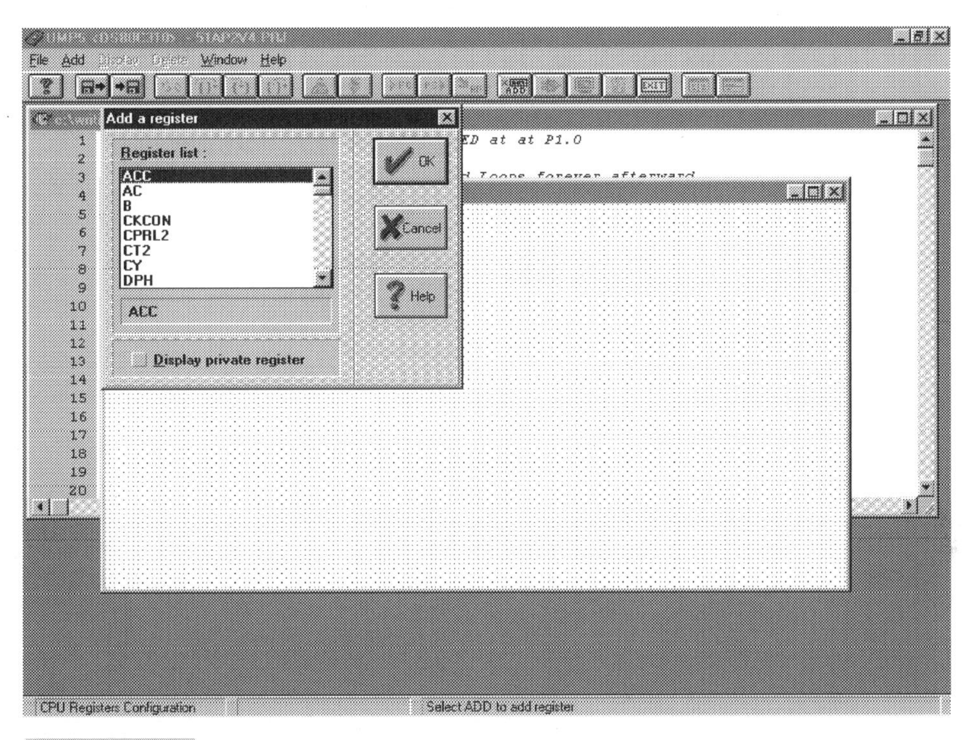

FIGURE 9-7 Selecting a register to monitor.

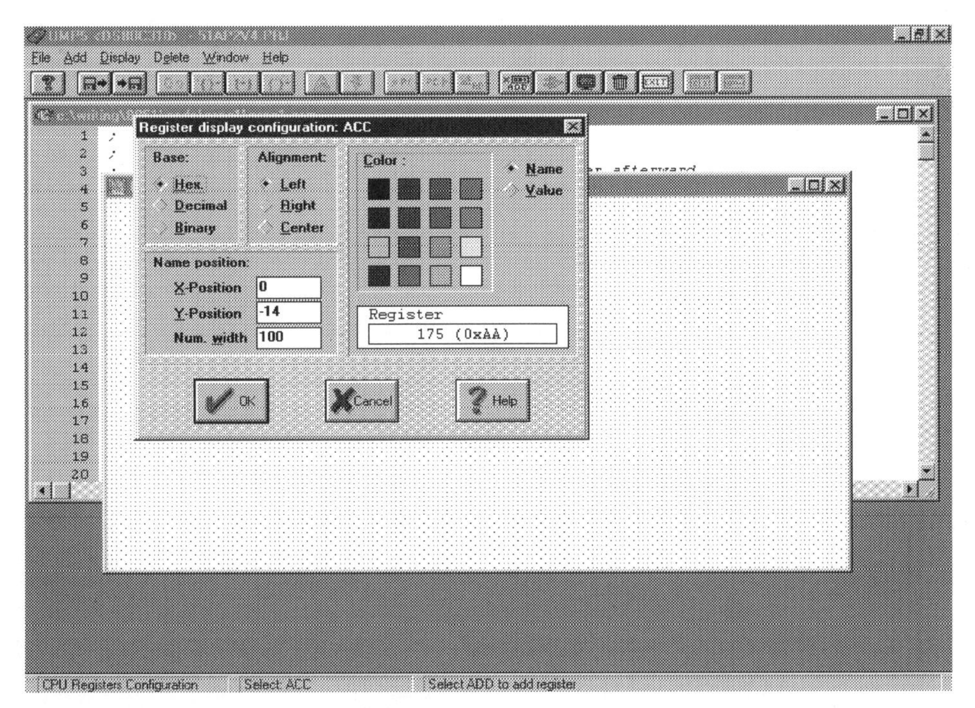

FIGURE 9-8 Selecting the register data format.

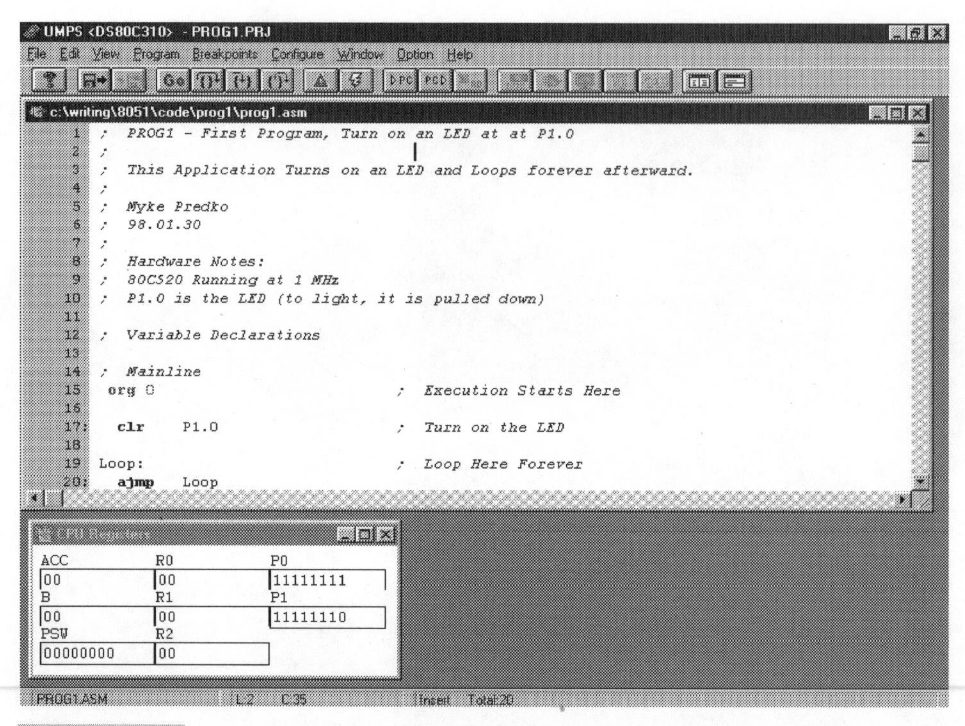

FIGURE 9-9 Placing the CPU registers onscreen.

the PSW and I/O pin registers as bits and the accumulators and bank registers as hexadecimal. This convention for register definitions will be used for all the experiments in this chapter.

In a typical IDE, this is where you would start running the simulation. The simulated processor in UMPS, on the other hand, can be interfaced to simulated hardware. This is done by clicking on `Configure` and then `Resources`. *Resources* are hardware devices (including LCDs, serial and parallel data sources, as well as a fairly wide variety of different discrete devices such as buttons and LEDs).

Once the Resource window is enabled, the devices can be added to the window by clicking on the Add icon. From here, you are given a list of different devices to include. For this program, I am going to add `LED (Red, Green, Yellow)` and `Push Button` (Fig. 9-10).

With the devices selected, I have to wire them to the simulated processor as well as choose their configurations or types. For the LED, I want to wire it in the same manner that I will in the hardware. This means that the simulated 8051 (like the real device) cannot source current, and I want the LED to light when the software pulls the pin down. To do this, I will connect the `Anode` (positive connection) to `Always "1"` and the `Cathode` to the I/O pin (Fig. 9-11). This is done by clicking on the icon on the tool bar that looks like an electrical plug, selecting the pin from the list, and then specifying the register that the pin is connected to.

The switch is a bit more complex to specify in UMPS. That is because two major types of switches are available in UMPS (with both being normally on or off). The switches are a matrix type of switch (which will allow you to implement a keyboard) and a push-button

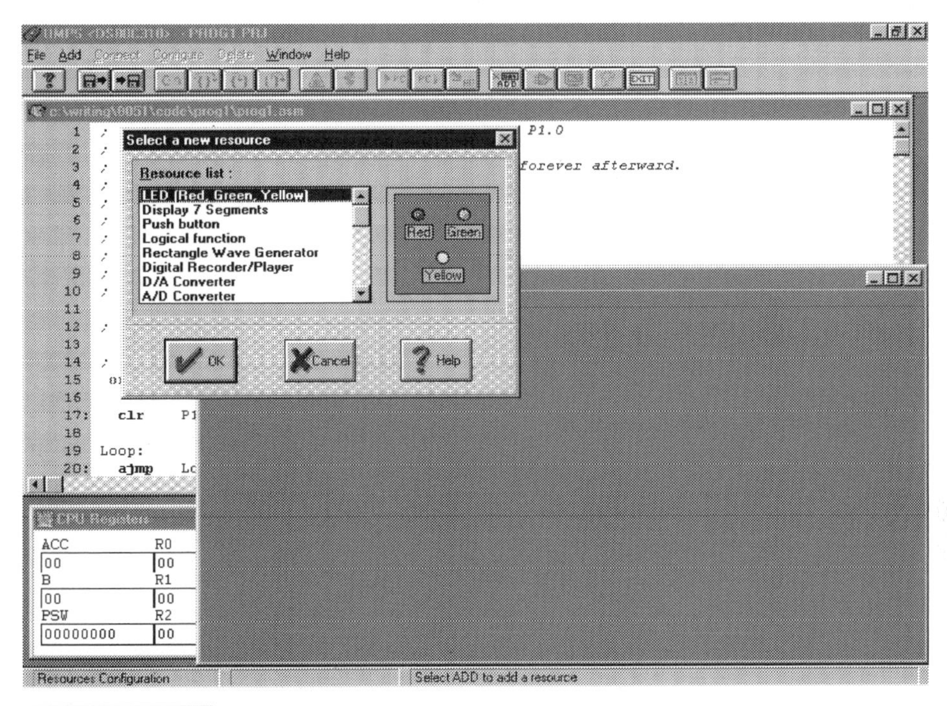

FIGURE 9-10 Selecting resources to add.

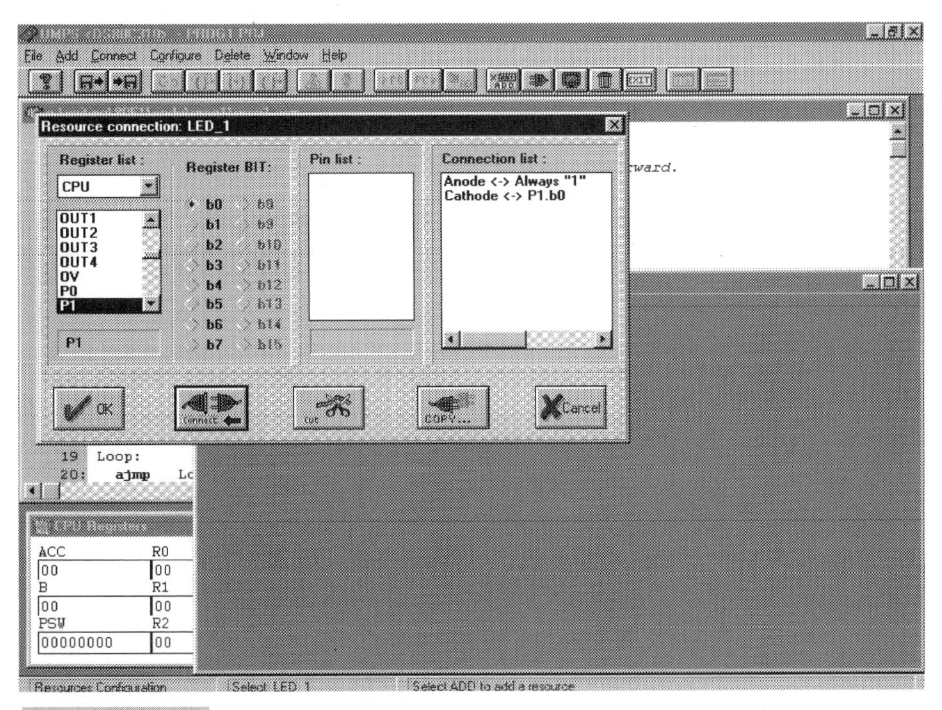

FIGURE 9-11 Specifying connections.

switch that pulls a signal to Vcc or Ground. For the first program, we will deselect `Push Button (matrix use)` in the configuration window that comes up when the CRT icon is selected for the button. With the configuration shown in Fig. 9-12, the button will be a momentary On switch pulling the signal to ground.

When I connect the switch, I connect the `Out` pin to `P1.1` (bit 1 of P1) and the `In` pin to `Always "0"`. In doing this, the pin will pull down to ground when it's pressed.

In UMPS, the simulation is more at a logic level and not an actual hardware level. This means that the current limiting transistors needed for actual LEDs cannot be added. As will be pointed out later, it also means that some connections required for running the program cannot be simulated as well.

With the hardware resources specified, I usually make the resource window as small as possible (like with the CPU register window) and put it along the bottom of the UMPS window as a standard way of setting everything up.

If you are following along with me and you compare Fig. 9-13 to where you are now, you'll notice one big difference between what you see on the screen of your PC and what is shown in the figure. That is the first line (`clr P1.0`) is highlighted in the figure and not on your PC's screen.

To make the program ready to run the program, click `Program` and then `Compile`. The program will be assembled, and if there are any errors in your code, an error message will appear. To show what this looks like, I have inserted the text "garbage" after the `org 0` statement and clicked on `Program` and `Compile` to assembling PROG1.

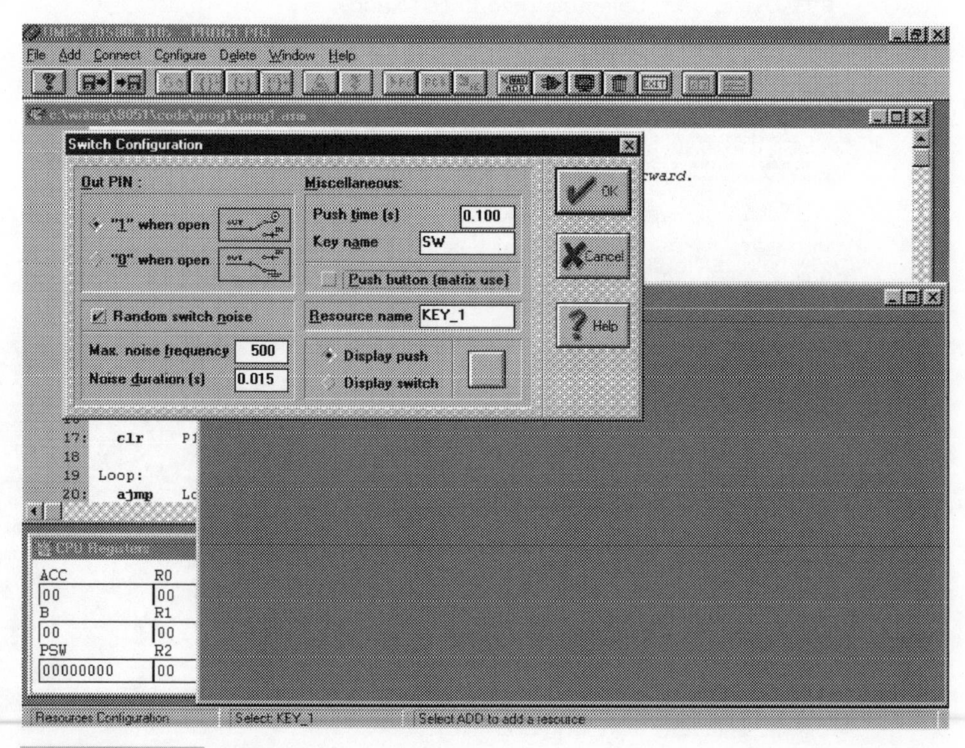

FIGURE 9-12 Specifying the switch type.

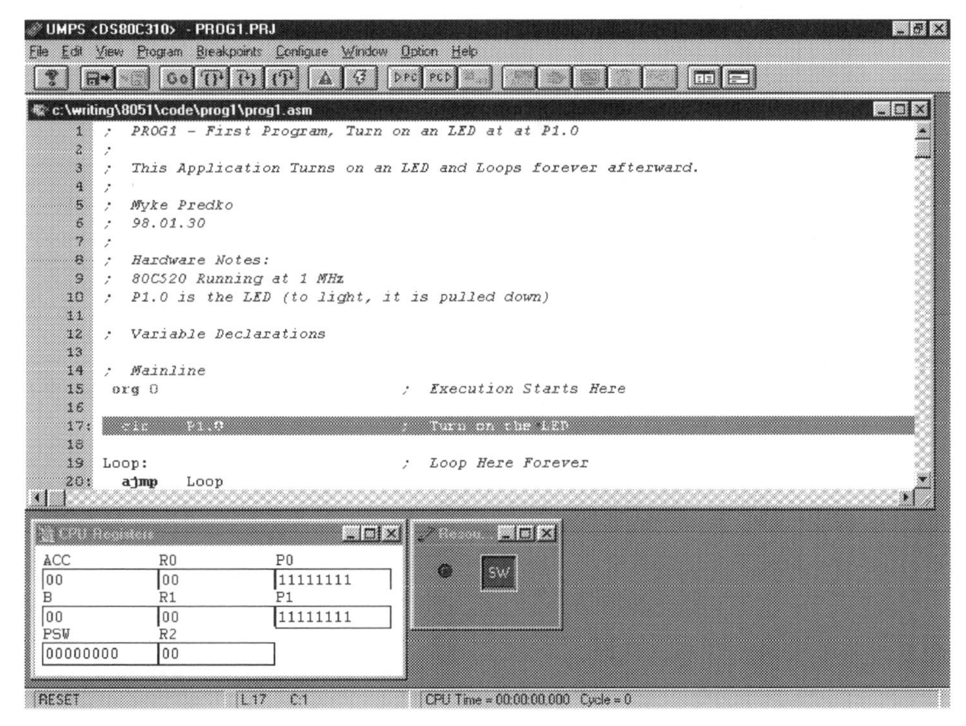

FIGURE 9-13 Ready to execute PROG1.

After clicking on `Ok` in the error window (for a normal assembly, no window will appear), the failing line will be highlighted in red (Fig. 9-14). The UMPS assembler will stop on first error and display it for you to fix it.

Once you have a clean compile (i.e., no error in the ASM Compilation Status window comes up during assembly), you are ready to start simulating your program. To make your screen identical to Fig. 9-13, click on `Program` and `Reset`. By doing this, UMPS has reset to the initial conditions of the microcontroller.

Once this is done, click the single-step icon (the arrow going into the two braces), and notice that the CPU register P1 changes from 011111111b to 011111110b, which indicates that bit 0 (and pin 0) of I/O port P1 has been changed to a zero value. When this happens, you should notice the LED will become brighter.

If you continue clicking on the single-step icon, you will see that you are in an endless loop. Rather than stop here, why don't you click on the Go icon and start program execution. The CPU Time and Cycles counters will start to increment.

While the program is running, click on the button and notice that bit 1 of P1 will change to "0" for a few seconds before changing back. You will see that, in the CPU Registers window, P1 will change to 011111100b and then back to 011111110b after a few seconds. This is showing that the button press is affecting the hardware registers inside the simulated 8051.

Now, what I have shown you here is only a tiny fraction of what UMPS can do for you. As I go through more of the experiments, I will show how additional information

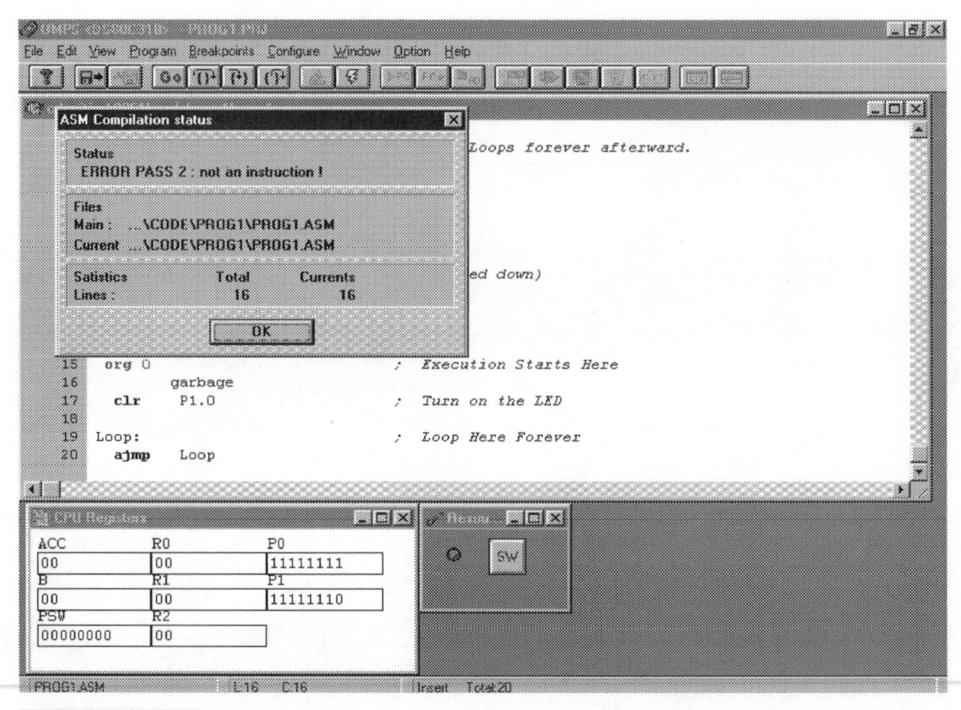

FIGURE 9-14 UMPS assembly error window.

can be displayed, the contents of registers and memory locations changed, and breakpoints are set.

PROG2: Arithmetic Operations

With the capability of being able to assemble source code in UMPS and simulate it's operation, we can now go ahead and start learning how the 8051 executes instructions and programs. In this first experiment, I will show you how arithmetic operations are carried out and point out a few quirks to look out for.

For this experiment (and a number that follow), you can continue to use the CPU Register and Resource windows that were created in PROG1 (this is why I put in the extra registers and hardware resources that weren't required for PROG1).

One of the important concepts to understand when you are doing arithmetic operations is how data moves. To implement the high level statement:

```
Var1 = Var2 + Var3
```

the accumulator (`A` or `ACC` in instructions that require an address) is used as temporary storage for the intermediate result of the statement. The previous statement would be converted (or compiled) into the following instructions:

```
mov       A, Var2          ; Load accumulator with Var2
add       A, Var3          ; Add Var3 to contents of accumulator
                           ;  and store the result back into
                           ;  the Accumulator
mov       Var1, A          ; Store the contents of accumulator
```

For PROG2, I will go through each line individually to explain what is going on. In later programs, I won't do this and rather focus on the new features and concepts that I am going to introduce.

PROG2 is shown in Listing 9-2.

LISTING 9-2 **PROG2.**

```
;  PROG2 - Arithmetic operations
;
;  This application, using the accumulator (A) and the bank registers
;   demonstrates various arithmetic operations.
;
;  Myke Predko
;  98.01.30
;
;  Hardware notes:
;   This program is only meant to run on the simulator.

;  Variable declarations
;   R0, R1 and R2 used as variables

;  Mainline
 org 0                    ; Execution starts here

;  Initialize registers

  mov     R0,#137         ; Load bank register R0 with 137

  mov     R1,#33          ; Load bank register R1 with 33

  mov     R2,#86          ;  Load bank register R2 with 86

;  Execute the data processing instructions

  mov     A,R1            ; A = 33
  add     A,R0            ; A = A + 137
                          ;   = 170 (0AAh)

  add     A,R2            ; 170 + 86 = 256, A = 0, C and OV flags set

  mov     A,R2            ; 86 - 33 = 53, Positive result
  clr     C               ;  NOTE: Clear carry explicitly
  subb    A,R1

  subb    A,R2            ; A = 53 - 86 = -33
                          ;   NOTE: Carry was reset from previous operation

  mov     A,R2            ; Repeat first subtract without clearing carry
  subb    A,#33           ; Result = 86 - 33 - (c) = 52

  mov     A,R1            ; Now, do multiplication
  mov     B,R1
```

LISTING 9-2 PROG2 *Continued.*

```
    mul     AB              ; Do 33 squared with:
                            ;  High byte of result in B
                            ;  Low byte of result in A

    mov     A,R2            ; Get mod 10 of 86
    mov     B,#10
    div     AB              ; Result in "A", remainder in "B"

    xrl     A,R0            ; XOR the bits with values in other registers

    rlc     A               ; Shift the value in accumulator 1x to the
                            ;  left and put the previous carry value in
                            ;  bit 0 and load carry with bit 7

    rr      A               ; Shift the value in the accumulator to the
                            ;  right, but avoid the carry flag

    swap    A               ; Swap the nybbles in the accumulator

  Loop:                     ; Loop here forever when finished
    ajmp    Loop
```

The first three instructions of PROG2,

```
mov     R0,#137         ; Load bank register R0 with 137
mov     R1,#33          ; Load bank register R1 with 33
mov     R2,#86          ; Load bank register R2 with 86
```

are simply used to load three bank registers with initial values. In later experiments, I will present more information on moving data to and from registers and RAM.

I should also point out something that you might have noticed when you looked over PROG2: I don't store the results of these operations anywhere (other than leaving the results in the accumulator). The purpose of PROG2 is to show how arithmetic operations execute, along with explaining the results. Loading and storing data in registers will be explored in later experiments.

The first two instructions of the actual experiment are:

```
mov     A,R1            ; A = 33
add     A,R0            ; A = A + 137
                        ;   = 170 (0AAh)
```

The `mov A,R1` instruction loads the accumulator (which is `A` in the instruction) with decimal 33 (which has been stored in R1).

Next, the contents of R0 are added to the contents of the accumulator. The result of this is 170 decimal (0AAh).

The next instruction:

```
add     A,R2            ; 170 + 86 = 256, A = 0, C and OV flags set
```

will add 86 (decimal) to the current value. The result in the accumulator will be zero, but the important changes are taking place in the PSW.

After the first add instruction, the PSW carry and auxiliary carry flags don't change. After the second add instruction, both the CY (carry) and AC (auxiliary carry or digit carry) flags are set. These bits indicate that the result for each hex nybble has overflowed and incremented the next higher nybble (for the most significant nybble, the carry flag is used to indicate that the next higher byte should be incremented as well). The OV (overflow) flag is set to the same state as the carry flag.

Subtraction works very straightforwardly except for one wrinkle: There is only a subtract with carry (borrow) instruction (addition can take place with or without the carry flag). This means that, when subtracting, you have to be conscious of the state of the carry flag at all times and make sure it is correct before executing a subb instruction.

The first subtraction operation handles this properly by clearing the carry flag before executing the subtract instruction:

```
mov        A,R2           ; 86 - 33 = 53, Positive result
clr        C              ;  NOTE: Clear carry explicitly
subb       A,R1
```

Before subtracting the contents of R1 from the accumulator (which has been loaded with the contents of R2), the carry flag is cleared so it won't affect the result.

The next instruction performs a subtraction that results in a negative result:

```
subb       A,R2           ; A = 53 - 86 = -33
                          ;   NOTE: Carry was reset from previous operation
```

Because I knew the previous instruction would reset the carry flag, I didn't bother explicitly resetting the carry flag before executing the subb instruction.

Is this good coding practice? I would say no, because over time, the code might be modified during maintenance and upgrading to include operations in front of the subtract instruction that have set the carry flag.

To avoid any chance of problems, you can use the "sub" macro:

```
MACRO csub(Parm)
 clr       C
 subb      A,Parm
ENDMAC
```

This macro will execute all the parameter options available for subb, as is shown in the sample invocations:

```
 csub(#12)                ; Subtract decimal 12 from the accumulator
 csub(R2)                 ; Subtract the contents of R2 from the
                          ;  accumulator
 csub(@R0)                ; Subtract the contents of the scratchpad RAM
                          ;  register pointed to by R0 from the
                          ;  accumulator
```

The next two instructions are a repeat of the first subtraction without the clr c instruction before the subb to show what happens when the carry is inadvertently left set:

```
mov        A,R2           ; Repeat first subtract without clearing carry
subb       A,#33          ;  Result = 86 - 33 - (c) = 52
```

The `subb` instruction also uses an explicit value for the subtraction (just to vary things a bit). There are a number of different addressing modes available for each of the arithmetic instructions (as I've pointed out in chapter 3).

Just as an aside, my number-one semantics error is forgetting to put in the "#" (pound) symbol before the explicit value. If you forget the "#" symbol, then the contents of the register at the address of the constant is used instead of the constant. When writing an application, I find it to be a good idea to either use numeric values or if I'm going to use a label for a value, I define it like:

```
DEFINE ConstA #041h      ; Define the letter "A" as a constant
```

By defining the value as a string starting with "#", I can always guarantee that the constant will be loaded correctly.

After executing the `subb` instruction, you will discover that your result is 52 (not 53 as expected and is correct). This is due to the carry being set. If you were to put the `csub` macro in place for all the 8-bit `subb` instructions, then there would never be a problem because of carry being wrong for the operation of the subtract instruction.

Following subtraction, the multiplication instruction is demonstrated. This instruction takes two 8-bit numbers and multiplies them together to get a 16-bit result, which is stored in B:A. This means the high byte is stored in the B register and the low byte is stored in the accumulator.

```
mov      A,R1          ; Now, Do Multiplication
mov      B,R1
mul      AB            ; Do 33 squared with:
                       ;  High byte of result in B
                       ;  Low byte of result in A
```

After loading A and B with 33 (decimal), the `mul AB` instruction is executed and puts the result back into A (containing 041h) and B (containing 004h) as expected.

Division works very similarly to multiplication, although you must keep the dividend straight with the divisor. For the `div AB` instruction, the accumulator is loaded with the number to be divided, and B is loaded with the divisor.

The mental mnemonic that I use to remember how this instruction works is to think of:

```
div      AB
```

as

```
div      A/B
```

The result (the number of times the divisor can be taken away from the number) is located in the accumulator after the instruction. B is loaded with the remainder (modulus) of the operation.

In appendix C, I will go through how the multiply and divide instructions can be used with 16-bit numbers.

Bit operations (`ANL`, `ORL`, and `XRL`) are used to carry out bitwise operations. In PROG2, I do a bitwise XOR (`XRL`) with a register value:

```
xrl      A,R0          ; XOR the bits with values in other registers
```

Like the add and subb instructions, the bitwise instructions modify the contents of the accumulator with the parameter data. But, the bitwise instructions do not affect the "AC", "C" or 0V frags.

There are two different types of rotate instructions (each type capable of rotating to the left, or up, and to the right, or down). One rotates through the carry flag, while the other doesn't.

```
rlc        A              ; Shift the value in accumulator 1x to the
                          ;  left and put the previous carry value in
                          ;  bit 0 and load carry with bit 7
rr         A              ; Shift the value in the accumulator to the
                          ;  right, but avoid the carry flag
```

The first rotate shifts the contents of the accumulator (which is 081h at this point after the div and xrl instructions) and loads the carry flag with the most significant bit (which is set to 1 after the rlc instruction).

The next rotate instruction shifts the contents of A back down without involving the carry flag. This means that the result of shifting up (and storing the most significant bit in the carry flag) and then shifting down external to the carry flag will effectively clear the most significant bit (bit 7) of the initial value.

The last instruction is the accumulator nybble swap instruction:

```
swap       A              ; Swap the nybbles in the accumulator
```

This instruction swaps the high and low nybbles of the accumulator. This instruction is most useful in situations like converting a byte to two hex (0-9 and A-F) characters.

With this experiment, there are a number of instructions that I haven't introduced you to (including the increment and decrement and bit ANDing and ORing). As well, it is impossible to go through every possible input state data combination and permutation for even the instructions that I have presented here. If you are unsure of the operations of these instructions, add them to PROG2, re-assemble the source, and see what happens when they execute.

PROG3: Direct Bank Register Addressing

When I first went through the 8051 architecture, I felt like the 8-byte bank registers weren't that useful. This feeling was based on my experience with other microcontrollers and the need for more than 8 bytes of primary RAM (actually, if R0 or R1 is used as index registers, you have less than 8 bytes), which would be used for fast instructions that required less space. As I introduced in chapter 8, I tried to look at the 8-byte bank register construct as a cache RAM that should only be the most often used variables (along with some scratchpad bytes for temporary storage and operations). For this purpose, 8 bytes are acceptable as a cache for small, embedded microcontroller applications.

In PROG3 and PROG4, I will fully introduce the operations that are possible with the bank registers (which were started in PROG2 and showed how the bank registers were initialized and used in arithmetic operations). In PROG3 (Listing 9-3), I will focus on direct

LISTING 9-3 PROG3.

```
;  PROG3 - Direct bank register addressing
;
;  This program shows the different methods that bank registers (8
;   register addresses in the "current" bank of the data space) can be
;   accessed directly.
;
;  This program will run in the UMPS simulator for a DS80C320.
;
;  Myke Predko
;  98.02.03
;
;  Hardware notes:
;   This program is only meant to run on the simulator.

;  Variable Declarations
;   R0, R1 and R2 used for data/variable storage

;  Mainline
 org 0                          ;  Execution starts here

  mov  R0,#123                  ;  Load bank register 0 with 123

  mov  A,#12                    ;  Load bank register 1 with 12
  mov  R1,A

  mov  A,R0                     ;  Load R2 with the contents of R0
  mov  R2,A

  mov  A,R1                     ;  Swap the contents of R1 and R2
  xch  A,R2

;  Change the banks

  setb        PSW.3             ;  Jump to bank 1

  mov  R1,A                     ;   Store the exchanged value

  mov  A,R0                     ;  Load R2 with the contents of R0
  mov  R2,A                     ;   NOTE: The contents of the bank
                                ;   register has changed

  clr  PSW.3                    ;  Go back to bank 0

Loop:                           ;  Loop here forever when finished
  ajmp       Loop
```

bank register addressing. In the next section (PROG4), I will introduce you to indirect addressing using the bank registers.

The first instruction of PROG3:

```
mov   R0,#123                   ; Load bank register 0 with 123
```

is simply initializing the bank 0 (which is the default) R0 register with a constant. An initialization with a constant (or the same value) can also be accomplished by first loading the accumulator as shown in the next two lines:

```
mov   A,#12                    ; Load bank register 1 with 12
mov   R1,A
```

In these two instructions, the accumulator is first loaded with a constant and then the contents of the accumulator are put into the specified register.

Having already been presented with a single instruction for this register initialization, it's probably not clear why you'd want to initialize a register (or memory location in the first 256 addresses) with two instructions. A single instruction takes fewer clock cycles and fewer bytes.

The reason why you'd want to use the two instruction initialization is if you had to load multiple registers (or RAM) locations with the same constant and it didn't make sense using a loop (i.e., if only two or three locations had to be loaded with the same value).

Finally, the contents of the R0 are saved into the accumulator and then moved into R2.

```
mov   A,R0                     ; Load R2 with the contents of R0
mov   R2,A
```

This is the code that would be generated for the high-level language statement:

```
R2 = R0
```

Before going too far, I should point out that as I went through the code for both this experiment and the next one, I changed the UMPS screen slightly. To display the CPU Register window and the first 256 memory locations, I have added the Internal RAM window to my UMPS window (Fig. 9-15). This is selected from the "View" pulldown in the IDE.

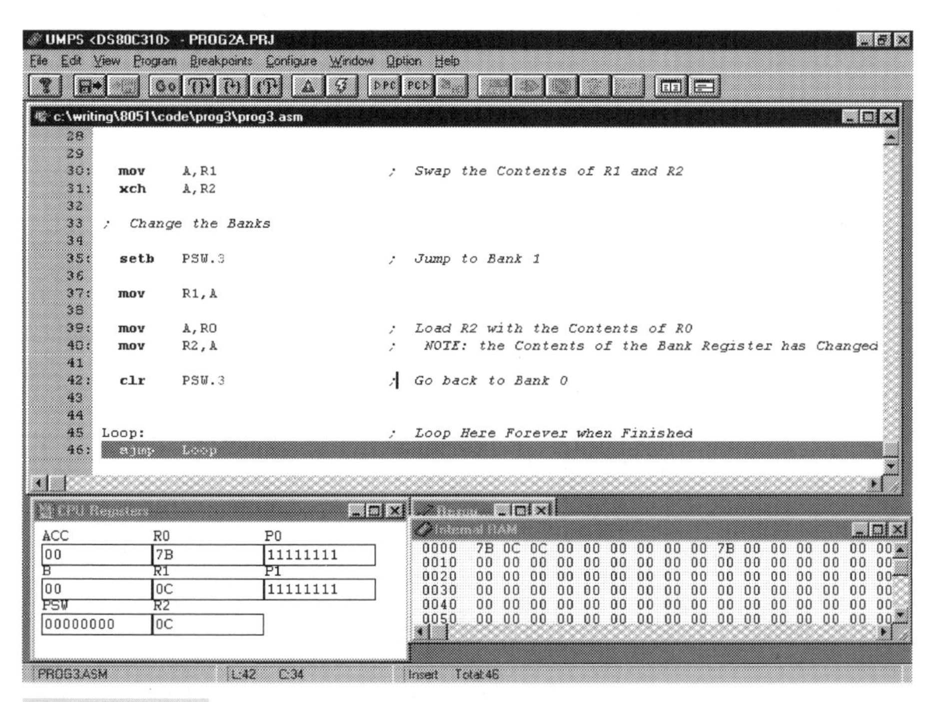

FIGURE 9-15 Adding the Internal RAM window.

Note that, in Fig. 9-15, I have resized the Internal RAM window down to the minimum size required for the two programs (to make sure the referenced locations will be displayed) and put it in the lower right-hand corner of the screen. Now, when I run the two programs, I can monitor what is in the bank registers in the CPU Register window as well as the complete register set including these bank registers.

The next two statements are used to exchange the contents of two registers without requiring the use of a third register.

```
mov    A,R1                    ; Swap the contents of R1 and R2
xch    A,R2
```

Before returning the exchanged value into R1, I play a dirty trick on the program and change the current bank:

```
setb   PSW.3                   ; Jump to bank 1
```

This instruction changes the base address of the bank registers from 0 to 8. When you execute the next instruction:

```
mov    R1,A                    ; Store the exchanged value
```

you will see that the contents of the accumulator (07Bh) will be stored at internal RAM address 9, rather than 1 (as it would be for the default bank).

What I'm trying to show here is, while being able to change the register bank focus is really nice, it can also have some pitfalls if you have a value to save and have lost awareness of what bank is currently executing.

The next two instructions:

```
mov    A,R0                    ; Load R2 with the contents of R0
mov    R2,A                    ;  NOTE: The contents of the bank
                               ;  register have changed
```

show that copying the contents of R0 into R2 is accessing completely different values than if R0 and R1 were in bank 0.

The program ends with the current bank restored to the default and the program going into an endless loop. With this program, there are two final comments that I should make about accessing bank registers and memory locations inside the 8051.

The first is, when the bank register set was changed and the contents of R2 was set to equal the contents of R0, the value transferred was 0. This means that the contents of memory location at bank 1 R0 is 0.

While this is true for the UMPS simulator, this is very rarely true for actual, executing hardware!

Except for the processor registers (such as PSW) with definite power-up values, you cannot assume anything about the state of a register or RAM location (even though, on power up, UMPS shows them as all being set to zero). Chances are that the registers and RAM will not be zero.

So, before reading *any* register or RAM location, make sure that it has been previously initialized. Personally, I like to initialize all my variables at the start of the program, right after I have defined them, to make sure I don't forget to do any.

The second comment I want to make is which bank you normally select when you are running. I like to keep execution running from bank 0 at all times. For this reason, after I complete executing code that runs in another bank, I immediately switch back to bank 0, so I always know where I am executing from.

PROG4: Indirect Register Bank Addressing

A very important part of the 8051 bank registers is their ability to access any indirect RAM byte in the scratchpad RAM. As I have pointed out earlier in the book, the stock 8051 has 128 bytes available for the scratchpad RAM, while other devices (such as the "8052") are capable of accessing 256 bytes, with the upper 128 only being accessible *indirectly*. This means that they are only accessible using the index registers. Directly addressing them will result in the program actually accessing the special-function registers (from 080h to 0FFh).

PROG4 (Listing 9-4) shows how the R0 and R1 registers of the currently active bank can be used for accessing scratchpad RAM. Later in this chapter (and in the book), I will show some examples of using the index registers for accessing tables.

LISTING 9-4 **PROG4.**

```
;  PROG4 - Indirect bank register addressing
;
;  This program shows how bank registers R0 and R1 can be used as 8-bit
;   indirection registers.  Note that for "true" 8051s, reads past
;   0x07F are invalid.
;
;  This program will run in the UMPS simulator for a DS80C320 (which has
;   indirect RAM from 0x080 to 0x0FF).
;
;  Myke Predko
;  98.02.03
;
;  Hardware notes:
;   This program is only meant to run on the simulator.

;  Variable declarations
;  R0, R1 and R2 are used for variable storage
General equ 044h                ;  Variable in RAM area

;   Just bank registers used for data storage

;  Mainline
 org 0                          ;  Execution starts here

  mov   R0,#General             ;  Load bank 0 R0 with general address

  mov   R1,General              ;   Common error, forgetting the "#"

  mov   @R0,#012h               ;  Load "General" with a value
```

LISTING 9-4 PROG4 *Continued.*

```
  mov  A,@R0                       ;  Get the value pointed to by R0

  mov  R1,#ACC                     ;  Load R1 with the offset of A

  mov  R2,A                        ;  Store the current value of "A"

Loop:                              ;  Loop here forever when finished
  ajmp       Loop
```

At the start of PROG4, I attempt to initialize the R0 and R1 registers of the current bank to the address of `General`. There are a few things that I should talk about here.

The first is the declaration of the `General` variable:

```
General equ 044h                  ; Variable in RAM area
```

When you have programmed other devices (such as a PC), declaring a variable has usually meant specifying the space needed for the variable so that code and other programs will not write over the space needed for the variable. This is usually required in Princeton architected processors, but not really required in Harvard architectures, because the code and data segments are separate and only the executing software has to make sure it does not overwrite the variable spaces.

The next two instructions are (as I said before) attempting to initialize the R0 and R1 registers with the address of the `General` variable:

```
mov     R0,#General               ; Load bank 0 R0 with general address
mov     R1,General                ;  Common error, forgetting the "#"
```

The reason why I use the term "attempting" when I refer to initializing R0 and R1 is because the initialization of R1 will not happen as I wanted it to. This is because I forgot the pound sign ("#") before the variable label, which puts the *address* of the label into R0 and not like R1 where the *contents* of the byte at address `General` are loaded in.

This probably seems like a pretty obvious mistake when you see it along side the correct method of loading the address. However, when you have program that doesn't work, you can spend hours staring at this and not seeing the problem.

This problem can become even harder to find if the application needs the variable initialized with zero. In the simulator (which has all the memory initialized to zero), you can't see the problem. However, when you burn a device and try to execute it . . ..

With the variable declarations and initializations out of the way, PROG4 can now begin to transfer data back and forth:

```
mov     @R0,#012h                 ; Load "General" with a value
```

This instruction stores the value 18 (decimal) into the address pointed to by R0 (which was initialized to `General` earlier). The data is passed to the data pointed to by `General` without passing through the accumulator.

```
mov     A,@R0                     ; Get the value pointed to by R0
```

Next, the accumulator is loaded with the contents of `General` (which was set to 18):

```
mov     R1,#ACC                     ; Load R1 with the offset of A
```

With this instruction, R1 is loaded with the address of the accumulator (which, if it is not explicitly defined as part of the instruction, is known as "ACC").

```
mov     R2,A                        ; Store the current value of "A"
```

Finally, R2 is loaded again with the contents of the accumulator. The reason that I put in this instruction was to show that the contents of the accumulator aren't affected by the instructions that bypass it. This is important to remember when you are writing programs and you don't want to necessarily change the contents of the accumulator.

Along with experimenting with the operation of the bank register index registers, I have also shown how data can be transferred behind the accumulator's back. The ability to move and manipulate data without accessing the accumulator is a very powerful feature of the 8051 and one that can simplify and speed up your application. In many other microcontrollers (and processor) architectures, this is not possible because the accumulator must be accessed to read and write data.

PROG6: RAM Direct Addressing

Accessing the scratchpad RAM for variables has been introduced in the previous experiments. In this program (Listing 9-5), I wanted to go through reading and writing from the scratchpad RAM as well as doing some processing on data contained within it.

LISTING 9-5 **PROG6.**

```
;  PROG6 - Accessing scratchpad RAM
;
;  This program reads and writes to the 128 bytes of scratchpad RAM
;   available at the start of the 8051's data space.
;
;  This program will run in the UMPS simulator for a DS80C320.
;
;  Myke Predko
;  98.02.09
;
;  Hardware notes:
;   This program is only meant to run on the simulator.

;  Variable declarations
;   R0, R1, and R2 used for data/variable storage
Var1 EQU 8                          ;  Define variables to be used in the
Var2 EQU 9                          ;   program

VarR0 EQU 0                         ;  This is the same as R0 in bank 0

;  Mainline
 org 0                              ;  Execution starts here
```

LISTING 9-5 PROG6 *Continued.*

```
  mov    R0,#123                      ;  Load bank register 0 with 123
                                      ;   to initialize the value for later

  mov    Var1,#12                     ;  Initialize the RAM variables
  mov    Var2,#34

;  Var1 = Var1 | Var2                - Create assembler code for this
;                                      statement

  mov    A,Var1                       ;  Load accumulator with first parameter
  orl    A,Var2                       ;   bitwise OR accumulator with contents
  mov    Var1,A                       ;   of Var2 and store the result

  mov    VarR0,#Var1                  ;  R0 now points to Var1

  mov    @R0,#0                       ;  Clear values that R0 is pointing to

Loop:                                 ;  Loop here forever when finished
  ajmp   Loop
```

The final result of this program will show you how you must understand how the bank registers interact as part of the scratchpad RAM and how changing one may affect the other.

With already explaining how variables are initialized, I can skip on to how the program works with scratchpad RAM. I like to create a hierarchy of data usage for my 8051 software that will make them as efficient as possible. For variables that are used often, they are located in the current bank registers. For individual byte and 16-bit variables, they are placed inside the scratchpad RAM. Finally, the external memory is used for large memory structures.

The instructions used to execute the high-level language statement:

```
;  Var1 = Var1 | Var2                - Create assembler code for this
;                                      Statement
  mov    A,Var1                      ; Load accumulator with first parameter
  orl    A,Var2                      ;  Bitwise OR accumulator with contents
  mov    Var1,A                      ;  of Var2 and store the result
```

show how easily scratchpad variables can be used instead of bank registers for processing data. The only drawback of using the scratchpad variables for this is the extra instruction cycle and byte needed to process them.

The next two instructions show what happens if you aren't careful with keeping the bank registers separate from the scratchpad RAM variables:

```
mov      VarR0,#Var1                 ; R0 now points to Var1
mov      @R0,#0                      ; Clear values that R0 is pointing to
```

In these two instructions, `VarR0` is defined to be at address 0, which is also the address of bank 0 register R0. When `VarR0` is written to, the contents of R0 are lost, making the next instruction execute incorrectly (R0 was originally loaded with 123 decimal).

PROG7: Bit Addressing

The ability to manipulate bits is one of the aspects of the 8051 that really sets it apart from other microcontrollers (and microprocessors). Thinking about the bit commands can really help make your applications much more efficient and easier to write.

PROG7 (Listing 9-6) is really an introduction to the bit operations. While I don't plan on burning it into a 8051, it is a very useful application for seeing how hardware can interact in the 8051.

To run PROG7, I put the Internal RAM window behind the switch and button I put into the resource window when I first set up UMPS (Fig. 9-16).

LISTING 9-6 **PROG7.**

```
;  PROG7 - Accessing bits
;
;  This program reads and writes individual bits in the bit addressable
;   area of the scratchpad RAM (Addresses 020 to 02F) and the I/O      .
;   ports.
;
;  This program will run in the UMPS simulator for a DS80C320.
;
;  Myke Predko
;  98.02.09
;
;  Hardware notes:
;   This program is only meant to run on the simulator.
$define LED P1.0                   ;  Virtual (real?) LED
$define Button P1.1                ;  Button that will be used for LED

;  Variable declarations
Bit1 EQU 7                         ;  Bit 7 of address 020h
Bit2 EQU 9                         ;  Bit 1 of address 021h

;  Mainline
.org 0                             ;  Execution starts here

  clr    Bit1                      ;  Reset/set bit 7 at address 020h
  setb   Bit1

  clr    C                         ;  Clear the carry flag

  orl    C,Bit1                    ;  OR the carry flag with Bit1 and save
                                   ;   result in carry

  mov    Bit2,C                    ;  Save the result
                                   ;   to initialize the value for later

Loop:                              ;  Loop here forever when finished

  mov    C,P1.1                    ;  Read the button value
  mov    P1.0,C                    ;  If pressed, light the LED

  ajmp   Loop
```

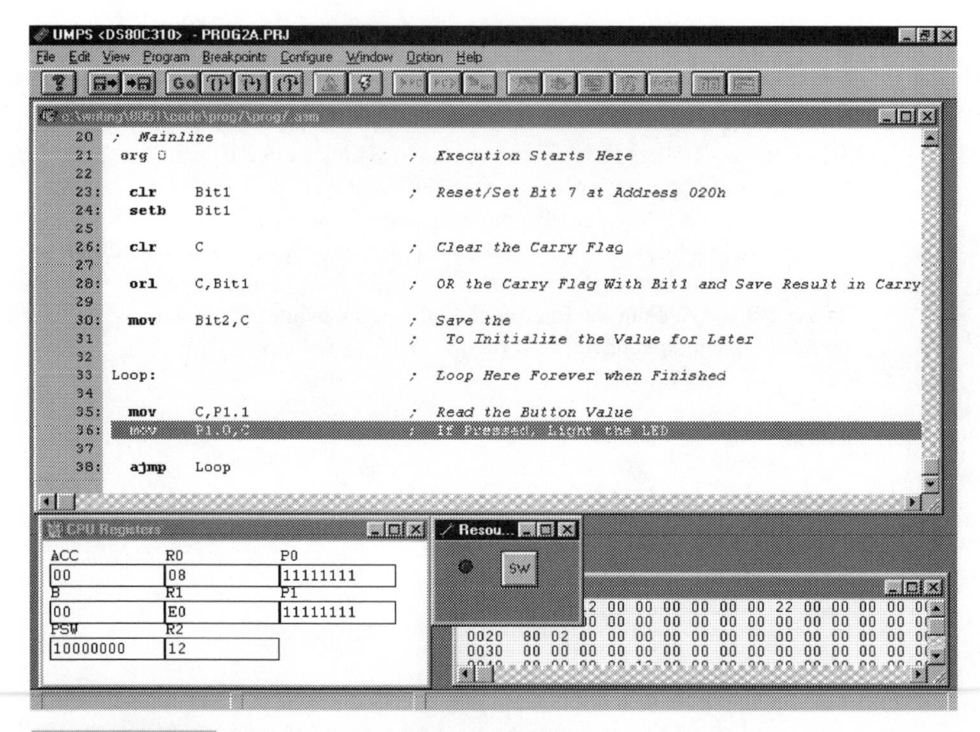

FIGURE 9-16 UMPS arrangement for PROG7.

The first five lines of the program:

```
clr      Bit1                    ; Reset/set bit 7 at address 020h
setb     Bit1
clr      C                       ; Clear the carry flag
orl      C,Bit1                  ; OR the carry flag with Bit1 and save
                                 ;  result in carry
mov      Bit2,C                  ; Save the result
                                 ;  to initialize the value for later
```

simply show how bits can be set or reset in the bit area of the scratchpad RAM (addresses 020h to 02Fh). The first two instructions reset and then set a bit. This sequence would probably be useful for toggling an I/O bit.

The next three-instruction sequence shows how bit ANDs and ORs can be used to modify the value of a bit. In the 8051, the carry flag can be used for temporary storage of a bit (this will be shown to greater advantage in the next two instructions), and it is first loaded with a value that is ANDed with another bit. The result of this operation is stored into the next bit.

After these instructions, you'll see the data line:

```
0020 80 02
```

in the Internal RAM window, which indicates that bit 7 (located in scratchpad address 020h) is set along with bit 1 of scratchpad address 021h. Typically, in all assembly language programming, data is displayed in right-to-left formats (as regular numbers are displayed). However, in cases like this, this really isn't appropriate. I personally think of the bit space in the 8051's scratchpad RAM as a contiguous set of bits, starting at 0 and going to 07F, with no byte boundaries.

So, for this program, with bit 7 and bit 9 set, I try to visualize the data as:

```
Bit Space: 1000 0000 0000 0010 0000 ...
```

I group the bits in sets of four to make it easier for me to understand where I am. However, if you look at the bits, you will see that the relationship between bits 7 and 9 is much easier to understand and might help you when trying to work with the bits in your own applications.

In PROG7, I have changed the endless loop at the end of the program:

```
Loop:                              ; Loop here forever when finished
 mov     C,P1.1                    ; Read the button value
 mov     P1.0,C                    ; If pressed, light the LED
 ajmp    Loop
```

In this loop, I continually poll the button (at bit P1.1) and store its contents into the carry flag (using it as temporary storage) and then moving it to the LED. This means that, when the button bit of P1 is pulled low (when it is pressed), this value is transferred over to the LED bit so that, when the button is pressed, the LED will light *without* affecting any other bits.

This ability to move bits of data in the 8051 is extremely useful and can make many microcontroller programming tasks easier. Often, in a microcontroller, you will have to transfer a bit to a new location, this could be represented as:

```
Bit2 = Bit1
```

In the 8051, as you saw earlier, this is simply implemented as:

```
 mov     C,P1.1                    ; Read the button value
 mov     P1.0,C                    ; If pressed, light the LED
```

In other microcontrollers, to move a bit without affecting other bits, the process becomes much more cumbersome. For example, in the 8051, if you didn't have bit operations and wanted to move P1.1 to P1.0, you would have to use the following code:

```
 mov     A,#1                      ; Set P1.0 high before polling bit
 orl     P1,A
 mov     A,P1                      ; Get P1 and shift bit 1 over to load
 rr      A                         ;  into bit 0
 anl     A,#1                      ; Isolate bit 0 so it can be loaded
                                   ;  into P1
 anl     P1,A                      ; Now, set bit 0 to the value read from
                                   ;  bit 1
```

This is actually pretty close to a general case, which means that most operations will be quite complex and difficult to optimize to fewer bytes or instruction cycles.

Understanding when and how to use the bit instructions is something that you have to get into the habit of because designing your application to use this feature of the 8051 can improve your code immensely. In the later experiments and example applications, I will point out situations where using the bit operations has simplified the application.

PROG8: The DPTR Pointer Register and External Memory

One of the most confusing features of the 8051 is the DPTR register. For some reason, it does not seem to be very well documented. DPTR is a 16-bit index register that provides access to external memory.

One of the features of the Dallas Semiconductor 8051s HSM processors is the inclusion of two DPTRs (the two 8-bit registers located at addresses 084h and 085h).

PROG8 (Listing 9-7) shows how writing to external memory using the DPTR register is accomplished.

Setting the value of DPTR is accomplished by loading the DPL and DPH registers (at addresses 081h and 082h, respectively) or by using the:

```
mov     DPTR,#Constant
```

instruction, which loads the two 8-bit registers with a 16-bit constant value.

LISTING 9-7 **PROG8.**

```
;  PROG8 - Using the DPTR to access data
;
;  This program uses the 16-bit DPTR register to access scratchpad RAM.
;
;  This program will run in the UMPS simulator for a DS80C320.
;
;  Myke Predko
;  98.02.09
;
;  Hardware notes:
;   This program is only meant to run on the simulator.

;  Variable declarations
;   No explicit variables are used for this program.

;  Mainline
 org 0                                   ;  Execution starts here

  mov    DPTR,#044h                      ;  Point the DPTR to a location for data

  mov    A,#12                           ;  Store a sample message
  movx   @DPTR,A                         ;   in external memory

  mov    A,#34

  movx   A,@DPTR                         ;  Retrieve the message from memory

Loop:                                    ;  Loop here forever when finished
  ajmp   Loop
```

To run this example program, you will have to View the external RAM in UMPS. When you single-step through, arrange the External RAM window so that the contents of address 044h can be seen easily. (I put it in the lower right-hand corner of the UMPS window, over the Internal RAM and LED/switch Resource windows).

PROG8 can be described pretty simply. First, DPTR is initialized to point to address 044h:

```
mov     DPTR,#044h              ; Point the DPTR to a location for data
```

Next, this address in *external memory* is written to with a 00Ch (12 decimal):

```
mov     A,#12                   ; Store a sample message
movx    @DPTR,A                 ;  in external memory
```

Note that the movx instruction is used instead of the mov instruction to write to external memory. While it is only possible to use movx with DPTR (which allows up to 64K of external memory), movx is available for creating a shadow scratchpad RAM that is accessible by the bank register indexes.

If you viewed the External Memory window, you can see that the contents of address 044h in the window were changed to 00Ch after this instruction.

The accumulator is given a new value:

```
mov     A,#34
```

so that, when the contents of the address pointed to by the DPTR register are read back into the accumulator:

```
movx    A,@DPTR                 ; Retrieve the message from memory
```

the values written to the accumulator cause the value already in there to be overwritten.

As an aside, when you're writing and reading back a register/byte/whatever, it's always a good idea to make sure that the accumulator's contents are always reset between the write and the read. This way, if the read happens successfully, you can see the change in the new value in the simulator, if the read could not be accomplished, then the accumulator will have the wrong value.

As I said before, the movx instruction is also available with one of the 8-bit current bank registers as the index to external memory. To demonstrate this, I modified PROG8 (Listing 9-8) to use R0, rather than DPTR. PROG8A (as I called this new program) will run almost exactly the same way as PROG8, except that the external memory address that I used was 043h instead of 044h as in PROG8. If you were to run this program on actual hardware, you'd find that it would take the same number of cycles to access the external memory as in PROG8.

It is important to note that this program could access the current bank registers or variable addresses in the scratchpad RAM and not read from or write to the scratchpad RAM. When I first described the 8051 and the variants that actually had 384 addresses in the first 256 addresses of the processor, I should have really pointed out that, with the movx instruction and external RAM, there are actually *640* addresses available in the first 256 addresses of the 8051.

LISTING 9-8 PROG8A.

```
; PROG8A - Using the bank registers as indexes into external memory
;
; This program uses R0 as an 8-bit index register to access scratchpad
;  RAM.
;
; This program will run in the UMPS simulator for a DS80C320.
;
; Myke Predko
; 98.02.12
;
; Hardware notes:
;  This program is only meant to run on the simulator.

; Variable declarations
;  No explicit variables are used for this program.

; Mainline
 org 0                          ; Execution starts here

  mov    R0,#043h               ; Point to different address in
                                ;  external memory

  mov    A,#12                  ; Store a sample message
  movx   @R0,A                  ;  in external memory

  mov    A,#34

  movx   A,@R0                  ; Retrieve the message from memory

Loop:                           ; Loop here forever when finished
  ajmp   Loop
```

While this seems confusing, it really isn't. The first 128 addresses of the 8051 data space can be accessed either through direct or indirect addressing (using `mov` in both cases) and through external RAM indirect addressing (using `movx`). The second 128 addresses of the 8051 data space are accessible as special-function registers using direct addressing (`mov A,`*`Register`*), 128 addresses of RAM using indirect addressing (`mov A,@R0`) and 128 addresses of external RAM (using the `movx` instruction).

PROG8 and PROG8A are really not about the DPTR register and its operation. They are really demonstrating the `movx` instruction, which is used to execute an indexed read or write or external memory and how that memory is architected.

PROG5: Jumping Around and Control Store Pages

When I work with my 11-year-old son on his science projects, one of the biggest problems I have is to get him to understand that experiments are not designed to discover something; they are designed to *confirm* a hypothesis or provide additional insight to the experimenter about how something works. A properly thought-out experiment does not deliberately lead the experimenter to a new train of thought (unless the theory behind the expected values of the experiment was faulty and a new theory has to be developed).

For this reason I do not feel immodest when I say that, for pretty well all the experiments presented in this chapter, the behavior of the 8051 that I have expected when creating the experiment is what I observed. However, this isn't to say that, for some experiments, I haven't received a few surprises and a better understanding of how things actually work.

PROG5 has an example of how creating the 8051 experiment has lead me to a better understanding of how UMPS works. This shouldn't be surprising at all because, for complex software products like the UMPS IDE, there are bound to be some issues that you can't predict or plan for and the documentation won't explain.

For 8051, when doing straight (non-conditional) "gotos" to new addresses, there are generally two ways of allowing the programmer to input the `jmp` instructions. There are three different non-conditional jumps in the 8051 instruction set: `sjmp`, which can jump +127 or –128 from the address of the next instruction; `ajmp`, which can jump anywhere within the current 2K code page; and `ljmp`, which can jump anywhere in the 8051's program space (up to 64K addresses).

The first method is to allow the programmer to just specify a `jmp` with the assembler figuring out which instruction (`sjmp`, `ajmp`, or `ljmp`) is appropriate. In my mind, this is the best way of programming the 8051 because the programmer is not burdened with the task of figuring out which jump operation is appropriate and, as the program grows and shrinks, the correct jumps are inserted by the software without the programmer's knowledge or concern.

The second method, you've probably guessed, is to make the programmer specify the `jmp` instruction to be used to change execution to a new address. While in the previous paragraph, I've probably made it sound like it's a lot more complex and error prone to have the programmer specify the `jmp` type, it's really not that bad.

It's really not that bad because 95% of the time, you will be using the `ajmp` instruction because that will allow jumps anywhere within the current 2K instruction address page. Normally this instruction is always used, except when you have to jump outside the current 2K page (in which case the `ljmp` instruction is used).

PROG5 (Listing 9-9) is an experiment that I have created to show how the different `jmp` instructions work. I wanted to see if my hypothesis that, if an `ajmp` instruction was given the offset in another code page, it would go to the wrong address. The PROG5 code is the final version of the experiment, with the modifications that I had to make to prove my `ajmp` hypothesis.

LISTING 9-9 **PROG5.**

```
;  PROG5 - Jumping around
;
;  This program demonstrates how jumping is done in the 8051.
;
;  This program will run in the UMPS simulator for a DS80C320.
;
;  Myke Predko
;  98.02.03
;
```

LISTING 9-9 PROG5 *Continued.*

```
;  Hardware notes:
;   This program is only meant to run on the simulator.

;  Variable declarations
;   No registers of variables are required.

;  Mainline
 org 0                            ;  Execution starts here

  sjmp       Short_Label          ;  Jump to a short label

 org 10

Back_Label:

  ajmp       Valid_Label          ;  Jump within the page

 org 20

Short_Label:                      ;  Showing how to jump to different
                                  ;   locations - short jump back a bit

  sjmp       Back_Label

 org 30

Valid_Label:                      ;  "ajmp Valid_Label" will come here

;  ajmp      Invalid_Label        ;  Invalid jump to next page
;  db     021h                    ;  Make my own jump to 0100h
;  db     000h

  ljmp       Invalid_Label        ;  Valid jump to next page

 org 0100h

;  #### - End up here, wrong address for "ajmp", should have used "ljmp"

Bad_Loop:                         ;  In the wrong section of code
  ajmp       Bad_Loop             ;   Have to look at your "jmp" statement
                                  ;   again

 org 0900h                        ;  In the next code page (over 2048)

Invalid_Label:                    ;  This is only accessible via "ljmp"

Loop:                             ;  Loop here forever when finished
  ljmp       Loop
```

The first three instructions, even though they do not execute in order, demonstrate how short jumps work. This will be important for understanding the operation of the instructions in the next experiment.

The short jump adds the offset to the program counter value of the address following the instruction. This offset can be from –128 to +127, so:

```
 org 0                                  ; Execution starts here
 sjmp        Short_Label                ; Jump to a short label
  :
 org 20
Short_Label:                            ; Showing how to jump to different
                                        ;  locations - short jump back a bit
```

will add the offset of `Short_Label` to the end of the current `sjmp` instruction to the contents of the program counter. The value of the offset is 18 because the end of the `sjmp` instruction is address 2 and `Short_Label` is at address 20 (as defined by the `org` statement).

If you were to look back at the `sjmp` instruction definition, you would see that for this case, you would get the machine code bytes:

```
080h 012h
```

which is the same as what the UMPS assembler will create for `sjmp Short_Label`. This can be confirmed by viewing the CPU Code window in UMPS (Fig. 9-17).

From `Short_Label`, program execution jumps back to `Back_Label`:

```
 org 10
Back_Label:
 ajmp        Valid_Label                ; Jump within the page
 org 20
```

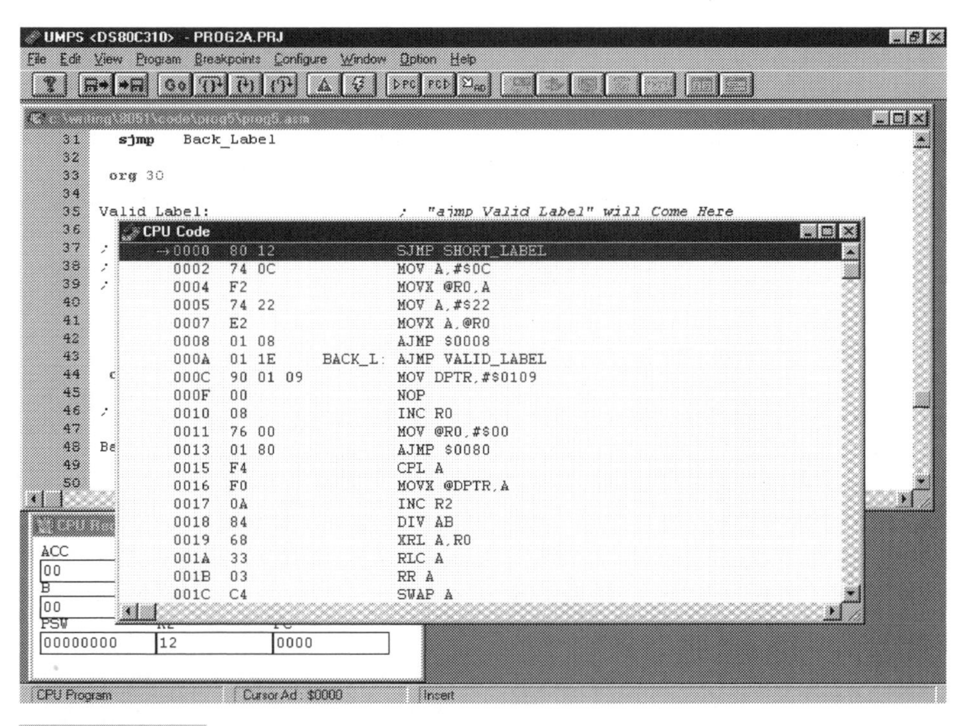

FIGURE 9-17 UMPS CPU Code window with `sjmp`.

```
Short_Label:                    ; Showing how to jump to different
                                ;  locations - short jump back a bit
 sjmp       Back_Label
```

Now, if you were to figure out what would be the correct offset to add to the program counter to jump to `Back_Label`, you'd find that it's a bit harder. This is because a negative number is added to the program counter to get the new value.

After reading in the `sjmp Back_Label` instruction, the 8051's program counter will be at 22. However, `Back_Label` is at address 10. To get to `Back_Label`, –12 has to be added to program counter.

To get the instruction opcode, you will have to convert –12 decimal to –12 hex. This can be done using the formula:

Negative = (Positive ^ 0FFh) + 1

So, to get the value for –12:

–12 = (12 ^ 0FFh) + 1
= (00Ch ^ 0FFh) + 1   // I've converted 12 decimal to C hex
= 0F3h + 1
= 0F4h

which means the machine code bytes are:

```
080h 0F4h
```

This is exactly what you'll find if you look at the instruction starting at address 20 decimal (014h) in the CPU Code window.

After reading all this, you're probably wondering why I went through this level of detail for figuring out the offset when the assembler will do it for you.

There are two circumstances that you might need this skill of understanding how to generate your own `sjmp` offsets. The first circumstance is when you are generating a program counter table and using the `movc A,@A+PC` and you have to figure out the offset from a specific location. The second is when you are patching code by hand to be loaded into the device.

You'll probably also wonder why I went through all this with the next comment I am going to make: I almost never use the `sjmp` instruction in my 8051 programming.

The reason is pretty simple: It's not that useful when compared to `ajmp` (both instructions take the same amount of code bytes and the same number of instruction cycles), but `ajmp` allows jumping to much wider address range. If I have to jump over a page, while `sjmp` will allow it, `ljmp` will make sure there's no confusion.

So why did I go through all this? I did it because the conditional jumps (which are explored in the next experiment) use the short jump format. By going through the short jump instruction here, I can explain how the conditional jumps work without worrying about how the short jump works.

With short jumps out of the way, it's now time to turn our attention to the `ajmp` instruction. This is by far the most common jump that you will see when you look at other people's 8051 code.

The reason for this is simply that there is much less of a danger of going outside the range of the offset within the current page, than there is of going outside the range of the `sjmp` instruction.

So, after showing how the `sjmp` instruction works at `Back_Label`, I show how the `ajmp` instruction can be used for a simple jump from a location to another.

Things begin to get hairy at this point, and you probably are seeing things that are unfamiliar and don't make any sense:

```
 org 30
Valid_Label:                     ; "ajmp Valid_Label" will come here
; ajmp        Invalid_Label      ; Invalid jump to next page
; db          021h, 000h         ; Make my own "ajmp" to 0900h
 ljmp    Invalid_Label           ; Valid jump to next page
 org 0100h
; #### - End up here, wrong address for "ajmp", should have used "ljmp"
  :
 org 0900h                       ; In the next code page (over 2048)
Invalid_Label:                   ; This is only accessible via "ljmp"
```

The purpose of the first four instructions after `Valid_Label` (three of which are commented out in this code) was to see what happens if `ajmp` is used to jump to a label in another execution page. In each of the three cases, I am attempting to jump to `Invalid_Label`, which is at offset 0100h in the code page starting at address 0800h.

In the first instruction, I tried to observe what the problem was by taking the comment off `ajmp Invalid_Label`, but the UMPS assembler returned the error message: `ERROR PASS 3: DATA value out of range!`. This means that the assembler recognized that the label to jump to was out of range and would not allow the program to assemble.

This is a good thing and something that all 8051 assemblers should do. However, it was a bit unexpected because I wanted to show how the `ajmp` instruction would jump to the same offset in the current page. My solution to the problem follows.

Next, I commented out the `ajmp` instruction and allowed the `ljmp` (long jump) instruction to be used as the source. As was expected, when I stepped through the program, I found that it worked as it should. Program execution jumps into the next page.

This left me in a bit of a quandary as to how to show that the `ajmp` instruction would change execution to the wrong location (right offset, wrong code page). I decided to bypass the assembler all together and the insert hand assembled code:

```
db          021h, 000h          ; Make my own "ajmp" to 0900h
```

If you look at the instruction's bit pattern, you will see that the lower 8 bits of the goto address are stored in the second byte of the instruction, while bits 10 through 8 are stored as the most significant three bits of the first byte of the instruction.

The previous instruction will change execution to offset 0100h in the current code page, which is correct, because address 0900h is at offset 0100h in its code page. If the `ajmp` instruction is given the offset to an address outside of its current code page (and the assembler doesn't give you an error message), then you will jump to the 11-bit offset within the new page in the current, which will just about always be in an error and will probably cause the execution of the 8051 to lock up or work erratically.

As for hand assembling instructions and putting in the bytes as data and letting the 8051 execute them, is this any kind of practice that I would suggest for you?

No way. I have used this technique quite a bit in the past when the 80386 first came out and the compilers/assemblers of the time could not create protect mode (32 bit) 80386 instructions. Hand assembling and inserting code manually into a program is only appropriate in situations where the assembler doesn't support an instruction (as in the 80386 situation) or you want to force a failure (as in this case).

PROG9: Conditional Branching

Conditional branching in the 8051 can be very powerful and can vastly simplify the programming of applications. However, to really take advantage of these capabilities, you have to understand the 8051 very well and have enough knowledge to look at things in different ways to see opportunities to improve the performance of your code.

Going through PROG9 (Listing 9-10) will not give you the information and skills necessary to really create funky control structures in the 8051, but it will give you some exposure to program execution changes based on different conditions. As I go through the example applications, I hope I can give you some insight into how to look to improve how you use conditional operations.

LISTING 9-10 PROG9.

```
;  PROG9 - Conditional branching
;
;  This program uses the simulator to demonstrate how conditional
;   branching can be accomplished in the 8051.
;
;  This program will run in the UMPS simulator for a DS80C320.
;
;  Myke Predko
;  98.02.09
;
;  Hardware notes:
;   This program is only meant to run on the simulator.

;  Variable declarations
;  R0 in bank 0 is used as an index register.

Temp EQU 020h                       ;  Temporary variable (in bit memory)

;  Mainline
 org 0                              ;  Execution starts here

  mov    Temp,#084h                 ;  Init variable for later checking

  clr    A                          ;  Do a jnz if "A" is not zero

  jnz    NZLabel                    ;  If A != 0 goto NZLabel

  nop                               ;  End up here if "A" == 0
```

LISTING 9-10 PROG9 *Continued.*

```
NZLabel:
  mov     R0,#Temp                 ;  Have R0 point to Temp
  cjne    @R0,#0,CNZLabel          ;  Jump to CNZLabel, if @R0 != 0
  nop                              ;  Execute this NOP if @R0 == 0
CNZLabel:
  cjne    A,#0,CNELabel            ;  Jump if "A" does not equal zero
  nop                              ;  Execute this if "A" does equal zero
CNELabel:
Loop:                              ;  Loop here forever when finished
  ajmp    Loop
```

Before going through PROG9, I should point out two things. In the 8051, there are two types of conditional branches. The first is the branches based on the current state of the PSW, the accumulator, or bits, while the second type compares one value to another. The branches based on the current state of the processor not only use the PSW and accumulator, but also the state of bits elsewhere in the processor. I'll demonstrate this latter type of conditional branch later in the experiments.

When I go through the code here, I will only really discuss the different branch instructions. To show whether or not the branch was taken, I have placed `nop` instructions after the branch instructions so that a change in execution can be observed.

After a scratchpad RAM variable and the accumulator are initialized, the first conditional branch instruction encountered is:

```
jnz      NZLabel                 ; If A != 0 goto NZLabel
```

This instruction compares the contents of the accumulator to zero and branches if it is not equal to zero. The instruction before this one clears the accumulator, so the branch will not be taken. This instruction is indicative of the first type described earlier, where the current state of the processor is used for determining whether or not a change in program execution should be made.

Along with jumping based on whether or not the contents of the accumulator are equal to zero, branches can also be based on the state of the carry flag as well as bits in the 128 bits in the scratchpad RAM from 020h to 02Fh and in the special-function registers.

Next, the program loads R0 of the current bank registers with the address of `Temp` and compares the contents of `Temp` (as pointed to by R0) to zero:

```
cjne     @R0,#0,CNZLabel         ; Jump to CNZLabel, if @R0 != 0
```

When this instruction compares the contents of `Temp` to zero, no registers or bits in the 8051's processor are affected. This is partly due to the lack of a zero flag in the PSW (I have continually debated over whether or not this is a good thing).

Not having a zero flag means that doing a compare and then loading up the accumulator before making a conditional branch is not possible. However, with the compare on external values, I'm not sure if I have any reason to complain. If I wanted to do a jump on the results of a comparison that took place before the accumulator was loaded with a different value, I could store the result of the comparison in the scratchpad RAM before changing the contents of the accumulator and then executing the compare and jump-if-not-equal instructions.

If I didn't want to change the value of the accumulator, I could compare the contents of the accumulator directly.

```
cjne     A,#0,CNELabel              ; Jump if "A" does not equal zero
```

In this last compare, I am checking to see if zero is equal to a constant value (in this case zero). Now, a `jnz Label` could be used instead of the `cjne`, but that is only for this case where I am comparing the contents of the accumulator to zero.

PROG10: Loop Control

One of the conditional branch instructions that I have not explored in the 8051 is the loop-control instruction (`djnz`). This instruction is used to allow you to count down inside of a loop and execute repeatedly until the count is equal to zero.

This instruction allows pretty simple implementation of "for" loops and are pretty easy to understand. In PROG10 (Listing 9-11), I first turn on the LED and then enter into the delay loop:

```
Dlay:                          ; Loop here 17x while LED is on
; #### - Inside loop code can go here
 djnz        R0,Dlay           ; Decrement the loop counter and,
                               ;  if not zero, loop around again
```

LISTING 9-11 **PROG10.**

```
;  PROG10 - Looping
;
;  This program uses the simulator to demonstrate how a set number
;   of loops can be simply accomplished in the 8051.
;
;  This program will run in the UMPS simulator for a DS80C320.
;
;  Myke Predko
;  98.02.09
;
;  Hardware notes:
;   This program is only meant to run on the simulator.
```

LISTING 9-11 PROG10 *Continued.*

```
;  Variable declarations
;  R0 in bank 0 is used as a counter.

;  Mainline
 org 0                          ;  Execution starts here

  clr    P1.0                   ;  Turn on the LED for the loop

  mov    R0,#17                 ;  Want to loop 17x

Dlay:                           ;  Loop here 17x while LED is on

;  #### - Inside loop code can go here

  djnz    R0,Dlay               ;  Decrement the loop counter and,
                                ;   if not zero, loop around again

  setb    P1.0                  ;  Turn off the LED

Loop:                           ;  Loop here forever when finished
  ajmp    Loop
```

In this loop, when the `djnz` instruction is encountered, R0 of the current bank is decremented. If the result of the decrement is not zero, then the branch is taken. In PROG10, R0 was initialized with decimal 17, which means that the `djnz` will branch 17 times before it will skip the branch and continue on with the program (which means turn off the LED and go into the endless loop).

PROG11: Stack Operations

For the next few experiments, I will show you how the stack works in the 8051. After reading through the `push`, `pop`, and subroutine call instructions, using the stack seems pretty simple. It actually is (hopefully I didn't scare you with this build up). However, there is one thing that you have to watch out for: Make sure the stack does not write over program variables stored in the scratchpad RAM.

Even if you've never written assembler before, understanding how pushes and pops work is pretty straightforward. A `push` puts a parameter onto a stack and increments the stack pointer to the next available stack element. A `pop` decrements the stack pointer to the previous parameter put on the stack and returns it.

In the 8051, the stack is located in the scratchpad RAM. This means that stack data can be pushed or popped anywhere in scratchpad RAM. Upon power up, the stack pointer is set to address 007h.

In PROG11 (Listing 9-12), a variable is set up with a specific pattern and then 8 bytes of data is pushed onto the stack. While stepping through this program, you might want to view the internal RAM while you step through the program.

In `StackLoop`, an incrementing value is pushed onto the stack. Pushing this data onto the stack writes over `TestVariable` after the third `push` (which has a parameter of 002h).

LISTING 9-12 PROG11.

```
;  PROG11 - Stack operations
;
;  This application uses the simulator to demonstrate how the 8051's
;   stack can be used and some pitfalls to watch out for.
;
;  This program will run in the UMPS simulator for a DS80C320.
;
;  Myke Predko
;  98.02.12
;
;  Hardware notes:
;   This program is only meant to run on the simulator.

;  Variable declarations
;  R0 in bank 0 is used as a counter/GP variable.
TestVariable EQU 00Ah           ;  Value that can be overwritten by
                                ;   stack

;  Mainline
 org 0                          ;  Execution starts here

  mov    TestVariable,#055h     ;  Load test variable with a dummy value

  clr    A                      ;  Start saving data from the stack

  mov    R0,#8                  ;  Just loop around 3x to make sure
                                ;   "TestVariable" is trashed
StackLoop:                      ;  Push contents of accumulator

  push   ACC                    ;  Push the contents of the accumulator

  inc    A                      ;  Change the value of "A"

  djnz   R0,StackLoop           ;  Loop around 3x

  mov    A,TestVariable         ;  Note that TestVariable HAS changed

Loop:                           ;  Loop here forever when finished
  ajmp   Loop
```

When you are developing software for the 8051, I find that it's a good idea, as one of the first operations that you perform, to set the stack pointer *above* all the variable scratchpad RAM (i.e., setting the stack pointer to 06Fh will mean that there are 16 bytes available for the stack [from 070h and 07Fh] in the 128 bytes of scratchpad RAM). This will mean that there is very little chance of your stack overwriting variables (although, if you go beyond the end of the scratchpad RAM, you will loose the data pushed onto the stack) unless you are pushing data onto the stack in an endless (or very large) loop without popping any data back off.

Actually, if I had a program running amok, I find that locating a problem with a subroutine return not being correct is easier to debug than finding a problem where the stack has overwritten variables. The stack going into RAMless addresses will fail much more consistently and logically than a variable being overwritten with a value that might or might not be invalid.

To prevent the stack from overwriting data or going beyond the bounds of memory, make sure that, when you specify the data locations, you understand how large the stack can grow to, in the worst case, and make sure there is sufficient space for it.

PROG12: Stack Arithmetic Operations

Using the stack for saving parameters for an operation is often used in high-level languages where statements like:

```
i = j + Increment( k ) / 7;
```

are common.

Using the stack to store arithmetic parameters can also be used in assembler programming, with you acting as the compiler. While this sounds somewhat weird, it is, nevertheless a very useful way of programming very complex operations (like the previous one).

LISTING 9-13 PROG12.

```
;  PROG12 - Stack arithmetic operations
;
;  This application uses the simulator to demonstrate how high-
;   level language ("C") statements can be implemented in an 8051.
;
;  Note, that comments starting with ";;" are "C" source code with the
;   8051 assembler source following.  With this virtual HLL, I am
;   assuming a word size of 8 bits.
;
;  This program will run in the UMPS simulator for a DS80C320.
;
;  Myke Predko
;  98.02.12
;
;  Hardware notes:
;   This program is only meant to run on the simulator.

;  Variable declarations
;  These declarations are modelled on the "C" variable declarations.
;;int i, j;                     //  Define some 8-bit counters
;;char Output;                  //  Output character
i EQU 0                         ;  Define "i" and "j" in the bank
j EQU 1                         ;   registers

Output EQU 030h                 ;  Location for saving a result

Temp EQU 03Fh                   ;  Have a temporary register for
                                ;   intermediate values

;  Mainline
 org 0                          ;  Execution starts here

;;  i = 4 - 3

  mov    i,#1                   ;  Initialize "i" (This was optimized)
```

LISTING 9-13 PROG12 *Continued.*

```
;;  j = i + 7

  mov    A,#7                         ;  First, load the second Value (which
  push   ACC                          ;   is added to "i") - Note that "ACC"
                                      ;   is being pushed and not "A".

  push   i                            ;  Add the second value for the
                                      ;   assignment to the stack.

  pop    ACC                          ;  Get the first value to operate on
  pop    Temp                         ;  Get the second value to operate on
                                      ;   and put in a variable that can
                                      ;   access "ACC"
  add    A,Temp                       ;  A = A - Temp
                                      ;    = i + 7
  push   ACC                          ;  Save the result for the next
                                      ;   operation
  pop    j                            ;  Retrieve result and store in the
                                      ;   destination

;;  Output = ( i * j ) + 'A';        //  Now, have a compound Operation

  mov    A,#'A'                       ;  Put the character value onto the
  push   ACC                          ;  stack

  push   j                            ;  Store the parameters for the
  push   i                            ;   operations

  pop    ACC                          ;  Get the numbers to multiply
  pop    B

  mul    AB                           ;  Perform the multiplication operation

  pop    Temp

  add    A,Temp                       ;  Add ASCII 'A' to the multiplication
                                      ;   result
  mov    Output,A                     ;  Store the result (Optimized)

Loop:                                 ;  Loop here forever when finished
  ajmp   Loop
```

In PROG12 (Listing 9-13), the first "C" assignment:

```
i = 4 - 3
```

is shown as a simple assignment of "1" to `i`, which can be done with a single instruction. The non-optimized version of this would be:

```
mov    A,#1                          ; Get the initialization value
push   ACC
pop    i                             ; Store the stack value in "i"
```

Any decent compiler should avoid this process all together and directly move the constant into `i`.

The next "C" statement:

```
j = i + 7
```

could be optimized very similarly to the first one (an add with two parameters is not an extremely difficult pattern for a compiler to look for), but I went through the full operation, which could be characterized as:

1. Push 7 onto the stack.
2. Push the contents of `i` onto the stack.
3. Pop the top of the stack into the accumulator.
4. Pop the (new) top of the stack into a temporary register.
5. Add the accumulator to the temporary register and store the result in the accumulator.
6. Push the contents of the accumulator onto the stack.
7. Pop the top of the stack into `j`.

This seems pretty complex, and you're probably (rightly) thinking that it could be optimized significantly. However, in this series of operations, there are some things I want to point out.

The first being the order in which I push parameters onto the stack. Note that I push the rightmost parameter first. The reason for this is to accommodate the subtraction operation exactly the same way that I handle addition.

To subtract two numbers, I can execute the code:

```
pop     ACC                 ; I have gotten the leftmost Value
pop     Temp
clr     C
subb    A,Temp              ; sub instruction works like add
push    ACC                 ; Put the result back onto the stack
```

This is useful in situations where I might be pushing the instructions onto the stack (such as in an interpreter) along with the data values. By making sure data is always pushed onto the stack in the same order, I can simply substitute the appropriate instruction (add, subtract, multiply, or divide) instead of having to handle each one differently. I should point out that, in this case, rather than using a `Temp` variable as I have done here, the B register should be used instead.

The order of what's pushed onto the stack for processing is something that has been written about in many books about compilers. It is shown in the last assignment:

```
Output = ( i * j ) + 'A';    // Now, have a compound operation
```

In this assignment, a decision has to be made on the order of operations. By convention, the code that is inside parentheses is executed first, which means that you have to push it onto the stack *last* (so that it will be first values pulled off of the stack).

When I created the code for this statement, I realized that I should have used the B register instead of creating the variable `Temp`. If I was going to go back and develop a compiler (or interpreter), I would make sure that I would make this change.

If you do decide to use this method of processing complex operations, there is one thing that you do have to watch out for: Make sure your don't have an unequal number of push and pop instructions. Not only will an extra push or pop result in an incorrect result after a series of instructions, but it will also cause the stack pointer to creep into a RAMless or variable area.

PROG13: Subroutines

Calling subroutines in the 8051 is actually quite simple, with nothing to look out for or worry about that I haven't already addressed. In PROG13 (Listing 9-14), I simply call a subroutine after making sure that I don't use any RAM greater than the initial value of the stack pointer.

LISTING 9-14 **PROG13.**

```
;  PROG13 - Calling subroutines
;
;  This application increments a 16-bit value in a subroutine.
;
;  This program will run in the UMPS simulator for a DS80C320.
;
;  Myke Predko
;  98.02.12
;
;  Hardware notes:
;   This program is only meant to run on the simulator.

;  Variable declarations
i EQU 0                                 ;  Define the 16-bit value to increment

j EQU 2                                 ;  8-bit counter

;  Mainline
 org 0                                  ;  Execution starts here

  ajmp   Mainline

Increment:                              ;  Increment the 16-bit value starting
                                        ;   at "i"

  inc    i                              ;  Increment the low byte of "i"

  cjne   R0,#0,Inc_Skip                 ;  Now, check it to see if it is zero
  inc    i+1                            ;  Low byte is == 0, increment high byte
Inc_Skip:

  ret                                   ;  Return to caller

Mainline:                               ;  Mainline, initialize "i" and then
                                        ;   increment it 32x

  mov    i,#0F0h                        ;  Initialize it to 0x012F0
  mov    i+1,#012h

  mov    j,#32                          ;  Want to increment the number 32x
```

LISTING 9-14 PROG13 *Continued.*

```
MLLoop:

  acall    Increment              ;  Now, increment "i"

  djnz     j,MLLoop               ;  Do 32x

Loop:                             ;  Loop here forever when finished
  ajmp     Loop
```

In the subroutine, I increment a 16-bit variable (`i`).

Because subroutines are so simple in the 8051, I decided to demonstrate how a 16-bit variable would be declared and accessed in the program.

When you're declaring a 16-bit variable, it's important to note that the issues aren't with the variable that you are declaring. They're with the variables that are encountered *after* the variable. This is why `j` is declared at address 2 (with `i` at address 0).

When I increment `i`, I first increment the lower 8 bits and then test the result to see if it is equal to zero (which means the number has overflowed and I have to increment the upper 8 bits.

When you look at the code for testing the lower 8 bits, you'll probably feel like I cheated because I used R0 (as the `cjne` instruction requires) rather than `i`. When I set up the variable, I knew that I wanted to use this instruction, so I made sure that I declared `i` as a bank register variable.

Now, if you wanted to use an arbitrary scratchpad register, you could move the result into the accumulator and test the result for being equal to zero.

I like to do all my arithmetic operations from the current bank registers. By following this rule, I can use the maximum number of instructions. This is shown to be advantageous if I wanted to *decrement* a 16-bit number.

In most microprocessors, you would have to subtract 1 from the lower 8 bits and then check the carry flag to see if a borrow was made before decrementing the high 8 bits.

In the 8051, a 16-bit decrement (using `i`) can be accomplished by:

```
 dec      i                        ; Decrement the low byte of "i"
 cjne     R0,#0FFh,Dec_Skip        ; Now, check to see if it rolled over
 dec      i+1                      ; Low byte is == 0, increment high byte
Dec_Skip:
```

To see if the upper 8 bits have to be decremented, this code simply checks to see if the result of decrementing the lower eight bits is 0FFh.

PROG14: Register Parameter Passing

Passing parameters to subroutines is usually reserved for high-level languages, but it does have its place in assembler programming as well. Many programmers, when developing subroutines, access parameters either in registers that are passed along with execution to the subroutine (which means data is usually passed in the accumulator and B register) or modify a global variable that is hardcoded into the program.

The disadvantage of passing data in the accumulator, B register, and any other registers is that this limits the amount of data that can be passed.

Creating global variables and using them to pass data back and forth is actually very simple when just a few variables and subroutines are used, but become unmanageably complex very quickly.

In PROG14 and PROG15 (Listings 9-15 and 9-16), I will demonstrate a number of different ways to carry out passing parameters to subroutines in the 8051 that do not have this limitation.

LISTING 9-15 **PROG14.**

```
;  PROG14 - Passing subroutine parameters via the bank registers
;
;  This application increments the 16-bit value in a subroutine.  Note
;   that there are two increment subroutines: one increments the value
;   and returns it, the other increments an index to the value.
;
;  This program will run in the UMPS simulator for a DS80C320.
;
;  Myke Predko
;  98.02.12
;
;  Hardware notes:
;   This program is only meant to run on the simulator.

;  Variable declarations
i EQU 020h                          ;  Define the 16-bit value to increment

j EQU 022h                          ;  Want to see the value roll

;  Mainline
 org 0                              ;  Execution starts here

  mov    i,#0F0h                    ;  Initialize it to 0x012F0
  mov    i+1,#012h

  mov    j,#16                      ;  Want to increment 32 times

MLLoop:

  mov    R0,i                       ;  Load up 16-bit value for incrementing
  mov    R1,i+1
  acall  Increment                  ;  Now, increment R0/R1
  mov    i+1,R1                     ;  Return the incremented values
  mov    i,R0

  mov    R0,#i                      ;  Increment the pointer to the value
  acall  Increment_Ptr

  djnz   j,MLLoop                   ;  Keep looping around

Loop:                               ;  Loop here forever when finished
  ajmp   Loop
```

LISTING 9-15 PROG14 *Continued.*

```
Increment:                         ;  Increment the 16-bit value in
                                   ;   "R0"/"Ri"
  inc     R0                       ;  Increment the low byte of "i"

  cjne    R0,#0,Inc_Skip           ;  Was the result == 0?
  inc     R1                       ;  Low byte is == 0, increment high byte
Inc_Skip:

  ret                              ;  Return to caller

Increment_Ptr:                     ;  Increment index to the 16-bit value

  inc     @R0                      ;  Increment the low 8 bits of the value

  cjne    @R0,#0,IP_Skip
  inc     R0                       ;  Increment high 8 bits of the number
  inc     @R0
  dec     R0
IP_Skip:

  ret
```

In PROG14, there are two increment subroutines, each of which use the bank registers as a method of passing the initial value to increment.

The first increment routine:

```
Increment:                         ; Increment the 16-bit value in
                                   ;  "R0"/"Ri"
 inc      R0                       ; Increment the low byte of "i"
 cjne     R0,#0,Inc_Skip           ; Was the result == 0?
 inc      R1                       ; Low byte is == 0, increment high byte
Inc_Skip:
 ret                               ; Return to caller
```

has the advantage of being very simple; however, to call it, R0 and R1 must be loaded with the value to increment. Then, after the subroutine is called, they have to be restored.

```
mov      R0,i                      ; Load up 16-bit value for incrementing
mov      R1,i+1
acall    Increment                 ; Now, increment R0/R1
mov      i+1,R1                    ; Return the incremented values
mov      i,R0
```

If I were to look at this operation as a "C" assignment statement, it would be:

```
  i = Increment( i );
```

The first increment routine ends up being much more complicated than the second method of executing the increment:

```
 mov      R0,#i                    ; Increment the pointer to the value
 acall    Increment_Ptr
```

```
  :
  :
Increment_Ptr:                    ; Increment index to the 16-bit value
 inc     @R0                      ; Increment the low 8 bits of the value
 cjne    @R0,#0,IP_Skip
 inc     R0                       ; Increment high 8 bits of the number
 inc     @R0
 dec     R0
IP_Skip:
 ret
```

As a "C" statement, this would be written as:

```
Increment_Ptr( &i );
```

In this method, I pass a pointer to an increment routine, which then increments the value being pointed to. In this method, I don't move the data directly, instead I manipulate it via the bank register index (or pointer). By passing a pointer, I'm really only setting up one parameter and operating on the data directly.

Which method is superior?

Neither method is superior to the other because they are both better in different circumstances. The first method essentially creates an automatic variable for the subroutine. An automatic variable is one which is created for storing parameters and temporary variables for a subroutine. If the input parameters are going to be modified inside the subroutine, these changes should not be made to the "global" values.

The second method is faster and has the advantage of writing directly to the global variable, rather than loading it and saving it after the subroutine. Obviously, this is a problem if you don't want the input value to change. However, for some cases (such as the one in PROG14), this is an advantage.

PROG15: Stack Parameter Passing

In many 8051 compilers, parameters are passed on the stack. This allows data to be passed to a subroutine without changing the global location as well as passing data of an arbitrary size.

While you can pass data parameters on the stack in 8051 assembler, it is actually quite complex and can be confusing. PROG15 (Listing 9-16) shows how a 16-bit value is passed on the stack to a subroutine.

In the `MLLoop`, the 16-bit variable `i` is incremented 32 times by calling `Increment`. Before calling `Increment`, `i` is pushed onto the stack. Afterward, the result is popped into `i`.

The `Increment` code is probably very hard to understand:

```
Increment:                      ; Increment the 16-bit value starting
                                ;  at "i"
 pop     3                      ; Pop the return address to access the
 pop     4                      ;  values
 pop     1                      ; Load the 16-bit counter into the bank
 pop     0                      ;  storage registers
 inc     0                      ; Increment the low byte of "i"
 cjne    R0,#0,Inc_Skip         ; If the inc result == 0
```

```
 inc     1                    ;  Increment high byte
Inc_Skip:
 push    0                    ; Store the values back onto the stack
 push    1
 push    4                    ; Restore the program counter
 push    3
 ret                          ; Return to caller
```

LISTING 9-16 PROG15.

```
;  PROG15 - Passing subroutine parameters via the stack
;
;  This application increments a 16-bit value in a subroutine.  The
;   value to be incremented is passed and returned on the stack.
;
;  This program will run in the UMPS simulator for a DS80C320.
;
;  Myke Predko
;  98.02.12
;
;  Hardware notes:
;   This program is only meant to run on the simulator.

;  Variable declarations
;  Address 0 and 1 used for temporary values in "Increment".

i EQU 044h                      ;  Define the 16-bit value to increment

j EQU 043h                      ;  8-bit counter

;  Mainline
 org 0                          ;  Execution starts here

  ajmp   Mainline

Increment:                      ;  Increment the 16-bit value starting
                                ;   at "i"

  pop    3                      ;  Pop the return address to access the
  pop    4                      ;   values

  pop    1                      ;  Load the 16-bit counter into the bank
  pop    0                      ;   storage registers

  inc    0                      ;  Increment the low byte of "i"

  cjne   R0,#0,Inc_Skip         ;  If the inc result == 0

  inc    1                      ;   Increment high byte

Inc_Skip:

  push   0                      ;  Store the values back onto the stack
  push   1

  push   4                      ;  Restore the program counter
  push   3

  ret                           ;  Return to caller
```

LISTING 9-16 PROG15 *Continued.*

```
Mainline:                          ;  Mainline, initialize "i" and then
                                   ;   increment it 32x

  mov     i,#0F0h                  ;  Initialize it to 0x012F0
  mov     i+1,#012h

  mov     j,#32                    ;  Want to increment the number 32x

MLLoop:

  push    i                        ;  Save the parameter for the call
  push    i+1
  acall   Increment                ;  Now, increment "i"
  pop     i+1
  pop     i

  djnz    j,MLLoop                 ;  Do 32x

Loop:                              ;  Loop here forever when finished
  ajmp    Loop
```

The reason for this is that I pop the return value off the stack before I can access the passed parameters. The return value is saved in scratchpad RAM locations 3 and 4.

A much more elegant way of carrying this out is to create a pointer from the stack pointer that points to the start of the passed parameters.

To demonstrate this, I created PROG15A, which has a different `Increment` subroutine:

```
Increment:                       ; Increment the 16-bit value starting
                                 ;  at "i"
 mov      A,SP                   ; Load R0 to point to the start of the
                                 ;  parameters
 add      A,#0FDh                ;  Take 3 away from the start of the
 mov      R0,A                   ;  stack pointer to point to parameters
 inc      @R0                    ; Increment the low byte of "i"
 cjne     @R0,#0,Inc_Skip        ; If the inc result == 0
 inc      R0                     ; Increment pointer to the next byte
 inc      @R0                    ;  Increment high byte
Inc_Skip:
 ret                             ; Return to caller
```

This subroutine is obviously a lot more efficient in terms of program and variable space usage and is probably much easier to understand. Rather than copying data from the stack (using the `pop` instruction), I create a pointer to the stack area and modify the stack contents. This code has the immediate advantages of not having to put in the instructions to copy the contents of the stack over (which means that the parameters are actually saved twice) and not requiring memory in scratchpad RAM to shadow the parameters.

In the past two experiments, I have presented you with four different ways of passing and handling parameters in 8051 assembly language.

Which one is the best?

There really isn't any one "best" way of passing parameters to a subroutine. Each of the different methods demonstrated has advantages over the others in different situations. If you are just passing a couple of bytes, you will probably find that passing them in the accumulator and B register is the most efficient method in terms of instruction cycles, scratchpad RAM, and instruction bytes required.

PROG16: Implementing Variable Arrays

A very important data type in all programming languages is the variable array. Single-dimensional arrays are used for many different purposes in applications, including storing character strings, data records, and subroutine parameters in some high-level language implementations.

In PROG16 (Listing 9-17), I will show you how a data array can be implemented in 8051 assembler and how different elements of the array can be accessed.

LISTING 9-17 **PROG16.**

```
;  PROG16 - Creating variable arrays
;
;  This application creates a variable array in the scratchpad RAM and
;   then accesses it with reading and writing.
;
;  This program will run in the UMPS simulator for a DS80C320.
;
;  Myke Predko
;  98.02.17
;
;  Hardware notes:
;   This program is only meant to run on the simulator.

;  Variable declarations
;  R0 is used as an index pointer to array.
;  R1 is used as for temporary storage of values when moving string
;   values.

Array EQU 070h            ;  Define array in scratchpad RAM

; Mainline
 org 0                    ;  Execution starts here

  mov  Array,#'H'         ;  Initialize the array to the
  mov  Array+1,#'E'       ;   ASCIIZ string "HEllo"
  mov  Array+2,#'l'
  mov  Array+3,#'l'
  mov  Array+4,#'o'
  mov  Array+5,#0         ;  Terminate the string

;  Array[ 0 ] = Array[ 0 ] - 'A' + 'a'      //  Convert 'H' to lower case

  mov  R0,#Array          ;  Reset R0 to point to start of array

  mov  A,@R0              ;  Get the character
  clr  C
  subb A,#'A'             ;  Convert to lower case
  add  A,#'a'
  mov  @R0,A
```

LISTING 9-17 PROG16 *Continued.*

```
;  Array[ 1 ] = Array[ 1 ] - 'A' + 'a'      //  Convert 'E' to lower case

  mov  R0,#Array          ;  Reset R0 to point to the start of the array
  mov  A,R0               ;  Change index to point to second character
  add  A,#1
  mov  R0,A

  mov  A,@R0              ;  Convert 'E' to lower case
  add  A,#('a'-'A')       ;  Calculate the value here
  mov  @R0,A

;  Array[ 1 ] = Array[ 5 ]                //  Convert the string to just "h"

  mov  R0,#(Array+5)      ;  Reset R0 to point to the start of the array

  mov  A,@R0              ;  Get the character pointed to by the index
  mov  R1,A               ;   and save it for later

  mov  R0,#Array          ;  Reset R0 to point to the start of the array
  inc  R0                 ;  Jump to index 1

  mov  A,R1               ;  Move the contents of Array[ 5 ] to
  mov  @R0,A              ;   Array[ 1 ]

Loop:                     ;  Loop here forever when finished
  ajmp       Loop
```

When you load this program into UMPS, you should have the Internal RAM window enabled so that you can watch the array being built and modified.

The first six instructions initialize the `Array` variable with the ASCIIZ string "HEllo".

```
mov  Array,#'H'          ; Initialize the array to the
mov  Array+1,#'E'        ;  ASCIIZ string "HEllo"
mov  Array+2,#'l'
mov  Array+3,#'l'
mov  Array+4,#'o'
mov  Array+5,#0          ; Terminate the string
```

This initialization is what I would consider a brute-force approach. In PROG17 (which follows this section), I will present another way of doing this that will take less space (for larger initializations), is easier to understand when you look at the program, and is able to accommodate changing strings without requiring that the actual code be modified.

With a string set up in scratchpad RAM, I am now able to begin executing the program and access it.

The first operation I perform on `Array` is to convert the first character of the string into lower case:

```
mov  R0,#Array           ; Reset R0 to point to start of array
mov  A,@R0               ; Get the character
clr  C
subb A,#'A'              ; Convert to lower case
add  A,#'a'
mov  @R0,A
```

The first instruction loads the address to be read from. Next, the character is converted into a lower case character by subtracting "A" from it and then adding "a". When this has completed, the result is stored back into its position in the array.

The next block of instructions converts the second character of the string stored in `Array` ("E") to a lower case character.

```
mov   R0,#Array           ; Reset R0 to point to the start of the array
mov   A,R0                ; Change index to point to second character
add   A,#1
mov   R0,A
mov   A,@R0               ; Convert "E" to lower case
add   A,#('a'-'A')        ; Calculate the value here
mov   @R0,A
```

This series of instructions does essentially the same operations as the previous group of instructions, but there are a few things I want to point out.

The first is: After I load in the `Array` starting address, I add the offset of the element that I want to access. This format is standard through this program and all the array programming that I do. Before executing an array read or write, I make sure the address is correct.

Arrays are an excellent method of finding out how effective a high-level language's optimizer is. If the optimizer was really on the ball, it would realize to set R0 to the address of `Array[ 1 ]`. It would just have to execute the instruction:

```
inc   R0
```

This instruction works because the previous set of instructions left R0 pointing to `Array[ 0 ]`, which is one address away from `Array[ 1 ]`.

This group of instructions also demonstrates an optimization of the upper to lower case operations:

```
mov   A,@R0
add   A,#('a'-'A')
mov   @R0,A
```

In this case, I've combined the subtract "A" (which returns the letter to be changed from upper case's offset from "A") and the add "a" (which gives the offset to the lower case letters).

When you're developing your own software, you have to look for opportunities like this.

The last operation:

```
Array[ 1 ] = Array[ 5 ]
```

moves the contents of the sixth element (which is "\0") into the second element. This essentially changes `Array` to a pointer to the string "h".

I've tried to make the operation optimized reasonably well, but there is one area I will bring to your attention, when the index pointer changes from `Array[ 5 ]` to `Array[ 1 ]`:

```
mov   R0,#(Array+5)       ; Reset R0 to point to the start of the array
mov   A,@R0               ; Get the character pointed to by the index
mov   R1,A                ;  and save it for later
```

```
mov   R0,#Array          ; Reset R0 to point to the start of the array
inc   R0                 ; Jump to index 1
mov   A,R1               ; Move the contents of Array[ 5 ] to
mov   @R0,A              ;  Array[ 1 ]
```

Yes, I could have changed the two instructions loading R0 to Array[1] to:

```
mov   R0,#(Array+1)
```

However, there's something else I could have done to improve the code more effectively.

The `mov R1,A` and `mov A,R1` instructions before and after the setting up of the `Array` address could be deleted. This is possible because the instructions used to set up R0 to `Array[ 1 ]` do not change the contents of the accumulator.

Loading the contents of `Array[ 1 ]` with the contents of `Array[ 5 ]` could be accomplished by:

```
mov   R0,#(Array+5)      ; Point to Array[ 5 ]
mov   A,@R0              ; Load accumulator with Array[ 5 ]
mov   R0,#(Array+1)      ; Point to Array[ 1 ]
mov   @R0,A              ; Store accumulator in Array[ 1 ]
```

which is quite an improvement in the number of instructions and cycles required to execute the operation.

As I've pointed out in chapter 8, there are a number of opportunities to improve the execution of your code, if you look for cases where instructions can execute without affecting the contents of the accumulator.

PROG17: Control Store Tables

In PROG16, there is one function of code that I don't really like. That was the initialization of the `Array` variable. Rather than hard coding the initialization instructions (i.e., `mov Array+n,#'`*`character`*`'`) into the program, I prefer using the `mov @A+PC` command to initialize a string from another string that has been put into control store.

This program will load any arbitrary sized string into the scratchpad RAM (not just "Hello\0"). This can be advantageous when strings of arbitrary lengths are to be accessed or during debug when user messages have to be expanded upon or added.

To load the `Array` variable with an ASCIIZ string of arbitrary length, the code in PROG17 (Listing 9-18) has a number of features that I would like to go through.

The first is reviewing how the instruction:

```
movc  A,@A+PC
```

works.

I think it's pretty easy to think of it as:

```
A = ControlStore[ A + PC ]
```

LISTING 9-18 PROG17.

```
;  PROG17 - Control store tables
;
;  This application initializes the creates a variable "Array"
;   in the Scratchpad RAM and then accesses it with
;   reading and writing.
;
;  This program will run in the UMPS simulator for a DS80C320.
;
;  Myke Predko
;  98.02.17
;
;  Hardware notes:
;   This program is only meant to run on the simulator.

;  Variable declarations
;  R0 is used as an index pointer to Array.
;  R1 is used as the source location for the control store table.

Array EQU 078h                ;  Define Array in scratchpad RAM

;  Mainline
 org 0                        ;  Execution starts here

  mov  R0,#Array              ;  Point to destination variable Array
  mov  R1,#(InitTable-InitGetTable) ;  Point to control store
                              ;   source string
InitLoop:                     ;  Loop around here until '0' is loaded
  mov  A,R1
  movc A,@A+PC                ;  Get the table value
InitGetTable:                 ;  Table Offset to Display
  mov  @R0,A                  ;  Save the Value
  inc  R0                     ;  Point to the next location in the
  inc  R1                     ;   strings
  jnz  InitLoop               ;  If the value saved wasn't zero, loop
                              ;   again

;  #### - Put code to process the array here

Loop:                         ;  Loop here forever when finished
  ajmp Loop

InitTable:                    ;  The control store table for
  db   'HEllo',0              ;   initializing "Array"
```

However, when you start trying to code it in assembler, you discover it's not quite as easy as you would have thought it is.

This instruction returns the byte in control store pointed to by the current program counter added to the contents of the accumulator.

The line:

```
mov  R1,#(InitTable-InitGetTable)
```

carries out this operation. With the `InitGetTable` label being defined after the `movc` instruction.

This instruction loads R1 with the byte address offset between the start of the table and the instruction following the `movc` instruction. Everything I'm writing about here should make perfect sense, except for the reason why I would use the address of the instruction *after* the `movc`.

In all processors, no instruction starts executing until it is completely loaded in. This means that, when execution starts, the program counter is now pointing to the *next* instruction when `movc A,@A+PC` actually starts executing.

One very important rule on placing the table is that it must be as close as possible to the `movc A,@A+PC` instruction and not before it.

The reason for placing the table after the `movc A,@A+PC` instruction is a function of how the instruction operates. In determining the actual address to be read from, an 8-bit value (in the accumulator) is added to the 16-bit program counter. If the start of the table was above (had a lower address than) the `movc A,@A+PC` instruction, then an 8-bit two's complement negative number would be added to a 16-bit number like a positive value. When this happens, the wrong address is actually calculated.

For example, if the `Table` was put at address 020h and the `movc A,@A+PC` instruction was put at address 030h, the address to read from the `Table` would be calculated as:

```
  mov  A,#(Table-GetTable)     ; 020h - 031h = 0EFh
                               ;             = -17 Decimal
  movc A,@A+PC
GetTable:
```

-17 or 0EFh would be added to the program counter at `GetTable`, which would mean that the address to be read from is 0120h rather than 020h, as would be desired.

Keeping the `Table` as close to the `movc A,@A+PC` is important because, the further away the table is, the smaller a maximum size it can have.

To maximize the space available in `InitTable` in PROG17, you might want to re-order the code to:

```
  mov  R0,#Array               ; Set up source and destination indexes
  mov  R1,#(InitTable-InitGetTable)
  goto      InitGet
InitEnd:                       ; Jump here when finished
  goto Next
InitLoop:                      ; Loop around here for each character
  mov  @R0,A                   ; Store the character in accumulator
  inc  R0
  inc  R1
  jnz  InitEnd                 ; Is a zero in the accumulator?
  mov  A,R1                    ; Now, read the character in the table
  movc A,@A+PC
InitGetTable:
  sjmp InitLoop                ; Jump back to do the compare/increment
InitTable:                     ; Table to initialize from
  db   'HEllo',0
```

By re-ordering the code in this way, only 3 bytes are used to read the table and jump back to the loop. This leaves a maximum of 253 bytes available for the table.

At the end of the program, when the table is defined, note that I put the initialization values on one line as a word with a zero at the end. This will reserve 6 bytes and store the

ASCIIZ string “HEllo” into it. I could have also reserved these 6 bytes using 6 “db” statements, but I find it useful to keep string definitions in this easily readable format.

PROG18: State Machines

For some applications, you can apply what is known as *non-linear programming*. In the next chapter, I will demonstrate one form of non-linear programming with real-time operating systems; however, in this chapter, I will demonstrate another form: the *state machine*.

In a state machine, a variable is used to keep track of the current state of the program and help an execution-control subroutine decide what to execute next. In PROG18 (Listing 9-19), I have created a program that can be used to control the state of traffic lights.

LISTING 9-19 **PROG18.**

```
;  PROG18 - State machines
;
;  This application demonstrates how a set of traffic lights could
;   be controlled in the 8051 using a state machine and table jumps.
;
;  This program will run in the UMPS simulator for a DS80C320.
;
;  Myke Predko
;  98.02.18
;
;  Hardware notes:
;   This program is only meant to run on the simulator.
;    It could be modified to run in hardware by making "Dlay" more than
;    a dummy routine
;
;  North/south lights
;   NSGreen  - P1.0
;   NSYellow - P1.1
;   NSRed    - P1.2
;
;  East/west lights
;   EWGreen  - P1.4
;   EWYellow - P1.5
;   EWRed    - P1.6

;  Variable declarations
State EQU 020h                    ;  The current state variable
StateDir EQU 004h                 ;  The direction bit - bit 4 of 020h
;  Bit states
;   Bit 4: _EW/NS - This bit is set for NS, reset for east/west
;   Low nybble: Current light state

;  R0 is used as an index pointer to Array
;  R1 is used as the source location for the control store table

Array EQU 0B0h                    ;  Define Array in scratchpad RAM

;  Mainline
 org 0                            ;  Execution starts here
```

LISTING 9-19 PROG18 *Continued.*

```
  mov   State,#0                    ;  Start with E/W "red"

  mov   P1,#0BBh                    ;  Turn on the red lights

  mov   DPTR,#LightTable

Loop:                               ;  Loop back here for each light state

  acall NextLight

  sjmp  Loop

NextLight:                          ;  Now, turn off the current light and
                                    ;   turn on the one for the next state
  mov   A,State                     ;  Get the state x2
  rl    A
  anl   A,#0DFh                     ;  Clear the _EW/NS flag

  jmp   @A+DPTR                     ;  Jump to the appropriate response

LightTable:

  sjmp  GreenLight                  ;  Have red light go to green
  sjmp  YellowLight                 ;  Have a green light go to yellow
  sjmp  RedLight                    ;  Have a yellow light go to red

GreenLight:                         ;  Turn the light to green

  mov   A,#0EBh                     ;  Are we turning on the E/W light?
  jnb   StateDir,GL_EW
  mov   A,#0BEh                     ;   No, do the N/S
GL_EW:
  mov   P1,A                        ;  Now, turn on the light

  mov   R0,#0                       ;  Delay for a green light period
  acall Dlay

  inc   State                       ;  Increment to the next state

  ret

YellowLight:                        ;  Turn the light to yellow

  mov   A,#0DBh                     ;  Are we turning on the E/W light?
  jnb   StateDir,YL_EW
  mov   A,#0BDh                     ;   No, do the N/S
YL_EW:
  mov   P1,A                        ;  Now, turn on the light

  mov   R0,#1                       ;  Delay for a yellow light period
  acall Dlay

  inc   State                       ;  Increment to the next state

  ret

RedLight:                           ;  Turn the light to red
```

LISTING 9-19 PROG18 *Continued.*

```
  mov    P1,#0BBh                ;  Turn everything to red

  mov    R0,#2                   ;  Delay for a red light period
  acall  Dlay

  mov    A,#0F0h                 ;  Reset back to green for next light
  anl    State,A

  cpl    StateDir                ;  Change which light is active

  ret

Dlay:                            ;  Delay a set amount of time for each
                                 ;   light

;  #### - Put delay code here

  ret
```

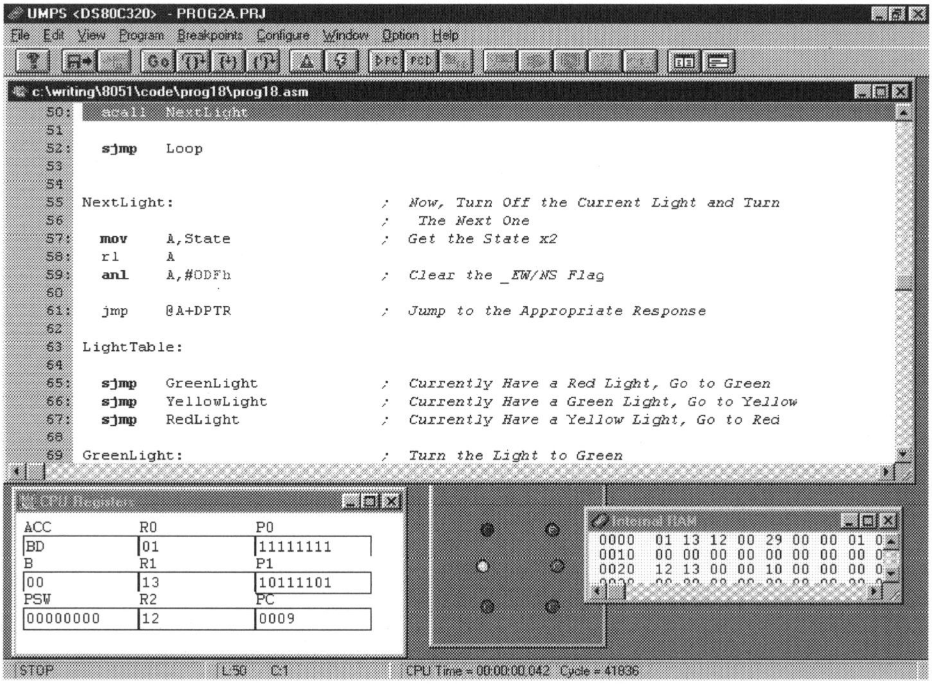

FIGURE 9-18 PROG18 UMPS display.

For running the program in UMPS, I have changed the resource window to include six LEDs (two each colored, appropriately enough, red, yellow, or green) to represent the traffic lights for north/south and east/west. (See Fig. 9-18.)

The heart of the program is the NextLight subroutine:

```
NextLight:                            ; Now, turn off the current light and
                                      ;  turn on the one for the next State
  mov  A,State                        ; Get the state x2
  rl   A
  anl  A,#0DFh                        ; Clear the _EW/NS flag
  jmp  @A+DPTR                        ; Jump to the appropriate response
LightTable:
  sjmp        GreenLight              ; Have red light go to green
  sjmp        YellowLight             ; Have a green light go to yellow
  sjmp        RedLight                ; Have a yellow light go to red
GreenLight:                           ; Turn the light to green
```

This program uses the State variable as an index into a control store table (the start of which is pointed to by DPTR). To jump quickly to the appropriate location, I have set jumps to each of the three different responses to the different LED conditions (red, green, or yellow).

Before executing the jump, I multiply the offset by two because each of the sjmp instructions requires 2 bytes. If I have a State value of 1, I want to jump to YellowLight, not to the byte after the first one in the sjmp GreenLight instruction. For a ljmp instruction, I would have to multiply the State variable by three to get the correct offset.

At the end of each of the light change subroutines, I have put in a ret instruction to return to the mainline that called NextLight. Even though we are jumping from NextLight to a light-handling routine, the call level has not been changed. You might wonder why I don't use a call instead of a sjmp instruction in NextLight; however, if I did, when a routine returned, it would return to NextLight and execute the next subroutine in the row (which is *not* how we want this application to run).

I could have written this application two different ways. The first way was to have six different light changing routines (i.e., EW_Green, EW_Yellow, EW_Red, NS_Green, NS_Yellow, and NS_Red), but I chose instead to have three different routines for each of the different light states and each routine determine whether or not the east/west or north/south routines are to be accessed.

The direction is accessed by bit 4 of the State variable. I deliberately put State at address 020h, so I could use the bit commands to access the current state.

I didn't put any code into the Dlay subroutine because this is purely an UMPS exercise for this book. If you would like to use this application for real (such as for a model train layout), you can add an appropriate delay for each of the different lights. This is why I load R0 with a light specific value before calling the delay.

This is actually quite a complex program. It uses a number of the different features that I have demonstrated in previous experiments. Before going on to the next experiments and attempting to run your applications in hardware, you should make sure you understand what I've tried to demonstrate in each of the experiments leading up to this one.

You might also want to try out an experiment or two on your own that can run in UMPS. I am suggesting this because, when you start writing your own applications and burn them into hardware, you might find that the 8051 just sits there and doesn't output anything or respond to input (or if it does, it seems very strange).

UMPS is an excellent tool for trying out your application before committing it to hardware and begin to trying to figure out why it won't run.

PROG1: Running a Program in Hardware

You're now ready to start playing with hardware. Actually, it's quite a leap going from the emulated code to a burning an application into a device and then seeing if it runs. Along with finding out whether or not your programmer works, you have to correctly wire up the 8051 to run.

To keep these variables down, my first program for a new microcontroller is always simply turning on an LED. For the 8051, I want to go back to the first program I presented you with (Listing 9-1).

Yes, I know there's not much to PROG1; however, if it's the first time you're trying to program an 8051 (or Atmel AT89C2051), you're going to have a lot more luck with this program than any other.

To run this program on a true 8051 (or a Dallas Semiconductor 87C520, as I used), you can use the circuit shown in Fig. 9-19. This program is only marginally different from the 8051 core diagram I presented at the start of the chapter. Only a resistor and LED have been added to the circuit.

So, what happens when you burn PROG1 into an 8051, wire the circuit, and apply power? If all these actions were successful, you should have an LED lit. If you can't believe that the 8051 is actually running, you can put a logic oscilloscope probe to pin 19 and see the oscillator running.

Not very exciting, is it?

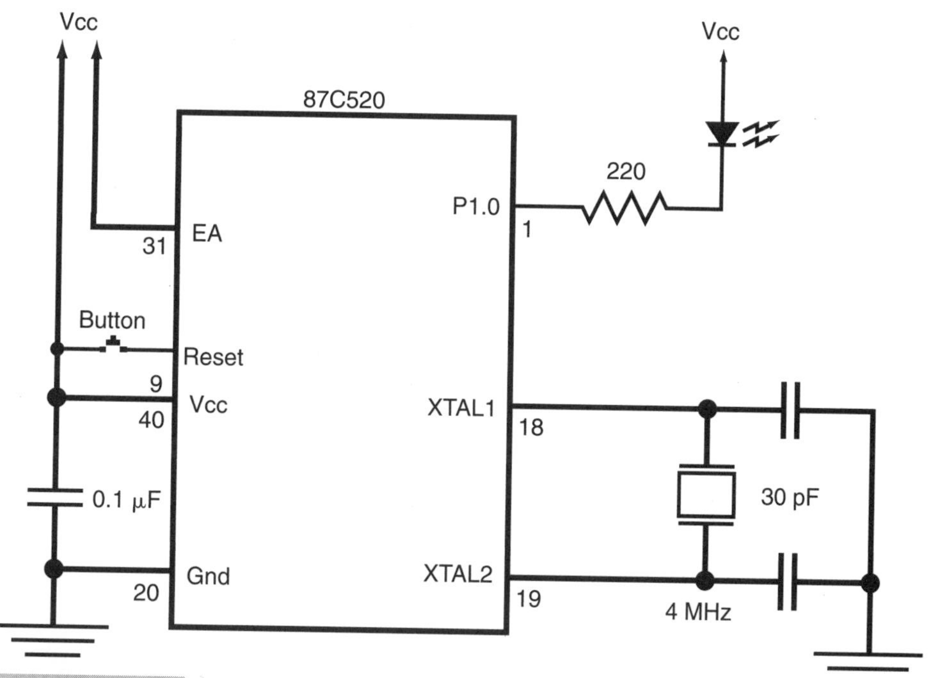

FIGURE 9-19 87C520 PROG1 schematic.

However, what happens if the LED doesn't light.

Chances are that this is what happened. Now you have to try to figure out what exactly the problem was. The process that I follow when I am faced with this situation is outlined in this section.

The first thing you have to check is whether or not the device was properly programmed. When I first ran this experiment, I used a Dallas Semiconductor DS87000 Programmer. Rather than pulling the Wall Wart that came with the unit out of the box, I decided to be lazy and use one that I had lying around. When I tried to program the DS87C520, I found that it wouldn't verify properly.

To determine if your programmer and its control software are running properly, try the following checks. First, do a "blank" check on the device. If this is the first time it's used, or if it is fresh from the UV eraser, it should pass the test. Next take the device out of the socket and re-run the test. If it passes, you might have a defective programmer, or the software isn't running properly.

Next, put your blank part back into the programmer, load the test software, and do a verify on the contents of the 8051's EPROM. This should fail (because you haven't burned anything into the device yet).

If it does fail, then try programming the 8051. After programming, run the verify again, both on the device and an empty socket. If the verify doesn't pass on the device and fails on the empty socket, then you have a problem.

If you have the device programmed and verified and the LED still doesn't come on, you're going to have to check the circuit. The first thing to check is whether, when you turn on the switch, power is getting to your prototyping board? I always put an LED indicator on my boards that lights when +5 V is applied, so I don't have to go looking with a DMM.

If power is going to the device and the LED still is dark, check all your wiring, including the polarity of the LED (which I've put in the wrong way more than once and then spent hours looking elsewhere for why the LED won't light). Another common mistake that I make with the 8051 is to forget to tie EA (pin 31) to Vcc.

You should check the oscillator pins (18 and 19) to see if the built-in oscillator is running. If it isn't, then chances are you have a problem with the crystal. Make sure you are using a parallel-cut crystal. I've found that strip-cut crystals do not always start up (although they can start oscillating when you put an oscilloscope probe on them, which makes things very confusing). I have put this step after checking the wiring because some wiring mistakes will prevent the oscillator from running and end up causing you hours trying different crystals and terminating resistors in parallel or series, and the problem turns out to be that the Reset or EA pin was incorrectly wired to Vcc or ground.

Despite all these suggestions for your debugging your circuit, chances are that you made a wiring mistake and found it about four paragraphs ago. The reason why I've put in this level of detail is to show you the steps that I feel will maximize the chances that you will find the problem.

To summarize them:

1 Make sure the device was programmed properly.

2 Make sure power is going to the circuit.

3 Check your wiring.

  3.1 Check the polarity of all the components on the board.

4 Check to see that the oscillator is running.

Before going on to any of the following experiments, I highly recommend that you make sure you have this one running. Yes, it's simple in terms of hardware and software, but it's simple for a reason. This experiment is relatively easy to debug, and once you have gotten the circuit debugged, you should use it for getting your execution core up and running and making sure that you can get the parts that you have on hand up and running. Believe me, spending the time getting this application up and running will save you hours of pulling your hair out later.

PROG19: Polling a Button

I realize that I have introduced the bit instructions of the 8051 earlier in this chapter. However, before starting to get hot and heavy into the operation of the 8051, I want to show how effective they can be for passing data back and forth between bytes (or, in this case between I/O bits). This experiment should also give you an idea of how flexible the I/O pins are, despite their simple design.

PROG19 (Listing 9-20) polls a button connected to one of the I/O pins (which pulls the line to ground from its normal pulled-up state) and passes the state of the pin to another one and lights an LED when the button is being pressed. (See Fig. 9-20.)

There is one thing I should note about adding LEDs to 8051s. In many microcontrollers, sink current is limited to 20 mA, which makes the job of adding LEDs to a microcontroller very easy. This is not true for many 8051s, which do not have an internal current limit or might have a boost circuit to make the pin state change faster. Currents damaging to the microcontroller or the LED can very easily be passed through the I/O pin. For this reason, *always* put in a 220-Ω current-limiting resistor when you are driving LEDs with an 8051.

LISTING 9-20 PROG19.

```
;  PROG19 - Poll a button and turn on an LED
;
;  This application polls a button and turns on an LED
;   when the button is pressed.
;
;  Myke Predko
;  98.02.18
;
;  Hardware notes:
;  80C520 running at 4 MHz
;  P1.0 is the LED (to light, it is pulled down)
;  P1.1 is connected to a momentary On switch that pulls down the line.

;  Variable declarations

;  Mainline
 org 0                              ;  Execution starts here

Loop:                               ;  Loop here forever

  mov  C,P1.1                       ;  Get the button state
  mov  P1.0,C                       ;  Display button state on LED

  sjmp   Loop
```

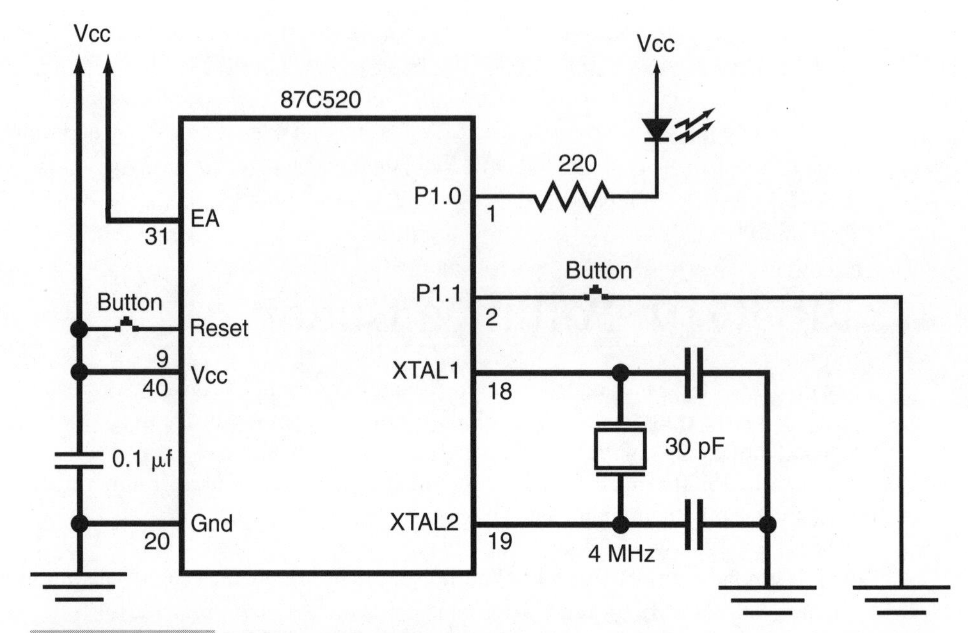

FIGURE 9-20 **87C520 PROG19 schematic.**

In most over microcontrollers, passing a single bit from one position to another or to another position in another byte is quite difficult. However, as you can see in PROG19, this isn't the case in the 8051.

This program only consists of three instructions. The program simply moves the button bit value into the carry flag and then moves the carry flag into the LED bit. That's really all there is to it!

Along with the power of the bit instructions, this simple program would not be possible without the flexibility of the I/O pins.

Most of my previous experience with microcontrollers has been with the PIC and Motorola 68HC*xx* series of devices. The I/O pins in these devices have a tri-state enable control register built in for each group of I/O pins. To change a pin from input to output, this register must be changed (and others, if optional pull-ups or pull-downs are to be enabled). The 8051 pins are a marvel of simplicity, and they do work extremely well in a variety of situations.

PROG45: Inadvertently Changing an I/O Bit

When I work with new devices and try to understand how they work, I will theorize how they work and then try it out. PROG45 is the result of one of these experiments that didn't quite work out, but I did end up with a better understanding of the 8051.

As I was working through the instructions, I started thinking about the:

```
anl  Pn,#0xxh
```

instruction, which ANDs the contents of the I/O port register (P0 through P4) with a constant and puts the result back into the I/O port register. This can be shown as:

```
Pn = Pn & 0xxh
```

which means that, if an input bit was being pulled low when another bit was being ANDed with zero, then the input driver would be loaded with zero at the end of the instruction. This could be a potential problem if input bits, which could be pulled low by an external circuit, were part of the same port as output bits.

The previous experiment, PROG19 had an button pull-down with an LED on the same I/O port and could be used to test this hypothesis with the source code.

When I first ran PROG45 (Listing 9-21) on UMPS, when the button was pressed, the LED would light and, when the button was released, I expected that, the LED would stay because, when the button was pressed down, the `anl` instruction would reset the button bit and it would never be able to go high.

It was a good hypothesis, but it turns out it is wrong. When the button was released, the LED turned off. My initial thought was that UMPS had been programmed wrong. I burned a DS87C520 and tried out the application.

The LED turned off when the button was released.

I also tried this application in an AT89C2051, and still the LED turned off when the button was released.

LISTING 9-21 **PROG45.**

```
;  PROG45 - Poll a button and turn on an LED
;
;  This application turns on an LED when a button is pressed.
;
;  Myke Predko
;  98.06.09
;
;  Hardware notes:
;  80C520 running at 4 MHz
;  P1.0 is the LED (to light, it is pulled down)
;  P1.1 is connected to a momentary On switch that pulls down the line.

;  Variable declarations

;  Mainline
 org 0                              ;  Execution starts here

Loop:                               ;  Loop here forever

  jb   P1.1,LEDON                   ;  If the button is NOT pressed, jump

  anl  P1,#0FEh                     ;  Turn on the LED (pull line down)

  ajmp Loop

LEDON:                              ;  Button is pressed, turn off LED

  orl  P1,#001h

  ajmp Loop
```

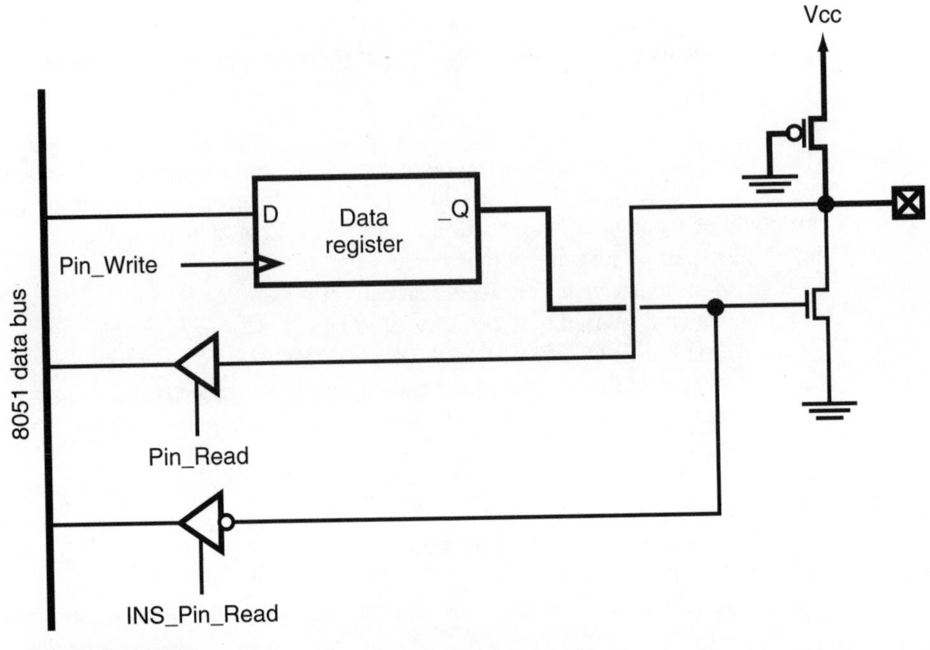

FIGURE 9-21 Actual 8051 parallel I/O pin circuitry.

I created a number of experiments to try to understand what the actual values read were and finally decided that there are two data paths from each pin. One path is used if the pin is to be read back (which is what I always expected there to be), and the other was a path used when the pin value was an intermediate value.

Going to my original drawings for the 8051 I/O pins, I decided that they had to be modified with the INS_Pin_Read (Fig. 9-21). This update to the pin layout gives a secondary path for intermediate values to make sure problems like I was anticipating don't happen.

While the two devices that I tried PROG45 on do work consistently, I would not bet my bottom dollar on this feature working for all 8051s. Instead, I would make it a rule to use the `clr` and `setb` bit instructions to change the I/O pins and avoid gambling on whether or not the INS_Pin_Read circuit and data path is present in the device.

PROG20: Button Debouncing

Debouncing a button input is a prime operation of microcontrollers. By adding a debounce routine in software, you can add button interfaces to your application without having to add any hardware to the circuit. For PROG20 (Listing 9-22), I used the same circuit as in PROG18 and added debounce code for toggling the LED's state each time the button is pressed.

The basic state (press or release) debug code can be modelled as:

```
Debounce:
  while ( Button == High );    // Wait for button press
  for  ( Dlay = 20msec; ( Dlay != 0 ) && ( Button == Low ); Dlay-- );
                               // Poll button while waiting for 20 msec
  if ( Dlay != 0 )             // Did we get a bounce?
    goto Debounce
```

LISTING 9-22 PROG20.

```
;  PROG20 - Do a debounce on a button and toggle an LED
;
;  This application polls the button and toggles the LED
;   state when the button is pressed (and has been "debounced").
;
;  The button is assumed to be debounced if it does not change
;   state for 20 msecs.
;
;  Myke Predko
;  98.02.20
;
;  Hardware notes:
;  80C520 running at 4 MHz
;  P1.0 is the LED (to light, it is pulled down)
;  P1.1 is connected to a momentary On switch that pulls down the line.

;  Variable declarations
;  Just "A" and "B" are used for this program.

;  Mainline
 org 0                             ;  Execution starts here

Loop:                              ;  Loop here forever

PressWait:                         ;  Wait for the button to be pressed
  jb   P1.1,PressWait

  mov  A,#191                      ;  Setup 20 msec delay
  mov  B,#10

PressDebounce:                     ;  Now, see if the button is held down
  jb   P1.1,PressWait              ;   20 msec - If released, start delay
  djnz ACC,PressDebounce           ;   over
  djnz B,PressDebounce

;  #### - Button press is debounced

  cpl  P1.0                        ;  Toggle the LED state

ReleaseWait:                       ;  Now, wait for the Button to be
  jnb  P1.1,ReleaseWait            ;  released

  sjmp       Loop
```

This code is converted to 8051 for the press as the instructions:

```
PressWait:                          ; Wait for the button to be pressed
  jb   P1.1,PressWait
  mov  A,#191                       ; Setup 20 msec delay
  mov  B,#10
PressDebounce:                      ; Now, see if the button is held down
  jb   P1.1,PressWait               ;  20 msec - If released, start delay
  djnz ACC,PressDebounce            ;  over
  djnz B,PressDebounce
```

The 8051 instruction set really makes this operation very efficient.

This is *not* repeated for the button release, to make sure it is debounced as well. Instead, I just wait for the button to be released. Rather than waiting for the bouncing to stop, I assume that, after the press has become validated (i.e., by being debounced), when the switch is at the released state for any reason, it means that the user is releasing the button.

Going through the code for the `Debounce`, the first label and `jb` instruction wait for the button to be pushed (and pull the I/O pin low). This will continue looping until this bit goes low.

I use the accumulator and B register as a 16-bit counter to time the delay. Using these registers, instead of the bank registers, is deliberate because I'm going to access the scratchpad registers in a later experiment, and I don't want to change their contents.

The wait for debounce loop code probably isn't intuitive (especially with the previous values being put into the accumulator and B registers). The loop code will jump back to the wait for the initial condition if the button changes state (i.e., bounces) or will fall out if the button state doesn't change for 20 msec.

```
PressDebounce:                      ; Now, see if the button is held Down
  jb   P1.1,PressWait               ;  20 msec - If released, start delay
  djnz ACC,PressDebounce            ;  over
  djnz B,PressDebounce
```

Once you see how this code works, it's pretty easy to set up. The only difficult part is how you determine the initial count values to get the delay. This code is stolen from my general case, mainline 16-bit delay routine (presented in appendix D, "Useful Routines") and just has the bit test condition placed before the delay code.

To figure out the delay, I broke this code into two cases. The first case is starting with the accumulator equal to zero (so that execution stays in the first `djnz` instruction) and timing this.

The timing delay works out to 8 cycles until the result of decrementing the accumulator to zero and then 12 cycles for the next case.

So, for each value of B:

$$(8 * 255) + 12 = 2052$$

cycles are executed.

By adding the clock period and the number of clocks per instruction cycle, we can calculate the time it takes to work through this loop. By setting B to a specific value, we can delay a set period of time:

$$Dlay\_Period = clock\_period * (B - 1) * 2052$$

I have put in the *B-1* because, if you go ahead and calculate the number for B, you will have a problem because executing this for B equal to zero (the expected minimum delay) will actually cause the longest delay possible!

When the second `djnz` instruction is executed, the result will be 0FFh, which will cause the code to run around 255 more times than you want.

In this case, if you don't want the full (outside) loop to execute because the desired delay is shorter than the minimum time through the loop, you should take it out of your code.

Once I have the number of full loops specified, I can calculate the remaining number of cycles for the inside, or first, loop to waste before starting to execute the full loop. To calculate the initial value for A, I simply divide this number by 8 (which is the number of cycles required for each loop).

I have two comments before finishing this experiment. The first is: The reason why I used a 4-MHz clock for this application is because this makes the instruction cycle period equal to 1 μsec and simplifies my calculations. When playing around with timing delays in your initial applications, I recommend that you use a base clock that will give you a 1-μsec instruction cycle period (for the true 8051, this would mean you should use a 12-MHz clock).

The second point I want to make is that I usually don't go to this much trouble in timing my delays (to the cycle in this case). Instead I usually round up the outside loop.

In this case, when I timed the outside loop, I plugged the formula I presented earlier into the specified delay:

$$\begin{aligned} 20 \text{ msec} &= 1 \ \mu\text{sec} * (B - 1) * 2052 \\ B - 1 &= 20 \text{ msec} / 2052 \ \mu\text{sec} \\ &= 9.747 \\ B &= 10 \end{aligned}$$

Normally for debouncing, I want a minimum of 20 msec delay, so I would set B to 11 and just clear the accumulator.

With B equal to 11, the delay is 20.5 msec, which is 2.5% over the exact timing. What I'm trying to say here is that going with the simplified, rounded up delay will not be that much more inaccurate than going through the entire process of getting the precise delay if it's not needed.

PROG21: Christmas Lights

Every year, around November, the electronics hobbyist magazines give a few projects that show LEDs blinking in a seemingly random pattern. Many of these decorations are designed with a microcontroller to drive the LEDs.

So far, with the experiments, you now have enough knowledge to develop one of these displays for yourself. The circuit (Fig. 9-22) should be pretty easy for you to understand, 16 LEDs are used for the display with the 8051 driving the outputs. The code, while very simple, does require some explaining.

In this code, I have created a simple pseudo-random number generator by treating the bits of P1 and P3 like a 16-bit shift register. Using a single tap, I produce a *cyclical redundancy check* (usually described as a CRC, for obvious reasons). This could be implemented as the circuit shown in Fig. 9-23.

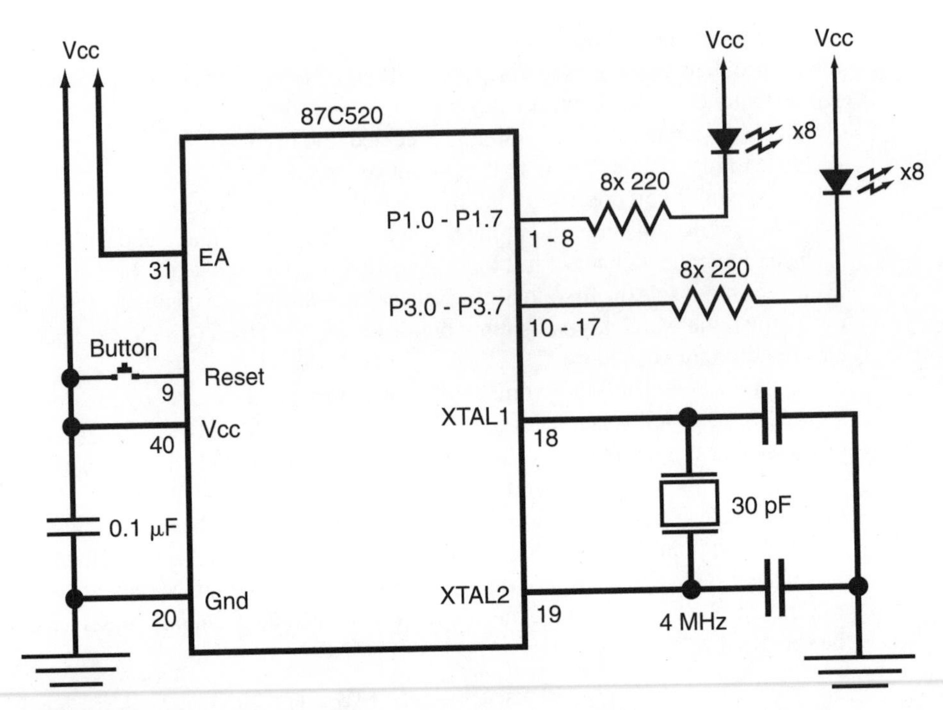

FIGURE 9-22 87C520 PROG21 schematic.

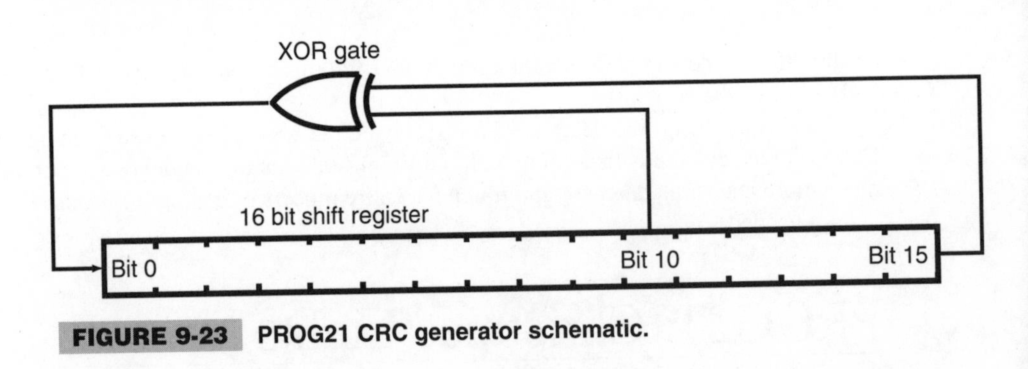

FIGURE 9-23 PROG21 CRC generator schematic.

In PROG21 (Listing 9-23), at power up, all 16 bits are set (equal to "1"). Going through the virtual XOR, I perform a pseudo-random operation on them.

The XOR code ORs together the results of bits 10 and 15 ANDed together to produce a result that is set when one or the other bit is set, but not both. When I presented the bit arithmetic operations and noted that I wished there was a bit XOR, this is one of the applications I was thinking about using for this instruction.

When this application is running in the 8051, you would probably be surprised to find that this simple CRC-based random-number generator will not repeat for over 4 hours and 14 minutes (assuming 61,000 unique combinations before repeating and a quarter second

LISTING 9-23 PROG21.

```
; PROG21 - Pseudo-random LED display
;
; This application displays a pseudo-random 16-bit number
;  generated by a single tap CRC at bit 10 (XORed with bit 15).
;
; The LSB is bit 10 ^ bit 15
;          = ( !bit 10 & bit 15 ) ¦ ( bit 10 & !bit 15 )
;
; This algorithm will give about 61K different patterns.
;
; A quarter-second delay is placed between every new LED pattern.
;
; This display is suitable for doing christmas lights.
;
; Myke Predko
; 98.02.22
;
; Hardware notes:
; 80C520 running at 4 MHz
; P1.0-7 connected to LEDs
; P3.0-7 connected to LEDs

; Variable declarations
; Just "A" is used for this program.

; Mainline
 org 0                        ; Execution starts here

Loop:                         ; Loop here forever

  mov   C,P3.2                ; Get the bits to AND together
  cpl   PSW.7                 ;  invert bit 13
  anl   C,P3.7                ;  AND it with bit 15
  mov   ACC.0,C               ; Save it for later

  mov   C,P3.7                ; Now, Get bit13 * !bit15
  cpl   PSW.7
  anl   C,P3.2

  orl   C,ACC.0               ; OR both values together in carry

  mov   A,P1                  ; Now, update the 16 LED display
  rlc   A
  mov   P1,A
  mov   A,P3
  rlc   A
  mov   P3,A

  acall Dlay                  ; Leave like this for 500 msec

  sjmp  Loop                  ; Loop around again

Dlay:                         ; Delay 263 msec

  clr   A
  mov   B,#0                  ; Longest delay possible
```

LISTING 9-23 PROG21 *Continued.*

```
DlayLoop:                           ;  Loop here for 500 msec
  djnz   ACC,DlayLoop               ;  Inside loop
  djnz   B,DlayLoop

  ret
```

between each transition). So, if some know-it-all says he sees the same pattern and it's been less than four hours, you can call him a liar with complete confidence.

Now, when you run this in the UMPS simulator, you might want to expand upon the CPU registers that you are monitoring. I added I/O port 3 (P3), the stack pointer, and the DPTR 16-bit value.

If you set the P1 and P3 (the two I/O ports used for this application) displays to binary and step through the program, you'll probably see that the value is being shifted through, and it really doesn't look all that random.

For example, if you run UMPS for about five minutes, you'll start seeing the data shown in Table 9-1. In the binary format, you can clearly see the data being shifted; however, when you look at the hex values, the shifting isn't quite so obvious.

When I first wired this up, I found that I could clearly see the shifting progression of the LEDs (I wired them in bit order initially). However, when I randomized them a bit, this shifting pattern became lost very quickly and the LEDs appeared to be turning on and off randomly. So, if you are going to use this circuit for a decoration, make sure the LEDs are randomized to some extent and are not put in any kind of row order where the shifting can be observed.

By rearranging the LEDs (in the past, I created a Frosty the Snowman, which had LEDs in his hat and scarf, based on a PICMicro running this algorithm). With the wiring of the LEDs randomized, the effect is pleasing, and nobody would guess that it is not a truly random pattern.

TABLE 9-1 THE PSEUDO-RANDOM DATA

P3/P1 BINARY	P3/P1 HEX
11011111 10000000	DF80
10111111 00000000	BF00
01111110 00000000	7E00
11111100 00000001	FC01
11111000 00000010	F802
11110000 00000101	F005
11100000 00001011	E00B
11000000 00010111	C017
10000000 00101111	802F
00000000 01011111	005F
00000000 10111110	00BE
00000001 01111100	017C
00000010 11111000	02F8

As a hint, when you wire this, buy a couple of 10-digit LED Bar Graph displays and a few 220-Ω DIP resistor packages. It will simplify your wiring immensely. If you go with individual LEDs and resistors, it will take an unreasonably long time to wire the LEDs for this experiment.

However, if you do spend the time wiring all the LEDs, it will really pay off for the next few experiments.

PROG34: Oscillators

For most of these experiments, I have used a crystal for providing the timing for the 878C520 and AT89C2051 example circuits. The crystal circuit (which includes two external 30-pF capacitors) is reliable and fairly cheap. The only real downside of the crystal is that is can be somewhat fragile in high-vibration environments.

However, it isn't the only option available for controlling the speed of an 8051. Another device that I use a lot of is the *ceramic resonator*. The ceramic resonator is somewhat more expensive than the crystal and doesn't provide as accurate a signal as the crystal, but it is a lot more robust and, with integrated capacitors, it can be simpler to wire into an application. For these reasons, I will tend to use ceramic resonators instead of crystals in applications like the 51Bot presented in chapter 12.

To demonstrate how the ceramic resonator works, I slightly modified the 16 LED circuit that we have been using (Fig. 9-24). For this circuit, I have used an ECS brand ZTT 4-MHz resonator with internal capacitors.

The program that I use, PROG34, is kind of an interesting one and will be of use in the next experiment.

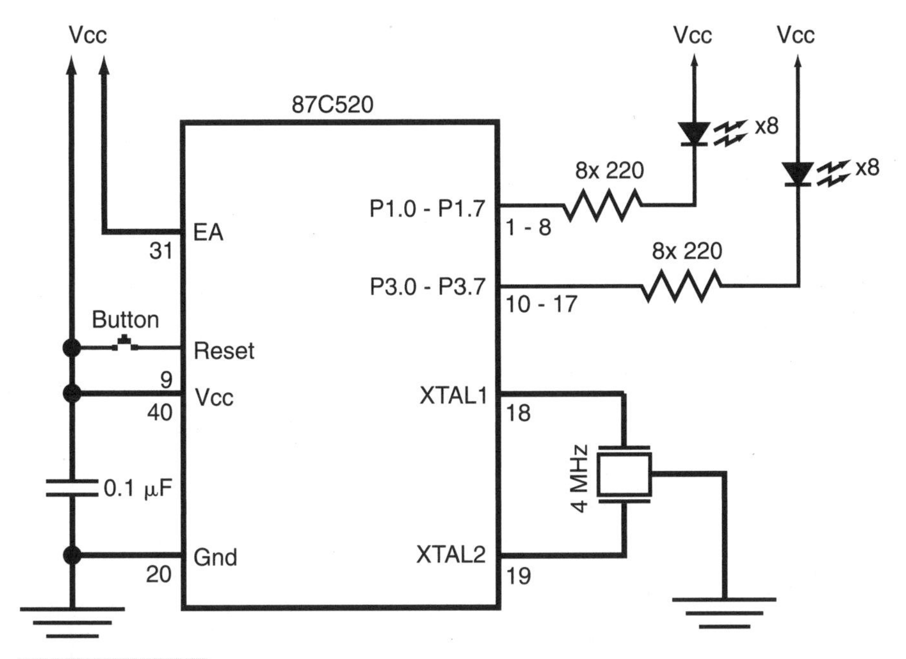

FIGURE 9-24 **87C520 PROG21 schematic with CR.**

PROG34 flashes the LEDs on and off. The lighted LEDS seem to shift across the display.

This application can be run using both a crystal as well as a ceramic resonator, and you should not see any differences.

As I have noted earlier in the book, ceramic resonators typically have a 0.5% accuracy, which should not be noticeable as you execute the program. Actually, you could run the program first with the crystal and then replace it with the ceramic resonator on either the 87C520 or AT89C2051. I doubt you'll see any differences between the two clock sources.

PROG34: Decoupling Capacitors and Power Supplies

PROG34 (Listing 9-24) was designed for a specific purpose: to provide as large as possible current transients in the device. By turning on and off the LEDs in varying groups, I have created a program where the current sunk by the microcontroller can be varied and the results monitored.

LISTING 9-24 PROG34.

```
;  PROG34 -  Testing the decoupling capacitor
;
;  This application "marches" a series of LEDs across P1 and P3 to try
;   to force a reset or device lock up.
;
;  Myke Predko
;  98.03.10
;
;  Hardware notes:
;  80C520 running at 4 MHz
;  P1 is connected to a set of LEDs
;  P3 is connected to a set of LEDs

;  Variable declarations
;  R0 - R1 is used for the program delay
;  020h - 021h is used for the LED bit states

 org 000h

  mov   020h,#0FFh                ;  Set the LEDs to all oFF
  mov   021h,#0FFh

Loop:                             ;  Loop here for each LED set

  mov   P1,#0FFh                  ;  Turn Off ALL the LEDs
  mov   P3,#0FFh

  acall Dlay                      ;  Dlay 1/4 second before doing turn-on

  mov   C,15                      ;  Complement bit 15 for low-order bit
  cpl   C
```

LISTING 9-24 PROG34 *Continued.*

```
  mov   A,020h              ;  Shift up the lower 8 bits
  rlc   A
  mov   P1,A                ;  Store in the output register
  mov   020h,A              ;  Save the new value

  mov   A,021h              ;  Shift up the upper 8 bits
  rlc   A
  mov   P3,A                ;  Store in the output register
  mov   021h,A              ;  Save the new value

  acall Dlay                ;  Delay another 1/4 second

  ajmp  Loop                ;  Loop around

Dlay:                       ;  Delay 1/4 second

  djnz  R0,Dlay             ;  Do a simple loop

  djnz  R1,Dlay

  ret
```

This program was created with the idea in mind that I wanted to experiment with power decoupling.

Primarily, I wanted to see what happened when I removed the decoupling capacitor while this program ran. The decoupling capacitor is used to help filter the incoming power in case the 8051 requires more current at a given time (i.e., at the transient of turning on or off a number of LEDs).

When I first ran this program on the Dallas Semiconductor 87C520 and pulled the decoupling capacitor, nothing changed. The program continued to run without any problems. When I tried it out on the Atmel AT89C2051, I found that the program would only run to five or six LEDs before it locked up.

While the results might point to a decoupling capacitor not being required for the 87C520, I highly recommend that you put a decoupling capacitor on all active components in your applications. I ran this experiment with a rather limited sample (only two devices). However, for one part number, having decoupling was absolutely critical for this application to run.

When you are designing an application of your own, I highly recommend that you provide decoupling for all active components, with the value being 0.01 µF to 0.1 µF, and keeping one end of the capacitor as close as possible to the positive terminal of the device.

PROG22: Reset

I've discussed how the 8051 powers up and reacts to a reset. Before writing your first application, you should ask yourself whether or not you really understand how the 8051 powers up and comes out of reset.

This is an important question because many applications are written with certain assumptions (that are generally wrong) that are reinforced by simulators. In most simulators, RAM is initially displayed as all zeros.

The biggest mistake you can make with any microprocessor is assuming that RAM is powered up in a certain state.

In the next two experiments, I will give you some idea of how the 8051 powers up and what its initial values are. Both these experiments use the circuit set up for PROG21.

In PROG22 (Listing 9-25), I have created a simple program that checks the first four locations in memory and sees if they match a set pattern. If they don't, then I turn on the eight LEDs on P1. If they do match, then I assume that the reset button was pressed while power was still on, and the LEDs on P3 are lit.

LISTING 9-25 PROG22.

```
;  PROG22 - Reset
;
;  This application checks to see if the 8051 had been reset by
;   checking the first four memory locations and if they are set
;   to R0 = 0AAh, R1 = 055h, R2 = 0F0h and R3 = 00Fh.
;
;  Myke Predko
;  98.02.23
;
;  Hardware notes:
;  80C520 running at 4 MHz
;  P1.0-7 connected to LEDs
;  P3.0-7 connected to LEDs

;  Variable declarations
;  R0, R1, R2, and R3 used for checking for the initial values

;  Mainline
 org 0                              ;  Execution starts here

  cjne   R0,#0AAh,NotReset          ;  Check 1st, if != 0AAh, then not reset
  cjne   R1,#055h,NotReset          ;   Repeat for the other three values
  cjne   R2,#0F0h,NotReset
  cjne   R3,#00Fh,NotReset

  mov    P3,#0                      ;  Turn on the LEDs at P1

RstLoop:                            ;  Reset loop
  ajmp   RstLoop

NotReset:                           ;  Set P3 On for Reset is On

  mov    R0,#0AAh                   ;  Initialize the reset positions
  mov    R1,#055h
  mov    R2,#0F0h
  mov    R3,#00Fh

  mov    P1,#0                      ;  Turn on the LEDs at P3

NRLoop:                             ;  Loop Forever
  ajmp   NRLoop
```

When you run this application through UMPS, the first time, you'll see it fail the four checks and turn on the LEDs on P1 (i.e., set P1 equal to zero) after setting these check values. If you reset UMPS and run it again, the program will not branch on the values not being equal and turn on the LEDs at P3.

Burning an 8051 and trying this application will give you similar results. The first time through, the LEDs at P1 will light. Pressing the Reset button (which pulls the Reset pin of the 8051 to Vcc) will cause the program to restart without changing the contents of the scratchpad RAM and then turn on the LEDs at P3.

When you first look at the test for the 8051 executing before and want to reduce the compare instructions by changing the code to:

```
mov A,R0                        ; Everything XORed together is zero
xrl A,R1
xrl A,R2
xrl A,R3
jnz NotReset                    ; If result is != 0, then not setup
```

these instructions will set the accumulator to zero if the four bank registers are set to the expected values. The problem with this code is that the accumulator will be set to zero in a lot of other cases, which makes the possibility of the code detecting a zero condition much more likely than in the code I used.

In the code used in PROG22, there is only one bit pattern out of 4.3 *billion* possibilities, which makes the chances for the power-up values being something that will appear as if the 8051 has had a reset while power was on much more remote.

In case you're wondering, I get the 4.3 billion number from the fact that there are 32 bits that I am checking and each bit has a 50% chance of being in one state or another. That means that there is 2^{32} different combinations possible, which is 4,294,967,296.

PROG23: RAM Contents on Power Up

In the previous experiment, I made the assumption that you could not guess the contents of the scratchpad RAM on power up, that it would be non-deterministic (or random). In PROG23 (Listing 9-26), I'm going to test that assumption.

Using the wiring that I created for PROG21, I have disconnected one LED (at P3.7) and put in a momentary On push button that pulls the pin to ground (Fig. 9-25).

In PROG23, the power-up value for each memory location will be shown along with the scratchpad RAM location. When you are finished with the contents of this memory location, you press the switch, and the contents of the next memory location along with it's address are displayed.

The button routine was cut directly from PROG20 and is really the most complex aspect of PROG23. Other than that, the program simply displays the byte pointed to by R0.

One thing to notice is that I use P3 to display the current address. You might be surprised to see that I simply decrement the register. This can be done in this application because it powers up with 0FFh. With this value, all the LEDs are turned off. Decrementing this will give the appearance to the user that the address is being incremented.

LISTING 9-26 PROG23.

```
;  PROG23 - Scratchpad RAM values on power up
;
;  This application outputs the value for each scratchpad RAM location
;   from 0 to 32 using the LEDs from PROG21 and a switch to increment
;   the counter.
;
;  Myke Predko
;  98.02.23
;
;  Hardware notes:
;  80C520 running at 4 MHz
;  P1.0-7 connected to LEDs - Displays the values
;  P3.0-6 connected to LEDs - Displays the address
;  P3.7 connected to a pull-down momentary On switch

;  Variable declarations
;  R0, once it has been read is used as an index into the internal RAM

;  Mainline
 org 0                              ;  Execution starts here

  mov    A,R0                       ;  Get the first address
  cpl    A                          ;  Invert the value
  mov    P1,A                       ;  Store it

  mov    R0,#1                      ;  R0, now starts to point to other
                                    ;   locations
MainLoop:                           ;  Loop here for each address

ReleaseLoop:                        ;  Wait for the button to be released
  jnb    P3.7,ReleaseLoop           ;   Loop while the button is low

PressWait:                          ;  Wait for the button to be pressed
  jb     P3.7,PressWait

  mov    A,#191                     ;  Setup 20 msec delay
  mov    B,#10

PressDebounce:                      ;  Now, see if the button is held Down
  jb     P3.7,PressWait             ;   20 msec - If released, start delay
  djnz   ACC,PressDebounce          ;   over
  djnz   B,PressDebounce

  dec    P3                         ;  Show address of byte being displayed

  mov    A,@R0                      ;  Get the value pointed to by R0
  cpl    A
  mov    P1,A                       ;   and display it inverted

  inc    R0                         ;  Point to the next address

  mov    A,P3                       ;  Are we at the end of the display?
  anl    A,#07Fh                    ;   Check if lower 7 bits are reset
  jnz    MainLoop                   ;   If not, can loop again

Loop:                               ;  Done - Loop forever
  ajmp   Loop
```

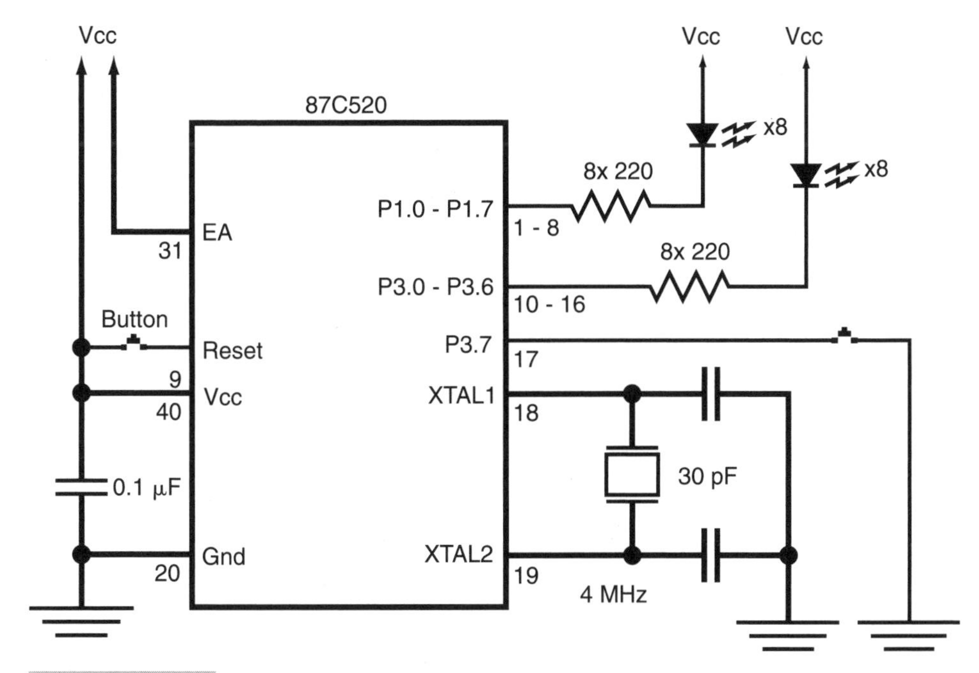

FIGURE 9-25 **87C520 PROG23 schematic.**

I could have passed the address value of R0 to the accumulator, inverted it, made sure bit 7 was set (because this bit is used as a pull up to the incrementing switch), and saved it into P3. However, when I designed this circuit, I planned to be able to use P3 as a counter that would display the correct next bit value by simply decrementing it.

To run this program in UMPS, I created a new push button resource with the output driving bit 7 of P3. When I ran through UMPS, I put a breakpoint at the `dec P3` instruction.

I ran the program in two different 87C520s: one with week code 9729 and the other with week code 9617. The 9729 date code was run twice, with power turned off for five minutes between each run. When I re-ran the program by simply pressing the reset button, I found I got the same results for the previous pass except for R0, which was set to the address I reset the previous pass at.

The results of executing the program for the first 32 RAM locations are shown in Table 9-2.

While there are many similarities between different locations at different power-up intervals and different 87C520s, they definitely are not something that you can count on. I was surprised to see the blocks of zeros that came up. However, as you'll notice at address 00Dh, this isn't something you can always count on getting.

Looking over this experiment, I can conclude two things. One, the data on power-up cannot be depended upon to be the same value each time and to be the same across different devices.

The second conclusion goes back to PROG22. In it, I made the statement that XORing four consecutive scratchpad RAM locations together and getting zero was a lot more likely

TABLE 9-2 THE RESULTS OF EXECUTING THE PROGRAM FOR THE FIRST 32 RAM LOCATIONS

ADDRESS	9729 PASS 1	9729 PASS 2	9617 PASS 1
000h	0FFh	0FFh	0DFh
001h	0FFh	0FFh	0FFh
002h	081h	009h	025h
003h	05Ch	058h	012h
004h	001h	000h	077h
005h	07Ah	01Ah	002h
006h	003h	002h	020h
007h	0D5h	084h	091h
008h	08Dh	0CDh	0AEh
009h	0D7h	0FFh	0B6h
00Ah	000h	000h	000h
00Bh	000h	000h	000h
00Ch	000h	000h	000h
00Dh	000h	020h	000h
00Eh	000h	000h	040h
00Fh	011h	013h	000h
010h	007h	057h	04Ch
011h	049h	049h	038h
012h	000h	000h	000h
013h	000h	000h	000h
014h	000h	000h	000h
015h	000h	000h	000h
016h	000h	000h	000h
017h	000h	000h	000h
018h	000h	000h	000h
019h	000h	000h	000h
01Ah	0FFh	0FFh	06Eh
01Bh	0F5h	0FFh	06Fh
01Ch	045h	001h	041h
01Dh	028h	028h	078h
01Eh	0D0h	0D0h	018h
01Fh	008h	018h	000h

than checking each memory location and making sure that each byte is set to an expected value. When I made this statement, I had not yet run or even written PROG23.

Looking over the data, I can see a number of locations where XORing four consecutive scratchpad RAM locations would result in a zero and the program assuming that the microcontroller had been reset without power going off. If this was the case, then I can see a real potential for disaster in any application that made this assumption.

Before I wrote this program, I wondered if I should have used a different display for the data. Using 15 LEDs for the data and address is not the most user-friendly output device possible. If I were going to develop an application that was going to output data such as this one, I would probably use a liquid crystal display or 7-segment LED displays and develop the code to display the data in simple numerics.

PROG24: Timer Interrupt with Context Saving

Interrupts in the 8051 are very easy to set up. In PROG24 (Listing 9-27), I have set up an interrupt that will occur every 262,144 clock cycles (or 65,526 instruction cycles) using Timer0. While this timer is running, approximately every second, I decrement the LEDs at P1.

LISTING 9-27 PROG24.

```
;  PROG24 -  Timer interrupt handler with context saving
;
;  This application decrements the value in P1 once per second
;
;  Myke Predko
;  98.03.01
;
;  Hardware notes:
;  80C520 running at 4 MHz
;  The 80C520's clock will be running at 1/4 second "ticks"
;  P1 is connected to a set of LEDs
;  P3 is connected to an external LED display

;  Variable declarations
;  R0 is used as the interrupt count.
;  R1/R2/R3 are used for the external display count.

org 0
 ajmp   Mainline

org 0Bh                          ;  Timer0 (real time) interrupt

  push  ACC                      ;  Save the accumulator
  push  PSW                      ;  Save the status register

  djnz  R0,IntSkip               ;  Count 4x before decrementing P1

  mov   A,P1                     ;  Decrement the LEDs
  dec   A
  mov   P1,A

  mov   R0,#16                   ;  Reset the loop counter
```

LISTING 9-27 **PROG24** ***Continued.***

```
IntSkip:                          ;  Return from interrupt

  pop     PSW                     ;  Restore registers pushed onto stack
  pop     ACC

  reti

 org 020h                         ;  Program mainline
Mainline:

  mov     R0,#15                  ;  Change value once every 15 interrups

  mov     CKCON,#%00001000        ;  Use internal /4 clock for Timer0
  mov     TMOD,#%00000001         ;  Timer0 - Uses internal clock
                                  ;         - Run in mode 1
  mov     TCON,#%00010000         ;  Start Timer0 running

  mov     IE,#%10000010           ;  Enable the Timer0 interrupt

;  Now, do the looping output

MainLoop:                         ;  Come back here after decrementing P3

  mov     R1,#0                   ;  Setup for the delay
  mov     R2,#0
  mov     R3,#4

DlayLoop:                         ;  Loop here for delays
  djnz    1,DlayLoop
  djnz    2,DlayLoop
  djnz    3,DlayLoop

  dec     P3                      ;  Decrement the LED output

  ajmp    MainLoop                ;  Wait to decrement the next time
```

In the mainline, I also run a similar loop, but its delay is carried out manually using `djnz` instructions.

In this code, I have deliberately written the interrupt handler exactly as I would in a regular processor. I save all the registers that are modified within the interrupt handler before executing the code inside the interrupt handler.

Yes, this example is a bit artificial because I load the contents of P1 into the accumulator, decrement it, and then save the new value back when all that is required is `dec P1`. However, I wanted to demonstrate how most microcontrollers would carry out the task.

In the mainline, I have taken the loop I first demonstrated for the button debounce and trimmed it down to a straight delay loop. Each inner loop takes four instruction cycles (16 clock cycles) while the outside loop takes 256 times the inner loop (four instruction cycles) plus four additional cycles for the outside loop. This gives me a total of 1028 instruction cycles for each outside loop. When I set the first two 8-bit counters to zero, then I have a delay of 267,040 instruction cycles.

To get a full delay of approximately a second, I loop around four times (which is why I put 4 into R3).

There is one thing to notice about this code: I have used the bank register scratchpad RAM addresses instead of referencing the bank registers themselves. I did this because the `djnz Rn,Label` instruction requires three instruction cycles, while the `djnz direct,Label` requires four to execute. This is something to remember because there might be times when you want to use one `djnz` over the other for different cases to allow more precise timing.

PROG25: Timer Interrupts without Context Saving

In PROG24, I introduced you to the timer interrupt. As I indicated earlier, the code was written how I would do it if I was writing it for a processor other than the 8051, the code itself is actually quite complex:

```
org 0Bh                        ; Timer0 (real time) interrupt
  push  ACC                    ; Save the accumulator
  push  PSW                    ; Save the status register

  djnz  R0,IntSkip             ; Count 4x before decrementing P1
  mov   A,P1                   ; Decrement the LEDs
  dec   A

  mov   P1,A
  mov   R0,#16                 ; Reset the loop counter
IntSkip:                       ; Return from interrupt

  pop   PSW                    ; Restore registers pushed onto stack
  pop   ACC
  reti
```

However, in PROG25, I looked at how I was doing it and was able to optimize the interrupt handler to:

```
org 0Bh                        ; Timer0 (real time) interrupt
 djnz   R0,IntSkip             ; One second up?
 dec    P1                     ; Decrement the timer display
 mov    R0,#15                 ; Reset the counter
IntSkip:
 reti
```

You're first reaction might be: "What happened?" Less than half the instructions of the original interrupt handler is required.

In PROG25, I took advantage of the features of the 8051. While in some cases, I lament the lack of a zero flag in the PSW, for situations like this, not having a zero flag is an advantage. In an interrupt handler, if I don't modify the carry flags, then I don't have to push the PSW register onto the stack.

Comparisons can be carried out using the `cjne`, `djnz`, or `jb` instructions, which don't modify the carry flags.

The 8051 has the capability of checking different registers as well as changing their values without involving the accumulator or the PSW registers. By using instructions that do not require or modify these two registers, they don't have to be saved on the stack.

Without having to save the accumulator or status register on the stack, chances are that you will be able to keep your interrupt handler within the 8 bytes that is allotted for it in the 8051's architecture.

Keeping your interrupt handler very short also has one other important advantage, it keeps down the time taken away from the mainline (and other interrupt handlers), which means there is less system overhead. This might seem like a nit (especially in a system that runs at several MHz), but there are real advantages to be gained.

For example, if you watch PROG24 run, you'll notice that the mainline loop updates the LED display about once every 1.05 seconds (or looking at it another way, several thousand instructions are unnecessarily executed in the interrupt handler which slows down the mainline). In PROG25, with the improved interrupt handler, this value is reduced slightly by several hundred cycles.

This might not seem like a lot; however, in a complex application with many interrupts happening within a given time, this improvement can become very significant very quickly.

PROG26: Button Debouncing using Interrupts and Timer

Button debouncing can also be very effectively carried out using an interrupt on I/O port going low (_Int0, _Int1, etc.) and using a timer interrupt.

The advantage of this method is that the button debouncing (and potentially, responding to the input) can be placed entirely inside interrupt handlers, which frees up the mainline to carry out its own execution without having to wait for the button debounce to complete.

The interrupt/timer-driven debounce code could be shown as the C code in Listing 9-28.

The code at the beginning of this section is not actually how I would block out the application. My version is PROG26 (Listing 9-29).

LISTING 9-28 **The C code for the interrupt/timer-driven debounce.**

```
ButInt() {                       //  Interrupt on the button pressed

  SetTimer( 20msec );            //  Set the timer to delay 20 msec

  ResetTimerOverflow();          //  Reset the timer's overflow

  EnableTimerInt();              //  Enable timer interrupts

}  //  End ButInt

TimerInt() {                     //  Interrupt on timer overflowing

  DisableTimerInt();             //  Disable the timer interrupt

  if ( Button == Low )           //  If the button is still pressed,
    ButtonFlag = 1;              //   indicate the button was pressed
```

LISTING 9-28 The C code for the interrupt/timer-driven debounce *Continued.*

```
} // End TimerInt

main() {                        // Mainline of the button routine

  TimerSetup();                 // Timer0 in mode 1, /4 input

  EnableButtonInt();            // Enable the button interrupt

  while ( 1 == 1 ) {            // Loop forever

    while ( Button == 0 );      // Wait for the button to be released

    ButtonFlag = 0;             // Reset the button flag

    while ( ButtonFlag == 0 );  // Wait for the button to be pressed

    LED ^= 1;                   // Toggle the LED

  }

} // End main
```

LISTING 9-29 PROG26.

```
; PROG26 - Using interrupts to debounce a button
;
; This application toggles the LED at P1.0 after the button on
;  P3.2 (Int0 pin) has been debounced.
;
; Myke Predko
; 98.03.01
;
; Hardware notes:
; 80C520 running at 4 MHz
; The 80C520's clock will be running at 1/4 second "ticks"
; P1.0 is connected to a set of LEDs
; P3.2 is connected to a button that pulls down the pin

; Variable declarations
Flags EQU 020h                  ; When bit 0 set, then button pressed

 org 000h
  ajmp   Mainline

 org 003h                       ; Int0 pin

  mov    TH0,#2*256/3           ; Wait 20 msec for key to be active

  setb   IE.1                   ; Enable timer interrupts

  clr    TCON.5                 ; Make sure overflow is turned off

  reti

 org 00Bh                       ; Timer0 (real time) interrupt
```

LISTING 9-29 PROG26 *Continued.*

```
                                    ;  When this is executed, set flags.0
                                    ;   and turn off interrupts

  clr     IE.1                      ;  Enable JUST button interrupts

  jb      P3.2,IntT0_End            ;  If button not pressed, don't mark it

  setb    0                         ;  Set the button interrupt

IntT0_End:                          ;  End of T0 overflow interrupt

  reti

 org 020h                           ;  Program mainline
Mainline:

  mov     CKCON,#%00001000          ;  Use internal /4 clock for Timer0
  mov     TMOD,#%00000001           ;  Timer0 - Uses internal clock
                                    ;         - Run in mode 1
  mov     TCON,#%00010001           ;  Start Timer0 running
                                    ;   Make edge triggered interrupts

  mov     IE,#%10000001             ;  Enable button interrupts

;  Now, do the looping output

MainLoop:                           ;  Come back here after inverting LED
  jnb     P3.2,MainLoop             ;  Loop waiting for button to be pressed

  clr     0                         ;  Clear the button press flag

ButtonLoop:                         ;  Wait for the button to be pressed
  jnb     0,ButtonLoop              ;  Wait for the bit to be set

  cpl     P1.0                      ;  Invert the LED state

  ajmp    MainLoop                  ;  Wait to decrement the next time
```

In the past few routines, I have gotten away from going through each line of code. To explain PROG26, I will go through each line to help explain what is happening (and reference the C example code as well). PROG26 is, by far, the hardest program so far to understand because not only is it demonstrating a pretty difficult concept, but it also uses some instructions and features that are not immediately obvious.

After the timer is set up:

```
Mainline:
 mov      CKCON,#%00001000        ; Use internal /4 clock for Timer0
 mov      TMOD,#%00000001         ; Timer0 - Uses internal clock
                                  ;        - Run in mode 1
 mov      TCON,#%00010001         ; Start Timer0 running
                                  ;  Make edge triggered interrupts
 mov      IE,#%10000001           ; Enable button interrupts
```

I enable interrupts to wait for the button to be pressed.

In the button interrupt routine:

```
org 003h                        ; Int0 pin
 mov     TH0,#2*256/3           ; Wait 20 msec for key to be active
 setb    IE.1                   ; Enable timer interrupts
 clr     TCON.5                 ; Make sure overflow is turned off
 reti
```

I wanted to use only 8 bytes of control store because the Timer0 interrupt address immediately followed the interrupt 0 button address. When you look at the three instructions between the `org` and `reti` statements, I'm sure none of them really make sense.

The first operation—loading the high 8 bits of Timer 0 with two-thirds of 256—probably doesn't make any sense, but it's actually a way to get an approximate delay.

As I've mentioned earlier in the book, to debounce a switch, generally you have to wait for 20 msec for the button to be valid in one state. With Timer0 running at 4 MHz (the clock frequency) with a divide-by-four counter, the full range of the timer is 65.536 msec, which is roughly three times what we want.

The timer counts up and interrupts when the TL0 and TH0 registers overflow. So, to get a specific delay, I have to load the timer with the full value *minus* the delay I want.

Keeping everything simple, I decided upon going with a third of the full value (which is approximately 22 msec), which meant that I had to work through:

$$
\begin{aligned}
&65.536 \text{ msec} - 1/3 * 64.536 \text{ msec} \\
&= 65.536 \text{ msec} * (1 - 1/3) \\
&= 65.536 * 2/3 \text{ msec}
\end{aligned}
$$

As I've pointed out, in this case, 65.536 msec is the full range, so I would start with TH0 equal to 256 and TL0 equal to 0. To calculate this value, I simply multiplied 256 (for TH0) by two-thirds to get the value to be loaded in.

When I first created this program, I also loaded TL0 with 0, but I found that it used up too many instruction bytes to keep the Int0 handler within the 8-byte limit. Not setting the line is not a problem because the lower 8 bits could only affect the delay by a maximum of 256 μsec, which meant that, even in the worst case, I would have at least 20 msec of waiting for a bounce.

The next line:

```
setb    IE.1                    ; Enable timer interrupts
```

will set the IE register to 0b10000011 or have the interrupts enabled with both the Int0 and Timer0 Interrupts active. Actually, at the time, IE will be set to 0b00000011 and will set the most significant bit (the global interrupt enable) active after the `reti` instruction.

The last line before the `reti` instruction:

```
clr     TCON.5                  ; Make sure overflow is turned off
```

turns off any pending timer interrupts. This has to be done because the timer will overflow once every 65.5 msec. In the application, chances are that the button will not be pressed before the timer initially overflows (or it will have overflowed between button presses). This line will stop any pending Timer0 interrupts until the 20 msec has gone by.

The first time I created this program the Int0 handler was no where near this efficient. In fact, what's presented here is about the third pass of the code before I got it to both work properly and be within 8 instruction bytes. Actually, when I first did this, I discovered a bit of a problem with the UMPS assembler.

If you overwrite memory, it does not return with an error. This is a problem for code that has to be kept within a set limit. When I first started simulating PROG26, I had a number of problems that I couldn't really understand.

The Timer0 interrupt handler might seem simple, but it also has a number of features that had to be worked through in the UMPS simulator before I was able to work efficiently with it.

```
 org 00Bh                          ; Timer0 (real time) interrupt
                                   ; When this is executed, set flags.0
                                   ;  and turn off interrupts
 clr     IE.1                      ; Enable JUST button interrupts
 jb      P3.2,IntT0_End            ; If button not pressed, don't mark it
 setb    0                         ; Set the button interrupt
IntT0_End:                         ; End of T0 overflow interrupt
 reti
```

The first line resets the Timer0 interrupt request line. After executing this instruction, a timer overflow can no longer cause an interrupt.

The jump on bit set (`jb P3.2,IntT0_End`) instruction is needed because, in the 8051, interrupt 0 is only requested when the pin is pulled down. If it goes back up, then the change will not be detected. So, I only set the button flag (bit 0) if, at the end of the 22 msec time-out, the button is still low.

There is one more case I should mention: what happens if, after the button is pressed, it bounces up and then back down.

In this case, after the button is pressed, the timer interrupt is enabled and the timer is loaded with a value that will cause it to interrupt after more than 20 msec.

When the button bounces up, the program doesn't notice it.

However, when it goes back down, the Int0 interrupt is requested again, which resets the timer value for another 20 msec.

If the button bounces up and doesn't bounce back down, then in the Timer0 interrupt, the button-up condition is noted and the button-pressed flag is not set. When the button goes back down, then the Int0 handler is requested again, and the whole process starts over and the timer waits for another 20 or more msec.

I realize that PROG26 is quite complicated and is the result of several iterations to get it right. As I created this program, I used UMPS to make sure that it would work properly, and I didn't burn a part until I was absolutely satisfied that it would work without any problems. In doing this simulation, I extended the UMPS CPU register display to show the timer registers (TCON and TMOD) as well as the interrupt enable register (IE) and stepped through the program repeatedly.

The good news is that this program worked perfectly the first time I tried it out in the DS87C520. However, with the problems that I had and the issues that I didn't foresee as being a problem (such as no interrupt on the rising edge of the _Int0 pin or that, for this program to work, I had to select edge triggered interrupts instead of level triggered) using the UMPS simulator was very important to get the program to work properly to avoid trying to figure out what was wrong in actual hardware.

There's one thing I haven't mentioned about the desirability of this method of debouncing inputs as opposed to the method shown in PROG20.

This method is very processor independent and will run in a true 8051 processor core as well as in the Dallas Semiconductor HSM parts without modification because it is based on the clock cycles and not the instruction cycles (as PROG20 is).

Developing application software to be processor independent is important because you might have to replace the processor at some point in the future for reasons that you can't foresee (such as the power requirement of your application is too high and going with the HSM parts will reduce the power required by an application running a true 8051) or maybe parts that you have specified won't be available.

Making the application processor independent will probably pay huge dividends because you won't be called in to work on a project you thought you finished years before.

PROG27: Memory-Mapped I/O

The last two experiments will require a device with an external bus, which really means a full 8051 (and not a 20-pin device). The external bus also means that some wiring is required to do the application. For both experiments, I have kept to the wiring that you have done so far (with the LEDs on P1 and P3) to minimize the amount of prototyping you will have to do. Despite this, I recommend that each of the last two experiments be started at the beginning of an evening or learning session to keep the wiring errors down.

I also recommend that you use a spring-contact prototyping system because you will probably make errors. Rather than hand soldering or wire wrapping, this type of system will make it easier to find and fix the problems (although it won't seem like it when you have to do it at the time).

So, with this warning, let's start wiring buses to the 8051.

Actually, the bus interface of the 8051 is a tremendous feature. To demonstrate it, I created PROG27 (Listing 9-30) to demonstrate reading eight switches at one of two bus addresses. The circuit is very simple when you see it as a block diagram (Fig. 9-26), but it does have significantly more wiring than any circuit I've introduced to you up to this point.

LISTING 9-30 **PROG27.**

```
;  PROG27 -  Polling 8 data pins on the data bus
;
;  This application reads the 8 switches at address 00000h (or address
;   00100h) on the address bus and displays them on the LEDs connected
;   to P1.
;
;  Myke Predko
;  98.03.09
;
;  Hardware notes:
;  80C520 running at 4 MHz
;  P1.0 is connected to a set of LEDs
;  P3.2 is connected to momentary On pull-down button
;  P0/P2 I/O bus wired to a 74LS245 with 8 input switches
```

LISTING 9-30 PROG27 *Continued.*

```
; Variable declarations

 org 000h
  mov   DPTR,#0                ; Initialize DPTR to the I/O address

Loop:                          ; Loop polling port and outputting data

  mov   A,P3                   ; Get the specified address

  mov   C,ACC.2                ; Save button value in carry

  clr   A                      ; Clear the accumulator
  mov   ACC.0,C                ; Put the button value into the acc.
  mov   DPH,A                  ; Update the DPTR to the address

  movx  A,@DPTR                ; Read the external I/O port
  cpl   A                      ; Complement read in value so LEDs are
                               ;  on for "highs"
  mov   P1,A                   ; Store the result in "A"

  ajmp  Loop
```

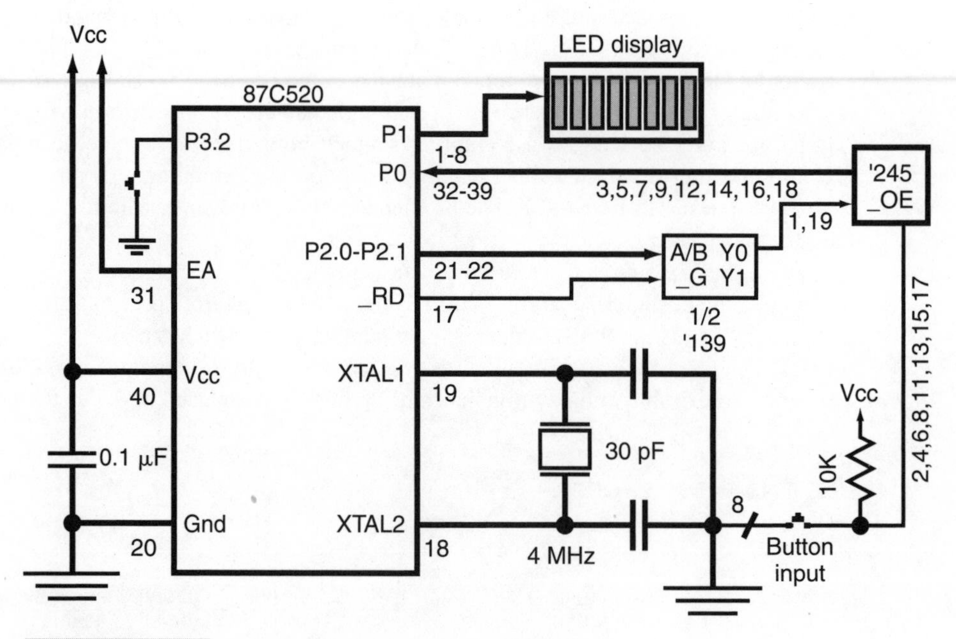

FIGURE 9-26 Memory-mapped I/O in the 87C520.

Before going through PROG27, I want to discuss a few aspects of the circuit first. You should notice that I have added a momentary On push button on P3.2. This button is used to select between reading addresses 00000h and 00100h.

The hardware at these addresses are known as *memory-mapped I/O*, and I don't see a lot of 8051 applications that use it, although it really offers some advantages in terms of allowing you to expand the I/O of your application significantly.

Address bits 8 and 9 are used to select which input address is used. When I first introduced you to the 8051's external bus, I showed that there was an address latch used to save the least significant 8 address bits (on P0) before doing the I/O (which uses P0). This circuit does not have it (to simplify it's wiring).

In this application, I only use 2 bits for selecting which device on the I/O bus to read from. This means that a maximum of four I/O devices can be accessed on the bus. The way it is wired, I can select from software address 00000h, 00100h, 00200h, and 00300 (I mentioned 00000h and 00100h earlier).

However, this really isn't true because I am only checking bits 8 and 9 of the address (with bits 0 through 7 being ignored). So, I can really select the ranges, 00000h to 000FFh, 00100h to 001FFh, 00200h to 002FFh, and 00300h to 003FFh.

To further confuse you, I also ignore the most significant 6 bits of the address, this means that addresses 00400h, 00800h, 00C00h, 01000h, and so on, all behave the same way as address 00000h. This is a very common method of selecting data addresses. In the IBM PC, even though 16 bits of I/O addresses are available, only 10 bits are used. To simplify your program, just work with 00000h, 00100h, 00200h, and 00300h and ignore the, uh, ignored address bits.

If you are doing memory-mapped I/O and not latching the least significant 8 bits, you are reducing the number of addresses you can access to 256. However, this is still pretty good considering how much it simplifies the circuit and is probably more than required in a device like the 8051.

The select circuit used is shown in Fig. 9-27 and is one half of a 74LS139. The _G input is connected to the 8051's _RD line so that, when it's low, the output select is asserted.

I typically make all my enable lines negative active, which fits in very well with the 8051 (because it cannot source current) and other logic devices such as the 74LS244 used in this application. A 74LS374 can also be used with the output of a negative active address selector, like the '139, connected to the clock pin (this will be shown later in the book).

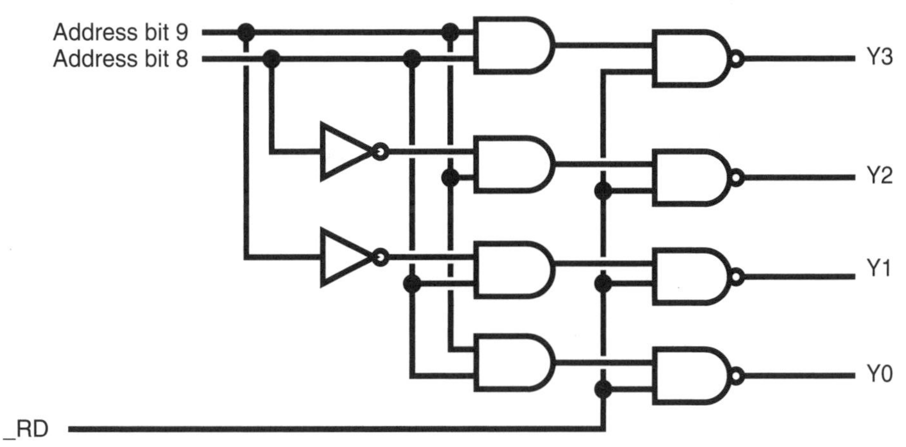

FIGURE 9-27 Memory-mapped I/O address selection.

PROG27 is a very simple program (only requiring 19 bytes of program space) and continually polls either address 00000h or 00100h and outputs the value onto the eight LEDs on P1.

This program shouldn't be difficult to understand, but there are two points I want to make from the program.

The first is that I only use the DPTR register to access the external device. This is normal. While the R0 and R1 bank registers can be used to access memory-mapped devices in the first 256 addresses of the 64K (16-bit) external address space, only the 16-bit DPTR register can be used for accessing external memory at address 00100h and above.

The second point is how I poll the address select bit (P3.2) and use it for specifying the external I/O address to be used.

The code:

```
mov     A,P3          ; Get the specified address
mov     C,ACC.2       ; Save button value in carry
clr     A             ; Clear the accumulator
mov     ACC.0,C       ; Put the button value into the acc.
mov     DPH,A         ; Update the DPTR to the address
```

first gets the value of P3 and then just loads the bit we're interested in (bit 2) into the carry flag.

With the bit saved, the accumulator is cleared and then the bit is stored back into the accumulator as bit 0. When the contents of the accumulator are put into the high 8 bits of the DPTR register (which is DPH), the address is selected between 00000h and 00100h.

I found that, with an 87C520, when P3.2 was left floating (pulled high by the pin), all zeros were read at address 00100h, which is what I expected because there is nothing driving the line.

When I pressed the button, the 8-bit DIP switch that I used for the device to poll was read.

I originally designed this circuit to include a '374 as a data latch and output values on LEDs driven by the '374. However, this turned into a real nightmare of wiring (and changed the experiment from learning how memory-mapped I/O could be implemented to learning how to wire a complex circuit). This circuit can be expanded to include four input sources and four output registers (using the second half of the '139).

PROG28: External Memory

In the last experiment, along with showing how the program reads and writes to external memory, I also wanted to show a few aspects of how the 8051 executes and how you can monitor externally how the program executes. Despite the fact that this program will take the longest to wire, I think you'll find the time spent going through it will be well worth the effort.

For this application, we start with the circuit that we have been working with for the last few experiments and then add a 74LS373 and a 6264 8K by 8-bit SRAM.

In Fig. 9-28, I have not included the pin numbers for the address latch (identified in the figure as the "'373") and the SRAM ("6264"). The pin-outs of these components are shown in Figs. 9-29 and 9-30.

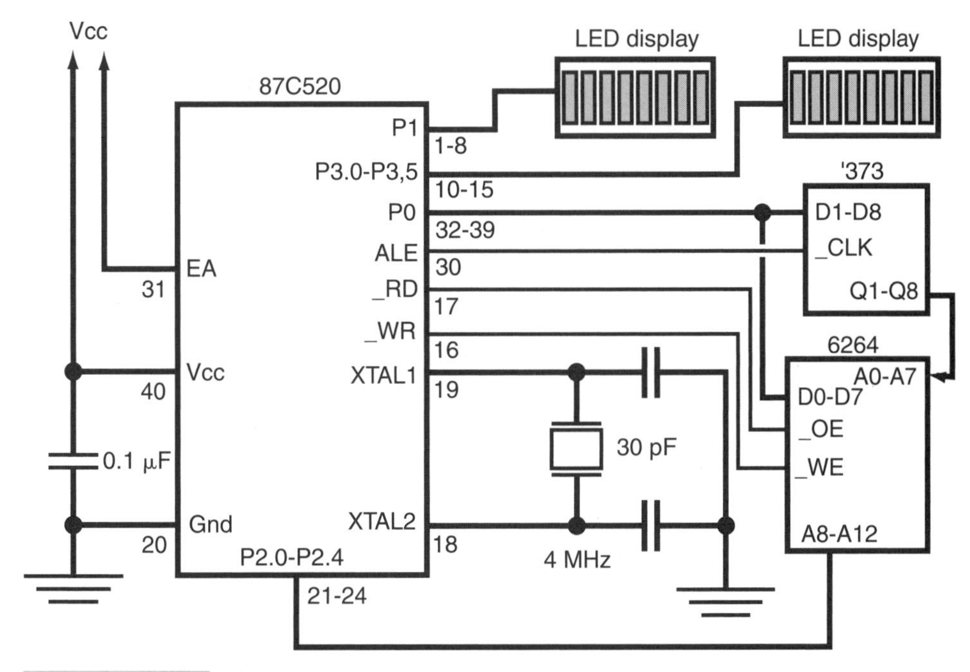

FIGURE 9-28 External memory wired to the 87C520.

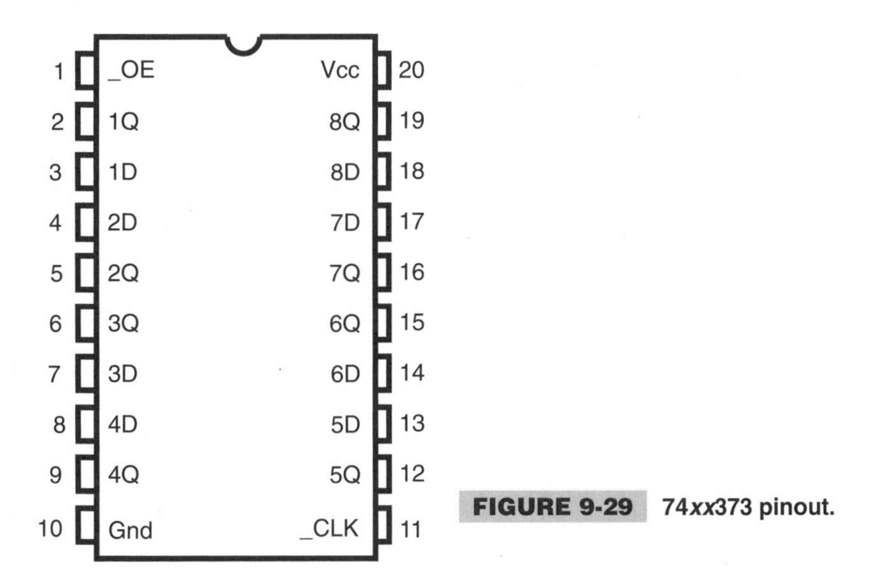

FIGURE 9-29 *74xx373 pinout.*

Both these parts are pretty run-of-the-mill, and you should be able to find them easily and quite cheaply. I used a 74LS373 because I had one lying around my work bench, and it will work fine at 4 MHz. For high-speed applications, a 74F373 is recommended.

You might be more familiar with the '374 as an 8-bit latch (and it's the part that I normally use). The reason why the '373 is recommended for the external memory latch is the

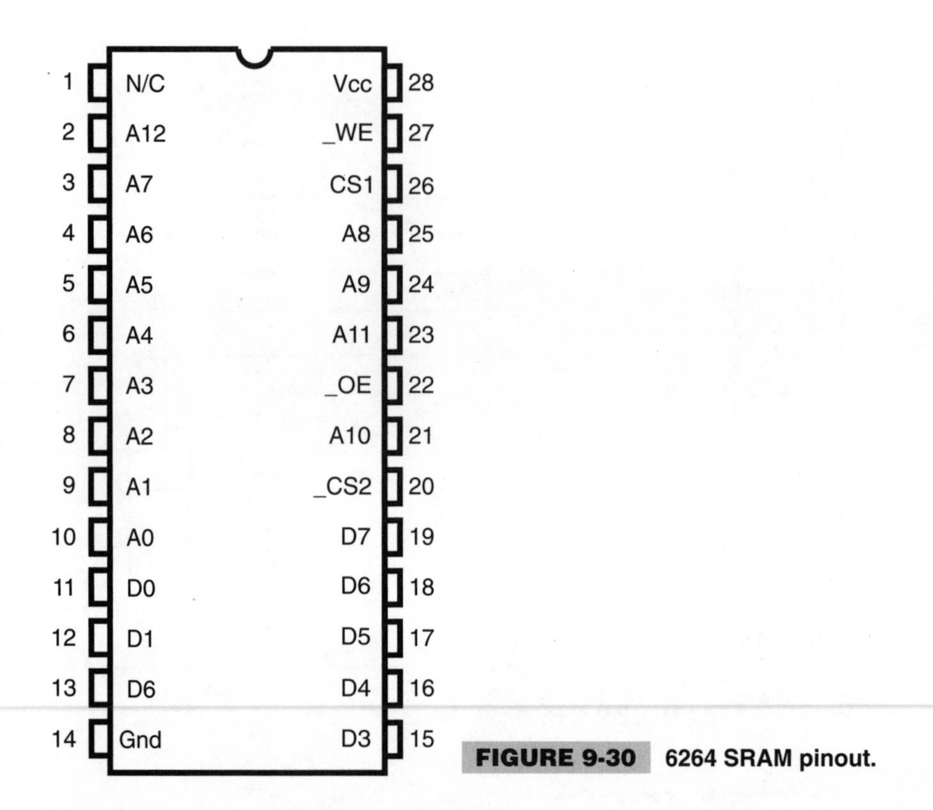

FIGURE 9-30 **6264 SRAM pinout.**

pass through feature of the device. When the Clock pin (pin 11) is high, the value at the input pins is passed directly to the output pins. In the '374, the new value shows up at the pins when the clock pulse returns to a low state. This method of operation gives a bit of additional time between when the address is available to the SRAM to when it is read or written to.

When I assembled this application, once I had the 87C520 ready (i.e., P0, P2, _RD, and "_WR" disconnected), I then placed the 74LS373 and 6264 on the protoboard and wired their power (with a 0.1-µF decoupling capacitor) and grounds. For the 74LS373, I tied the _OE pin (output enable, pin 1) to ground. The _CLK line was connected to the ALE pin of the 8051.

On the 6264, I tied CS1 (pin 26) to Vcc and _CS2 (pin 20) to Ground to ensure that when the _RD or _WR lines became active, the SRAM would respond. These lines could be connected to P2 lines to select the address the 6264 is active at. By adding these lines, the 6264 will be specifically addressed within an 8K block within a 32K (15 address line) address space. Connecting these select lines to the address lines of the 8051 can result in a simplified circuit because address decoders are not required.

After the chip selects and read/write lines were connected, I went through and wired P0 to the 74LS373's D1 to D8 pins and then to the 6264 as data lines D0 through D7. This is the multiplexed address/data bus. The first cycle of the I/O operation is to load the '373 with the least significant 8 bits of the address. The I/O operation is to take place in the next cycle.

This can be seen in Fig. 9-31.

The ALE (address latch enable) line is first asserted to load the '373 with address lines 0 through 7, and then the P0 bus is used for transferring data.

Once the input to the 74LS373 was wired, I then wired the 6264's A0 to A7 bits with the outputs (Q1 through Q8) of the '373.

With the least significant 8 address lines and the 8 data lines wired, I then finished off with the remaining 5 address lines, starting at P2.0, connected to A8 and going to P2.4 and connecting it to A12.

With the wiring behind you, you can now load the software (Listing 9-31). This application performs a "Data equals Address" test on the SRAM for each 256 address block in the SRAM. The 16-bit DPTR is used for addressing the external memory.

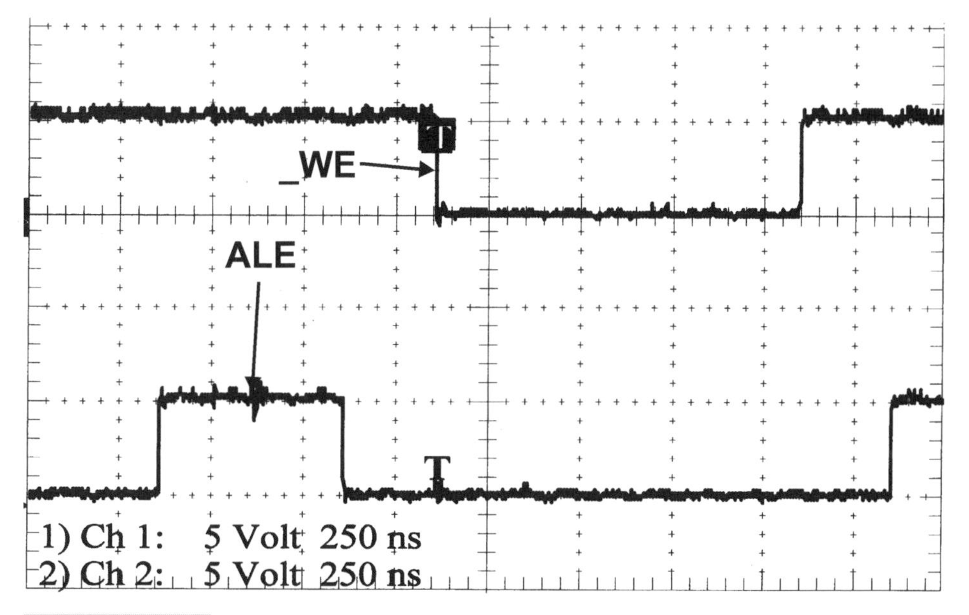

FIGURE 9-31 87C520 external memory waveform.

LISTING 9-31 PROG28.

```
;  PROG28 -  Interfacing to an SRAM
;
;  This application reads and writes to the 8K by 8 SRAM connected to
;   the P0/P2 bus.
;
;  Myke Predko
;  98.03.10
;
;  Hardware notes:
;  80C520 running at 4 MHz
;  P1.0 is connected to a set of LEDs
;  P3.2 is connected to momentary On pull-down button
;  P0/P2 I/O bus wired to a 6264 8Kx8 SRAM with a 74LS373 for address
;   buffering

;  Variable declarations
```

LISTING 9-31 **PROG28 *Continued.***

```
 org 000h
  mov   DPTR,#0                  ;  Initialize DPTR to the I/O address

Loop1:                           ;  Load the SRAM

  mov   A,DPL                    ;  Do an Data == Address test

  movx  @DPTR,A                  ;  Save the low 8 bits of the address

  inc   DPL                      ;  Point to the next address

  mov   R0,DPL

  cjne  R0,#0,Loop1              ;  Do until the first 256 bytes of SRAM
                                 ;   loaded

Loop2:                           ;  Now, compare the data

  movx  A,@DPTR                  ;  DPL is at 0 again

  xrl   A,DPL                    ;  Does the data match the address?

  jnz   Error                    ;  No, show the error address

  inc   DPL                      ;  Else, try the next address

  mov   R0,DPL

  cjne  R0,#0,Loop2

  inc   DPH                      ;  Now, repeat with next 256 addresses

  ajmp  Loop1

Error:                           ;  Miscompare, show the expected/actual
                                 ;   on the LEDs

  movx   A,@DPTR                 ;  Read the value again

  mov    P1,A                    ;  Display on P1's LEDs

  mov    P3,DPL                  ;  Display the expected

Loop3:                           ;  The read/write worked okay, just loop

  ajmp   Loop3                   ;  Now, loop forever
```

For each byte in a 256 address block, the least significant 8 bits of the address are written into the byte at that address. Once this is complete, I then read back the 256 bytes. If the address is correct, I then increment the high byte of the DPTR and repeat the process forever.

If there is a miscompare on the readback, I then display the actual and expected values on the LEDs on P1 and P3.

After wiring the application and burning PROG28 into an 87C520, you can now try it out. If you have wired everything correctly, the LEDs connected to P1 and P3 should remain out. If any of the LEDs light, then go back and look at your wiring. Chances are that there's a wire out of place (or my favorite, put on the same pin as another wire).

With this, you can start playing around with the application.

One of the first things you probably want to do is look at the signals going to and from the SRAM. Figure 9.31 shows the relationship between the ALE and _WR line. The lower 8 bits of the address are latched into the 74LS373 when the line goes low, and the data is latched into the SRAM when _WR goes high (which, in this example, is 1.5 μsec later).

If you were to look at the address lines themselves, you'd notice that A0 was different each time _WR goes low, A1 changes every two _WR cycles, A2 changes every four _WR cycles, and so on.

If you don't have access to an oscilloscope, you can see this relationship with a logic probe. If you put your logic probe into pulse mode, you'll see that at the lower addresses it screams, which indicates that the address is changing too fast for it to keep up. As you go to the higher addresses, you'll find that the changes in the address line slow down to the point where you can watch the high and low lights change at a rate of several times per second.

You can also check to see if the program is working properly by pulling one of the address lines from A0 to A7. If you pull any of the lines from A8 to A12, the program won't change because the least significant 8 address bits will never be in error (and that's all the program is really testing).

Before finishing this program, I did want to explore the operation of the external memory accesses a bit further. To get a better idea of what's going on, I expanded my oscilloscope view until I could see what was happening between two loops. To observe this, I moved back my time scale until I could see two _WE pulses (Fig. 9-32).

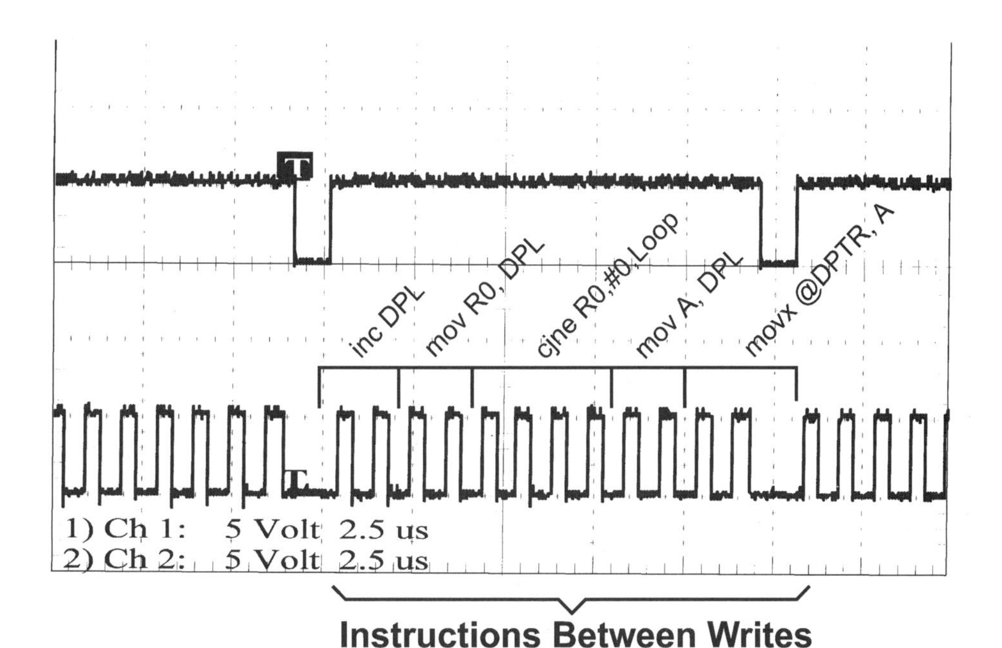

FIGURE 9-32 87C520 external memory waveform.

The most obvious thing to notice is that ALE pulses for each instruction cycle. This was a bit surprising to me because I expected it to only be active on an external address. However, as the oscilloscope shows, the ALE is active for each instruction cycle (for two clock cycles).

I then wanted to see if I could figure out what was happening. I took the scope picture and marked on it the cycles required for each instruction in the write loop ("Loop1" in PROG28).

When I did this, I discovered something very interesting; each instruction seemed to take the number of cycles specified in the documentation except for the `mov @DPTR,A` instruction. This instruction actually takes three instruction cycles, instead of the two specified in the documentation.

However, other than this, I am able to chart how the program executes. This could be useful as a tool for timing applications. By doing an external write at either end of something you want to time, you could observe exactly what is happening.

In Closing

In this chapter, I have gone through a lot about the 8051. As I have been writing the book, I have found other areas to experiment with, and I have gone back and added them to this chapter (which is why some PROG#s are out of order).

These simple experiments are invaluable tools for understanding how the 8051 works. I encourage you to make sure you understand exactly what is happening in each line of code and at each I/O pin of the 87C520 or AT89C2051.

By going through these experiments, you'll have a very good understanding of how the standard 8051 operates. What you might want to do is begin creating your own experiments to test hardware that is specific to devices that you are going to work with. For example, I really haven't investigated the extended timer modes of the 87C520 or the analog voltage comparators of the AT89C2051.

When you set up your own experiment plan out what you want to understand and what the results are that you expect. The experiments can be run from inside UMPS or in actual hardware with LEDs to indicate whether or not what you expected happened.

10

EMULATORS

CONTENTS AT A GLANCE

Over the past few years, I have tried to fix in my mind what is the best way to use emulators in application development. In my day-time job, all too often I discover young engineers using an emulator on an application that was written and designed quickly and sloppily, with the feeling that the emulator will find the problems, so there is no point in putting in a lot of up-front effort. This turns into a serious fallacy because, without properly planning out how an application is to work and designing with this information, the engineer will end up doing this work but will have lost much of the initial time doing the quick-and-dirty initial design.

Having said this, there are a number of cases where an emulator is the most useful tool you can get your hands on for unexpected or intermittent problems. In these cases, an emulator is an important debugging tool, but it should not be considered as a part of application development.

Types of Emulators

As I look through advertisements for 8051 products, I have found that there can be quite a difference in cost and capabilities between different "emulator" devices. I have put the word "emulator" in quotes because the different manufacturer's devices can be radically different. Knowing what the options are can make the difference between an expensive tool that you use all the time or one that will sit on a shelf gathering dust.

On the market today, there are four types of application debugging products that can be called emulators. These devices range from being quite inexpensive and limited in performance to very expensive and able to support all aspects of your applications. The primary differentiator is the speed (performance) at which the emulator is capable of executing. Many inexpensive emulators have features that rival that of the most expensive and capable tools.

The first type of emulator is a hardware simulator, which uses a hardware interface to the development system (the PC or workstation) to emulate the 8051 microcontroller (Fig. 10-1). The actual hardware is very simple and, for the 8051, could be as few as eight or nine standard TTL logic chips.

In this type of emulator, a PC program is used to write to the latch and read the pin states back via a tristate ('244) driver.

The advantage of this type of emulator is that it is very cheap. Often the hardware costs for the emulator will be less than $50 dollars (although the software costs can be much higher). Several product manufacturers will provide software tools to interface with assemblers, compilers, and source-code-level debuggers.

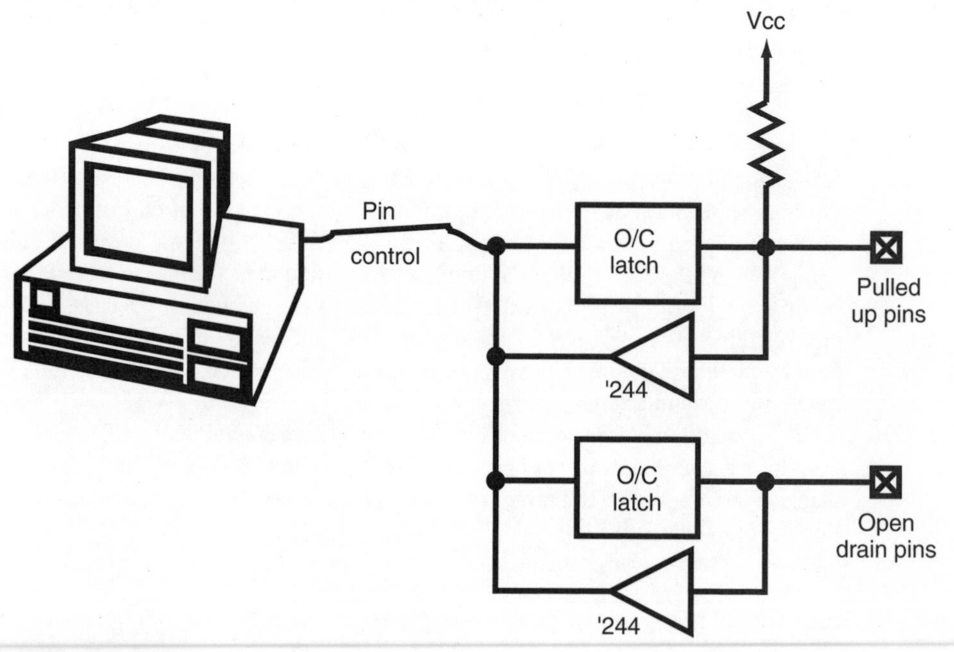

FIGURE 10-1 Simple I/O port emulator.

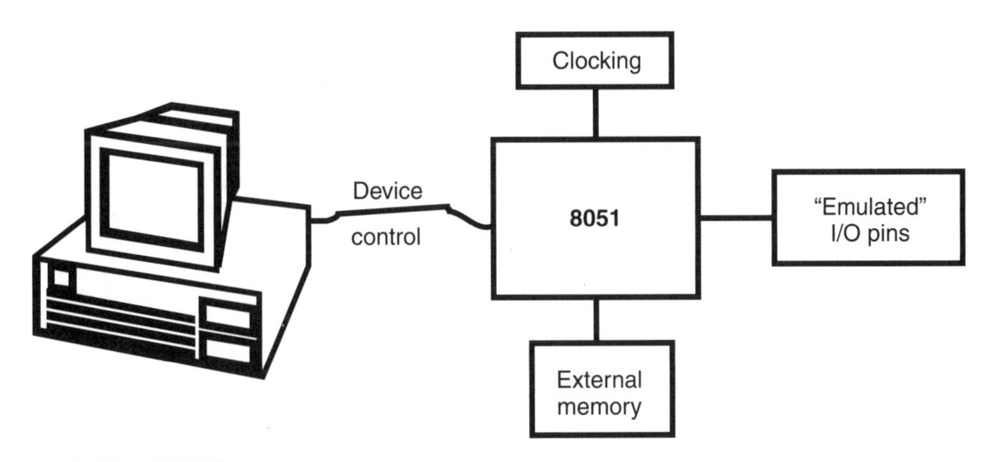

FIGURE 10-2 **8051 as the emulator.**

The big disadvantage of this type of emulator is that it can only run at slow speed (often only hundreds of instructions per second). Another drawback of this emulator is the total reliance on software to provide hardware features such as timers and serial ports, which means that some events can be missed or some features (such as serial I/O) cannot be properly tested out in the emulator.

This type of emulator is most appropriate for learning a device because it is an inexpensive way of seeing how the microcontroller device works.

In the next type of emulator (Fig. 10-2), the application to debug is downloaded into the 8051's external program memory and then executed directly out of the external program memory. In the example emulator application, I run the complete emulator code out of the DS87C520's built-in control store EPROM, but other devices might require a special application to be run on the PC to control the emulator (and provide source-code-level debugging).

This type of emulator eliminates many of the deficiencies of the first type, while still being quite inexpensive. Applications can run at high speed in this type of application, but this speed is not necessarily the speed that you want.

The speed issue could be resolved with a programmable clock, but this would required intelligence to change the host communications speed or have the software compensate for it. For this reason, in my application, I just stick with a constant speed and concentrate on coding the application in such a way that it uses the built-in timers for delays whenever possible and not the time required to execute instructions.

This type of emulator can also only do a subset device to the one used in the emulator. For my emulator project given later in this chapter, I have emulated a pin device with 15 I/O pins with a device with 32 I/O pins. As in the case later in this chapter, there might not be all the features available in the target device.

This subset emulator can lead to problems if features of the emulator microcontroller are accessed. For the emulator that I present at the end of this chapter, if the application writes to enhanced features (such as the Dallas Semiconductor ROM size register), the emulator might stop running and lock up. These problems can be found and eliminated (which is why you're using an emulator in the first place) by searching out the address where the problem occurs.

The third type of emulator (Fig. 10-3) is the most flexible because it's made up of discrete logic that mimics the operations of the microcontroller.

The biggest advantage of this type of emulator is its ability to mimic a wide range of parts by selecting different I/O features. This can be useful for a device like the 8051, where features will allow different applications to work in different devices.

This type of emulator tends to be the most expensive of all four types of emulators. The high cost is due to the need to develop custom ASICs for the processor and I/O functions.

Clocking tends to be an issue for most types of emulators (although this is less important for the last two because the controller can access the memory and control functions of the device directly). In the latter two types of emulators, the actual clock used in the system could be used, although in the third type (with the emulated 8051 ASIC), a programmable clock is often provided because its cost is not significant compared to the cost of the complete system.

The last type of emulator (Fig. 10-4) requires a great deal of forethought by the device designer. A bondout chip is used as a bridge between the development system and the application.

The bondout chip is typically an actual product microcontroller chip with pads and circuitry accessed that will allow external hardware to control its operation. Ideally, the bondout chip will simply be a repackaging of the actual chip (with the emulator features built onto all chips). The big advantage of this type of emulator is that there is no difference between the emulator and the actual product in the application. As well, the clocking circuit used in the application can be used in the emulator.

This type of emulator is not cheap (it usually costs a couple of thousand dollars or more), and each emulated part module containing the bondout chip would cost a few hundred dollars. However, if an application is wired wrong and the bondout chip is damaged, it can be repaired very easily by the user for just the cost of the bondout chip.

When choosing an emulator, remember to make sure that you understand which application development tools it works with. Using emulators for code-level debugging is especially useful and will simplify the effort required to debug an application. This is especially true if you are writing in a high-level language. If the emulator and development system are not capable of working at a source code level, you will have to print a listing

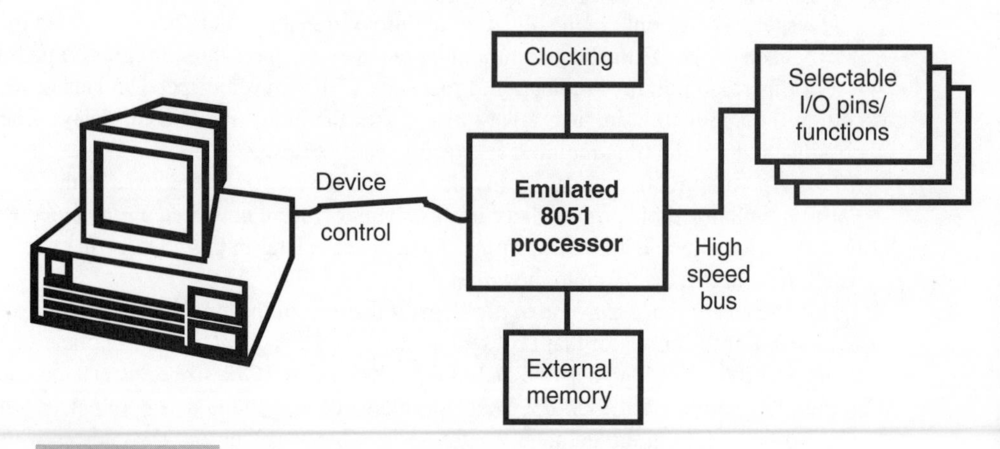

FIGURE 10-3 **8051 as emulator with changeable I/O.**

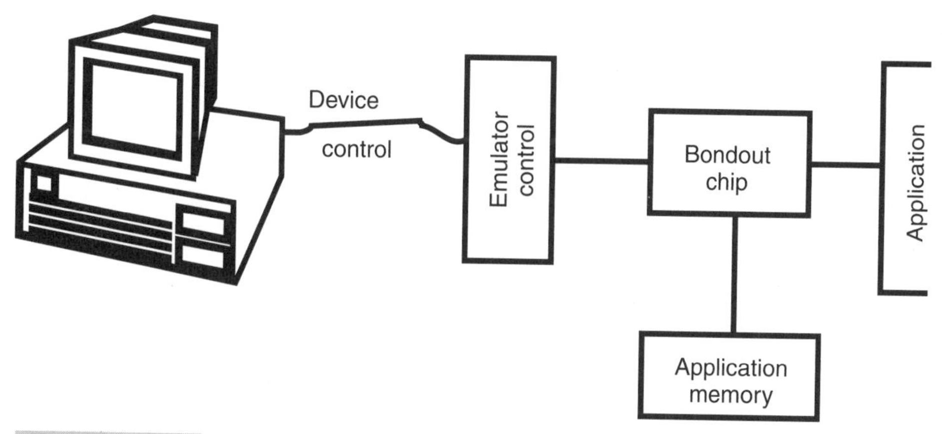

FIGURE 10-4 **8051 as emulator using a bondout chip.**

with the assembler statements and addresses. This will very quickly turn into an unreasonable amount of work, and you will find that your (potentially) very expensive tool is difficult to work with.

Monitor Programs

A monitor program for a computer processor is a program that is intended to be used in the debugging of applications. Hopefully this usage of the term *monitor* will be familiar to you and won't be too much of a surprise (i.e., you thought I was talking about an application that drives a video monitor). They seem to be somewhat unusual now, their function having been replaced by GUI-based simulators and emulators.

To finish off the previous paragraph and indicate that monitors were used all the time, I originally wrote "Although when I was in school..." which makes me sound like I went to school in the dark ages. It was actually the middle ages.

Recent advances have really improved the operation of developing and debugging applications, but there are some situations where a monitor program (with an integral debugger) is required.

One of these is the monitor/emulator project that I am presenting at the end of this chapter. Many of the features and implementation methods I'm going to present here will be used for this application.

A *monitor* is a program that controls the execution of an application program and gives a user an interface into what's happening. I like to think of it as a moat around the application, preventing access to the 8051 only in specific circumstances. To provide this control interface, a separate console system is provided. Usually this is a PC running a terminal emulator program (as I use in the monitor/emulator I present next). (See Fig. 10-5.)

The usual operations provided by a monitor include:

- Application loading
- Application execution control

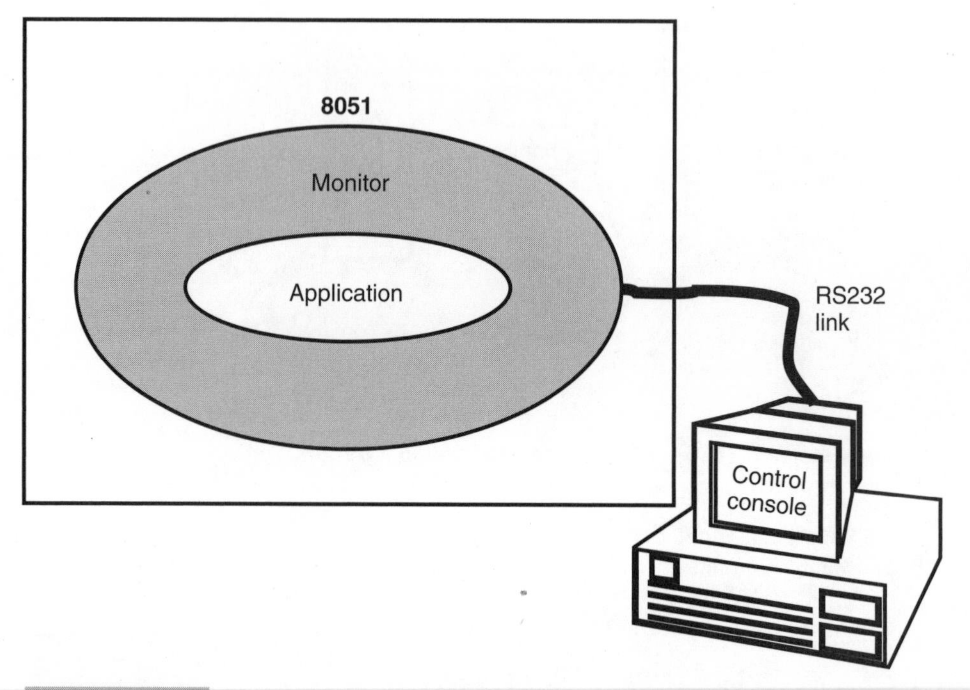

FIGURE 10-5 **Monitor/application execution.**

- Single stepping
- Set, reset, and list breakpoints
- Display application code
- Modify application code
- Display scratchpad RAM and registers
- Modify scratchpad RAM and registers

I will go through each of these operations and describe how they would be implemented in a monitor program. The base assumption is that the application to debug using the monitor would be stored in RAM connected to the external memory bus.

Loading an 8051 program is quite simple if the assembler/compiler produced a hex- or Intel-format object file. The hex file contains all the instructions and addressing information for an application. The data is simply put into RAM memory as an image of the start of control store. If you are using a monitor on an 8051 with 2K or less of control store, the application can be located anywhere in the control store space at the start of a 2K page. This is possible because the `ajmp` and `acall` instructions can access anywhere within a 2K page.

If the device can have more than 2K of control store space, you are going to have to either change all the `ljmp` and `lcall` instructions to the correct actual address inside the 8051 or somehow disable the built-in EPROM and move the RAM to the start of the control store space and start executing from there.

At first blush, executing an application probably seems very simple, but there are a number of issues to consider. The first is: *Where* is execution going to begin? To allow control over execution, I treat the monitor program like a *subroutine* to the application (similar to how an RTOS looks like a subroutine to the application). Conceptually, this might fly in the face of the Fig. 10-5, where I showed the monitor *outside* of the application, but the reason for this is to save the current program counter.

To begin executing at an arbitrary address (say 01234h), the following code could be used:

```
mov     9,#012h      ;  Save the high address byte
mov     8,#034h      ;  Save the low address byte
mov     SP,#009h     ;  Set the stack pointer to these addresses
ret                  ;   Start executing at 01234h
```

In this code, the stack is loaded with the address to execute at and then a `ret` (return from subroutine) instruction is executed, which pops the address I want to start executing at and loads it into the 8051's program counter.

Single stepping can be implemented two different ways in an 8051 monitor. The first way is a brute-force approach. The instructions following the current instruction could be overwritten with an `lcall` *`Monitor_Address`* instruction. This means that the instructions that were overwritten by the `lcall` instruction have to be saved and replaced before the next instruction is to execute.

There is one potential problem with this method. What if the `lcall` instruction overwrites table data to be used in the single step?

For example, a table read could consist of the following code:

```
  movc    A,@A+PC        ;  Read from "Table"
  ret
Table:
  db      ...
```

In this example, single stepping through the `movc` instruction will have a problem if the accumulator is loaded with one or two because the `lcall` instruction will overwrite the first two bytes at `Table` and will return the last two bytes of the `lcall` instruction.

Ideally, a single-byte long jump instruction to a specific address (also known as a *software interrupt*) would be available in the 8051 for this use.

This method probably seems pretty reliable, and for the most part it is. However, there are some cases where there can be problems that will complicate the situation. The most serious ones are jumps and conditional branches. For jumps, the destination will have to be replaced with the `lcall` instruction, rather than the next location and for conditional jumps. Both the next instruction and the destination will have to be set with the `lcall` instructions.

There is another way of implementing a single step that avoids the concerns: setting an unused timer to overflow after executing one instruction cycle and then jumping to an interrupt handler. This method is much cleaner than the first method, although interrupts have to be enabled for this to work.

A non-maskable single-step interrupt would be implemented in the processor in which case the monitor would appear as an interrupt handler (especially if there was a single-byte long jump instruction) instead of a subroutine. The two previous suggestions are actually very similar to how the 8086 microprocessor in the PC handles single stepping and breakpoints.

Breakpoints are often an extension of the first method I showed for implementing single stepping. An `lcall` instruction is inserted at the address that you would like to stop at. One problem with breakpoints is that not only do the three saved bytes of the application that `lcall` overwrote have to be saved, but the address of the breakpoint must also be saved so that the application can figure out where to replace the code.

Reading and writing scratchpad RAM and registers and memory is quite simple with only a few problem cases to watch for. Using R0 and R1 will give you access to just about everything in memory. However, before doing a read or write to any registers, you will have to check to see if the register to be accessed is a context register (such as the accumulator, bank, PSW, and the IE registers).

One serious consideration is how you are going to implement a read/write access to a SFR address. You could create subroutines to access each SFR, or you could use a bit of self-modifying code:

```
SFR_READ:
  mov DPTR,#SFRR_ADDR+1        ;  Have DPTR point to the register to read
  movx@DPTR,A                  ;   Save the address to read
SFRR_ADDR:
  mov A,012h                   ;  This can be any register
  ret
```

In this code, the address in the accumulator will be written into the instruction that reads from (or writes to) the SFT. Yes, self-modifying code is wrong; however, in cases like this, I feel it is appropriate as an advanced programming technique to carry out operations that cannot normally be executed by the instruction set. I would not recommend doing this in your applications (actually, I wouldn't recommend writing data into your stack space as I showed earlier for jumping to a specific address).

Allowing reading and writing to the control store is accomplished by ANDing the _PSEN and _RD lines together and having the application control store show up as external memory. If you could read anywhere in the application source (a 16-bit control store access instead of the 8-bit `movc` instruction) without having to modify large blocks of memory, you would probably just connect _PSEN. However, this isn't available, so you should just AND the two control pins together and avoid the hassle of writing overwriting code into the application RAM that will allow reads using the `movc A,@A+PC` instruction.

PROG29: AT89C*x*051 Monitor/ Emulator using the DS87C520

I had a both a lot of fun and a lot of frustration creating the emulator that I have presented in this book. The emulator (Fig. 10-6), as it's presented, consists of a DS87C520 wired to an external RAM for both external storage and control store as well as an interface to console and an application circuit. I ended up emulating an Atmel AT89C2051, although other 20-pin devices could be emulated using this device.

If you are planning on going through and building each project in this book, I recommend that you read through this one first and then go on and work through the other projects and then come back and build this one. This project is quite complex (actually the

FIGURE 10-6 AT89C2051 emulator.

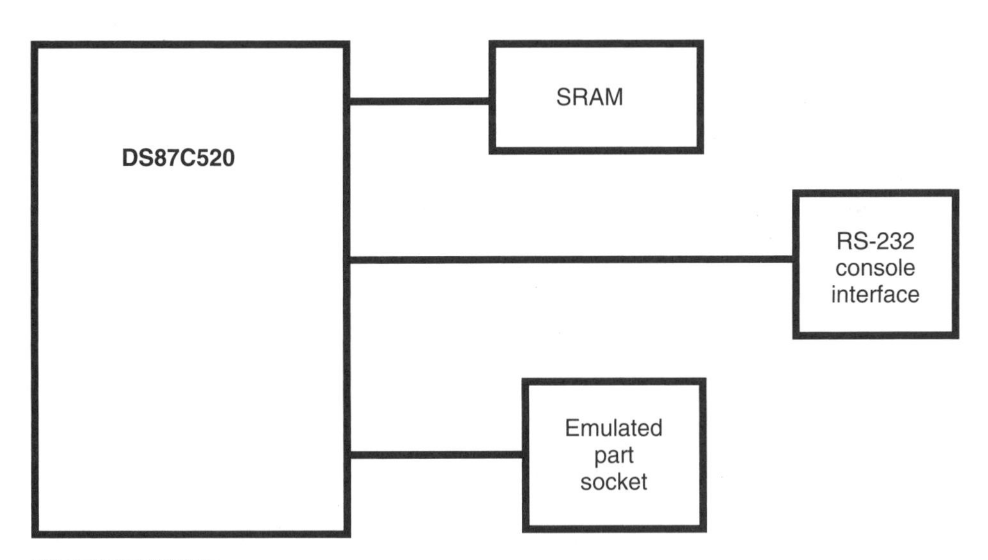

FIGURE 10-7 AT89C*x*051 emulator block diagram.

most complex application presented in this book) and uses code that I will explain in greater detail later in the book.

Before building this application, I highly recommend that you build a raw card (I show the foil patterns in this section and the board gerber files are on the CD-ROM) rather than attempt to build it using prototyping techniques. While the actual concept is quite simple, as is shown in Fig. 10-7, the actual implementation will seem quite complex (see Fig. 10-8), and it would be very difficult to wire wrap.

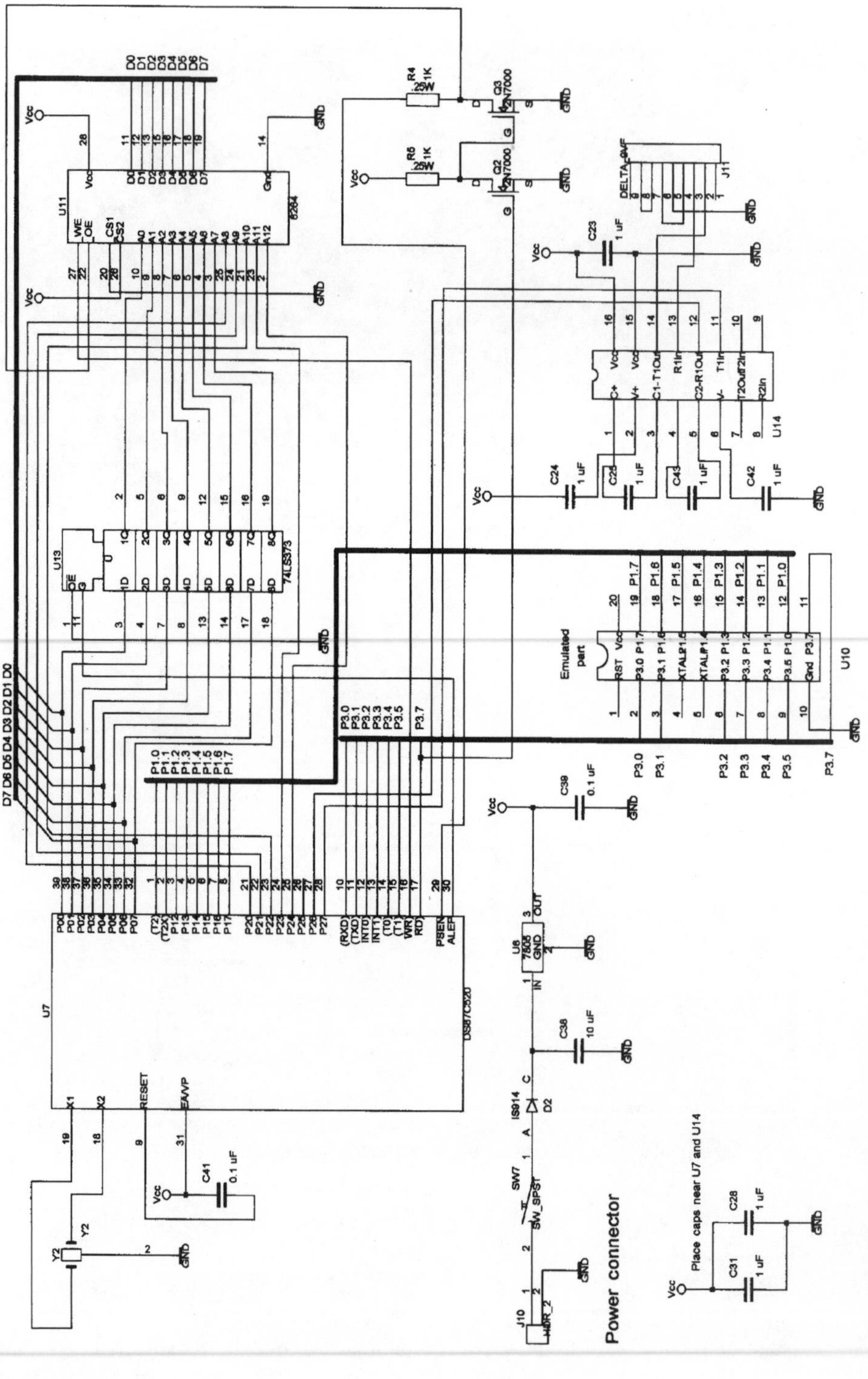

FIGURE 10-8 AT89Cx051 emulator schematic.

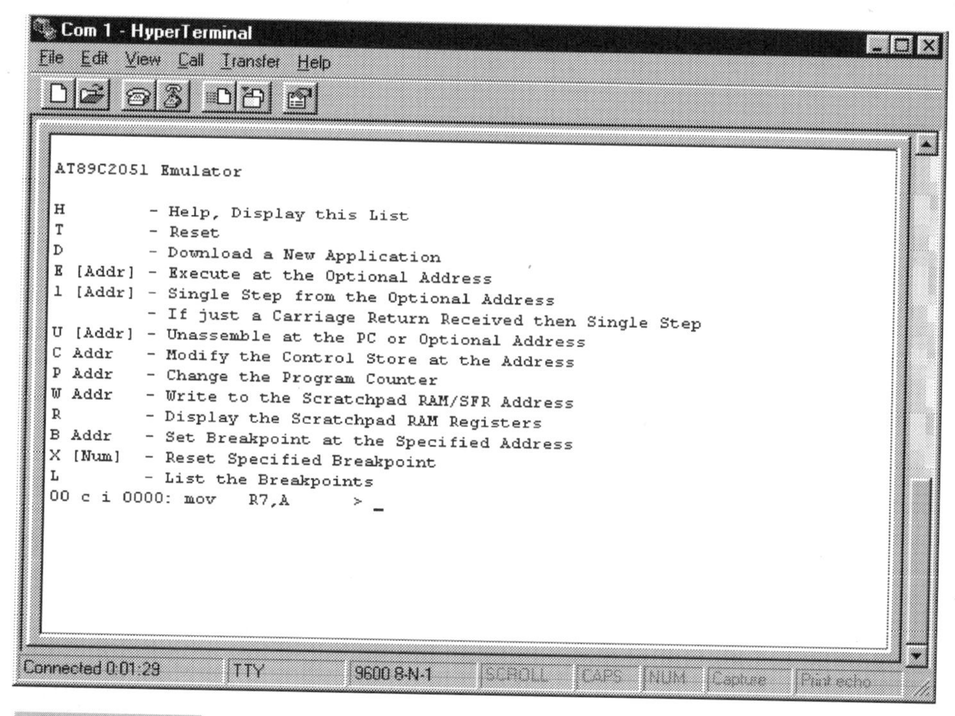

FIGURE 10-9 **AT89C*x*051 emulator power-up.**

The circuit itself centers around the DS87C520 HSM microcontroller running at 20 MHz. Other 40-pin 8051 compatibles could be used without changing any of the code except for the serial communications routines. One of the big philosophies that I wanted to use when developing this application was to not use any of the advanced (8052 or HSM) features present in the DS87C520. Instead, I worked through the application without using the built-in serial ports, Timer2, or any of the other features that could have made implementing the application easier, because I wanted these features to be available to the emulated part.

This application demonstrates one important idea that I like to implement in my microcontroller applications: taking advantage of the microcontroller's intelligence when interfacing with a host system. This emulator only requires a host system's serial port and a terminal emulator program; it does not require any custom PC or workstation software to run. By placing the user interface inside the microcontroller, I have created a product that can work with just about any hardware or software system on the market today. The host system's only requirement is that it has an RS-232 port.

The commands and entry screen are shown on my PC running Hyperterminal (at 9600 bps "8-N-1") in Fig. 10-9.

The RS-232 interface consists of a number of subroutines:

- `SendChar`—Send the character in ACC to the host.
- `SendDec`—Send the decimal number in ACC to the host.

- SendBit—Send the binary number in ACC to the host.
- SendHex—Send the hexadecimal number in ACC to the host.
- SendMsg—Send the message number specified in ACC to the host.
- GetChar—Return the incoming character from the host in ACC or set the carry flag after a 10 msec timeout.
- GetCommand—Get the one- or two-parameter command from the host.
- GetHex—Get a hex number from the host or set the carry flag if just CR is entered.

These subroutines are used to provide the basic serial interface to the host. If a serial port was to be used, then the SendChar and GetChar routines could be replaced with routines that simply interface to the built-in serial ports (i.e., read and write to SBUF).

The basic interface consists of the user entering a two-parameter string consisting of a command and, in some commands, an address (such as B 44). These two parameters are used to carry out the commands listed in Table 10-1.

If no command is specified (i.e., just Enter or Carriage Return is sent), then the interface assumes that a single-step command is being made.

Most of the commands are quite straightforward, but there are a few comments I would like to make. The first is how I store the current context information. While the emulator software is running, I need a number of registers for program execution. In the high 128 bytes of the scratchpad RAM (which is not normally accessed in the AT89C2051), I store the bank registers, the stack pointer, the DPTR, the interrupt control register (IE), the PSW, and the two accumulators (ACC and B). Most of these should seem quite straightforward, except for the IE register. In the emulator software, I keep interrupts disabled (to

TABLE 10-1 THE COMMANDS OF THE INTERFACE

COMMAND	DESCRIPTION
H	Display the help screen.
T	Reset the emulated microcontroller.
D	Download a new application.
E	Start execution at the current program counter.
E *Addr*	Start execution at the specified address.
1	Single step from the current program counter.
1 *Addr*	Single step from the specified address and update the program counter.
U	Disassemble the program store at the current program counter.
U *Addr*	Disassemble the program store at the specified address.
C *Addr*	Change the program store at the specified address.
P *Addr*	Change the program counter value.
W *Addr*	Write to the specified scratchpad RAM/SFR address.
R	Display all the scratchpad RAM registers.
B Addr	Set a breakpoint at the specified address.
X	Clear all the breakpoints.
X *Num*	Clear the specified breakpoint.
L	List the set breakpoints.

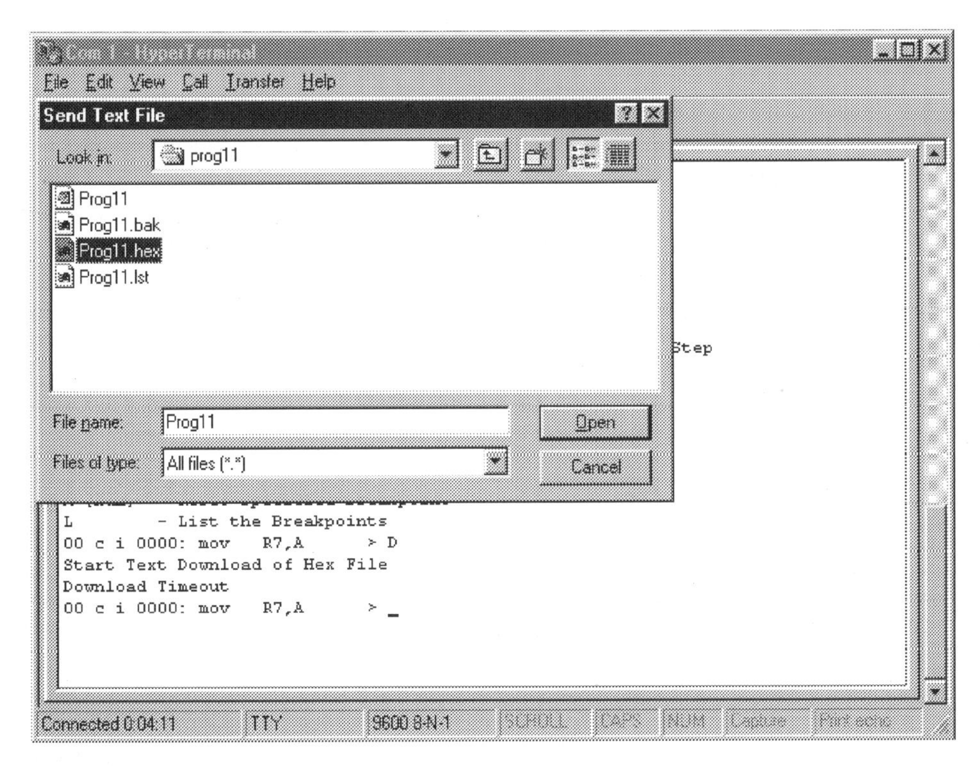

FIGURE 10-10 AT89Cx051 emulator application download.

avoid jumping to an interrupt while I've modified these registers), so I have to keep a copy of IE around.

The program counter is also saved in the upper 128 scratchpad bytes but is kept in a separate location, along with an input buffer, memory for saving code replaced for single stepping, breakpoints, and most importantly, a separate stack for the emulator (so the stack and information is not screwed up by the emulator's execution).

The Reset (command `T`) will reset only the registers available in the AT89C2051 along with the emulated program counter. Note that none of the 128 scratchpad RAM bytes are modified. I didn't modify any of them to help make missing initializations become more obvious.

Changing the scratchpad RAM or SFRs requires the `W` command, which will prompt you for the new byte (just pressing enter will not change the register) to be loaded at the address. This command only works with one byte at a time and is quite tedious if you have to change a large number of addresses.

To read back the scratchpad RAM and the AT89C2051 specific SFRs, you use the `R` command.

Loading application code into the emulator is simply accomplished by sending a text file from the terminal emulator program (see Fig. 10-10). After entering the `D` command, you will select `Send Text File` from your terminal emulator's `File Transfer` command menu and simply select on the .hex file that you want to download into the emulator.

In Fig. 10-10, at the bottom of the terminal emulator screen, you can see the message `Download Timeout`. If a file hasn't been received within one minute of entering `D`, then the emulator will timeout. This brings up a very important thing to note: Because I do not use the built-in serial ports and have interrupts disabled for the emulator, you cannot stop the operation of the emulator during execution (command `E`). This means that, if you forget to put in breakpoints or if your program never executes code at those addresses, you're going to have to hit the BRS ("big red switch") and start all over again.

In the future, I might have the DS87C520's reset controlled by DTR. When the terminal emulator is disconnected, the DS87C520 would then be reset to fix this deficiency.

You can either execute (`E`) or single step (`1` or just Carriage Return) the code. In either case, before executing, the stop addresses (breakpoints, using the `B` command, or the instruction after the current one) will be overwritten with the `ljmp` code snippet:

```
  jbc     IE.7,Step_Skip
  inc     DPL1                  ;  Mark that interrupts were enabled
Step_Skip:
  lcall   Step_Done             ;  Finished, load the code
```

Earlier in this chapter, I indicated that just a single `lcall` instruction could be used for breakpoints or single-step instructions. For some applications, this is true. Where there is a problem is when interrupts are enabled (remember I don't want to be interrupted in the emulator code). This code will make sure that interrupts are disabled when the monitor is entered.

One test that I put in before single stepping or executing is to see if any case of this code snippet is ever placed within 8 bytes of the program counter or another case of this snippet. If this test wasn't in place, then multiple snippet writes could overwrite each other, resulting in invalid code. If there is a problem, then an error message will be transmitted and the `1` or `E` command will be ignored.

Using this code, breakpoints are quite straightforward, but there is a problem with single stepping. There are three cases where the instruction to have the code snipped shown earlier is not the next one. They are jumps and calls, subroutine and interrupt handler returns, and interrupts.

For jumps and calls, I put the previous code after the instruction and at the destination. By doing this, I avoid any problems with conditional statements. Returns require me to go back through the stack and find the next address.

The real problem was supporting interrupts. At each interrupt handler, I put in the following code:

```
 org 023h                      ;  Serial port interrupt
_Ser:
  mov     TH2,A
  mov     A,TL2                ;  Get the operation type
  cjne    A,#1,_Ser_Step       ;  Is it a step or execute?
  mov     A,TH2                ;  Restore ACC
  ljmp    04023h               ;  - Execute the interrupt handler
_Ser_Step:                     ;  Now, return to the step location
  mov     A,#023h              ;  Put the interrupt handler address
  push    ACC                  ;   on the stack
  mov     A,#040h
  push    ACC
  ljmp    Step_Done
```

The `Step_Done` label replaces the three instructions inserted for the stepping and breakpoints.

In this code, I test to see whether or not the code is single stepping (`1` command) or executing freely (`E` command). If it is stepping, then the interrupt vector is saved as the next address to start executing at; otherwise, the interrupt is passed straight to the code in the external RAM.

The external RAM is wired as both external data memory and control store by ANDing the _RD and _PSEN lines together using the two transistors and resistors (see Fig. 10-11). This circuit will AND the two input signals together and output a low to the SRAM's output enable pin (_OE) when either one is low.

I wired a standard 8Kx8 SRAM to the DS87C520 for use as the emulated part control store using this circuit for getting data from it. When I was 'scoping the various signals, I discovered that the _PSEN goes low once every instruction cycle. This means that the P0 port should not be used for anything other than external memory unless address decoding is used.

I used a 100-nsec SRAM and didn't find any problems. At 20 MHz, the minimum access required is 200 nsec for external control store reads.

For the serial interface, I did use P2.6 and P2.7, but this is okay because these two bits are not required for decoding the external memory addresses. P2.5 is also available for other functions.

The level converter used for this application is a Maxim MAX232 (U14), which converts the 0-V to +5-V CMOS logic levels to the –12-V to +12-V inverted RS-232 logic levels. This device requires five 1-μF capacitors for internal charge pumps to generate the voltages required for RS-232 while only requiring 5 V. Later, in chapter 12, I will show you two other commercial RS-232 logic level translator chips that only require 5 V to provide a valid RS-232 interface.

The DS87C520's P1 and P3 ports are passed directly to the emulated part. For the most part, the pins from the emulated part can be connected to the application, but with a few differences that primarily center around clocking and reset. In the emulator, clocking is a function of the DS87C520, and the clock is not redriven to the application. This should not

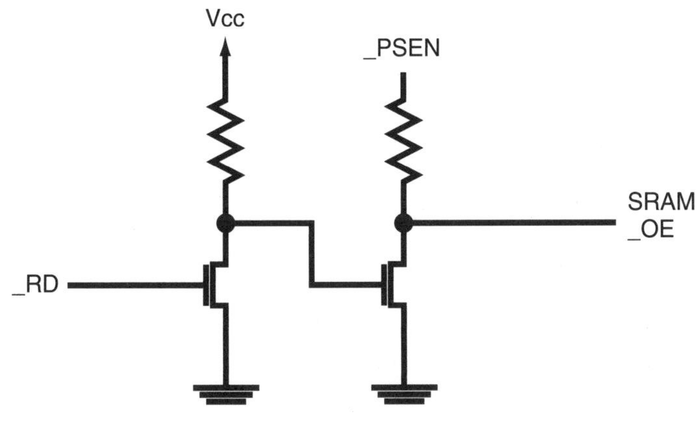

FIGURE 10-11 **AT89Cx051 emulator _RD AND gate.**

be a problem because the clock should be provided by either the built-in oscillator or an external circuit.

In terms of clocking, the DS87C520 HSM runs at 20 MHz. Now, an external source could be used, and attempts at matching the DS87C520's timing could be made to the actual source circuit. However, this would also involve changing the serial I/O subroutine's constants. Ideally, in the application, I would recommend using conditional code for changing delay constants between the 20-MHz HSM and the actual application.

Reset is controlled in the DS87C520. Some kind of control should be available from the circuit. In applications where the power down/idle is used, this could be useful. However, if I had to decide about whether or not to control Reset from the application or the Host's serial port, I would probably go with the host's serial port.

When debugging applications using this emulator, there are a few rules for writing the application code that I would recommend for you to follow. I went through the application code in chapter 9 and found that very few of them would run in the emulator without problems. To allow everything to run cleanly, here are a few suggestions.

To allow single stepping throughout the code, make sure that tables are at least 8 byte addresses away from the `movc A,@A+PC` instruction that loads data from them. Going with this, no jump/call destinations can be less than 8 byte addresses away from the originating instruction.

For example:

```
Loop:                        ;  Finished, loop forever
  ajmp       Loop
```

would have to be:

```
Loop:                        ;  Finished, loop forever
  nop
  nop
  nop
  nop
  nop
  nop
  nop
  nop
  ajmp       Loop
```

On the DS87C520, pins P3.6 and P3.7 are connected to the SRAM's _WE and _RD pins. Looking back, I should have masked P3.6 to _WE by P2.5 (if P2.5 is set, then P3.6 can access _WE). When the application is executing, I would reset P2.5, so the SRAM is never written to. As well, the `movx @DPTR,A` instruction should never be used in the application.

Going along with not accessing P3.6 or P3.7, you cannot use `lcall` or `ljmp` instructions because the application code to be debugged actually is located in external memory, starting at address 04000h instead of 00000h. Now, I do not decode addresses on the board, so in the DS87C520 emulator code, I could disable the internal RAM before executing the application code, but that would add a lot of complexity to the code.

The last rule to follow when writing an application to be debugged with this emulator is to make sure there are 4 bytes of stack space more than the application actually requires to

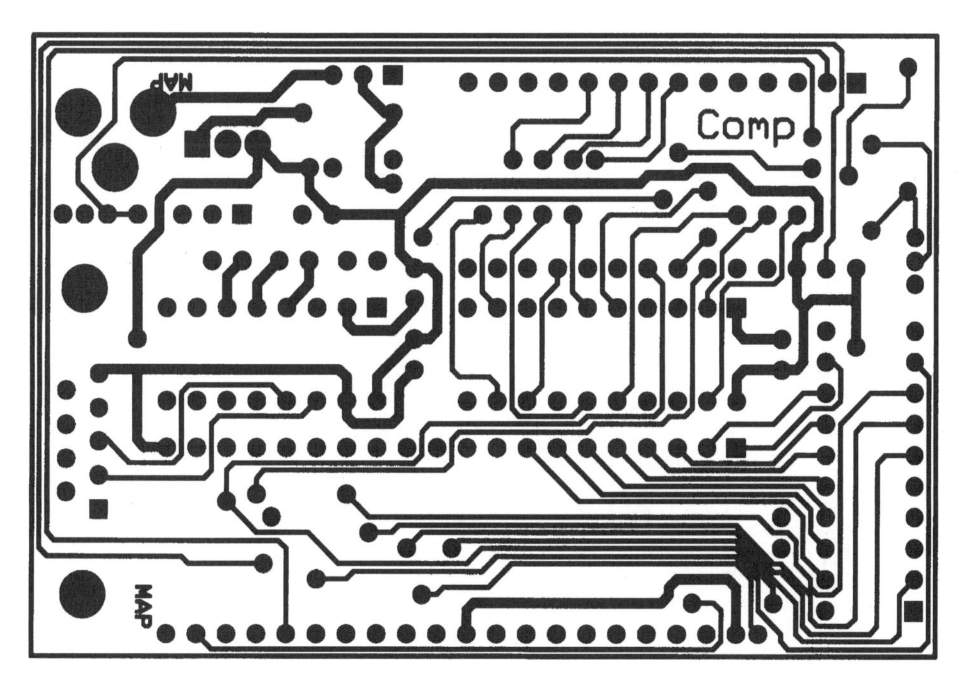

FIGURE 10-12 **AT89C*x*051 emulator top side.**

execute. These 4 bytes are used for saving the address when the emulator code is called; the worst case is after interrupts are disabled and the emulator code is still running. Not only is the execution return address still on the stack, but the current emulator address is also pushed onto it.

The final emulator application code is 5360 lines (and 7410 bytes) long. As good a programmer as I think I am, I'm sure a few problems have crept their way into this code. If you find any, I would appreciate you notifying me with sample source code that shows the problem. As well, you might want to look on my Web site to see if there are updates to the application code.

If you look through the emulator code, you'll see a large number of `lcall` and `sjmp` instructions. After reading through chapter 3, you'd probably feel like I don't think that they should be used. To a large extent, you'd be right.

This application is different from all the others in this book because of its size. As I was creating the application, I found that many subroutines would go over a 2K page boundary (and also within the subroutines, a label could be across a 2K page boundary). To avoid problems with page crossing, I have used `lcall` and `sjmp` instructions almost exclusively because they don't experience the problems that `acall` and `ajmp` do. I ended up making wholesale changes after `SendChar` went into the second 2K page.

For this project, I have designed an embedded card. Like the other embedded projects presented in this book, you can either reproduce the card from the drawings in the book (Figs. 10-12 through 10-14), or use the gerber files located on the CD-ROM.

If you are interested in getting a kit of the emulator, please check my Web site or drop me a line indicating your interest.

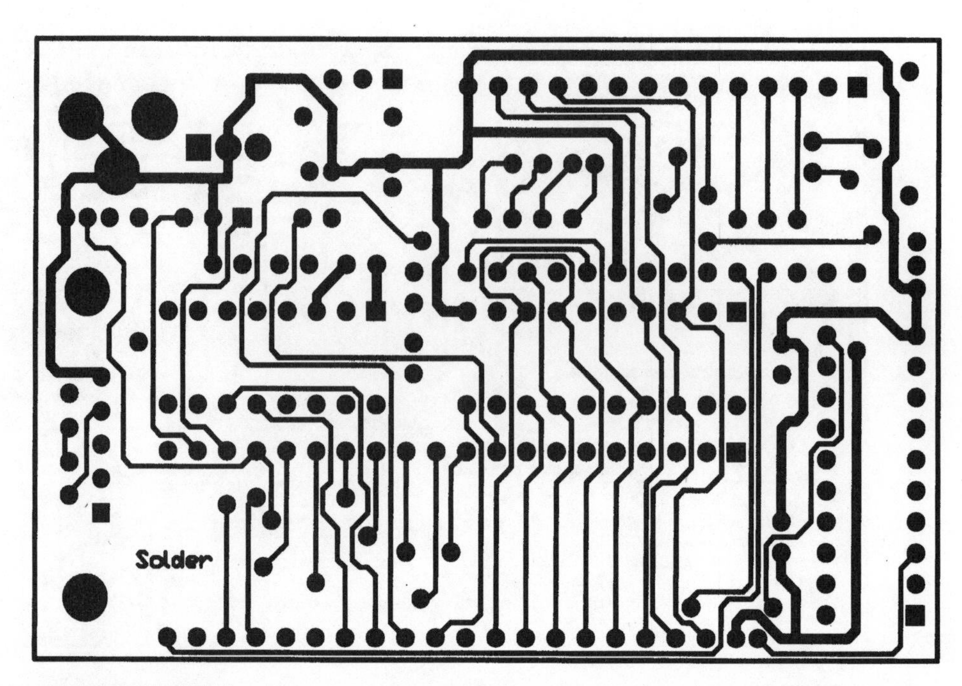

FIGURE 10-13 AT89C*x*051 emulator back side.

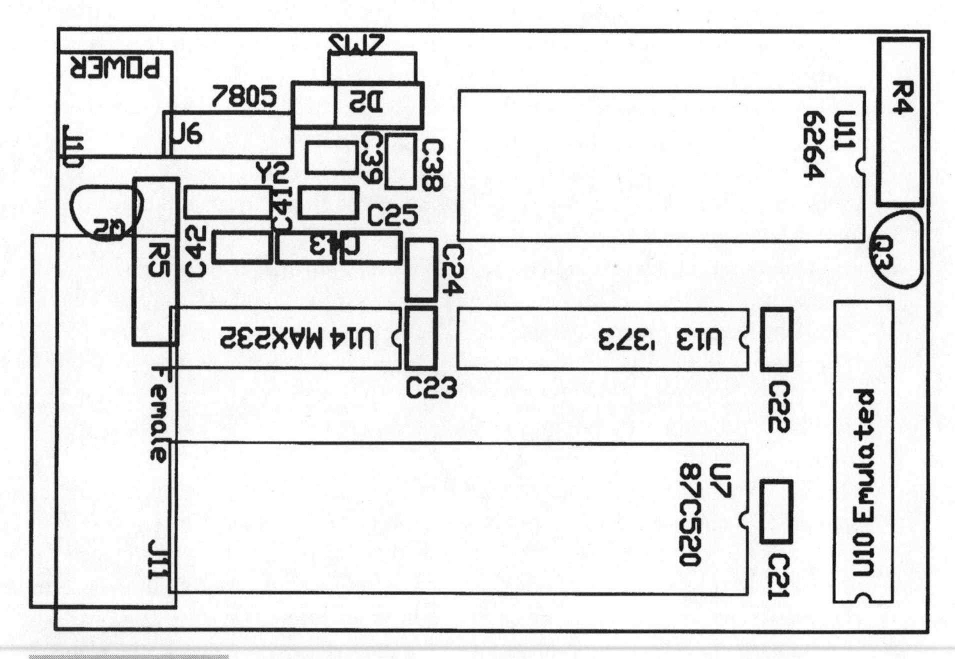

FIGURE 10-14 AT89C*x*051 emulator overlay.

11

REAL-TIME OPERATING SYSTEMS

CONTENTS AT A GLANCE

One of the most useful application programming environments that can be used in a microcontroller is the *real-time operating system* (RTOS). When properly used an RTOS can simplify application development. By compartmentalizing tasks, the opportunity for errors is significantly reduced.

Implementing an RTOS for the 8051 is not trivial and requires a good understanding of how the microcontroller works. This is why this chapter is located after chapter 9, the experiments chapter. Along with a good understanding of the 8051, an understanding of how a RTOS are implemented and how to write applications for them is very important.

In this chapter, I will introduce you to real-time operating systems and give you some example *kernels* that can be run on the 8051 in a variety of different situations.

RTOS Basics

So, what is a real-time operating system? The best definition I can come up with is:

> A real-time operating system is a background program that controls the execution of a number of application subtasks and facilitates communication between the subtasks.

An important point about *subtasks* (which are usually referred to as just *tasks*): If more than one task can be running at one time, then the operating system is known as a *multitasking operating system* (which I usually refer to as a *multitasker*). Each task is given a short time slice of the processor's execution, which allows it to execute all or part of its function before control is passed to the next task in line.

Before going on with this chapter, I should point out a few things. The first is that I usually call a *subtask* a *task* or *process*. I will also blur the difference between an RTOS and an OS (operating system).

The reason why I tend to lump both an OS and an RTOS together is because the central *kernel* (the part of the operating system that is central to controlling the operation of the executing tasks) is really the same for both cases. The difference comes into play with the processes that are loaded initially along with the kernel. In a PC's operating system, the console I/O, command interpreter, and file system processes are usually loaded with the kernel and everything has been optimized to run in a PC environment (which means respond to operator requests). In an RTOS, the actual application tasks are loaded in with the kernel with priority given to tasks that are critical to the operation of the application.

You might be a bit suspicious of an RTOS after what I've written so far. After all, you probably have a PC running Windows 95 or Windows NT, and it's not like these operating systems never crash or have problems.

Not only that, but these operating systems require literally hundreds of megabytes on a hard drive space to operate. With this background, you're probably wondering why I would suggest a multitasking operating system for an 8-bit microcontroller with only 4K of control store.

Continuing with this thread, you might ask: What does an 8051 have that a PC doesn't?

The answer to this question is really in what the 8051 *doesn't* have relative to the PC. The smaller system environment of the 8051 is what's important and makes RTOSs ideal for this type of device. The 8051 does not have a file system, arbitrary amounts of memory (including virtual memory models), or a user console (unless you want to provide one yourself). As well, in the microcontroller's case, you can specify the interfaces and hardware rather than come up with standards that might have to be bent when new technology becomes available.

To make matters worse for the PC, there are literally millions of programmers working on the Windows operating system, drivers, and applications. All of these individuals are humans. With the more code written, the greater the chance that incompatibilities between the system, drivers, and applications will exist.

In a small microcontroller, the team of people developing the application and the interface software is very small and the chances for these incompatibilities not being detected and being passed along to the end user is much more remote. The 8051 has some architectural features and processor instructions that I've really only hinted at to this point that can be used to help implement an RTOS more efficiently than is possible than in other microcontrollers.

A good example of how an RTOS works is in a nuclear power plant's process. Instead of having multiple computers located around the site, one computer could be used to control all the processes needed to run the reactor and get power out.

In Fig. 11-1, I have written all the tasks in boxes and overlaid them on a reactor's block diagram.

For the various tasks, I have labelled them as "high," "med"(medium), or "low." This reflects the priority of messaging and operation of the tasks. This means that, if high priority and medium priority tasks are waiting to execute, the high priority task will run first. So, if you had an emergency in the nuclear power plant, the 'Emergency "SCRAM"' task would override all other tasks until it completes.

Other events, such as non-critical water leaks (which require topping up from the reserve tank) or periods of overcapacity (in which water is circulated through the reactor and radiators to keep everything within operating temperatures) are medium, or regular, priority.

Low priority tasks are the interactions with the operators because, if a task takes a long time to execute (in computer terms), the humans operating the system probably won't notice any delays in the computer's operation.

There is one important point about task priorities. As will be shown later, the 'Emergency "SCRAM"' task will only execute if it is told to by a lower-priority task. If a high-priority task is allowed to execute without blocking (waiting for a message or control from another task), then none of the lower-priority tasks will be able to execute (which is known as *starvation*). Typically, systems are designed with high-priority tasks only being specified in cases where their execution will not be a problem if it causes starvation

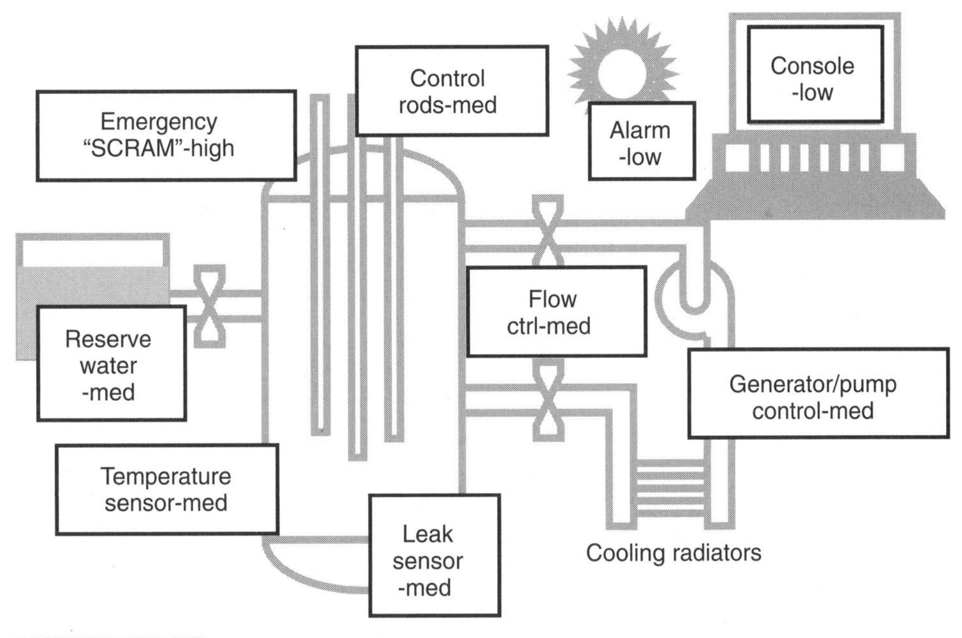

FIGURE 11-1 **Nuclear reactor schematic diagram with tasks.**

in lower-priority tasks. As will be shown in the nuclear reactor example, 'Emergency "SCRAM"' is only enabled when something bad happens and the reactor has to be shut down quickly. Starving the other tasks at this time is not a problem because ensuring there is not a meltdown is the overriding priority.

Before going too far, I have to say that developing a control system for a nuclear power plant (or many other processes and applications) is not something that should be done lightly, as I've shown here. The purpose of this example is to show how the different processes interact, not to be used as a primer for nuclear power plant control design. If you are designing systems that control dangerous processes or have human lives depending on them, please develop and test your software according to the appropriate specifications.

If you do fail to heed this warning and design a nuclear power plant control system based on what's presented here, please let me know so I can stay as far away as possible.

All of the tasks can be arranged in a diagram to show how they communicate via messages. An important concept about messaging in an RTOS is that each message is initiated by a task. Normally each task is waiting (blocking) on a message, waiting to respond to it and execute the request that is part of the message.

This can be seen in Fig. 11-2 with the two tasks sending messages to the "Timer" task. The "Timer" task will acknowledge the messages sent to it (which will allow the other tasks to execute their functions) when the requested time has passed.

After the task has executed its function (which is to monitor the state of the reactor), it then can respond to the conditions by sending messages to other tasks.

For example, the reactor "Temperature Sensor" task could execute the code in Listing 11-1. This task will monitor the temperature of the water, comparing it to what it should

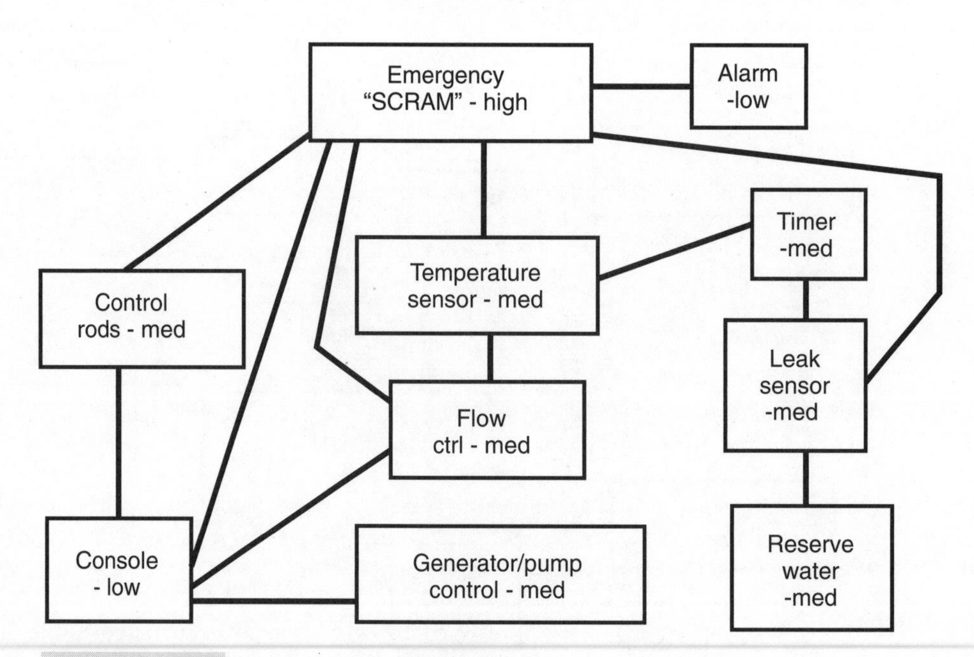

FIGURE 11-2 Nuclear reactor task diagram.

LISTING 11-1 **Code for the reactor "Temperature Sensor" task.**

```
ReactorTemp()                   //  Reactor temperature sensing task
{

int actualValue;                //  Value read in

external setValue;              //  Value the reactor should be running at

  while ( 1 == 1 ) {            //  Loop forever

    SendMsg( Timer, 1Min );     //  Delay one minute between measurements

    actualValue = ReadTemp();   //  Read the reactor's temperature

    if ( actualValue < setValue )
      SendMsg( FlowControl, close );        //  Reactor cooling down,
                                //   restrict water flow
    else
      if ( actualValue > setValue )
        if ( actualValue >= dangerValue )
          SendMsg( SCRAM, NOW );      //  Reactor is too hot, SHUT IT
                                //   DOWN!
        else
          SendMsg( FlowControl, open );    //  Reactor heating up, allow
                                //   more water flow

  } //  endwhile

} // End ReactorTemp
```

be and shutting it down if it gets too hot. The task might appear to run continuously, but the `SendMsg( Timer, 1Min );` statement will block the task on the acknowledgement from the "Timer" task. This task will delay one minute before responding to the `ReactorTemp()` task's message.

I should point out one aspect of tasks and processes in RTOSs that might be apparent: Tasks should only access one hardware device. So the temperature sensor in the previous example only reads the thermometer value inside the reactor vessel. To access other hardware in the system, it will send messages to the appropriate tasks that control the different hardware.

This is an important philosophical point about access and RTOSs, because applications obviously could be written that accessed multiple hardware interfaces. This allows code reuse or modification and allows multiple software developers to work on an application without having to understand how the other interfaces work. At first, it might be unnatural for you to think in terms of simple, single devices, but once you get the hang of it, you'll be amazed at how easy it is to work with an RTOS.

The last aspect of RTOSs that I want to present to you is the *semaphore*. The semaphore is a flag that is controlled by the operating system, which can be used to restrict access to a resource (which can be a hardware device, data, or even another task) until the owning task has finished with it.

Semaphores are used in situations like controlling access to an operator console. In the earlier reactor example, you might want to have a semaphore controlling what is written on the screen. If the temperature sensor is writing to the operator console while the leak sensor has detected a problem and has requested a SCRAM, the operator might get a message on his console like:

```
Temperature is 95 Degrees CelsiDANGER! DANGER! Evacuate the building!
us Nominal
```

which is obviously confusing. If a semaphore was used to serialize the writing to the console by the temperature task and then the SCRAM task, the message would come out like:

```
Temperature is 95 Degrees Celsius Nominal
DANGER! DANGER! Evacuate the building!
```

which is actually easier to read and looks less like a software error.

I have thrown a lot of information and concepts to you in this section that you have probably never seen before. If it doesn't make a lot of sense or if you can't see how something is implemented, don't worry. As I go through the rest of the chapter, I will show how these concepts are implemented in the 8051.

PROG30: 8051 Example RTOS

For the balance of this chapter, I would like to present to you two different RTOSs that I have written for the 8051 with an application. With both of them, I have tried to focus in on different performance aspects of RTOSs. These aspects include size and speed (RTOSLite) and something I consider a mix of speed and flexibility—basically just right (FullRTOS). As I go through each type, I'll focus in on the measurements that I consider important as well as features that I like. At the end of the chapter, I will present an example application that uses FullRTOS for implementing a digital LCD clock/thermometer.

For commonality, I am designing each RTOS to use the same interfaces from the application's perspective. These interfaces will be specified in a macro because of differences in how the actual RTOS interfaces are implemented. Some of these interfaces will be straight calls to the RTOS, while others will involve substantially more code in the application. The goal for using the macros is to allow applications to be run under any of the three operating systems and to provide code that may simplify the RTOS design.

These interfaces are listed in Table 11-1.

When you first see this list of tasks, many will probably seem pretty obvious, while others might seem trivial or unneeded. Actually, this is probably the minimum list of functions necessary for a complete RTOS. As I go through the test and example applications, I think you'll be surprised at how much everything is used.

The way I have structured the two RTOSs is to embed the application in their source as an include file. When the RTOS is assembled, the application is assembled along with RTOS. There aren't any problems with linking object files together (as would be required in a PC).

Just as a point of warning, I am in no way the ultimate authority on RTOS design and optimization. The RTOSs I have presented here have been created for this book and are really not bad general purpose RTOSs but you should look around at other RTOS's to make sure the one you want to use has all the features you need.

TABLE 11-1 THE RTOS INTERFACES

REQUEST	INPUT	OUTPUT (IN R0)	COMMENTS
ReadMsg	Message#	Pointer	Return a pointer to the specified message number.
		–1	If the message number is not available.
GetMsg		#	Returns the number of messages waiting for the task.
AckMsg	Message#	0	Acknowledge the specified message.
		–1	If the message isn't available.
StartTask	Pointer	Task#	Start the task pointed to.
		–1	No task information blocks available.
EndTask			The current task has outlived its usefulness. Make the task information block available for another task.
NextTask			Jump to the next executable task.
SendMsg	ptr, Task#		Block until Task# acknowledges the message.
WaitMsg		#	Block until a message has been sent to the task.
WaitInt	Int#		Block until interrupt "n" has executed
GetSem	Sem#		Block until the semaphore is allocated to the specific task.
RelSem	Sem#	0	Release the requested semaphore.
		–1	If the semaphore wasn't allocated to the specific task.

PROG30: RTOSLITE

Before embarking on a full RTOS, I wanted to make sure I had a good idea of how to implement an RTOS in the 8051 by creating as small an RTOS as I could. Creating it as small as possible does have some advantages in the areas of simplicity and ease of debugging. Keeping it small is also an advantage in devices like the Atmel AT89C*x*051 series, which has less control store than many other full-featured 8051s.

I was surprised to discover that there were a number of features I could pass along to the full RTOS to make it simpler.

For the RTOSLite, my initial targets were:

- Create an RTOS that used less than 256 bytes of control store.
- Minimal overhead scratchpad RAM would be required.
- Each task would be contained in a set of eight bank registers.
- Only four tasks would be supported in the RTOS (because there are only four banks). There will be an "AlwaysTask" used for allowing interrupts when the other tasks are blocked.
- No task priorities would be used.
- With a true 8051 running at 1 MHz, worst case task switching would take less than 500 msec.
- An interface that could be used with the full RTOS.
- Include simple interrupt handling for the _Int0 and _Int1 pins and the two basic timers.
- Only the scratchpad RAM could be accessed by each task. If the DPTR is to be accessed, then a semaphore flag should be devoted to it.

The result was RTOSLite in the PROG30 subdirectory of the CD-ROM. Along with the operating system, I have also put in a simple application that increments a counter and outputs the value to a set of LEDs on P1. I'll go through each of my design criteria and then offer a few comments on how successful I was along with some pointers on writing RTOS Applications.

The first two specifications I set out for myself are actually quite stringent and difficult to achieve. In particular, I had to go over RTOSLite three times before I was able to achieve the goal of writing an operating system that required less than 256 bytes of control store. There are a few factors that help with making this requirement possible.

The first factor is the use of a set of bank registers as all the context information for a task. This is the third point, and there are a few subtleties that make this restriction important to keeping the size down (and the performance up of the operating system). The big advantage to this data movement is kept to a minimum.

For each task, I have defined the bank registers as shown in Table 11-2.

The stack area for each task is actually the most significant four bank registers. By placing the stack information in the bank registers, I can very easily calculate where the stack pointer should be when I'm switching tasks. This reduces code space quite a bit.

As well, making the bank registers specific to a task eliminates the need for saving them during task switching. To resume execution of a task right where it was stopped, all I have to do is pop the task's accumulator and status register off the stack before returning.

The task block information register (R3) is used to specify the current operating status and characteristics of the task (Table 11-3).

The different values might seem clumsy, but they actually have been optimized to provide very short and quick responses to different events (such as passing messages). I will show how the different task block information register values are used for the different operations of the RTOS.

This might seem like cheating, but the operating system interface actually reduces the amount of control store bytes required for the operating system by using macros to carry out the interface operations of RTOSLite. These macros are a combination of code functions and subroutine calls that are invoked in the application code to carry out the various functions (Listing 11-2).

Saying that the macros are cheating a bit because they have code in them that should rightly be in the operating system itself is probably a bit inaccurate. The code in the macros, for the most part, is used to set up the parameters for the different operations.

TABLE 11-2 THE BANK REGISTERS FOR EACH TASK

REGISTER	TASK FUNCTION
R0	Message pointer/index/task address/GP register
R1	RTOS request number/index/task address/GP register
R2	GP task register/task address
R3	Task block information
R4	Low byte of return address
R5	High byte of return address
R6	Saved accumulator
R7	Saved PSW

TABLE 11-3 VALUES FOR R3

VALUE	TASK STATUS
000h	Task can execute.
1*nn*000b	Sent message to task "*nn*," waiting for ack.
001*xxx*b	Waiting for a message.
01*iiii*b	Waiting for interrupt number "*iiii*."
0FFh	No task loaded for register bank.

LISTING 11-2 **The macros to carry out the interface operations of RTOSLite.**

```
MACRO NextTask                  ;  Let the next task execute

  acall  _RTOSNextTask

ENDMAC

MACRO EndTask                   ;  End the current task

  mov    R3,#0FFh               ;  Set task block that it has ended

  acall  _RTOSNextTask

ENDMAC

MACRO StartTask Task            ;  Start the new task

  mov    R0,#(Task AND 255)
  mov    R1,#(Task SHR 8)

  acall  _RTOSStartTask         ;  Task number in Acc (or -1)

ENDMAC

MACRO WaitInt(Number)           ;  Wait for the specific interrupt number

  mov    R3,#(%00010000 OR (1 SHL Number))
                                ;  Wait for the specified interrupt

  acall  _RTOSNextTask          ;  Now, start waiting for the other tasks

ENDMAC

MACRO RelSem(Number)            ;  Release the semaphore number

  setb   Number                 ;  Just set the semaphore number

ENDMAC

MACRO GetSem(Number)            ;  Get the semaphore

@_RTOSGetSem:                   ;  Loop back here for each semaphore

  jnb    Number,@_RTOSGetSemGot  ;  Semaphore gotten

  acall  _RTOSNextTask          ;  Wait for the next task
```

LISTING 11-2 **The macros to carry out the interface operations of RTOSLite *Continued.***

```
  ajmp   @_RTOSGetSem          ;  Try to get the semaphore again

@_RTOSGetSemGot:               ;  Have the semaphore

ENDMAC

MACRO AckMsg(Task)             ;  Acknowledge the message from task #

  mov    ((Task * 8) + 3),#0   ;  Tell the task to stop waiting

ENDMAC

MACRO ReadMsg(Task)            ;  Read the message from the task

  mov    R0,#((Task SHL 3) + 3) ;  Make sure the correct address

  acall  _RTOSReadMsg

ENDMAC

MACRO SendMsg(Task,Ptr)        ;  Send the message pointed to

  mov    R3,#(%00100000 OR (Task SHL 3))  ;  Block the task

  mov    R1,#((Task SHL 3) + 3)
  mov    R0,#Ptr

  acall  _RTOSSendMsg

ENDMAC

MACRO WaitMsg                  ;  Wait for the message to be sent

  mov    R3,#%1000             ;  Make the task blocking

  acall  _RTOSWaitMsg

ENDMAC
```

With the macro interfaces, you should see that I only allow hardcoded (constant) values for task numbers when communicating. These could be changed, but you should note that, if you were to put in a label for a variable in scratchpad RAM, chances are that the code would compile correctly, but not execute correctly because the address of the label would be used rather than the actual value at the label. To correct this, the macros could be modified to check for defined values (which would be labels) and use the contents of the scratchpad RAM at the label value, rather than using the label value itself.

With the macro interface, you might be wondering why I didn't go even further and put in defines in the macros so that, after the application code, only the required operating system features would be loaded.

The problem with this method is that there is no simple way of seeing which features are used and working around it. If I did not have to worry about interrupts, I could structure the code as:

```
;  Macros defined at the start of the program
 org 0                  ;  Reset vector
 ajmp  Mainline         ;  Skip over the application Code
Application:
  :                     ;  Put in the application code here
Mainline:               ;  RTOS code follows
```

In each macro, I would add a define to indicate that it is being invoked and, inside the RTOS code, add an `$IFDEF` statement around the code:

```
MACRO WaitMsg                  ;  Wait for the message to be sent
$DEFINE _WaitMsg               ;  Define the wait message flag
  mov     R3,#%1000            ;  Make the task blocking
  acall   _RTOSWaitMsg
ENDMAC
 :                             ;  Skip over the application code
Mainline:                      ;  RTOS code follows
 :
$IFDEF _WaitMsg
_RTOSWaitMsg:
 :                             ;  Put the "_RTOSWaitMsg" code here
$ENDIF
```

This would work and would potentially reduce the amount of control store needed for the application, but it also would make the RTOS more complex and harder to debug for different applications. When I reviewed this, I felt I had a good balance between code size, readability, and not doing something I'll never be able to debug later.

In the operation of the RTOS functions, you'll see that I tend to use the R0 and R1 registers of the task that called the RTOS, rather than define separate registers for the RTOS to use. I used R0 and R1 because they can also be used as index registers into the scratchpad RAM. This means that, when you are calling an operating system function, you can't depend on the R0 and R1 registers being at the same value as when you called the function.

Having said this, I should point out that, if the task is interrupted, nothing will be lost.

A maximum of four tasks (as will be shown later, the maximum is actually three) might seem like an unreasonable restriction to tasks, especially when I say that each task should have a single hardware feature devoted to it. For many applications this limits the usefulness of RTOSLite.

What the limited number of tasks loses in terms of application flexibility, it does make up for in small code size and application speed. One of the requirements was to take no more than 500 μsec to switch a task.

The advantages of using the task registers is clearly shown in the NextTask function of the operating system. In the worst case, NextTask will be called from the currently executing task, check the other three tasks to see if they are capable of executing and then begin to execute again.

The code that does this is shown in Listing 11-3.

`_RTOSDoNexTask` increments the current bank register set (as pointed to in the PSW), then if the task has a status of 000h (task can execute), it starts it executing. In an 8051, this requires 40 instruction cycles to execute, which means, at 1 MHz, 480 μsec are required in the worst case (only one task able to execute) for the function. In a Dallas Semiconductor HSM device, 284 μsec are required to execute in the worst case.

LISTING 11-3 The code for NextTask.

```
_RTOSNextTask:                    ;  Jump to the next active task

  push  ACC                       ;  Save the task registers
  push  PSW

_RTOSDoNextTask:                  ;  Now, find the next task

  mov   A,PSW                     ;  Which task is active?
  add   A,#%00001000              ;  Just increment the bank register
  anl   A,#%00011000              ;  Make sure task isn't > %00011000
  mov   PSW,A

_RTOSDoNextTaskNextCheck:         ;  Check the task to see if it can execute

  cjne  R3,#0,_RTOSDoNextTask ;  If task information block == 0, then
                                  ;   execute it

_RTOSExecuteTask:                 ;  Now, the task can start executing

  mov   A,PSW                     ;  Get the correct stack pointer
  anl   A,#%00011000              ;  Get rid of extra bits
  add   A,#7                      ;  At the end of the register bank
  mov   SP,A

  pop   PSW                       ;  Restore the PSW
  pop   ACC

  reti                            ;  Return where program stopped
```

Four basic interrupts are handled within the operating system. Most 8051s and derivatives have two interrupt input pins and two timers. To handle an operating system, a task first sets up the hardware to request an interrupt and then invokes a WaitInt macro.

When the interrupt request is acknowledged by the 8051 hardware, the code in Listing 11-4 is executed.

If additional interrupts are required, then the RTOS would have to be modified, but I do have to caution you that, before adding additional interrupts, you need to fully understand how they operate. The two timers and two interrupt pins will reset the interrupt requesting hardware upon execution of the return from the interrupt instruction. Other hardware interrupts don't necessarily work the same way and might require explicit code to reset the interrupt.

There are a few things I want to point out with this code. A big one is how interrupt requests are allowed to be serviced.

I probably haven't emphasized the point enough, but it is important for you to realize that hardware interrupts *request* the software to service them. The software can refuse or put them off. In the 8051 RTOSLite, I want to enable interrupts only at sections of code that I know I can control. This is important because the 8051 works differently than other processors that you might be familiar with.

When a hardware interrupt is serviced, other interrupts can interrupt the handler unless you explicitly disable them (by resetting bit 7, EA, in the IE register). Note that this is the first thing I do in each of the interrupt handlers.

As well, the EA bit of IE should not be enabled unless you are going to disable it before any operating system functions. If a hardware request is serviced from within an operating system function, the stack that should be contained within the current set of bank registers will overwrite data outside of the bank registers, potentially changing the execution of another task.

If a task overwrites the bank registers, additional execution problems could arise because the stack pointer value is calculated in the code and is not saved for each task. If the task is interrupted during an operating system function, then the actual stack address of the task is lost forever. This could cause problems that would be just about impossible to find.

You might have figured out that this architecture means that tasks cannot call subroutines. This might be a problem for some applications (although FullRTOS does not have this deficiency).

LISTING 11-4 **The code that is executed when the interrupt request is acknowledged by the 8051.**

```
org 000Bh                   ; Timer0 interrupt

 clr   IE.7                 ; Disable global interrupts

 push  ACC                  ; Save the accumulator

 mov   A,#0FDh              ; Have Timer0 interrupt

 ajmp  _RTOSInt             ; Handle the interrupt

 :                          ; Skip over the other interrupts that are
 :                          ;  handled the same way

_RTOSInt:                   ; Handle the interrupt requests

 anl   IE,A                 ; Disable requested interrupt and global ints

 push  PSW                  ; Save the task's PSW

 cpl   A
 orl   A,#%00010000         ; In "A", have expected interrupt wait value

 xrl   A,3                  ; Now, is task 0 waiting for this?
 jnz   _RTOSI1              ;  No, check task 1
 xch   A,3                  ; Mark that the task can now run

_RTOSI1:                    ; Restore the accumulator
 xrl   A,3
 xrl   A,00Bh               ; Check the next task
 jnz   _RTOSI2              ; It's not, check task 2
 xch   A,00Bh

_RTOSI2:                    ; Restore the accumulator
 xrl   A,00Bh
 xrl   A,013h
 jnz   _RTOSDoNextTask      ; It's not, jump to the next task code
 xch   A,013h               ; It is, enable the task
 ajmp  _RTOSDoNextTask
```

In most RTOS, there is a scheme for allowing different tasks to run at different priority levels. I have not put in this type of mechanism into RTOSLite simply because I don't believe that it's necessary. The very fast task switching of the RTOS along with the need to keep the size of the code short really has eliminated any need for this.

Messaging is a bit unusual in this operating system because I pass a single byte pointer to a message. Often, RTOS will pass several bytes, some of which are a command and some of which are a pointer to data (or the data itself).

I have not used this format for RTOSLite because I wanted fast execution. After space is required for the stack, the task block information, and the RTOS function parameters, I didn't want to use up any more registers than were absolutely required.

Before discussing the application, I want to point out one important aspect of an RTOS that I haven't fully covered yet that is important to correct operation of an application. This is the need for having a single task that doesn't do anything but allows interrupts to take place.

In RTOSLite, I have put in the following code that has two purposes. The first is to start the application as well as allow interrupts to take place within the operating system in a controlled manner.

The code for this task is:

```
_AlwaysTask:                          ;  The always available task
  StartTask Application               ;  Start the application task
_AlwaysTaskLoop:                      ;  Loop here forever
  clr  IE.7                           ;  Disable global interrupts
  NextTask                            ;  Look at the next task
  setb       IE.7                     ;  Allows interrupts to happen
  ajmp       _AlwaysTaskLoop          ;  Loop around (with interrupts enabled)
```

From this task, the application tasks are started by invoking the StartTask macro. Once they have started, this task runs continuously although each time it is allowed to execute, it allows pending interrupt requests to be serviced and then immediately passes up execution. This might seem strange, but it is actually a very efficient way to allow interrupt requests to be serviced while the other tasks are blocked. If this task wasn't available, then I would have had to come up with specialized code to enable interrupts and respond to them.

`_AlwaysTask` is the only place where interrupts should be globally enabled. As I've discussed, the reason for this is because the stack is actually part of a set of bank registers. If an interrupt occurs during another task when it is calling the operating system functions, then the stack will go beyond the task's bank registers and overwrite data that is used for other purposes (such as by other tasks).

The application itself (Listing 11-5) is pretty simple and simply sends an incrementing value to eight LEDs connected to P1. If you have gone through chapter 9, you should have no trouble understanding what the code does.

The first task application, after initiating task `App2`, ends itself. This might seem a bit unusual, but I wanted to make sure the code worked properly.

In true RTOS fashion, each hardware element used is controlled by a single task. Timer0 is controlled by `App2`, and `App3` receives messages from `App2` and outputs the passed value via P1.

Message passing should be pretty self evident. Task `App2` simply invokes `SendTask` and is blocked by the operating system until the message has been acknowledged by the re-

LISTING 11-5 **The code for the application.**

```
Application:                      ;  Simple test application

  StartTask App2                  ;  Start a second task

  EndTask                         ;  End this task

App2:                             ;  Second task

  mov    CKCON,#%00001000         ;  Use internal /4 clock for Timer0
  mov    TMOD,#%00000001          ;  Timer0 - Uses internal clock
                                  ;         - Run in mode 1
  mov    TCON,#%00010000          ;  Start Timer0 running

  mov    021h,#0                  ;  Use Address 21 as the data byte

  StartTask App3                  ;  Start the third application

  mov    022h,#15                 ;  Set the counter for a one second loop

Application_Loop:                 ;  Loop around here

  setb   IE.1                     ;  Enable the Timer0 interrupt

  WaitInt(1)                      ;  Wait for the interrupt

  djnz   022h,Application_Loop    ;  Wait for counter to count to 1 second

  SendMsg(2,021h)

  inc    021h                     ;  Increment the output value

  mov    022h,#15                 ;  Reload the counter

  ajmp   Application_Loop

App3:                             ;  Wait for messages from App2

  WaitMsg                         ;  Wait for a message

  ReadMsg(1)                      ;  Read the message from task 0

  mov    R0,A                     ;  Read and output the message
  mov    A,@R0
  cpl    A                        ;  Invert the data before displaying
  mov    P1,A

  AckMsg(1)                       ;  Acknowledge the message

  ajmp   App3
```

ceiver. App3, which is the message's receiver, simply waits for the message, reads it, and when it's finished with it, it acknowledges it.

Before moving on, I should point out that messaging uses absolute task numbers exclusively. If you were going to use this operating system in an application, you might want to save the task number that is returned to the initiating task in a scratchpad RAM byte and add code to call the messaging subroutines using the value in scratchpad RAM, rather than using absolute values. By doing this, if the program's execution changes the task numbers used by the different tasks, you won't discover that, at some point, the application stops working and you're left high and dry trying to figure out why.

If you were to load PROG30 into UMPS and run it, you should notice that, most of the time, nothing is executing. The program is continually running `_AlwaysTask` and waiting for an interrupt.

This is normal for most RTOS applications. Ideally, you want all the tasks to be blocked waiting on an external event. When the external event happens (which, in this case is Timer0 overflowing), then it can be responded to very quickly. An RTOS differs from a multitasking operating system used in large systems in that it takes priority in servicing outside requests. A large computer's operating system's priority is to finish the currently executing number-crunching applications as quickly as possible.

When I originally outlined out this chapter, I created the initial specifications to hobble the final operating system and show how limited the minimum operating system could be. In this task, I fear that I didn't really accomplish that. The operating system here is actually quite capable and very efficient.

In fact, in FullRTOS, I have taken many features from this one to help make it more efficient for application execution.

I realize that three (or four, if you're willing to modify `_AlwaysTask`) tasks is quite restrictive. However, if you are working with an 8051 that has limited hardware features and a very limited amount of control store, you might want to consider using this RTOS. Despite being extreme in a number of areas, it is very efficient and fast and consists of all the features required in an operating system. I would recommend it for use in an application that is going to be put into a device like the AT89C1051, which has very little control store available (as well as quite limited internal resources).

PROG31: FULLRTOS

With the experience of the small RTOS, I wanted to create a much more substantial RTOS that could be used for more general cases but still have the advantages of the original RTOS in terms of speed and flexibility. I also identified a few areas that I felt were deficient in the original RTOSLite that I would like to improve upon. The resulting FullRTOS is less than 50% larger than RTOSLite but should be usable in a wide variety of cases.

In creating a full-featured RTOS, I wanted to add the following features:

- More than four tasks.
- Tasks are able to call subroutines.
- Each task is given a priority level.
- Keep part of the bank registers as well as context registers in the task block.
- Pointers to messages,
- Simplify interrupt operation.

The first four of these new requirements meant I could no longer just use the 8 byte register blocks for the task information blocks. Instead, I would have to reserve memory for the task information blocks.

TABLE 11-4 DESCRIPTION OF THE TASK INFORMATION BLOCK SCHEME

BYTE	DESCRIPTION
0	Task status 00000000—Task can execute 0001*nnnn*—Send message to task "*nnnn*" 00100000—Wait for a message 010000*ii*—Waiting for interrupt "*ii*" 11111111—No task loaded for TIB
1	Task priority ("0" highest/"0FFh" lowest)
2	Task stack pointer
3-15	Task context registers in the stack

At this point, I encountered my first design challenge. I wanted the RTOS to work within the 128 bytes of scratchpad RAM of the 8051. As I worked through this operating system, I realized that most full-featured RTOSs would use external memory for storing the task information blocks (which would largely be the context registers for each task and the stack pointer to them).

To avoid having to move data in and out of external memory (and losing those I/O pins and the bandwidth required for the data movement), I wanted to come up with a task information block scheme that would allow the scratchpad RAM to be used comfortably in an application.

The scheme I did come up with was to use a 16-byte task information block as described in Table 11-4.

When each task stops executing, the task context registers are stored in the order:

1. PCH—16-bit program counter
2. PCL
3. IE—Interrupt control register
4. ACC—Accumulator
5. PSW—Status register
6. R0—Bank registers unique to the task
7. R1
8. R2

These eight registers define where the task was executing. By saving the interrupt enable register, you can create high-priority tasks that have interrupts disabled to allow a timing critical operation to take place.

This structure resulted in 11 bytes minimum being required for saving a task and, in a 16-byte block, gave 5 bytes for saving data and calling subroutines. Looking at this structure, you might think that there is space for two subroutine calls (each of which saves a 16-bit program counter) and one variable pushed onto the stack. Realistically, there is only enough space for one subroutine call and one variable save in the application because there might be an interrupt inside an operating system call, before the global interrupt enable bit is masked (and after it has been pushed on the stack), which means that potentially up to 3 bytes can be placed on the task's stack as overhead that cannot be used for subroutine calls. This means that there are only 2 bytes, or one subroutine call is available on the task's stack.

While this seems like a restriction, it's really a feature because RTOSLite's problems with keeping track of the interrupt register status and only unmasking interrupts at specific points in the application are eliminated. In FullRTOS, I enable interrupts globally as well as the four basic interrupts (the two interrupt pins and two timer interrupts). So tasks that wait on these interrupts just have to enable the software. I found this much easier to write applications for.

Like RTOSLite, multiple tasks can wait on the same interrupt. If multiple tasks are waiting for a time delay, they can both wait on the same interrupt. This is rather than having a single task that requests for other tasks waiting on it. While this method works very well for the four primary 8051 interrupts (which only require the `reti` instruction to reset), it does not work as well for the serial port or other devices' hardware interrupts that require an action to be taken before the interrupt is acknowledged in hardware. In the case of the serial port, the data incoming interrupt is acknowledged and reset by reading the byte that was read in.

When the task is to be switched either through an interrupt or an operating system request, the context data is stored in the task information block's stack area using stack `push` instructions:

```
push       ie               ;  Save the Interrupt enable register
clr  IE.7                   ;  Disable global interrupts
push       ACC
push       PSW
push       0                ;  Save R0 - R2 in bank 0
push       1
push       2
```

As I indicated earlier, this seems to be a reasonable amount of data to save. If additional registers are to be unique to each task or more stack depth is required, then you have to change the `_TIBSize` constant to a larger value than the 16 that I used.

This task information block size might seem a bit arbitrary, but it does meet the initial requirements that I set out and means that each task information block could be displayed on a single line of the UMPS simulator.

This is actually important in debugging because it makes each task easier to find and read through. Now, you could change this value. However, when you are using UMPS or other simulators to debug your application, you will find that it will be hard to see where the different task information blocks start and stop. (In Fig. 11-3, you should note that I loaded the internal RAM with 044h so that I could see where the data was being written to.) If you do change the block size, you might want to change it to 32 bytes so that you are always starting on a new line. It might seem like I'm being finicky; however, when you have been debugging an application for several hours, being able to quickly pick up the task information and not search for it is a big plus.

Like RTOSLite, FullRTOS has placed the task information blocks in a chunk of Scratchpad RAM devoted to it (Fig. 11-4). To access different tasks, an index from the starting task is used.

In other RTOSs that I have seen, the task information blocks are arranged in queues of linked lists (Fig. 11-5). In this method of processing task information, pointers to the task

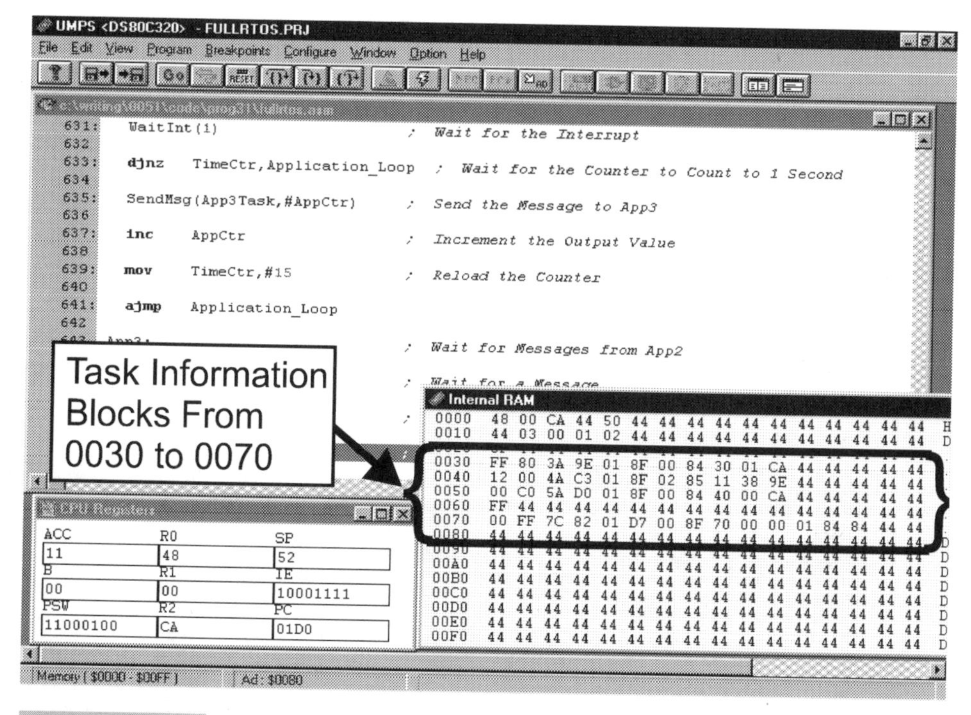

FIGURE 11-3 Task information block on a line.

	State	Ptry	SP	Context data
Address	0	1	2	3 - F
Task 0				
Task 1				
• • •				
Task n				

FIGURE 11-4 Task information block in RAM.

information blocks are moved between queues and the different queues are accessed at different times.

The different ways of executing this function are reflected in the different development systems and tools used. The method that I have used is best for assembly language programming for a device with limited memory. The queues of linked lists are best suited to high-level languages with lots of memory available.

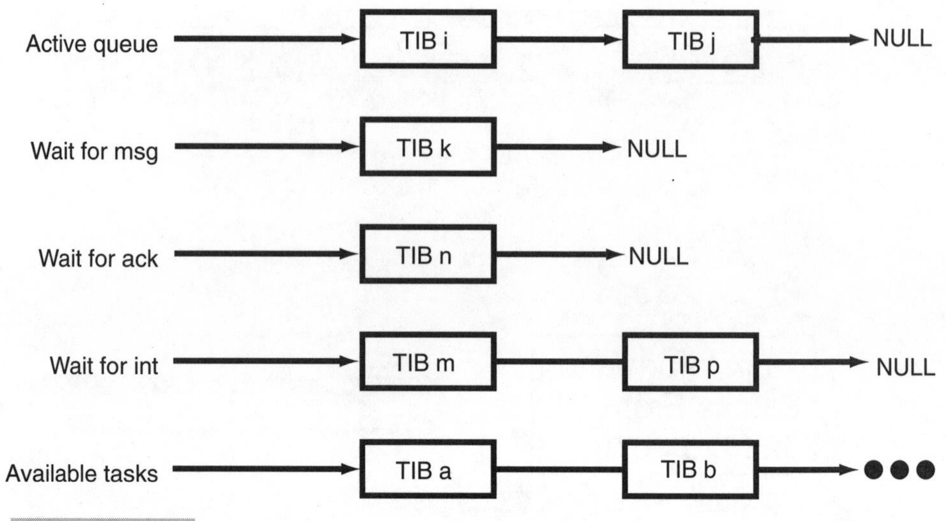

FIGURE 11-5 **Task information block in queues.**

The last comment I would make about how I implemented the task information blocks is to explain the obvious question. If you looked at the FullRTOS code in the PROG31 subdirectory, you might have wondered why the FullRTOS is better than RTOSLite if it only has space for five task information blocks.

The way I originally laid out the code was for the task and RTOS variables to be stored in register bank 0. Register banks 1, 2, and 3 were available for system variables while addresses 020h to 02Fh were available for semaphores, flags, and more system variables.

If more tasks are required for an application, the `_TaskStart` address can be moved to 010h from 030h (although, in this case, you cannot use any semaphores or bit variables and you are limited to 11 bytes for system variables), which would give you seven task information blocks. You could also use an 8052, like the Dallas Semiconductor HSM, which has 256 bytes of scratchpad RAM (and the high 128 bytes are only available via index registers).

The biggest difference between the RTOSLite and FullRTOS task information blocks is the inclusion of a priority value for each task in FullRTOS. This byte is used to differentiate between tasks and help the operating system determine which task should execute next. This is an improvement over RTOSLite that simply executes the next task available in the list.

In FullRTOS, when a task switch is requested, the first task with the highest priority is identified and begins to execute using the algorithm:

```
NextTask()
{
  curTask = AlwaysTask#;        //  Assume only "Always" is able to run
  curPrty = Priority[ curTask ];
  i = _CurTask;                 //  Start at the current task
  do {                          //  Loop through all the tasks
    if ( ++i >= TotalTask# )    //  Point to the next task
      i = 0;
    if ( Priority[ i ] < curPrty ) {    //  Have a higher priority task
```

```
        curTask = i;                    // Specify as the
        curPrty = Priority[ i ];        // executing task.
      }
    } until ( i == _CurTask );
}  //  End NextTask
```

At the end of this routine, `curTask` will have the next task able to execute at the highest priority. If other executable tasks that have the same priority are encountered, they will not replace `curTask`, to ensure that all tasks execute and none are starved for cycles.

While I have provided a full byte for each task's priority, I try to keep the priority values to the convention shown in Table 11-5.

With this, interrupt requesters are always serviced as quickly as possible, followed by simple polled input. Output tasks are at the lowest priority because they are the ones typically used to output information to the user. Even if these tasks are the slowest in the application to execute, then chances are the operator won't notice.

In FullRTOS, I have used the same macros wherever possible for the operating system request functions. The only differences are in the `StartTask` request (I have added the priority of the task to be created) and in the messaging functions.

In the messaging functions, I have created the message system so that a single byte (which can be used as a pointer to messages) is passed to the receiving task and the sending task is explicitly referenced by the receiver when reading and acknowledging the message. Changing to requesting the sender's task number made the messaging system easier to code and allowed me to pass the contents of a variable rather than a constant.

To make sure you know which task is which, when the `StartTask` command completes, the requester has the new task number in its R0 register when it begins executing again. After starting a task, the next instruction in the initiating task should be to save the contents of R0 in a variable that the different tasks can access.

For the most part, the task numbers will remain constant, so this might seem to be unnecessary. However, if you reuse the code or make changes, you'll probably find your hardcoded values have changed and your application no longer works.

The FullRTOS request macros are listed in Table 11-6.

To demonstrate the operation of FullRTOS, I have used the same test application as I created for RTOSLite as well as an LCD digital clock/thermometer as shown in the next section. In the test application presented along with FullRTOS in subdirectory PROG31, I have changed the application source slightly to use the updated RTOS request macros.

TABLE 11-5 THE CONVENTION USED FOR THE PRIORITY VALUES

PRIORITY	TASK TYPE
0	Tasks waiting for interrupts
040h	Input polling tasks
080h	General information processing tasks
0C0h	Output tasks
0FFh	"Always" task—Nothing else

TABLE 11-6 THE FULLRTOS REQUEST MACROS

FUNCTION	INPUT PARAMETERS	RETURN VALUES
StartTask	@Task, Priority	R0 = Task number R0 = 0FFh if no TIB's available
EndTask		
NextTask		
WaitInt	Interrupt#	
SendMsg	Task#, MsgPtr	
WaitMsg		
ReadMsg	Task#	R0 = Message pointer R0 = 0FFh if no message from Task#
AckMsg	Task#	
GetSem	Semaphore#	
RelSem	Semaphore#	

LCD DIGITAL CLOCK/THERMOMETER USING FULLRTOS

The first example application in this book is a digital clock and thermometer that runs on a DS87C520, with a Dallas Semiconductor DS1820 temperature sensor and a Hitachi 44780 2-line LCD (Fig. 11-6).

If you have read the *Handbook of Microcontrollers*, this application will probably look pretty familiar to you. I used this application to demonstrate how different microcon-

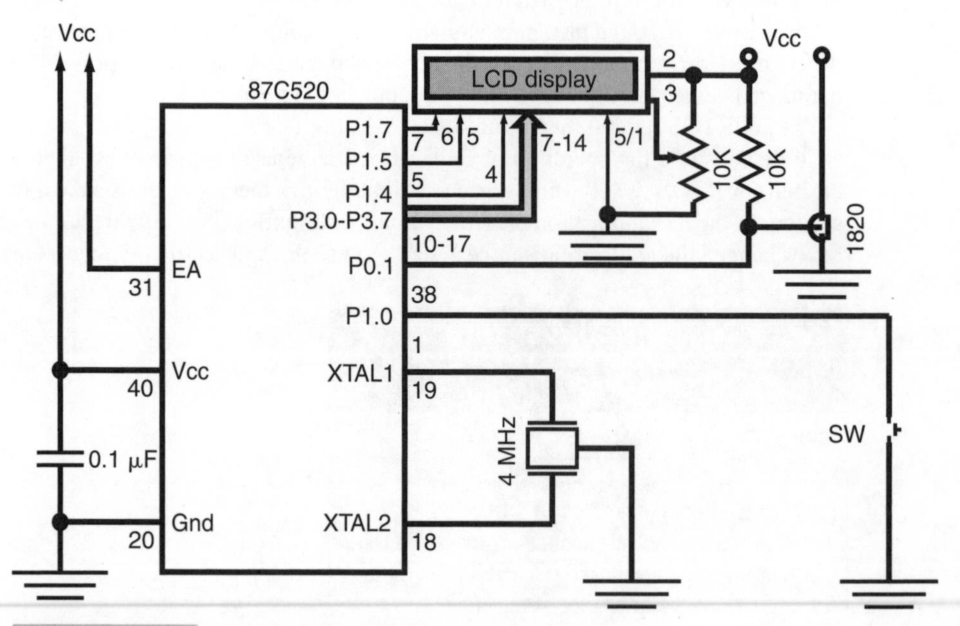

FIGURE 11-6 87C520-based clock and thermometer.

trollers would implement the same application. For this book, I am using this application to show how a typical microcontroller application can be implemented using an RTOS for controlling the application's execution. (The fact that I had working DS1820 and LCD interface code had nothing to do with this decision).

The application code interfaces with three hardware devices: the LCD, the DS1820, and an internal device, which is one of the timers built into the DS87C520. The purpose of the LCD and DS1820 temperature sensor should be pretty obvious, but I also include the DS87C520's timer as a peripheral because it is a hardware circuit that is responsible for the operation of the application.

This circuit, while it probably looks complex, actually is pretty easy to wire because I arranged the P1 and P3 bits to interface almost directly to the LCD's bits. The LCD's pin 14, if it lines up with pin 17 of the DS87C520, will allow mostly straight-through wiring to the LCD. The DS1820 has a 10K pull-up on its line and uses one of the I/O pins that is not internally pulled up to allow the DS1820 to send data to the DS87C520 without having to overcome the strong pull-ups when the DS1820 is allowed to send data.

When I created this application, I used a 4-MHz ceramic resonator (mostly out of laziness, because I had just finished chapter 9 and I still had the core circuit still wired). If you are going to use this application as a clock, you might want to put in a 4-MHz crystal (with external capacitors) to get the best timing accuracy possible.

With the circuit designed, I then looked at how I would architect the application software. Because this code would be running under the FullRTOS, I wanted to make sure that I could develop it without using any more than the available resources.

I always find creating a simple block diagram of how I expect the tasks to execute to be invaluable when I am designing an RTOS application. For this application, I was able to create the task block diagram shown in Fig. 11-7 and allow a total of only five tasks to be created.

The "Always" task is the "Always" task built into the RTOS, and its function is to start the "Application" task executing and provide a task that never blocks and is able to execute instructions that aren't masked for interrupts.

The "Application" task might seem to be a bit unusual in that it starts the other tasks and then ends. I have done this because I wanted to keep the functions of each task as specific as possible with regard to their functions and the hardware that they access. In this scheme, I have specified that "Application" is used to start the application tasks and nothing else. I could allow "Application" to keep executing (in a loop that continuously calls `NextTask`, as "Always" does). However, I already have a task that does this, and this keeps everything simple.

"Application" carries out one function that I haven't noted in Fig. 11-7, which is its setting up of the timer interrupt. The first thing that I do with "Application" is to set up Timer0 in mode 1 with interrupts enabled. This gives me an interrupt every 65,536 instruction cycles (each instruction cycle takes 1 μsec at 4 MHz in the DS87C520 HSM microcontroller), which occurs every 65.536 msec. This 65-msec clock is the timebase for the application.

In "Application," I should point out that I was careful with the order in which I started the other tasks and how they were handled. The first task initiated ("LCD" was chosen because it will have messages sent to it by the other tasks). To help speed up the loading of the application, while the LCD is being initialized, I use the `WaitInt` command to wait for the 65-msec timebase to provide the gross delays. During the `WaitInt` "LCD" task blocking, "Application" will load and begin executing "Time" and "Temp," both of

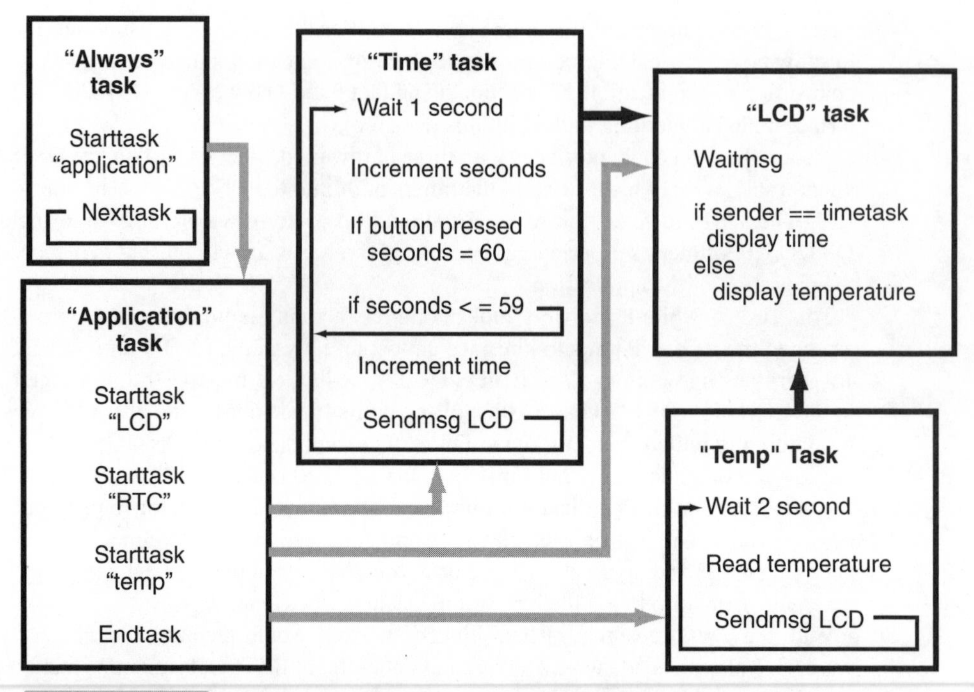

FIGURE 11-7 Digital clock/thermometer task diagram.

which will attempt to send a message to "LCD," which won't be read and acknowledged until "LCD" has finished initializing the LCD hardware.

The "LCD" task is really a simple modification of the 8-bit LCD code that I present in appendix D. As I mentioned earlier, instead of using a code delay, I use the Timer0 65-msec timebase delay for providing the delay to the LCD.

Along with using the 65-msec timebase, when I am polling the busy flag and waiting for the operation to complete, I have inserted a `NextTask` request so that, while data is being written to the LCD's, other higher-priority tasks are not blocked while the LCD is executing.

```
SendCHAR:                        ; Send character to the LCD
  setb   P1.5                    ; Set the RW line
  clr    P1.4                    ; Clear the RS line
  setb   P3.7                    ; Make sure the polling line is high
SendCHARLoop:                    ; Loop here until the LCD is active
  setb   P1.7                    ; Enable the "E" flag
  jnb    P3.7,SendCHARSkip       ; If bit 7 is low, send the character
  clr    P1.7
  NextTask                       ; Else, loop through the operation
  ajmp   SendCHARLoop
SendCHARSkip:                    ; Now, we can output the character
  clr    P1.7                    ; Turn off the "E" flag
  clr    P1.5                    ; Turn off the LCD read operation
  setb   P1.4                    ; Send out the character
  mov    P3,A                    ; Output the character to send
  setb   P1.7                    ; Toggle the "E" flag
  clr    P1.7
  ret                            ; Return to the caller
```

By getting rid of the 5-msec delay and checking other tasks for allowing them to execute, the "LCD" task only executes when it is processing information or writing to the LCD. This is an important feature of all tasks. Putting in polling loops or delays without checking other tasks can really impact the performance of a RTOS application. The only time polling loops or delays should be put in is for timing critical applications (which will be shown below).

When you read through the code, you'll see that the "LCD" task does the conversion of the time and temperature before displaying. When a message is passed to the LCD, the values to be displayed are pulled out of memory.

This is really in violation of how an RTOS should send messages. I would feel most comfortable with a pointer to the message data be passed to the "LCD" task and then the LCD task copy in the message data before acknowledging the message. By treating a message this way, I would be protecting the good copy of the data from being corrupted. If I had critically timed code, I would be returning to the calling tasks much more quickly.

As it works out, the time can be processed and displayed on the LCD before the next Timer0 tick, but this really wasn't the proper way of coding the application.

One aspect of the messaging to the "LCD" task that I am pleased with is the ability of the RTOS to check on messages from specific tasks. I found this to be very useful in this application because, if I hadn't put in this feature, I would have had to poll through each available message, which could be a problem with the "Timer" task because I would like to respond to that task first (which is why I do a `ReadMsg` on it first in "LCD") so that I will always respond before the next timer tick. If I miss a timer tick on the DS1820 controlling task ("Temp"), it is not a problem. So, I have prioritized "Time" to be responded to first in "LCD."

The real-time clock ("Time") task works very well, and it's operation is something that I am quite proud of (especially with regards to how the clock is set). Although it might seem like this is the first time I've created this task, I've actually implemented it about seven times in the past year, so I'm very familiar with its operation.

The algorithm used for keeping track of the current time runs off of the 4-MHz clock used in the microcontroller rather than a separate, dedicated crystal. In many real-time clocking applications, a 32.768-KHz crystal is used for the clock because this frequency is an even power of two. By dividing by 2_{15}, you will get a one-second clock.

In the "Time" task, the second variable is updated once per second by waiting through 16 of the 65,536 instruction cycle delays facilitated by waiting for Timer0 to interrupt the processor. This count is kept track of in a 24-bit second fraction counter. Each time Timer0 overflows and interrupts the processor, this value is incremented by 65,536 (which is accomplished by incrementing the high 8 bits of the fraction).

This statement is somewhat in error because I really wait for bit 20 of the 24-bit second fraction counter to be set. When this bit is set, then the 24-bit timer has a value of more than one million (actually 1,048,576), which means that more than a second has passed by.

Rather than clearing the entire 24-bit fraction counter, I subtract 1,000,000 (0F4240h) from it. The first time 1,000,000 is subtracted from the actual value, 48,576 is left over for the next second. After the second 16 cycles, 97,152 is left over, which is greater than the 65,536 added by each Timer0 cycle. So, for the third cycle, 15 cycles are required for the total to have bit 20 of the 24-bit counter set.

Obviously, the actual time taken for each second will be different; however, over a minute, it does even out very nicely. When the task sends a message to the "LCD" task to update the minute on the display, 916 Timer0 cycles have passed, which is 60,030,976 actual cycles, which is only 0.05% greater than the expected count of 60,000,000. As time goes on, this error becomes even smaller as to become infinitesimal.

The "Time" task can be represented in C as shown in Listing 11-6.

LISTING 11-6 The "Time" task written in C code.

```
TimeTask() {                    //  Real-time clock task

int  Increment = 1;             //  Value to increment minutes by

long Fraction = 0;              //  24-bit second fraction
                                //   counter

  Hours = Minutes = Seconds = 0;  //  Initialize the time variables

  while ( 1 == 1 ) {            //  Loop forever

    while ( Fraction < 0100000h ) { //  Wait for 1,000,000+ cycles

      WaitInt( 1 );             //  Wait for the timer to overflow

      Fraction += 65536;        //  Increment the Fraction counter

    }

    Fraction = Fraction - 1000000;  //  Take a million away from the
                                //   Fraction counter

    if ( Button == 0 )          //  If the Button is pressed
      Seconds = 60;             //   force an increment
    else
      Increment = 1;            //  Reset the increment value

    Seconds++;                  //  Increment the current Second

    if ( Seconds > 59 ) {       //  Have to increment the time

      Seconds = 0;              //  Reset the Seconds

      Minutes += Increment;     //  Increment the Minutes (with setting)

      if ( Minutes > 59 ) {     //  Increment the Hours

        Minutes = 0;

        if ( ++Hours > 23 )     //  If rolling over the Hours
          Hours = 0;            //   - Start a new day

      } //  endif

      if ( Button == 0 )        //  If the Button is pressed
        Increment = (( Increment < 1 ) + 1 ) & 0x03F;
                                //   increment to a maximum of 63
      else
        Increment = 1;          //  Else, Reset the increment value
```

LISTING 11-6 The "Time" task written in C code *Continued.*

```
      SendMsg( LCDTask, Something );  //  Notify LCD task that it must
                                      //   display a new time

    }  //  Finished displaying the new time

  }  //  endwhile

}  //  End TimeTask
```

In "Time" task, the setting button is polled once per second. If it is being pressed, then the seconds are automatically set to 60 (plus one in the later increment), then the minutes are incremented using the `Increment` variable.

Before sending a message to the "LCD" task to notify it that it's time to update the display, the button is polled again. If it is pressed, then it is increased by the next power of 2 (to a maximum value of 63) or reset back to 1. When it is reset back to 1, it will increment the minute counter by 1 when the clock is running normally.

By pressing the button, the `Increment` value is increased by a power of 2. So, after 1 second, it will increase the minute counter variable by 3 minutes (and not one). After 2 seconds, it will increase the minute counter variable by 7 minutes. After 3 seconds, by 15. When an `Increment` variable value of 63 is reached, `Increment` won't be incremented any more, and every second, the hour will be updated. Using this algorithm, it's possible to run through a total of 24 hours in less than 30 seconds with quite a bit of control over the value you are trying to arrive at.

Now, if you are a digital clock manufacturer, I hope you remember where you first saw this! I would be happy to license this algorithm to you for a modest fee.

The temperature sensor used in this application, the Dallas Semiconductor DS1820, is a rather interesting little beast. It is available in a 3-pin transistor-like package and only requires a 10K pull-up resistor for the driving signal. I have included a PostScript format of the data sheet on the CD-ROM, and you can see that the operation of the device is quite simple.

As you would expect, this device, even though it seems to be the simplest one used in the application, caused me the most trouble.

The data sent and received used a Manchester-encoded signal in which the controlling device (the DS87C520 in this case) pulls down the line for a specific amount of time to write to the DS1820. A "0" is a low pulse of 15 µsec to 60 µsec (with 30 µsec being typical), and a "1" is a low pulse greater than 1 µsec and less than 15 µsec.

To read data from the DS1820, the microcontroller pulls down the line for approximately 1 µsec and releases it, at which time the DS1820 holds the line low for a specific length of time to write a bit on the line. If the DS1820 holds the line low for less than 15 µsec, then a "1" is being sent. If the line is low for greater than 15 µsec, then a "0" is being transmitted. I found it best to poll the line 3 µsec or 4 µsec after the line is allowed to go high.

Figure 11-8 shows a read of a "0" followed by a "1", to give you an idea of what data looks like on the single line.

Data is passed between the controlling device and the DS1820 8 bits at a time with the least significant bit first. Before any command, a reset pulse of approximately 500 µsec is

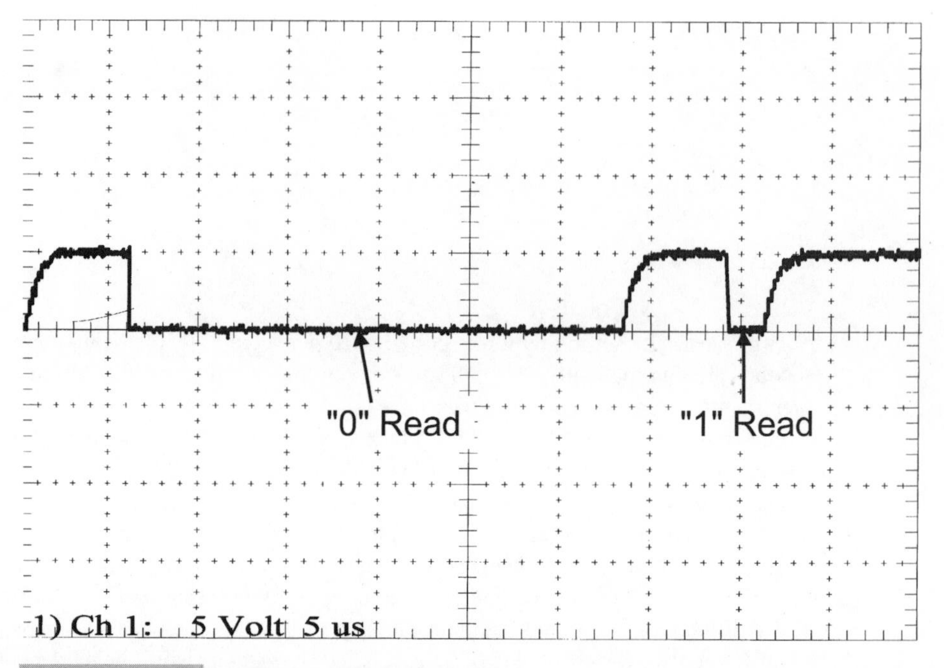

FIGURE 11-8 DS1820 "read" waveform.

output onto the line. When the line goes high, the DS1820 responds by pulling the line low for about 60 µsec. After the reset pulse, the DS1820 should not be accessed again for 1 msec.

When I first coded this application, I created three subroutines for operating the DS1820 from the DS87C520 from this application. The first routine was `DSReset`, which pulled the line down for 480 µsec and then waits 1 msec before returning to its caller.

The transmit routine, `SendDS` was coded as a simple subroutine with no hooks used for the RTOS (Listing 11-7).

There is one thing to note in both `SendDS` and `GetDS` (the data receive routine from the DS1820), and that is that the first thing that I do is to mask interrupts. This is important because, if an interrupt handler is requested and executes during the bit data transfer, the timing of the bit will be lost and incorrect data will be received, or in the case of the DS1820, the device will be reset and discard all the data received so far.

Reading data from the DS1820 was accomplished by pulsing the line low and then see how long it would stay low (Listing 11-8).

To read data from the DS1820, I carried out the following instruction process:

1 Reset the DS1820.
2 Send 0CCh followed by 044h to begin the temperature sense and conversion.
3 Wait 480 µsec for the temperature conversion to complete.
4 Send another reset to the DS1820.
5 Send 0CCh and 0BEh to read the temperature.
6 Wait 100 µsec before reading the first byte in the DS1820.

LISTING 11-7 **The transmit routine,** SendDS.

```
SendDS:                         ;  Send the byte in "A" to the DS1820

  mov     TempCount,#8          ;  Sending 8 bits

  clr     IE.7                  ;  Disable interrupts during operation

SendDSLoop:                     ;  Loop here for each bit

  rrc a                         ;  Put the LSB into the carry flag

  clr     P0.1                  ;  Drop the line low

  jnc     SendDSDlay            ;  Do we delay the line low?

  setb    P0.1                  ;  No, sending a "1"

SendDSDlay:                     ;  Delay 30 instruction cycles before
  mov     R0,#9                 ;   setting the pin
SendDSWait:
  djnz    R0,SendDSWait

  setb    P0.1

  mov     R0,#7                 ;  Wait 21 cycles before sending the
SendDSWait2:                    ;   next bit
  djnz    R0,SendDSWait2

  djnz    TempCount,SendDSLoop

  setb    IE.7                  ;  Enable interrupts again

  ret
```

LISTING 11-8 **The read routine,** GetDS.

```
GetDS:                          ;  Get 8 bits from the DS1820

  mov     TempCount,#8          ;  Sending 8 bits

  clr     IE.7                  ;  Disable interrupts during operation

GetDSLoop:                      ;  Loop here for each bit

  clr     P0.1                  ;  Toggle the I/O pin
  setb    P0.1

  nop

  setb    C                     ;  Now, see what is returned
  jb      P0.1,GetDSSave        ;  Is the bit still high?
  clr     C
GetDSSave:                      ;  Save the carry flag as the bit read
  rrc     a

  mov     R0,#5                 ;  Delay 16 cycles for DS1820 to reset
GetDSWait:
```

LISTING 11-8 The read routine, `GetDS` *Continued.*

```
djnz    R0,GetDSWait

djnz    TempCount,GetDSLoop

setb    IE.7                    ; Enable interrupts again

ret
```

7 Read the first, or SP0, byte of the DS1820.
8 Wait another 100 μsec before reading the second or SP1 byte of the DS1820.
9 Reset the DS1820.

The DS1820 has a unique serial number burned into it. This allows multiple temperature sensors (and other devices using the Dallas Semiconductor one-wire protocol) to be placed on the same pulled-up bus. To avoid first having to read the serial number out of the device and then referencing it each time a command is being sent to it, by sending a 0CCh, the DS1820 assumes that it is the only device on the network and executes each command without checking for a valid serial number being sent.

As I indicated earlier, the DS1820 gave me a few problems with the data transfers. The first problem that I had was with the transmit. I had originally coded `SendDS` to transfer data at the minimum values on the data sheet. This turned out to be too close for the operation to complete correctly. The problem demonstrated itself by having all 1s sent from the DS1820.

I found that, for the DS1820 to work properly, I had to use the typical values for the data writes. I originally used the minimums because I wanted to keep the time the microcontroller was running with interrupts masked to a minimum. However, after changing the timings to the typical values in the data sheets, the DS1820 seemed to respond properly. However, instead of a single long pulse for a "0," I received a short pulse followed by a longer pulse.

This caused me to really tear my hair out until I decided to go back and look at the DS87C520's data sheet. As I originally created the application, I had placed the connection to the DS1820 on pin P1.1, which has an active pull-up. When a "1" is written to this pin, an extra driver is enabled for 2 clock cycles to help pull up the pin. This driver seemed to be able to overdrive the DS1820.

The connection to the DS1820 was moved to P0.1 (the P0 port does not have internal pull-ups), and the application seemed to work without any problems.

This litany of problems might make it seem like I had problems with implementing the DS1820 in the RTOS, and I would have to disagree. The DS1820 code could be placed in a single task microcontroller application without any modifications. If anything, the RTOS allowed me to easily debug the application by turning off unneeded tasks while I was debugging specific ones.

Using the RTOS and making sure that most operations are keyed off of the timer, rather than instruction delays, means that it should be easily transferred to another 8051 device. The only issue that you would find in porting this RTOS application to another 8051 is with the DS1820 code. In these routines, I used the instruction/cycle timing of the

DS87C520 HSM to meet the specified values. If you were to run this application on a standard 8051, you would have to go through the DS1820 code and modify it for the other microcontroller.

In *Handbook of Microcontrollers*, I implemented this application in a Motorola 68HC05 in an RTOS that I had written for it. This application was actually a lot easier to implement in FullRTOS on the 8051 because of the ability of this RTOS to call subroutines. I found that this simplified the creation of the application and avoided having to create large and complex macros because subroutines cannot be called.

12

EXAMPLE APPLICATIONS

CONTENTS AT A GLANCE

Now that you understand how the 8051 works, understand how it is programmed, and have gone through a few example experiments, it's now time for the fun stuff: actual applications. The applications that I have selected are designed to show off different characteristics of the 8051 and how to interface to various devices. I have created raw card designs for several of the example applications that you can have built simply (see appendix I). This will simplify the task of being able to create the applications and take the drudgery out of wiring the circuits yourself.

These various example applications can be expanded upon or aspects of them taken and used in other applications. As well, I've tried to make the example applications as device independent as possible so that you can mix and match between different 8051 processor architectures (i.e., classic, HSM, XA, 151, etc.). In the interests of allowing you to easily program the devices, I have primarily used the AT89C*x*051 devices, for which I have provided the programmer circuit and software earlier in the book.

Many of these sample applications are classics in the sense that they are typical of what a microcontroller is called upon to do. In these applications, I think you'll be impressed at how efficient the 8051 code is for different applications as well as how simple the wiring is. These applications really show off the power of the 8051.

Please note that, while I am providing you with all the information needed to reproduce these circuits, I am not giving you permission to build these projects for commercial purposes.

I should also point out that none of the projects presented in this book are debugged to what I would consider a production-qualified level. While I have done a reasonable amount of testing, I have not exhaustively looked for cases where the projects won't work. For this reason, I have avoided presenting projects that could be potentially dangerous. All the projects given in this book are powered either from batteries or Wall Wart power supplies and have simple 5-V regulators on board.

For several of the projects produced in this book (namely the programmer, emulator, 51Bot, and Breakout Box), I might arrange to have kits made at a later time. If you are interested in these kits, please check my Web page or drop me a line.

Marya's Music Box

When I wrote *Programming and Customizing the PIC Microcontroller*, one of the most popular projects that I came up with was creating code and example hardware that would play simple tunes on the PICMicro. For this book, I wanted to expand upon this and fix a few of the problems that I was left with the PICMicro project. The result, given in this book, is a small project (Fig. 12-1) that will not drive you crazy as fast as the PICMicro project (playing "Frosty the Snowman" over and over again) and will give you an introduction to interfacing a matrix keyboard to the 8051.

After I created the PICMicro "Tunemaker," I went back and looked at what I felt could be improved upon. The original application ran with interrupts and required some funky code to make sure the notes would come out correctly. As well, to enter a song into the PICMicro, I wrote a simple compiler for the PC that took a script and converted it into data that was included with the PICMicro source at assembly time. Creating tunes for the application was quite labor-intensive.

This project was designed for my three-year-old daughter, and I wanted to make it a little less clunky and more robust than the projects I have made for her in the past. In light of this, I decided to use a matrix keyboard to allow her to select different songs. As well, I wanted to avoid making it complex to use, so I wanted the 8051 inside the unit to turn on or off via the keyboard and not require a separate power switch.

FIGURE 12-1 **Music box AT89C2051, speaker, battery, and keypad connections.**

My first experiments were to figure out a way that I could turn on and off the AT89C2051 (the microcontroller that I was planning on using) using its internal features (namely idle and power-down modes).

Looking at the documentation, in the PCON (power control) register, by setting bit one or zero, the microcontroller would be put in a low-power power-down or idle modes, respectively. The idle mode results in an 80% decrease in power consumption, while power-down mode reduces the power required by the microcontroller down to around 10 μAmps.

I decided to first attempt to use power-down mode and made a simple circuit to program and test it (Fig. 12-2). This circuit was designed to hold Reset high for a few cycles before P1.3 pulled the line low (by turning on the transistor). In the application, I wanted to set up the circuit so that, when the push button was pressed, the gate of the transistor would be pulled to ground, the transistor would turn off, and Reset would be high until P1.3 became active (high).

To test this circuit, I simply created a small circuit and used the code in Listing 12-1. Upon entry, this code checks the state of the first four bank registers. If they are equal to the expected values, then the warm-boot LED will be lit; otherwise, the start-up LED will light.

Looking at the code, you might think the warm boot test could be simplified to:

```
mov   A,R0
xrl   A,R1
xrl   A,R2
xrl   A,R3
jz    WarmBoot
```

because, when these values are XORed together, you will end up with zero (as I will in the code in Listing 12-1). The problem with the simplified code is that there are many possible combinations of power-up values that, when XORed together, result in zero.

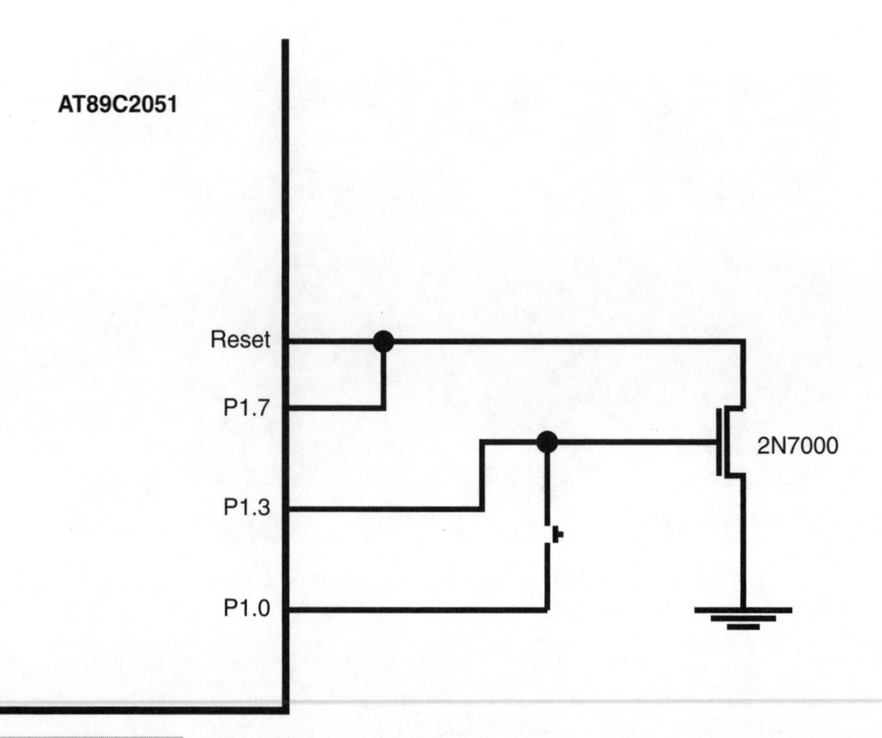

FIGURE 12-2 AT89C2051 Reset control.

LISTING 12-1 Code to test the Reset control.

```
;  PROG44 - Making music from an AT89C2051
;
;  Simple application to test power-down/power-up code.
;
;  This application after power up, places the AT89C2051 into
;   power-down mode and sees if the transistor pull-up circuit can
;   turn it back on.
;
;  Myke Predko
;  98.05.29
;
;  Hardware notes:
;  AT89C2051 is used as the microcontroller
;   - Oscillator speed is 10 MHz
;  P1.0 is used as a pull-down to ground
;  P1.3 is connected as a pull-up to the Reset transistor's base
;  P1.7 is used as a voltage source for Reset
;
;  P3.0 is the start-up LED
;  P3.1 is a reset From power-down indicator LED

;  Constant declarations

;  Variable declarations
```

LISTING 12-1 **Code to test the Reset control *Continued.***

```
;  Macros


;  Mainline
 org 0                          ;  Execution starts here
MainLine:                       ;  Mainline of the program

  clr     P1.7                  ;  Make sure Reset can't be set

  xrl     0,#055h               ;  Does R0 = 055h?
  xrl     1,#0AAh               ;  Does R1 = 0AAh?
  xrl     2,#000h               ;  Does R2 = 000h?
  xrl     3,#0FFh               ;  Does R3 = 0FFh?

  mov     A,R0
  orl     A,R1
  orl     A,R2
  orl     A,R3

  jz      WarmBoot              ;  If they are, then warm boot

  mov     R0,#055h              ;  Set up for the warm boot
  mov     R1,#0AAh
  mov     R2,#000h
  mov     R3,#0FFh

  clr     P3.0                  ;  Turn on the power-up LED
  setb    P3.1

  clr     P1.0                  ;  Enable the reset condition
  setb    P1.7

  orl     PCON,#%00000010       ;  Set the power-down bit in PCON

  ajmp    Mainline              ;  After power up, jump to mainline (if
                                ;   no reset)

WarmBoot:                       ;  Have executed here after reset

  setb    P3.0                  ;  Indicate the situation
  clr     P3.1

  clr     P1.0                  ;  Can we bounce back and forth?
  setb    P1.7

  orl     PCON,#%00000010       ;  Set the power-down bit in PCON

  ajmp    Mainline
```

By checking each bit, there is only one combination in *4.3 billion* that matches what I check for, which maximizes the probability that, if the result is equal to zero, then we are experiencing a warm boot.

This circuit works very well with the program it is executing. The problems arose when it was first powered up. I found that upon power up, the I/O pins were not in the state that I needed to get a valid Reset high signal to start the application running every time. I found that I had about a 25% success rate with powering up the application properly (most of the time it stayed in reset with both LEDs lit). After experimenting with different configurations and different ideas, I came to the conclusion that the power up with a button power-down circuit did not work as I envisioned. I was able to come up with some alternatives, but these all required additional circuitry and logic. My goal was to just have the 8051, a voltage regulator, and minimal extra components to keep the complexity and size to a minimum.

As an alternative, I played around with idle mode and found that I could turn it on and off very easily in the circuit shown in Fig. 12-3.

For the final application, I used an experimenter's box, keypad, and 9-V radio battery holder (PacTec model K-HM-9VB-SP). To get out of idle mode, a reset of interrupt request had to be made. This was easily implemented by using a key on the keyboard that pulls the _Int0 line low when it is pressed.

Idle mode doesn't provide the same extreme reduction in current, but I did find that the final application was still running after two months with a 9-V alkaline radio battery. To be fair, at the end of two months, the battery, when tested, indicated that it had to be replaced, but it would still power the circuit.

With the power-down decision made, I could concentrate on making the music. Pulling a speaker to ground would allow me to click it. By using a timer in mode 2, I was able to output some pretty decent tones out of an 8-Ω speaker wired as shown in Fig. 12-3.

If you look at the actual signal across the I/O pin, you would see that looks like Fig. 12-4. The spikes are caused by inductive effects within the speaker.

I decided to use both timers in the AT89C2051 to generate the music. Timer0 was used to output the tone, while Timer1 would time how long each note plays for.

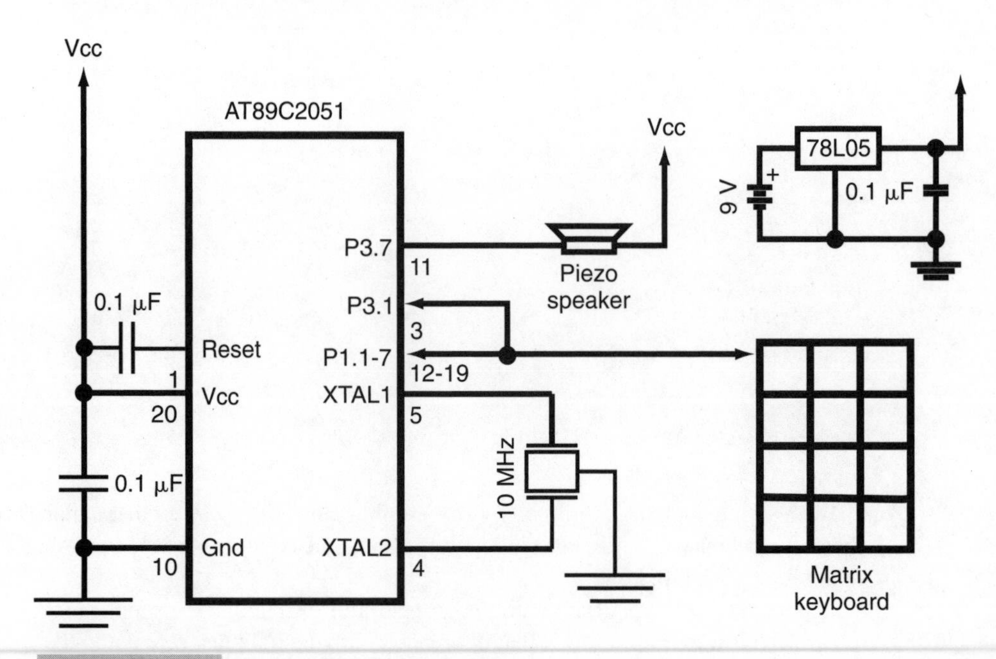

FIGURE 12-3 **Marya's music box circuit.**

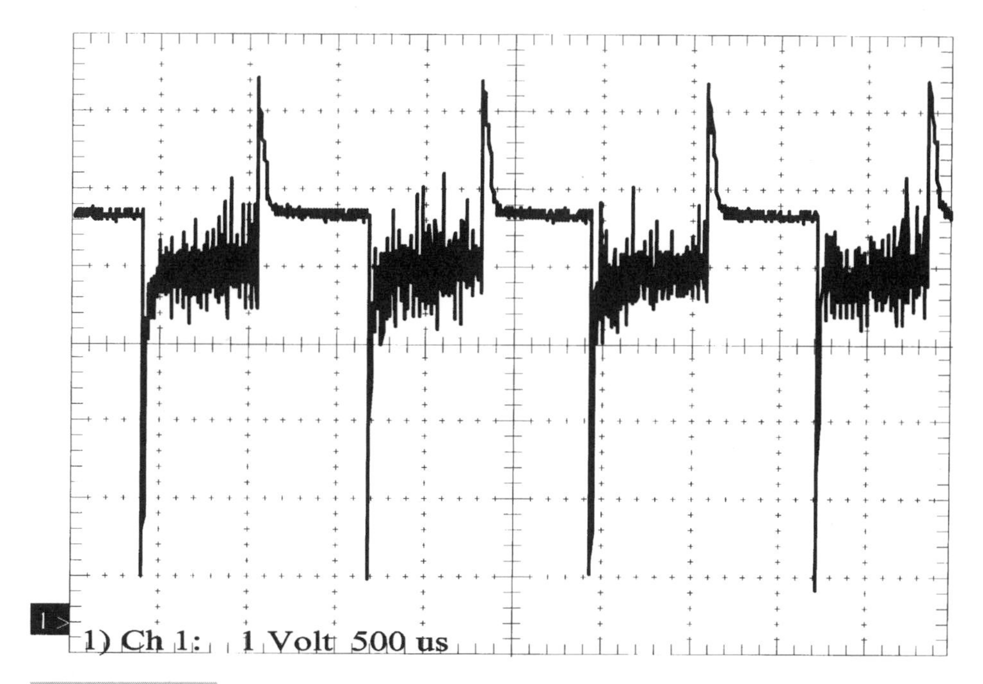

FIGURE 12-4 Music box speaker output waveform.

The Timer0 interrupt handler is very simple:

```
 org   00Bh
Tmr0:
  jb   0,Timer0End
  cpl  P3.7
Timer0End:
  reti
```

Each time it is invoked, if bit zero (bit 0 of scratchpad RAM address 020h) was reset, then the speaker bit would be complemented (toggled). If bit 0 was set, then the speaker bit wouldn't change. This was how I basically implemented a pause in the tune output.

The Timer1 interrupt handler is considerably more complex, and I first wrote it up in C as shown in Listing 12-2. The Timer1 interrupt handler is responsible for actually reading

LISTING 12-2 The Timer1 interrupt handler in C.

```
Timer1 Int()                  //  Timer1 interrupt handler, load next
{                             //   note or output a short "pause"

  if ( —BeatCount == 0 ) {    //  Is the note finished?

    BeatCount = 4;            //  Four beats == 0.5 seconds

    if ( —NoteDuration == 0 ) //  Is the note finished?
      if ( Tune[ CurTune ][ ++Note ] != 0 ) {
```

LISTING 12-2 The Timer1 interrupt handler in C *Continued.*

```
        if ( TL0 = TH0 = Freq( Curtune, Note )) == 0 )
                              //  Load the timer with the new note
          PauseFlag = 1;      //   If zero, "pause"
        else                  //  "PauseFlag" is bit 0
          PauseFlag = 0;      //  Not zero, play the note
        NoteDuration = Delay( Curtune, Note );
                              //  Load the duration of the note
      } else {                //  The song is over

        Curtune = Pause;      //  Just continue playing a short "pause"
        NoteDuration = 1;
        PauseFlag = 1;        //  Don't output the note

        TuneActive = 0;       //  Flag mainline that tune is complete

      }
  }
} //  End Timer1Int
```

the tune from the table that the tune is loaded into, processing it, and, at the end, just continually playing a short pause.

The note information is stored in a single byte with the least significant 5 bits being the note to play (with "0" being pause) and the top 3 bits being the duration (in eighth notes).

To simplify loading the note data, I created the `Note` macro, which takes the note and duration and combines them into a single data byte:

```
MACRO Note (Value,Duration)
  db      (Value+(Duration SHL 5))
ENDMAC
```

The `Value` is the note, and I created a list of defines from "A" below "middle C" to "D" one octave above "middle C." Table 12-1 shows the frequencies that I used and the note identifiers (defines).

Using this collection of notes, quite a wide range of songs can be played. You might want to expand the note range. By simply doubling (when going up) or halving (when going down) the frequencies of the notes in Table 12-1, you can get the appropriate frequencies for notes in other octaves.

The basic theory behind reading a switch matrix keyboard is quite simple. A signal is put onto a column, and rows are scanned to see if this signal appears.

The most usual way of doing this is to pull up the rows and selectively tie the columns to ground. When a button is pressed (and its switch is closed), a low voltage will be sensed at that row position.

In Fig. 12-5, I have shown a 2x2 matrix keyboard connected to an AT89C2051. In this diagram, I have assumed that the internal pull-ups of the microcontroller's pins are used for the row pull-ups, and I have shown explicit transistors for the pull-down controls, although in the actual application, I used the pin low function.

When I wired the keyboard to the AT89C2051, I deliberately chose pins that would allow me to pull up and down the lines without any additional components. Because of the

TABLE 12-1 THE FREQUENCIES AND NOTE IDENTIFIERS

NOTE IDENTIFIER	NOTE	FREQUENCY
LA	A	440 Hz
LAS	A#	466 Hz
LB	B	494 Hz
LC	middle C	523 Hz
LCS	C#	554 Hz
LD	D	587 Hz
LDS	D#	624 Hz
LE	E	659 Hz
LF	F	698 Hz
LFS	F#	740 Hz
LG	G	784 Hz
LGS	G#	831 Hz
HA	A	880 Hz
HAS	A#	932 Hz
HB	B	988 Hz
HC	C	1047 Hz
HCS	C#	1109 Hz
HD	D	1175 Hz

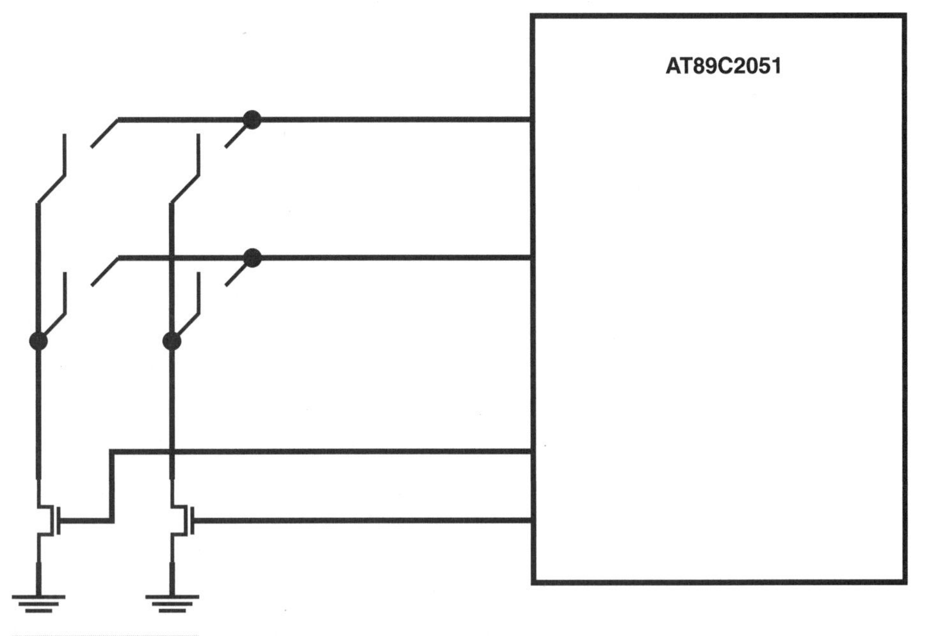

FIGURE 12-5 Switch matrix wiring to the microcontroller.

AT89C2051's pin arrangement, I wired the 4x3 keypad with P1.1 (which has no internal pull-up) as a column pull-down.

The mainline of the code polls the keypad and specifies the tune to be played. Writing it out as C pseudo-code, it looks like Listing 12-3.

This code waits while a tune is active (playing). Then, if it is the button that starts the idle mode (I used the one in the bottom right-hand corner), the AT89C2051 is put into this mode until the button is pressed again.

I deliberately kept the keypad scan code vague in the `main` routine in Listing 12-3. This is because you might want to use a keypad that's different from the one that I used (or can't find the PacTec keypad/box combination). Learning and scanning the keypad actually took me as much effort as coming up with the code to play the songs.

Before working through the scanning software to read a keyboard, I beep out the keyboard to see how it is wired. Very rarely is a keyboard laid out in such a way that it can be wired directly to the microcontroller. Usually the wiring is optimized so that a single copper layer is required for the keypad circuitry.

When I beeped through the PacTec keyboard using a DMM, I first numbered the keys from 0 to 11 by row (left to right) followed by column (top left is 0, bottom right is 11). This keypad has a single ribbon cable with seven conductors coming out from it. I arbi-

LISTING 12-3 **Code for the mainline in C.**

```
main() {

  Init();                         //  Initialize the timers and _Int0

  while ( 1 == 1 )
    if ( TuneActive == 1 ) {  //  Something is playing

      while ( TuneActive == 1 );      //  Wait for it to complete

      if ( LastButton == 11 ) {       //  Power down?

        Enable( _Int0 );              //    Yes
        Set( Button11 as _Int0 );

        Enable( IdleMode );     //  Power down

        CurTune = 11;           //  "Beep" when coming back up
        TuneActive = 1;
        LastButton = 0;         //  Mark with an invalid button

      }
    } else {                    //  Wait for key

      while( NoPress );         //  Wait for a key to be pressed

      while( Press );           //  Wait for the key to be released

      CurTune = LastButton = ButtonPressed;
      TuneActive = 1;

    }
}  //  End main
```

TABLE 12-2 THE CONNECTIONS FOR THE KEYBOARD

	COLUMN 1	COLUMN 2	COLUMN 3
Column 1	2–4	3–4	4–6
Column 2	2–7	3–7	6–7
Column 3	1–2	1–3	1–6
Column 4	2–5	3–5	5–6

trarily numbered the connectors from 1 to 7. The connections came out as shown in Table 12-2. From this, I determined that the rows were pins 1, 4, 5, and 7 while the columns were 2, 3, and 6.

To read the keyboard, I connected the seven pins directly to P1.1 through P1.7 of the AT89C2051. I connected P1.1 to pin 1 of the keypad (and used the columns to pull down to ground) to avoid problems with not having a pull up on the P1.0 and P1.1 pins.

In the `main` C pseudo-code, you'll notice that I don't wait for the key to be released before starting the tune. The reason why I did this was to avoid having to debounce the keystrokes. Because each tune takes at least a large fraction of a second (and some tunes take many seconds), debouncing is really not an issue.

The resulting application code is found on the CD-ROM as PROG44G.ASM in the PROG44 subdirectory. As I have written it, each eighth beat is half a second long for a tempo of 120 beats per minute. This is a pretty lively tempo that works well for most songs.

Creating a song consists of simply transposing the notes and their duration into a table of `Note` macros for each tune (or button selection). Valid durations are from one to seven. A zero or eight will result in the delay being 256 eighth notes, which will take about a second to work through. If this is too limiting for you, you might want to create two tables: one with the notes and one with the durations. Because I wanted to potentially use all 12 buttons, I used a 1-byte table to maximize the number of songs/notes I could put in.

In some songs, notes are repeated. In this program, a definite dead space can be heard between the notes. This is caused when the timer is reloaded (and already has a set value in it) and, in this case, is desirable. For other applications where the time out value is updated, you might want to come up with a different way of implementing the reload (such as not reloading TL0 when a new interval is loaded into TH0).

In terms of song length, you should be able to implement song tables of up to 254 notes. Make sure that you always end the song table with a 0 byte, and don't use it again until you are at the end of the song. I recommend trying to keep the songs fairly short (less than 20 seconds) because even the jauntiest rhythm will seem like a dirge after a while.

You might feel like the application is limited, because there is only one constant volume voice. However, if you look through beginner piano books, you should find some interesting and entertaining songs (listen to the "Hawaii Five-O" theme that I have included in this application to see what I mean).

PROG36: 51Bot—Getting the Wheels Turning

The next example application is actually a robot and the various subsystems required to get it running (Fig. 12-6). When you finish this application, you will have a robot that, using two electric motors, can navigate over a flat surface, be controlled by a TV infrared remote control, move at different speeds, or control motors via RS-232.

Robots really encompass a lot of different sciences and engineering disciplines. One of the reasons why I enjoyed doing this project is because, through the different interfaces, I ended up with something that was a lot of fun, and I really learned a lot about the 8051. If you go through with doing all the different subsystems of this project, you'll end up with a learning tool that I think you'll really get a lot out of. This robot could also be a good introduction to electronics and programming for children and teenagers.

The basis for the robot was the Tamiya Wall Hugging Mouse (Tamiya Item Number 70068).

This "toy" (for lack of a better term) consists of two electric motors driving different wheels with a wall sensor (also known as a piece of wire connected to a microswitch) that turns the motors on and off to keep the mouse running along a wall. The motors and sensor are mounted on a sturdy clear plastic base that you can drill or mount equipment on.

I actually liked the vacuum formed mouse top, so an important goal for me was to create the controller in such a way that it could be placed inside the original package without having to cut apart the top of the mouse. To accomplish this, I had to begin to make some assumptions about how this would be done.

The first assumption was how to run the motors. The motors supplied with the kit are designed for a 1.5-V "C" cell battery. When I was trying to decide how to run the robot, I tried

FIGURE 12-6 Wall Hugging Mouse with an AT89C2051 control card installed.

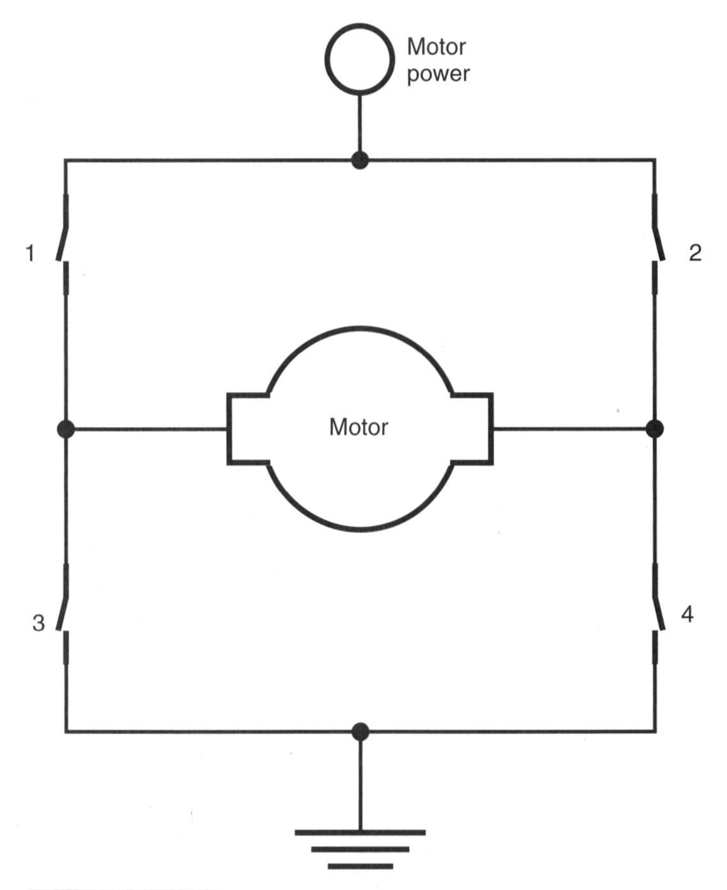

FIGURE 12-7 **"H" bridge motor driver.**

a 9-V radio battery and found that the motors raced and gave off a bit of a burning smell. To keep the motors from burning out, I decided to keep the motors running from as close to 1.5-V source as was possible. This was a concern because I was planning on running the electronics from a 9-V battery (hence the test), and this meant that I would have to use separate batteries for the motor supply. This was actually a good thing because it meant that the motor's power supply would be separate from the electronic's power supply.

How to control the motors was an issue because of the space required and the low power required. Often in small, simple robots, relays are used for the motor controls. Relays didn't seem to be appropriate for this application because of the power required to energize the coils, and I couldn't vary the robot's speed.

To control the motors electronically, I decided to use an "H" bridge (Fig. 12-7). This circuit is used to control the current path through the motor. In Fig. 12-7, if the top-left and bottom-right switches are closed, then the motor will turn in one direction. If the top-right and bottom-left switches are closed, the motor will turn in the opposite direction. Both switches on one side of the motor should never be closed, else current will pass right through the circuit without going through a load, which will probably short out the power supply.

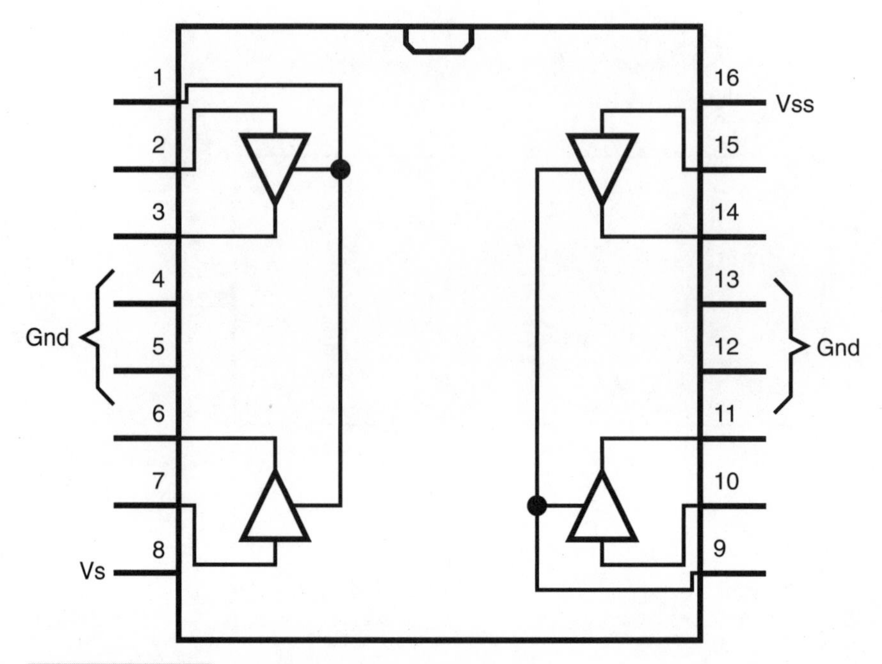

FIGURE 12-8 293D "H" bridge motor driver.

Instead of designing my own circuit to implement the "H" bridge, I decided to look for a commercial solution. I chose the Unitrode "293D" motor driver chip because it is simple to interface to and can control up to two Amps of current at a variety of voltages to motors, relays, or other magnetic components.

Not really shown in the 293D's pin-out (Fig. 12-8), the 293D can be used to control motors in an "H" bridge configuration by wiring two motor power leads to each half of the chip. The 293D has a few features that make it very desirable for motor control.

The first, and probably most important, feature is that each of the four drivers have clamping diodes to suppress back EMF when the motors are turned off. This is important because all magnetic devices produce a large voltage spike when the drivers are turned off. This spike is caused by the collapsing magnetic field when the current is turned off. A diode should be placed across the coil to prevent this back EMF from disturbing or damaging any local electronic components.

Another feature of the 293D is the enable line for each of the drivers. This line can be used to implement a pulse width modulation (PWM) speed control for the motors without having to change the driver controls.

The transistor controls of the 293D will cause a 1.5-V drop within the chip, which means that, if I wanted to run the robot motors at 1.5 V, I could do it with two 1.5-V batteries wired in series without any problems.

The last feature is the wide range of voltages that the chip can control and pass along through the drivers. For high current applications, the board that the chip is mounted on should be designed with a large, heat-sinking flood area around the 293D's ground pins. This flood area will give the chip some additional copper to allow current induced heat to be radiated away.

Using the 293D, I could create the circuit shown in Fig. 12-9 to control the two motors of the Wall Hugging Mouse. This circuit is the basis for the next five example applications. Rather than add to this circuit, I will just show the hardware enhancements and how they are wired to the AT89C2051.

The power provided to the AT89C2051 is provided by a 9-V battery regulated by a 78L05 and both the electronics (going to the AT89C2051 and 293D's logic control) and two "AA" batteries wired in series as I described earlier. Both power sources are controlled by two SPDT switches that can be turned on and off. I wasn't able to find a DPDT switch that was small enough to fit in the application.

Because I knew I wanted to put the control logic inside the Wall Hugging Mouse and I wanted to have the electronics inside the mouse body without having to modify it, I created a raw card (PCB) for the robot circuit. This card has a lot more features than just the AT89C2051 and 293D chips. Later in this chapter, I will present the printed wiring card that I designed for this robot.

With the circuit built, I then wanted to make sure that I could turn the motors on and off reliably. To do that, I created the program shown in Listing 12-4 that makes the mouse go forward for two seconds and then turn to the right (by reversing the right motor and keeping the left motor running forward) for one second.

This program was written in a very modular form so that the two motors could be controlled from various locations within the program.

Before you start modifying the Wall Hugging Mouse for the AT89C2051 card, you should at least build it and try it out. I found it took me about an hour to assemble it.

The mouse is really designed to solve a maze by keeping the feeler (the wire) in contact with the left wall at all times. If it looses contact with the wall, it will turn to the left until

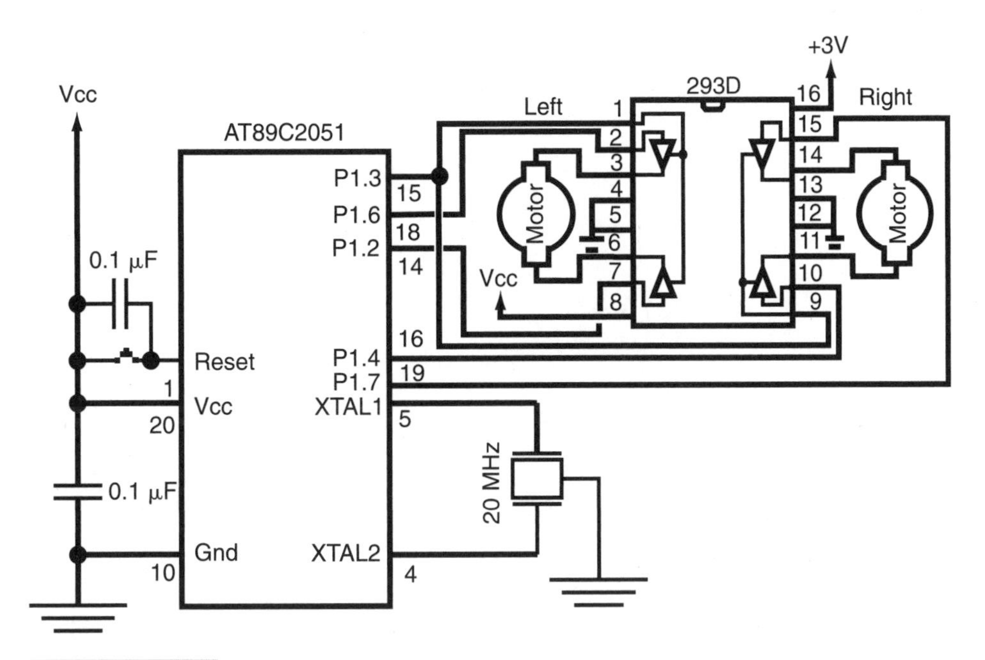

FIGURE 12-9 PROG36 motor control schematic.

LISTING 12-4 **Code for getting the mouse's wheels turning.**

```
;  PROG36 - Getting 51Bot's wheels turning
;
;
;  This is the first program of the "51Bot" and will be used to start
;   turning the wheels.  The robot will go forward for 2 seconds and
;   then turn to the right for one second and then repeat the going
;   forward for 2 seconds.
;
;  Myke Predko
;  98.03.22
;
;  Hardware notes:
;  AT89C2051 is used as the microcontroller
;    - Oscillator speed is 20 MHz
;  Wheel PWM control is at P1.3
;  Left motor control 1 is at P1.6
;  Left motor control 2 is at P1.2
;  Right motor control 1 is at P1.7
;  Right motor control 2 is at P1.4
;  LED control is at P1.5

;  Constant declarations

;  Variable declarations
DlayVar EQU 030h                  ;  Delay subroutine variables - 3 bytes

;  Macros


;  Mainline
 org 0                            ;  Execution starts here

Loop:                             ;  Loop around here

  acall   Forward                 ;  Go forward for two seconds

  acall   Dlay
  acall   Dlay

  acall   Right                   ;  Turn right for one second
  acall   LED_ON

  acall   Dlay

  acall   LED_OFF                 ;  Turn off the LED

  ajmp    Loop


Left:                             ;  Turn left

  mov     A,#%00100011            ;  Clear the P1 bits NOT doing motion
  anl     A,P1
  orl     A,#%11001000            ;  Enable the robot to go forward
  mov     P1,A

  ret

Right:                            ;  Turn right
```

LISTING 12-4 **Code for getting the mouse's wheels turning *Continued.***

```
  mov     A,#%00100011            ;  Clear the P1 bits NOT doing motion
  anl     A,P1
  orl     A,#%00011100            ;  Enable the robot to go forward
  mov     P1,A

  ret

Backward:                         ;  Go backwards

  mov     A,#%00100011            ;  Clear the P1 bits NOT doing motion
  anl     A,P1
  orl     A,#%01011000            ;  Enable the robot to turn right
  mov     P1,A

  ret

Forward:                          ;  Go forward

  mov     A,#%00100011            ;  Clear the P1 bits NOT doing motion
  anl     A,P1
  orl     A,#%10001100            ;  Enable the robot to turn right
  mov     P1,A

  ret

LED_ON:                           ;  Turn on the LED

  clr     P1.5                    ;  Clear the LED bit

  ret

LED_OFF:                          ;  Turn OFF the LED

  setb    P1.5

  ret

Dlay:                             ;  1-second delay (1,666,666 cycles)

  mov     DlayVar,#0              ;  Set the 1-second delay variables
  mov     DlayVar+1,#179
  mov     DlayVar+2,#13

DlayLoop:                         ;  Delay loop for 1 second
  djnz    DlayVar,DlayLoop        ;  Loop for 1,666,666 cycles
  djnz    DlayVar+1,DlayLoop
  djnz    DlayVar+2,DlayLoop

  ret
```

the feeler comes in contact with it again. During normal operation, the mouse will alternate between having the left and right motors on (only one can be on at any given time) and will follow a wall with a slight wobbling motion. You might find that the mouse gets stuck in inside (turning right) corners, and you might have to place a 35-mm film canister in the corner to avoid this problem.

The Wall Hugging Mouse is interesting to watch as it follows the contours of a wall. I tried it out in our kitchen, and it was interesting to watch it follow the side of our cabinets and obstacles I placed in its path.

However, it was only interesting for five minutes or so. You'll very quickly become bored with this and look forward to wiring in the AT89C2051.

The first thing I did when modifying the mouse was to remove the microswitch, wall-sensor assembly, and the power switch. Next, I modified the base plate to accept the electronics card.

To mount the electronics board, I drilled four holes for 1.5" 10/32" bolts (using a 0.125" drill bit) and cutting and filing a new hold in the bottom for the logic and motor power switches. I ended up putting the two SPDT switches side by side and using the original power switch and "C" cell battery holder screws for securing the switches to the base plate. (See Fig. 12-10.)

I put the switches on the side opposite the wall sensor's microswitch so that I could still use the microswitch without having to relocate it. To make the hole for the switches, I drilled a 0.375" diameter hole and then filed it out to allow the switches to operate. I then used a 0.031" bit to make the holes for the mounting screws.

With this done, the board could be put over the motors and bolted down. The AT89C2051 should be mounted in a socket to allow for it to be pulled and reprogrammed. The 293D should be soldered to the board to make sure that the heat-sinking features built into the board can be taken advantage of. The 293D can be socketed for this application. However, if you are going to use the board for controlling other robots that use different motors, I highly recommend soldering in the 293D.

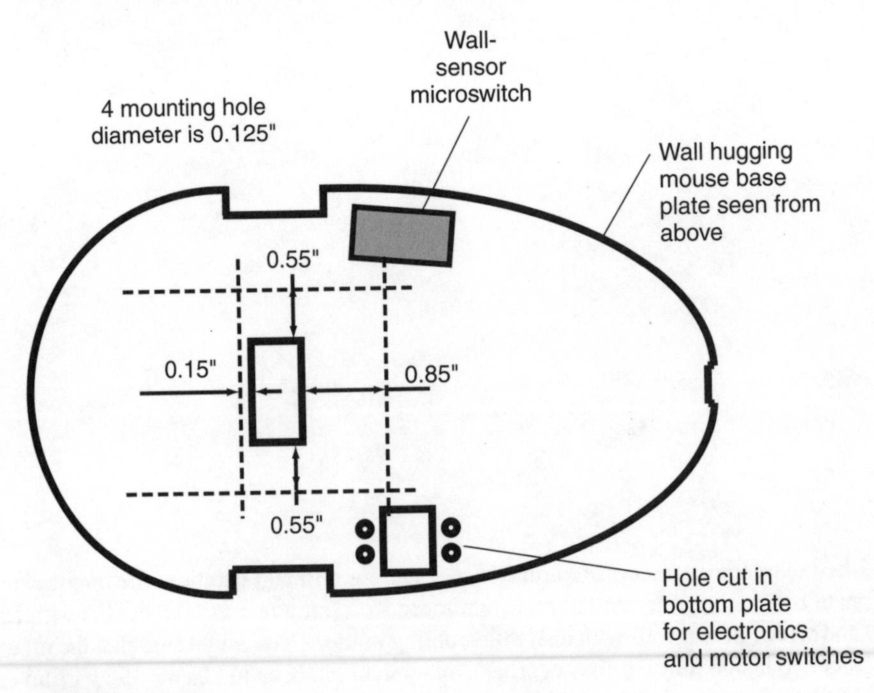

FIGURE 12-10 **51Bot bottom plate modifications.**

The battery and switch connections are soldered to the board. I have used screw terminals and two position connectors for the motor connections to allow the board to be pulled from the mouse without having to pull the motors along with them.

Depending on how the PWC is made, you might have to drill the electronics card mounting holes out a bit. The holes should be the correct diameter and location. However, with copper and solder plating on the inside of the holes, the actual diameter might be a little small.

To mount the screws, I found that I had to have 1.10" space from the top of the base plate to clear the motors. After putting in the bolts, I then put nuts on them to secure them to the base plate and then put 1" plastic standoffs over the bolts. This gave me the exact clearance I needed.

To distribute the weight of the batteries, the two "AA" batteries were put in front of the motors with the 9-V battery placed at the back of the robot (where the "C" cell was originally located). I have tried using cable ties to hold the batteries in place, but I found that simple clear adhesive tape works the best.

You can now run the robot with the AT89C2051 controlling it using the program in Listing 12-4. If everything is wired correctly, you will see the robot run forward for a couple of seconds, turn right for a second (my mouse turned more than 90°) and repeat the process.

It's really not very impressive. However, it does have a lot of potential, which I will start to exploit in the next few example applications.

PROG37: 51Bot—Remote Operation Control with an Infrared TV Remote Control

With the controller able to turn the 51Bot's wheels, I could start experimenting with controlling the robot with one of the most useful and often overlooked devices: the TV infrared (I/R) remote control. These controls can be purchased quite cheaply, and chances are you might already have one around the house. For the 51Bot controller, I used a universal remote that I bought a number of years ago. For controlling the 51Bot, I will use this remote control set to Sony TV encoded signals.

I/R remote controls might not work exactly as you would expect. When it is sending a code (i.e., a button is pressed), the signal is modulated. The Sony type I/R remote modulates the signal at 40 KHz. This modulation means that the signal cannot be received by a simple I/R diode. Instead a filter is required to demodulate the signal.

Fortunately, there are a number of different devices available commercially. For the 51Bot, I used the LiteOn 1060 as the I/R receiver in the circuit shown in Fig. 12-11.

Actually, I put this circuit in twice on the 51Bot's 89C2051 controller board. The outputs of the I/R receivers were directed to the interrupt pins (P3.2 and P3.3) while the power was filtered using the 100-Ω resistor and 10-μF capacitor network.

The RC network is used to filter the incoming power as well as isolate the 40-KHz onboard oscillator from other devices in the circuit. The output of the LiteOn 1060 can be connected directly to a digital input pin. (See Fig. 12-12.)

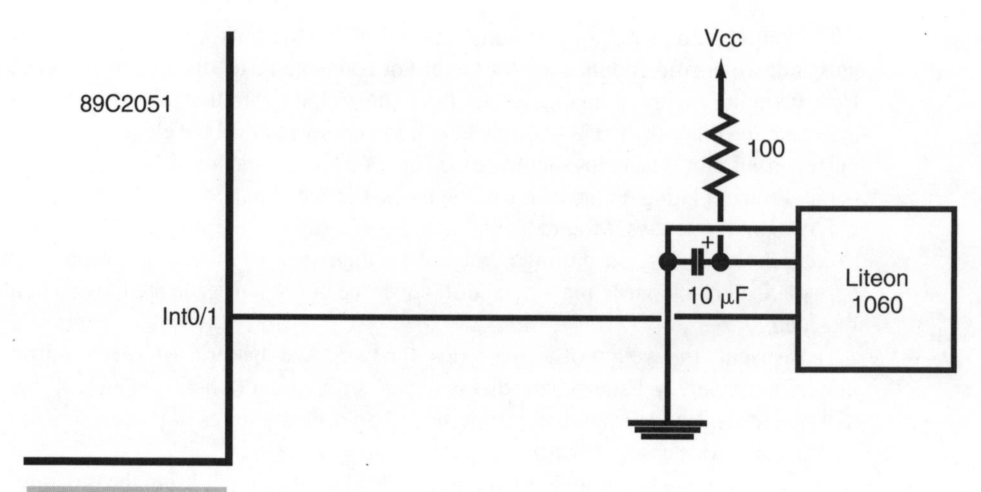

FIGURE 12-11 I/R receiver connection.

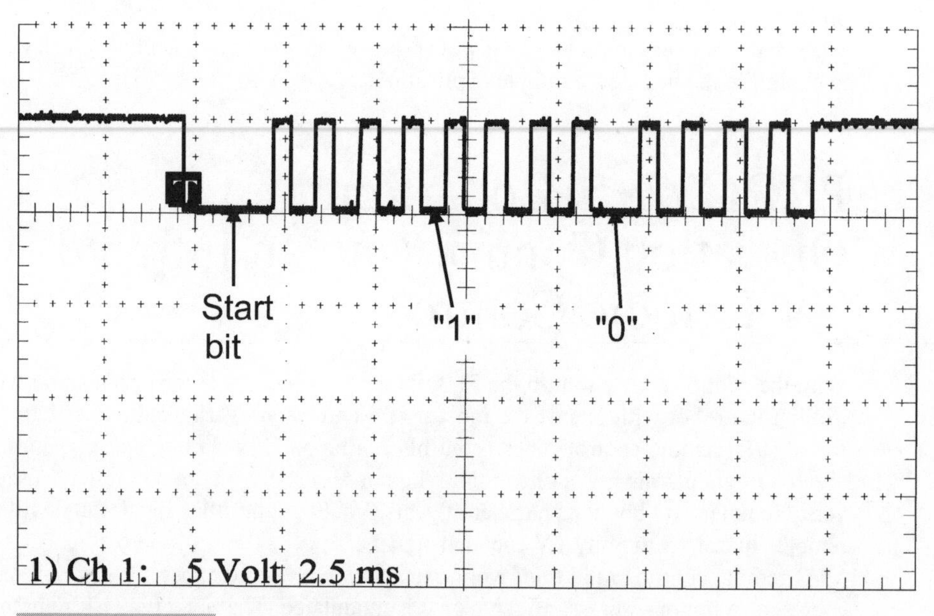

FIGURE 12-12 TV I/R remote-control input.

When the signal output from the I/R receiver is low, then it is receiving a modulated signal from the transmitter. Data is coded as different low lengths. By working with an oscilloscope, I was able to determine the bit values as shown in Table 12- 3.

These values are not absolutely precise because the I/R receiver might not pick up the first few pulses of the signal. If you were to look at this signal, you will find that the signal edges jump back and forth depending on when the I/R receiver is able to lock onto the 40-KHz carrier.

One data packet is sent every 25 msec while a button is being pressed on the remote control. The receiver is responsible for handling the signal and potentially debouncing or auto-repeating the character according to its own unique algorithm. For example, you might want to pass along the packet when the first one comes along and, afterward, send the packet on for processing once every 40 receptions for an auto-repeat once per second. By continually sending packets, the transmitter is simplified and the receiver can determine how they are to be processed without debouncing the initial button press.

Each data packet is 13-bits long, and each bit is preceded by a positive pulse. For the start bit, I have assumed there is a virtual pulse before it to keep things consistent.

The first bit is a start bit used to indicate that data is being transmitted. This bit is longer than the other bits, so the receiver can easily recognize it. It is also longer in case two I/R transmitters are operating at the same time. If an I/R transmitter begins sending a packet while another is partly through its own packet transmission, the start bit of the second packet will garble the first transmitted packet by obliterating the data bits. To detect when this happens, the receiver should have a time-out routine if a high-to-low transition is not detected when it is expected.

The 12 data bits are encoded using a Manchester encoding scheme. This data format consists of a "0" being a long low (1.3 msec) and a "1" being a short low (0.79 msec). By timing how long each pulse is, the data being transmitted can be easily decoded. When you look at the data signal, you would probably be surprised to see that the data packet varies in length. This is different from a typical asynchronous or synchronous serial data transmission scheme where each packet takes the same amount of time.

With understanding how the data is transmitted, I then had to come up with an algorithm to read the incoming data and interpret it in an 8051.

One of the first assumptions that I made was to use the interrupt pins for the data input and a timer module inside the 89C2051 for the hardware for decoding the incoming data. Timer0 was used to count execution cycles. If it timed out, the read would be reset and started over with the reception of the next packet.

As one more wrinkle, when I created the algorithm for the 51Bot, I decided to use LiteOn 1060s, one on each side of the Wall Hugging Mouse to maximize the probability that an incoming signal would be received no matter how the mouse was oriented with respect to the I/R transmitter (remote control). When I experimented with the LiteOn I/R receivers outside (where nothing would reflect the I/R signal), I found that they had the

TABLE 12-3 THE BIT VALUES FOR THE I/R REMOTE CONTROL

BIT	LENGTH
Start	2.62 msec
1	0.79 msec
0	1.33 msec

capability to read signals from a source in an arc of almost 180°. By putting two on the mouse, on opposite sides, I gave it almost 360° data reception.

This might be regarded as being overkill because infrared signals can bounce off a variety of different surfaces (although not my cat), and as long as the I/R receiver has an unobstructed field of view, only one might be regarded as sufficient. However, I decided to go with two to maximize the probability that the signal would be received as well as work through the challenge of working with two I/R receivers in one device.

When I compared the incoming signals from two I/R receivers, I found that sometimes one receiver would pick up the signal a carrier cycle (which is 25 μsec in length) or two before the other, which meant that I had to come up with an arbitration scheme to decide which signal I should work with.

The arbitration scheme I came up with was pretty simple: I would work with the first interrupt source that I got for the entire packet. As I was working through the logic, I realized that the code could execute entirely within interrupt handlers of the interrupt pins and Timer0. This made the code quite simple to query from within the mainline without having to code up with a complex input signal processing routine.

The I/R receiving code that I came up with can be modelled in "C" as shown in Listing 12-5. This code was used to test out the interrupt receiver. When I first ran it, the data read by the 89C2051 was output as synchronous serial data (Fig. 12-13).

LISTING 12-5 **The I/R receiving code in C.**

```
main() {

; #### - Initialize the timer and enable interrupts on Int0 and Int1

  :

  IRComplete = 0;              //  Wait for the incoming data
  while ( IRComplete == 0 );

  :

} //  End main

Int0() {                       //  _Int0 pin interrupt handler

  IE = Timer0 | Int0;          //  For the rest of the packet, just allow
                               //   INT0 as the source

  if ( IRRunning == 0 ) {      //  Do we have the start bit?

    IRComplete = 0;            //  Yes, start the process off
    IRRunning = 1;

    Timer0 = 3msecDlay;        //  Wait for start bit

    Count = 12;                //  Initialize the count
    Data = 0;                  //   and data storage

  } else {                     //  Not the first bit

    Data = Data << 1;          //  Shift up the data and add the bit
    if ( Timer0 < 0.8msec )    //  Do we have a "1" coming in?
      Data++;
```

LISTING 12-5 The I/R receiving code in C *Continued.*

```
    if ( —Count == 0 ) {        //  Have we finished with the data?

      IE = Int0 | Int1;         //  Yes. Now, wait for the next start bit

      IRRunning = 0;            //  Mark that the data has been received
      IRComplete = 1;

    } else                      //  Wait for the next bit
      Timer0 = 2.2msec;

  }

} //  End Int0 Handler

Int1() {                        //  _Int1 pin interrupt handler

//  #### - Same as "Int0" handler, except with references to "Int0"
//         changed to "int1"

} //  End Int1 Handler

Timer0() {                      //  Timer0 overflow interrupt handler
                                //   - Incoming data overflow

  IRRunning = 0;                //  Turn off the packet read

  IE = Int0 | Int1;             //  Begin waiting for the start bit again

} //  End Timer0 Handler
```

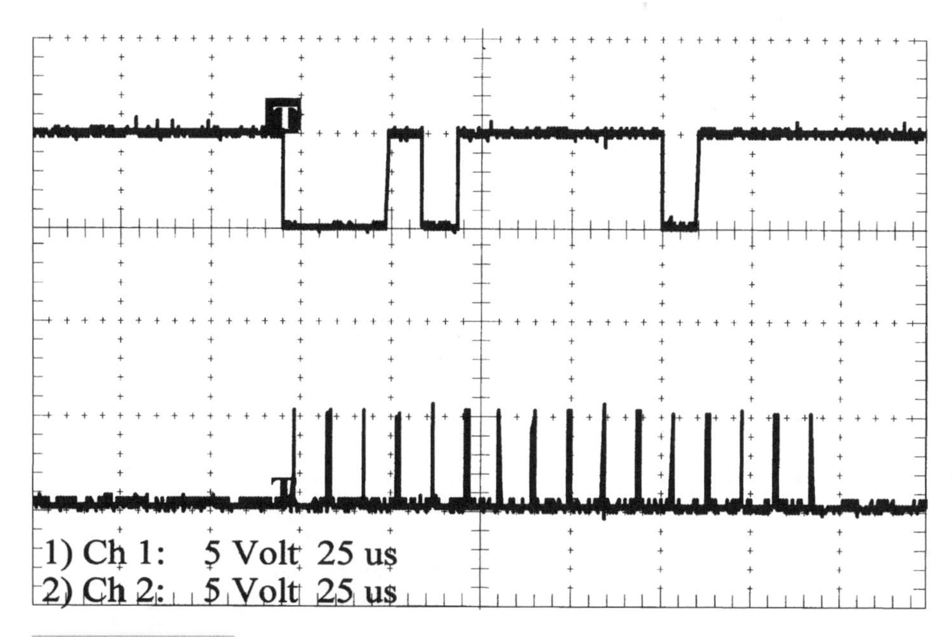

FIGURE 12-13 8051 TV I/R remote-control input data.

I output the value so that I could observe how the data was received and what its value was for different buttons being pressed.

I was surprised to find that this algorithm ran very reliably the first time I tried it out and without any misread data bits or missed characters. Actually, it ran so well that I was able to modify PROG36, add the reading information with PROG37, and run them together (as PROG37A) immediately on the Wall Hugging Mouse's control board without any problems.

If you've read through the previous code, you'll probably notice one thing that I haven't mentioned yet: why I only read 11 bits (including the start bit) when 12 bits are sent by the I/R transmitter.

The Int0 and Int1 pins can only request an interrupt on the falling edge of the incoming signal. I could have set an interrupt when the pin was low (not going low or falling, but when low), but I would have to poll this interrupt pin and time how long it was low. The method that I have used for this application is totally timer- and interrupt-based, which means that the overhead (in terms of cycles) required to implement it is quite low.

However, by going with a totally interrupt-based solution, I have had to ignore the length of the last bit (because there was no way to measure from the start of a bit's falling edge to the rising edge of the "low"). This wasn't a terrific handicap because I discovered that all data words sent by the Sony TV remote control ended in a "1." So, while I'm only working with two data bits, no potential data is lost.

With the I/R control of the robot in PROG37A, the Wall Hugging Mouse has become much more useful and interesting. It's something that you can really start playing with. I found it useful to spend a few minutes moving it around and watching how it moves to start planning the next experiments.

One of the interesting things that I did observe was that, when the robot was going forward, it turned to the right in about a 6' arc. This was caused by one motor turning at a greater speed than the other.

PROG41: 51Bot—RS-232 Interface

With the 51Bot now able to run along the floor with some control, I next turned toward creating the resources necessary to achieve the ultimate goal of the robot being a complex, programmable device. The next step was to figure out how to implement changing the speed that the robot's motors ran at.

Reading the previous paragraph, you probably feel like I mislabeled an application or something. You should be wondering why I've headed this section with "RS-232 Interface" when I introduce it as a way of controlling the robot's motor speed. The answer lies within how I am going to maintain resources within the AT89C2051 that I use to control the robot.

In the AT89C2051 that I am using for the robot's control, there are only two timers available. I have already allocated TMR0 to the I/R remote-control interface, leaving me with only TMR1 for the RS-232 interface and the PWM motor-speed control. Now, I could implement a bit-banging RS-232 interface, but I wanted to use the built-in RS-232 ports in the AT89C2051 to simplify the work required to develop the software. This meant that I had to share TMR1 between the serial UART and the PWM motor control and determine what the parameters would be that would allow both the UART and the PWM motor control to work together and not require changing of its parameters while it is running.

Fortunately, the requirements of the PWM and UART are not completely mutually exclusive. I decided to start from the PWM's side and decide that I wanted to start with a PWM frequency of 20 KHz. This frequency is above the range of human hearing, so when the robot is operating, you shouldn't hear any whine from it.

As I will show in the next section, I wanted to get four different duty cycles out of the PWM. I needed to actually run the PWM at four times that frequency, or at about 80 KHz.

With this information, I then wanted to see what would be an appropriate data rate that I could get out of the AT89C2051. To figure this out, I began working backwards from the serial data speed calculations.

Starting with the formula:

$$Data\ Rate = Clock * (2 \wedge SMOD) / (32 * 12 * (256 - Reload))$$

I could assume that I was starting with an 80-KHz clock interrupt frequency for the PWM. Putting this value into the formula I presented for figuring out the repetition rate for the repeating interrupt:

$$Interrupt\ Rate = Clock / (12 * (256 - Reload))$$

or:

$$Reload = 256 - Clock / (12 * Interrupt\ Rate)$$

For 80 KHz, This worked out to:

$$\begin{aligned} \text{Reload} &= 256 - 20\ \text{MHz} / (12 * 80\ \text{KHz}) \\ &= 256 - 20.83 \\ &= 233.17 \end{aligned}$$

Now, Putting this value into the previous UART formula, I find that, with an 80-KHz reload rate in a 20-MHz AT89C2051 and assuming that SMOD is equal to 0, I get a data rate of:

$$\begin{aligned} Data\ Rate &= 20\ \text{MHz} / (32 * 12 * (256 - 233)) \\ &= 2264\ \text{Hz} \end{aligned}$$

which is very close to 2400 bps, which is a standard serial data rate.

I then recalculated the interrupt rate based on a 2400 bps data rate, which came to a reload value of 22. This value gives an absolute data rate of 2367 bits per second, which has a 1.36% error. Over 10 bits, the full error rate is 13.6 percent, which means that there is no danger of the wrong data being read in because of the bit shifting.

For the PWM, I had a TMR1 reload (which will cause an interrupt in the next application) rate of 75.76 KHz.

With the timer values decided upon, I was able to add the RS-232 interface (Fig. 12-14).

The DS275 (the DS1275 is the follow-on part) is a wonderful device for interfacing a microcontroller to an RS-232 device. The chip will pass the negative voltage from the incoming RS-232 signal as the "1" output on the RS-232 line. The "0" will be +5 V from Vcc. I use it in this circuit to translate the positive voltage CMOS logic levels to negative voltage RS-232 logic.

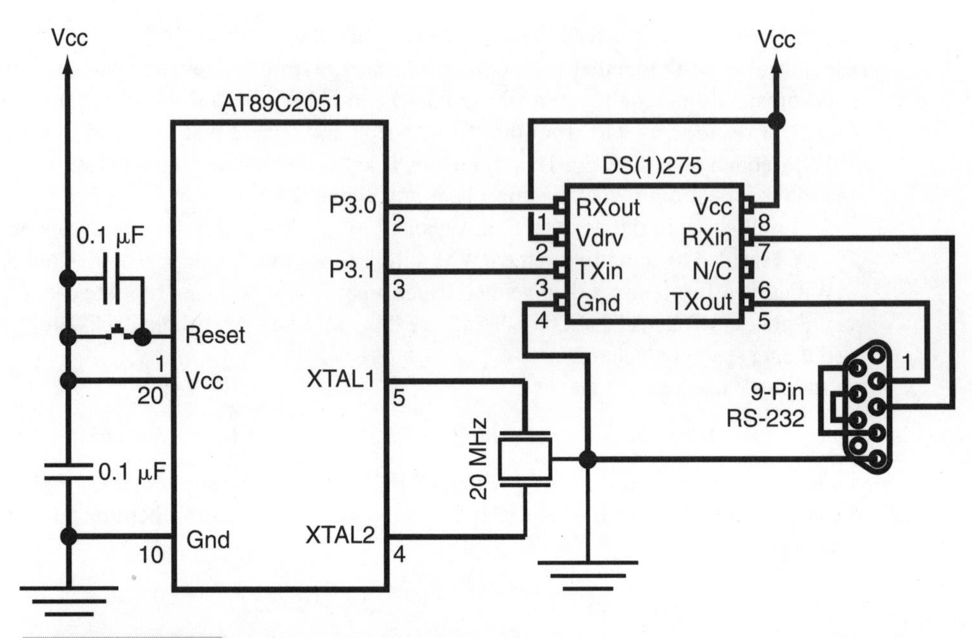

FIGURE 12-14 PROG41 RS-232 serial I/O.

I have also tied the RTS, CTS, DSR, and DTR lines together to fool the RS-232 device that will communicate to it (i.e., your PC) into assuming the handshaking lines are active for data to be sent to the AT89C2051.

The code used for this application is actually pretty simple as well, with data sent to the SBUF serial port I/O register once each second, followed by saving the last character received when the second delay was called (Listing 12-6).

LISTING 12-6 Code for the 51Bot RS-232 communications.

```
;  PROG41 - Start doing RS-232 communications with the 8051 in 51Bot.
;
;  This application will read and send RS-232 commands from the
;   AT89C2051 in the 51Bot.  Requirements for this application include
;   setting up Timer1 for use as a PWM interrupt source as well as the
;   timer for the RS-232 serial data transmission.
;
;  Timer1 interrupt rate: 75.76 KHz (for a 18.94-KHz PWM frequency)
;
;  Serial interface speed: 2400 bps (actually)
;
;  Myke Predko
;  98.04.19
;
;  Hardware notes:
;  AT89C2051 is used as the microcontroller
;   - Oscillator speed is 20 MHz
;  Serial output is on P3.1
;  Serial input is on P3.0

;  Constant declarations
```

LISTING 12-6 Code for the 51Bot RS-232 communications *Continued.*

```
TimerReload EQU 0EAh                 ;  Value to reload the timer with

;  Variable declarations
LastChar     EQU 028h                ;  Last character sent to the program
Dlay1        EQU LastChar+1          ;  24-bit value to delay 1 second
Dlay2        EQU LastChar+2          ;   between transmissions
Dlay3        EQU LastChar+3

;  Macros

;  Mainline
 org 0                               ;  Execution starts here

  ajmp    MainLine                   ;  Jump over to the mainline

 org 01Bh                            ;  Timer1 interrupt vector

Mainline:                            ;  Mainline code, set up Timer1 and
                                     ;   transmit "A" once every second or
                                     ;   last character received.

  mov     SCON,#%01010000            ;  Run serial port in 8-bit mode
  mov     TMOD,#%00100001            ;  Run the timers, Tmr1 in 8-bit reload
                                     ;   mode
  mov     TH1,#TimerReload           ;  Enable the timer
  mov     TL1,#TimerReload
  orl     TCON,#040h                 ;  Enable Tmr1 to run
  anl     PCON,#07Fh                 ;  Make sure SMOD is reset

  mov     LastChar,#'A'              ;  Set up the last character to transmit

Loop:                                ;  Loop here to transmit each character

  acall   Dlay                       ;  Wait one second

  clr     TI                         ;  Turn off the transmit completed flag
  mov     A,LastChar
  mov     SBUF,A

  jnb     RI,Loop                    ;  Was anything received?

  clr     RI                         ;  Yes, clear RI

  mov     A,SBUF                     ;  Save the character received
  mov     LastChar,A

  ajmp    Loop                       ;  Wait to transmit the next character

Dlay:                                ;  Delay 1 second

  mov     Dlay1,#0                   ;  Initialize delay values
  mov     Dlay2,#0
  mov     Dlay3,#00Ch

DlayLoop:                            ;  Loop here for 1 second
  djnz    Dlay1,DlayLoop
  djnz    Dlay2,DlayLoop
  djnz    Dlay3,DlayLoop

  ret                                ;  Return to the caller
```

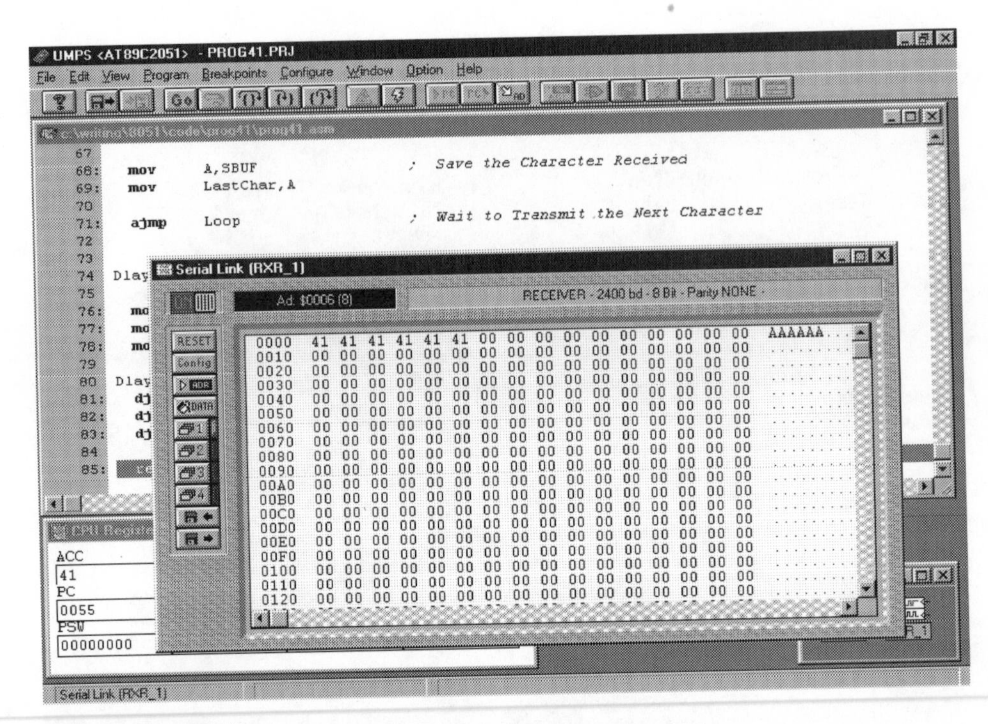

FIGURE 12-15 UMPS screen with serial resources.

To debug this application, I used the Serial Receiver and Transmitter virtual instruments available in UMPS. In Fig. 12-15, you can see the Serial Receiver virtual instrument with data received from the AT89C2051. The receiver and transmitter were connected to the RXD (P3.0) and TXD (P3.1) pins of the AT89C2051 in UMPS and used as a regular device.

PROG42: 51Bot—PWM Motor Control

When developing a large application like the 51Bot and developing it like I am doing in this book (working through each feature individually), you'll find that some features will primarily be software enhancements instead of hardware. Adding a PWM control to the motor is one such project. While no hardware changes are made to what we have already, I have made some significant changes to the software that drives the motors in the robot.

As a review, a Pulse Width Modulated (PWM) waveform is a dc logic waveform that provides a controlled power source to a device by sending a pulsed signal to the device to be controlled. The fraction of the total cycle the power is expressed as a percentage and known as the *duty cycle* and is a measure of how much power is available to the dc device. In Fig. 12-16, a 50% duty cycle is shown.

In PROG41, I created an RS-232 interface for the PC with an idea toward using Timer1 for both the RS-232 ports as well as providing a PWM clock signal. The PWM clock signal was designed to allow five states for the PWM. The PWM that I created could have a

0% (motor off), 25%, 50%, 75%, or 100% (motor full on) duty cycle from the 76-KHz Timer1 interrupt frequency (which was divided by four to provide the PWM's cycle period and the five different power outputs).

A significant issue for the PWM interrupt handling code was to make it execute in as few cycles as possible. This is an important issue because the final code would be used along with the infrared receiver code in the 51Bot, and I didn't want to have to retime anything already done.

```
Tmr1Int:
  cjne      OffCount,#0,PWMOff
PWMOn:                              ;  The PWM is on
  setb      P1.5                    ;  Output a high value
  djnz      OnCount,IntEnd          ;  If OnCount != 0, then loop around
  mov       OnCount,OnValue
  mov       OffCount,OffValue
IntEnd:
  reti
PWMOff:                             ;  Turn off the PWM
  clr       P1.5                    ;  Output a low value
  jb        10,IntEnd               ;  Is the value always low?
  dec       OffCount                ;  No, decrement the counter
  reti
```

This interrupt code was tested in PROG42, which simply controlled the 51Bot's LED using the PWM interrupt handler to show the LED off, with one-quarter, half, three-quarters, and full power.

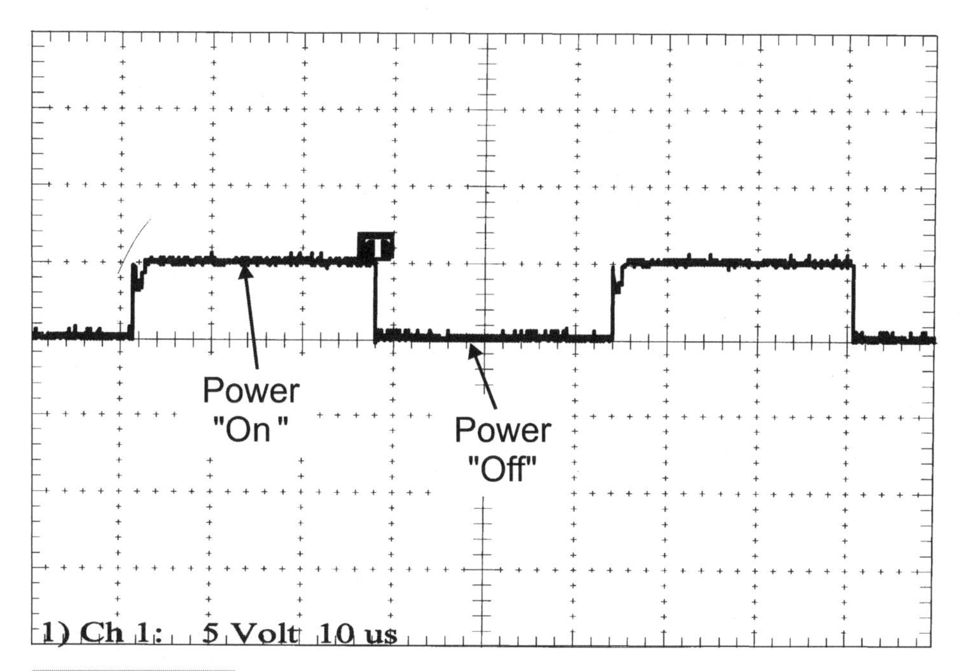

FIGURE 12-16 PWM 50% duty cycle waveform.

From a gross point of view, the interrupt handler could be described as:

```
Timer1 Int() {                      //  Timer1 interrupt handler
  if ( OffCount == 0 ) {            //  Turn on the signal?
    Output( on );                   //  Turn on the signal
    if ( —OnCount == 0 ) {
      OnCount = OnValue;            //  Reset the timer
      OffCount = OffValue;
    }
  } else {
    Output( off );                  //  Turn off the signal
    if ( OffCount != 4 )            //  Decrement the off ONLY if the PWM
      OffCount—;                    //   isn't off all the time
  }
}  //  End Timer1 interrupt handler
```

The code uses two Count variables that count the number of interrupt invocations to a total of four. When the four cycles have completed, the Count variables are reloaded from the Value variables (which contain the current PWM duty cycle information).

In the interrupt handler code, there are a couple of tricks that I used to make sure this code operated in as few cycles as possible.

The first was that I defined the Count variables as bank registers. This allowed data to be manipulated and tested without being saved in the accumulator. With this accomplished, not having to save the accumulator on the stack at the start and finish of the interrupt handler helped make it operate very quickly.

Another was my placement of `OffValue` at address 021h in the scratchpad RAM. By placing this variable inside the bit addressable space, I could do a bit test for it equal to 4, rather than loading up the accumulator and testing the value to see if I ever decrement the count value. The bit test is the `jb 10,IntEnd` instruction. Bit 10 is bit 2 of the second byte in the bit addressable area. When this bit is set, this byte (which is `OffValue`) has a value of 4 in it, and it is never to be decremented (i.e., the PWM signal is always off).

This code, along with the timer setup, was added to the PROG37 code (Listing 12-7).

LISTING 12-7 The interrupt handler code along with the timer setup.

```
;  51Bot42c - Reading a TV I/R remote control and running the
;   61Bot motors and controlling the speed of the motors.
;
;
;  This application will read the incoming I/R commands.
;
;  A Sony TV remote control is used.
;
;
;  I/R data (repeated every 45 msec):
;
;  -------+               +----+        +----+    +----+
;         |               |    |        |    |    |    |
;         |               |    |        |    |    |         o  o  o
;         +---------------+    +--------+    +----+
;
```

LISTING 12-7 **The interrupt handler code along with the timer setup *Continued.***

```
;         |                 |    |        |     |  |
;         |   Start bit     |    |"0" bit |     |"1" |
;         |    < 3 msec     |    | 1 msec |     |<1  |
;                           |High| > 1.5  |     |msec|
;                           |<0.5| msec   |
;                           |msec|
;
;
;  This application uses the 75.76-KHz interrupt signal for a motor PWM
;   output.
;
;  For PROG42, the LED will be used instead of a motor PWM control.
;
;  Timer1 interrupt rate: 75.76 KHz (for a 18.94 KHz PWM frequency)
;
;  Serial interface speed: 2400 bps (actually)
;
;  51Bot42a - Change the initial and "off" value to not use the PWM
;  51Bot42b - Run the PWM at 1/2 value to see how it works
;  51Bot42c - Add the PWM speed to the motors
;
;  I/R values:
IR2     EQU 017EFh             ;  "2" on remote control
IR4     EQU 013EFh             ;  "4" on remote control
IR6     EQU 015EFh             ;  "6" on remote control
IR8     EQU 011EFh             ;  "8" on remote control
IRPower EQU 0156Fh             ;  "Power" key on remote control
VolUp   EQU 01B6Fh             ;  Volume + key
VolDown EQU 0136Fh             ;  Volume - key
;
;  Myke Predko
;  98.05.22
;
;  Hardware notes:
;  AT89C2051 is used as the microcontroller
;   - Oscillator speed is 20 MHz
;  Wheel PWM control is at P1.3
;  Left motor control 1 is at P1.6
;  Left motor control 2 is at P1.2
;  Right motor control 1 is at P1.7
;  Right motor control 2 is at P1.4
;  LED control is at P1.5
;
;  Left I/R receiver At P3.2 (_Int0)
;  Right I/R receiver At P3.3 (_Int1)

;  Constant declarations
TimerReload EQU 0EAh           ;  Value to reload the timer with

;  Variable declarations
define OnCount=R3              ;  Use registers for PWM variables
define OffCount=R4

Flags       EQU 020h           ;  Current operation flags
OffValue    EQU 021h           ;  Bit 2 is used to check for 4
IRState     EQU 0              ;  Set when reading I/R data
IRComplete  EQU 1              ;  Set when finished with I/R read
```

LISTING 12-7 **The interrupt handler code along with the timer setup *Continued.***

```
IRButton    EQU 030h               ;  Value read in (2 bytes)
IRCount     EQU 032h               ;  Number of bits to read
OnValue     EQU 033h               ;  Reload value for the PWM "on"

;  Macros

;  Mainline
 org 0                             ;  Execution starts here

  ajmp    MainLine                 ;  Jump over to the mainline

 org 003h                          ;  Interrupt pin 0 - Left I/R receiver

  mov     IE,#%10001011            ;  Just Int 0 for incoming data

  ajmp    IRInt

 org 00Bh                          ;  Timer0

  clr     IRState                  ;  Don't have anything to do here
  mov     IE,#%10001101            ;  Just leave buttons ready for
                                   ;   interrupts
  ajmp    IRStop                   ;  Stop the motors

 org 013h                          ;  Interrupt pin 1 - Right I/R receiver

  mov     IE,#%10001110            ;  Just Int 1 for incoming data

  ajmp    IRInt

 org 01Bh                          ;  Timer1 (PWM and RS-232 interrupt
Tmr1Int:                           ;   handler)

  cjne    OffCount,#0,PWMOff

PWMOn:                             ;  The PWM is on

  clr     P1.5                     ;  Turn on the LED
  setb    P1.3                     ;  Turn on the motor

  djnz    OnCount,IntEnd           ;  If OnCount != 0, then loop around

  mov     OnCount,OnValue
  mov     OffCount,OffValue

IntEnd:

  reti

PWMOff:                            ;  Turn off the PWM

  setb    P1.5                     ;  Turn OFF the LED
  clr     P1.3                     ;  Turn OFF the motor

  jb      10,IntEnd                ;  Is the value always low?

  dec     OffCount                 ;  No, decrement the counter
```

LISTING 12-7 **The interrupt handler code along with the timer setup *Continued.***

```
  reti

IRStop:                         ;  Stop the motors

  mov     P1,#%00101011         ;  Turn off the motors

  reti

IRInt:                          ;  Handle the button interrupt

  jb      IRState,IRRead        ;  If IRState set, then executing
                                ;   interrupt
;  Else, have to set up for a new interrupt

  setb    IRState               ;  Set the I/R state interrupt
  clr     IRComplete            ;  Don't have a valid value

  mov     IRButton,#0           ;  Clear the I/R button
  mov     IRButton+1,#0

  mov     IRCount,#12           ;  Going to read 12 going lows

  mov     TL0,#0                ;  Going to delay 4 msec for end of
  mov     TH0,#LOW(256 - 26)    ;   start pulse

  clr     TCON.5                ;  Clear the Tmr0 overflow flag

  reti                          ;  Return to mainline

IRRead:                         ;  Now, read the value

  push     ACC                  ;  Save the accumulator

  clr     C
  mov     A,TH0                 ;  Load the timer value and set
                                ;   according to time remaining
  subb    A,#(256 - 4)          ;  Have we waited 8 or 11 cycles?
  mov     A,IRButton
  rlc     A
  mov     IRButton,A
  mov     A,IRButton+1          ;  Remember, it is a 16-bit value
  rlc     A
  mov     IRButton+1,A

  mov     TL0,#0                ;  Wait up to 2 msec for the next delay
  mov     TH0,#LOW(256 - 13)

  dec     IRCount               ;  Decrement the bit count

  mov     A,IRCount             ;  Is it equal to zero?
  jz      IRDone

  pop     ACC                   ;  Restore the accumulator

  reti                          ;  Return to executing

IRDone:                         ;  Have read the I/R value
```

LISTING 12-7 **The interrupt handler code along with the timer setup *Continued.***

```
  setb    C                        ;  Put in the last set bit
  mov     A,IRButton
  rlc     A
  mov     IRButton,A
  mov     A,IRButton+1
  rlc     A
  mov     IRButton+1,A

  clr     IRState                  ;  Have finished with the code
  setb    IRComplete               ;  Indicate that the button is ready

  mov     TL0,#0                   ;  Run for 1/10th of a second
  mov     TH0,#0

  pop     ACC

  reti                             ;  Return to the mainline

MainLine:                          ;  Mainline of the program

  mov     P1,#%00100011            ;  Turn off the motors initially

  clr     IRState                  ;  Clear the flags variables
  clr     IRComplete

  mov     OffValue,#0              ;  Start with full PWM
  mov     OffCount,#0
  mov     OnValue,#4
  mov     OnCount,#4

  mov     SCON,#%01010000          ;  Run serial port in 8-bit mode
  mov     TMOD,#%00100001          ;  Timer0 - Uses internal clock
                                   ;         - Run in mode 1
                                   ;  Timer1 - Uses internal clock
                                   ;         - Run in mode 2
  mov     TH1,#TimerReload         ;  Enable Timer1 to reload 32x 2400
  mov     TL1,#TimerReload
  mov     TCON,#%01010101          ;  Start Timer0 & Timer1 running
                                   ;   make edge triggered interrupts
  anl     PCON,#07Fh               ;  Make sure SMOD is reset

  mov     IE,#%10001101            ;  Enable Int 0 & 1 pin
                                   ;   AND Timer1 interrupt

Loop:                              ;  Loop around here

  jnb     IRComplete,Loop          ;  Nothing to handle

  mov     A,IRButton               ;  Now, do we have the number "2"?
  cjne    A,#LOW(IR2),Test4
  mov     A,IRButton+1
  cjne    A,#HIGH(IR2),Test4

  acall   Forward                  ;  Go forward

  ajmp    Loop

Test4:                             ;  Check for "4" pressed (and turn left)
```

LISTING 12-7 **The interrupt handler code along with the timer setup *Continued.***

```
  mov     A,IRButton              ;  Now, do we have the number "2"?
  cjne    A,#LOW(IR4),Test6
  mov     A,IRButton+1
  cjne    A,#HIGH(IR4),Test6

  acall   Left                    ;  Turn left

  ajmp    Loop

Test6:                            ;  Check for "6" pressed (and turn
                                  ;   right)
  mov     A,IRButton              ;  Now, do we have the number "2"?
  cjne    A,#LOW(IR6),Test8
  mov     A,IRButton+1
  cjne    A,#HIGH(IR6),Test8

  acall   Right                   ;  Turn right

  ajmp    Loop

Test8:                            ;  Check for "8" pressed (and go
                                  ;   backwards)
  mov     A,IRButton              ;  Now, do we have the number "8"?
  cjne    A,#LOW(IR8),TestVolPlus
  mov     A,IRButton+1            ;   - If not pressed, go back
  cjne    A,#HIGH(IR8),TestVolPlus

  acall   Backward                ;  Go backwards

  ajmp    Loop

TestVolPlus:                      ;  Check for the volume up key

  mov     A,IRButton              ;  Is the volume + pressed?
  cjne    A,#LOW(VolUp),TestVolMinus
  mov     A,IRButton+1
  cjne    A,#HIGH(VolUp),TestVolMinus

  mov     A,OnValue               ;  Can we increment the value?
  xrl     A,#4                    ;   If equal to 4, then "no"
  jz      TestVolPlusEnd

  mov     R5,#23                  ;  Use "R5" as a counter of the number
                                  ;   of repeats
TestVolPlusLoop:                  ;  Loop here until button has been
                                  ;   pressed 22x
                                  ;  Wait for the next packet to come in
  jb     IRComplete,TestVolPlusLoop

TestVolPlusLoop1:                 ;  Wait for the character input to
                                  ;   complete
  jnb     IRComplete,TestVolPlusLoop1

  mov     A,IRButton              ;  Is the volume + still pressed?
  cjne    A,#LOW(VolUp),TestVolPlusEnd
  mov     A,IRButton+1
  cjne    A,#HIGH(VolUp),TestVolPlusEnd
```

LISTING 12-7 The interrupt handler code along with the timer setup *Continued.*

```
  djnz    R5,TestVolPlusLoop     ;  If R5 is equal to 23, then increment
                                 ;   the count
  inc     OnValue                ;  Change the value
  dec     OffValue

TestVolPlusEnd:                  ;  Finished, return to loop
  ajmp    Loop

TestVolMinus:                    ;  Check for the volume down key

  mov     A,IRButton             ;  Check for volume - pressed
  cjne    A,#LOW(VolDown),Loop
  mov     A,IRButton+1
  cjne    A,#HIGH(VolDown),Loop

  mov     A,OffValue             ;  Can we increment the value?
  xrl     A,#4                   ;   If equal to 4, then "no"
  jz      TestVolMinusEnd

  mov     R5,#23                 ;  Use "R5" as a counter of the number
                                 ;   of repeats
TestVolMinusLoop:                ;  Loop here until button has been
                                 ;   pressed 22x
                                 ;  Wait for the next packet to come in
  jb     IRComplete,TestVolMinusLoop

TestVolMinusLoop1:               ;  Wait for the character input to
                                 ;   complete
  jnb     IRComplete,TestVolMinusLoop1

  mov     A,IRButton             ;  Is the volume + still pressed?
  cjne    A,#LOW(VolDown),TestVolMinusEnd
  mov     A,IRButton+1
  cjne    A,#HIGH(VolDown),TestVolMinusEnd

  djnz    R5,TestVolMinusLoop    ;  If R5 is equal to 23, then increment
                                 ;   the count
  inc     OffValue               ;  Change the value
  dec     OnValue

TestVolMinusEnd:                 ;  Finished, return to loop
  ajmp    Loop

;  Motor subroutines
Stop:                            ;  Stop the motors

  mov     P1,#%00101011          ;  Turn off the motors

  ret

Left:                            ;  Turn left

  mov     A,#%00101011           ;  Clear the P1 bits NOT doing motion
  anl     A,P1
  orl     A,#%11000000           ;  Enable the robot to go forward
  mov     P1,A

  ret
```

LISTING 12-7 **The interrupt handler code along with the timer setup *Continued.***

```
Right:                          ;  Turn right

  mov     A,#%00101011          ;  Clear the P1 bits NOT doing motion
  anl     A,P1
  orl     A,#%00010100          ;  Enable the robot to go forward
  mov     P1,A

  ret

Backward:                       ;  Go backward

  mov     A,#%00101011          ;  Clear the P1 bits NOT doing motion
  anl     A,P1
  orl     A,#%01010000          ;  Enable the robot to turn right
  mov     P1,A

  ret

Forward:                        ;  Go forward

  mov     A,#%00101011          ;  Clear the P1 bits NOT doing motion
  anl     A,P1
  orl     A,#%10000100          ;  Enable the robot to turn right
  mov     P1,A

  ret

LED_ON:                         ;  Turn on the LED

  clr     P1.5                  ;  Clear the LED bit

  ret

LED_OFF:                        ;  Turn OFF the LED

  setb     P1.5

  ret
```

If you compare 51Bot42d to PROG37, you'll see a number of changes that probably don't seem related to the PWM. The most important change is how I control the motors.

In the earlier programs, I turned off the motors by resetting the PWM bit. This is obviously impossible to do because the PWM bit is controlled by the Timer1 interrupt handler. So, this meant that I had to explicitly turn off the motors. This ended up complicating the Timer0 interrupt handler. Originally, I was able to keep all the Timer0 interrupt handler functions within 8 bytes, but explicitly turning off the motors meant that I had to write a byte explicitly to P1. This meant that I had to split the handler up into two parts to get the code in it.

In this code, I use the LED to indicate the current PWM level. If you burn PROG42 into an AT29C2051, you'll find that the LED does not light as you would probably expect for the different levels. The light levels do change noticeably, but not as much as you would expect until no power is output (at which point the LED turns off).

When I added 51Bot42d to my Wall Hugging Mouse, I found that, with two 1.5-V "AA" cells, the motors didn't get enough current except at a duty cycle of 100%. This was improved by substituting a 9-V battery for the power; however, at 25%, the motors wouldn't turn at all. I was able to improve this by connecting two 9-V batteries in parallel. With this power source, I began to notice that the 293D motor driver was getting quite warm. This means that, for some applications, very high current sources at reasonably high voltages will have to be made available to take advantage of the PWM.

One bit of code in the previous application that I would like to bring your attention to is how I handle the PWM speed change inputs. On the remote control, I decided to use the volume-control buttons. However, I knew that the remote control auto-repeats 23 times per second. This meant that I had to come up with a simple algorithm for allowing the user to control the PWM value.

The algorithm I came up with is actually quite simple. It polls the input data for one second before bumping the PWM control values:

```
if ( PacketIn == VolUp ) {          //  Increase the PWM duty cycle?
  for ( i = 0; ( i < 23 ) || ( PacketIn != VolUp ); i++ ) {
    while ( IRComplete == 1 );      //  Wait for the next packet
    while ( IRComplete == 0 );
  }
  if ( PacketIn == VolUp )          //  Increase the PWM duty cycle
    if ( OnValue != 4 ) {           //  But ONLY if it's not at the max
      OnValue++;
      OffValue-;
    }
}
```

Note that, in this code, when the on time increases, the off time decreases so that the total of the two always equals four. When changing the PWM value, in this code, you have to be careful to make sure that the on and off time totals never deviate from four, or you will end up with a different frequency for the PWM (and unpredictable behavior).

With regards to controlling the motors, there is one aspect of them that I would like to bring to your attention. You might have looked at the direction-control subroutines and wondered why I haven't optimized them.

For example:

```
Forward:                          ;  Go forward
  mov      A,#%00101011           ;  Clear the P1 bits NOT doing motion
  anl      A,P1
  orl      A,#%10000100           ;  Enable the robot to turn right
  mov      P1,A
  ret
```

could be optimized to:

```
Forward:                          ;  Go forward
  anl      P1,#%00101011          ;  Clear the P1 bits NOT doing motion
  orl      P1,#%10000100          ;  Enable the robot to turn right
  ret
```

This optimization will save 4 bytes and 4 instruction cycles in the applications. However, it will also turn off the motors after the `anl` instruction, which means that, when the direction key is pressed and autorepeats, the motors will be turned off after each packet is received. This will result in a 23-Hz PWM signal, which is unacceptable.

51Bot: Embedded Card

The 51Bot is actually quite a complex project. To help simplify building it, I have designed a raw card for your use. This card can be built from the diagrams shown in Figs. 12-17 through 12-19, from the gerbers on the CD-ROM, or by contacting me directly.

The actual design of the board has a couple of interesting features with the most important being that the I/R receivers can be placed on the top side or bottom side of the board.

Another feature that you should notice is that I have included 0.020" spacing for motor power terminals as well as a socket for an I2C serial EEPROM.

There are five vias on the top right edge of the card, these pins consist of:

- Vcc
- Gnd
- P3.3 (_Int0)
- P1.0
- P1.1

and can be used for interfacing the 51Bot controller card to external hardware.

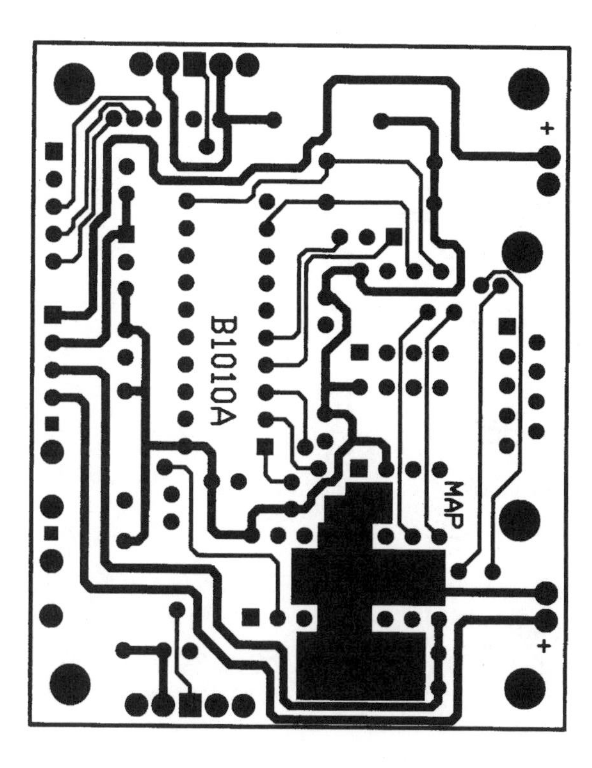

FIGURE 12-17 Top side of the 51Bot raw card.

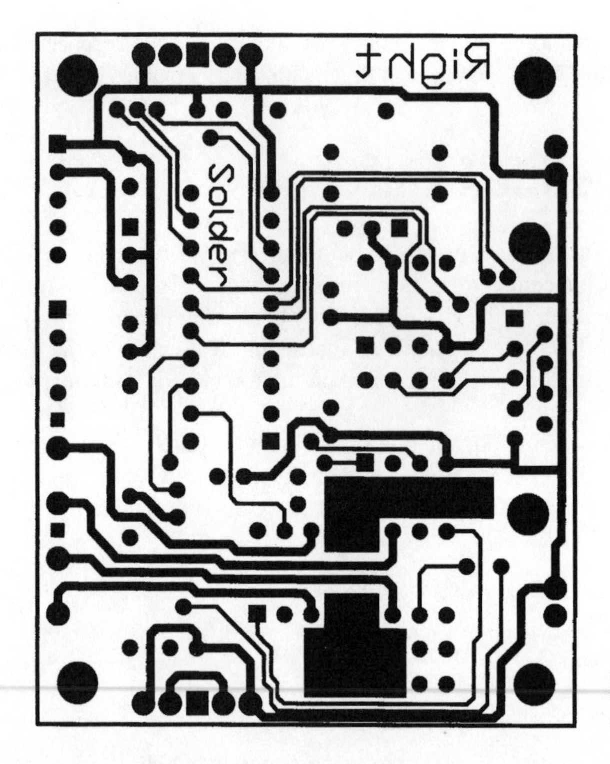

FIGURE 12-18 Bottom side of the 51Bot raw card.

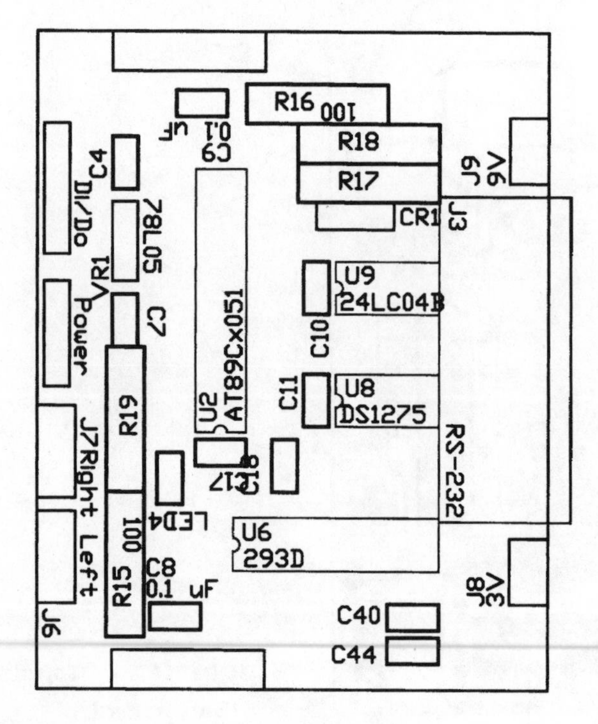

FIGURE 12-19 Overlay of the 51Bot raw card.

The actual design is quite flexible and will give you a lot of room for doing your own robot and motor control experiments.

PROG33: Aircraft Control Demonstration

During the writing of this book, my oldest son, Joel, had a simple machines project to do at school. To demonstrate how the direction of push/pull motion could be changed to a different direction (as is done in an airplane to move the control surfaces), I created a simple demonstrator model airplane using radio control servos, an old surplus joystick, and two Atmel AT89C2051s (Fig. 12-20). The radio control servos were used with standard model hardware to move the control surfaces as the joystick was moved.

This project is a bit like the 51Bot project. To get it to work, I had to actually work through a number of problems. The first was that I had to come up with application code that would allow me to control the servos. As well, I had to read the value of the potentiometers in the joystick. Because these two functions required similar 8051 hardware features, I decided to use a separate microcontroller for each function and come up with a simple communications protocol that could be used for sending the data back and forth.

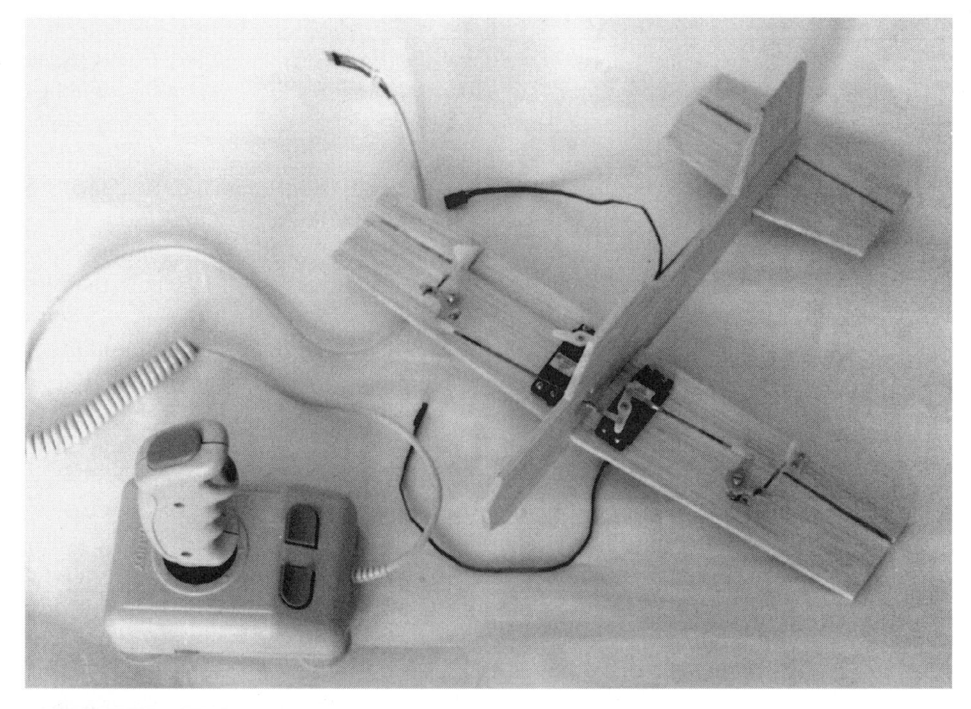

FIGURE 12-20 Joystick-controlled model airplane.

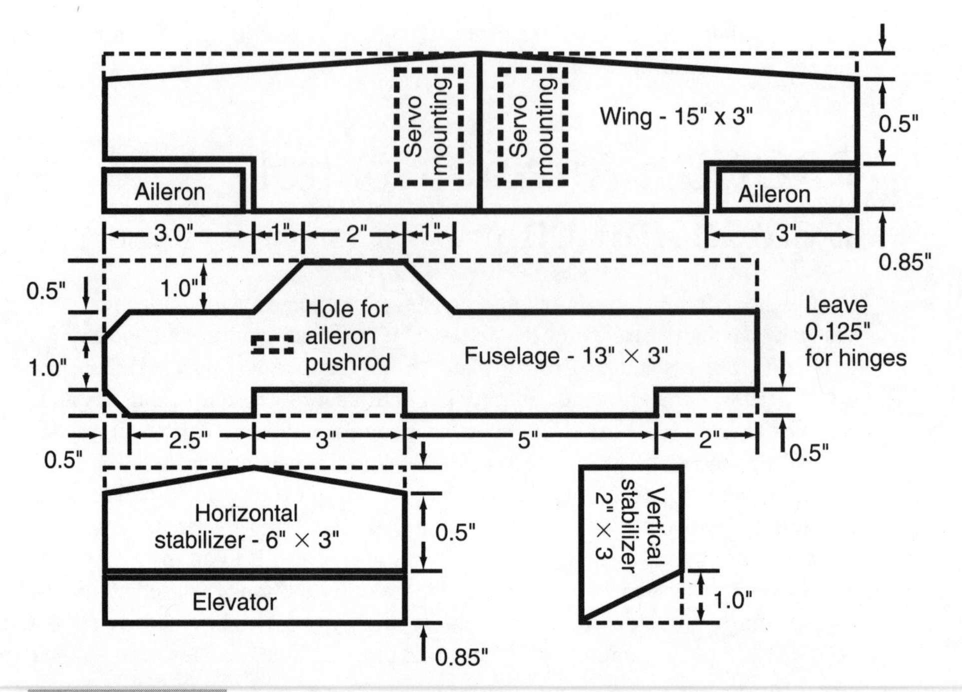

FIGURE 12-21 Cut-outs on 36" by 3" by 0.125" balsa wood.

Before I even started with developing the electronics hardware, I built the model airplane. I was able to do it for about $10 after going to a local hobby shop and buying a 36' by 3" of ⅛" (0.125") piece of balsa wood and some radio control hinges, control horns, bellcranks, and pushrods.

The design itself (Fig. 12-21) was the simplest one that I could come up with that would look like an airplane and allow simple mounting of the servo hardware.

The servos are placed into the roots of the wing using holes that are cut just large enough to hold them. You could put in pieces of hardwood to secure the servos, but I didn't and they held quite well for this application. Once the pieces of wood were cut, I glued the airplane together with Weldbond, using small nails to hold the pieces together while the glue hardened. I was able to cut the balsa wood and assemble the model in about an hour.

For the hinges, I cut a ⅛" away from the control surfaces to make space for them to move back and forth, then used pinless hinges (which is basically just a piece of plastic). Pinless hinges have the advantage of being cheaper than pinned hinges and are easier to install (i.e., lining them up isn't very critical).

With the design of the airplane, the two servos will extend past the bottom of the model and provide a stable base for it. You might have to make a small tail skid to keep the model level. (I left in one of the nails that I used for holding the horizontal stabilizer while the glue was drying down at the same level as the servos.)

If you aren't familiar with how radio-control models have their control surfaces connected, you can go to your local hobby shop, and they'll be happy to show you how the different parts are used to connect the servos to the airplane's control surfaces. You can also pick up one of the various magazines that are devoted to radio-control modelling.

For the servos, I used the cheapest servos that I could find (the Tower Hobbies model TS-51). These servos are controlled by a PWM pulse, which repeats once every 20 msec and has a 1-msec to 2-msec pulse range (Fig. 12-22).

One of the really nice things about servos is how simple they are to interface to (Fig. 12-23). All that's required is a CMOS/TTL PWM signal along with +5 V and Ground.

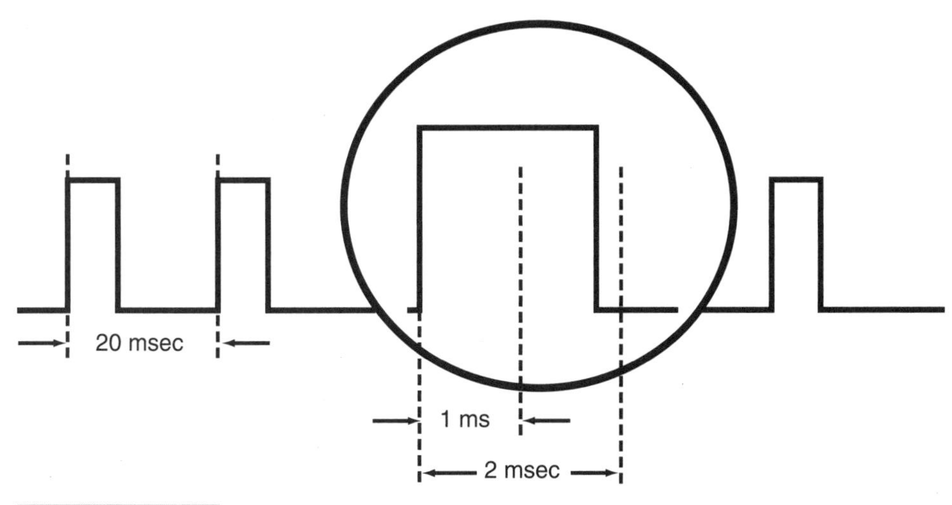

FIGURE 12-22 Servo PWM waveform.

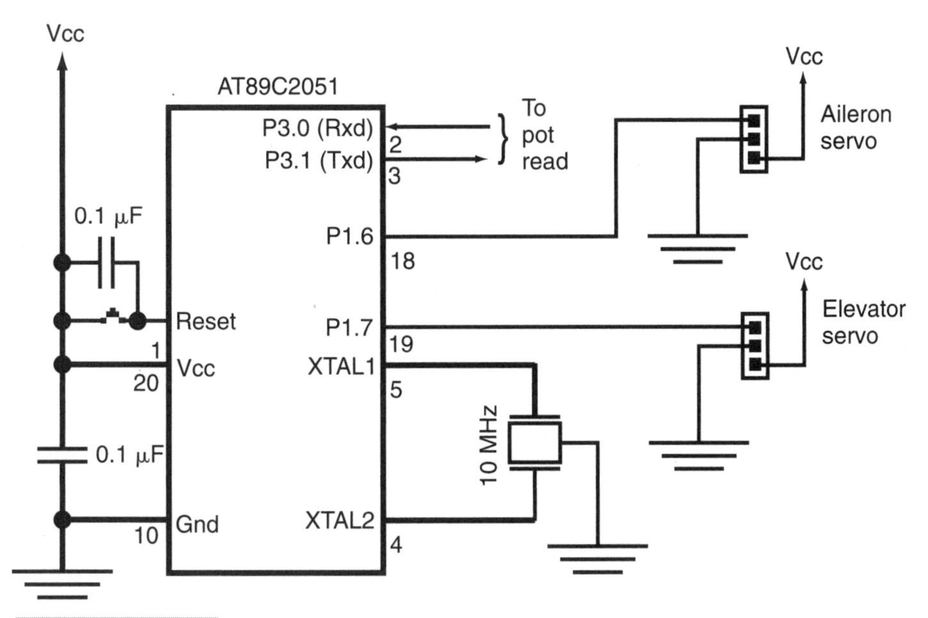

FIGURE 12-23 8051 servo positioner.

After building the model with the servos and control surfaces connected and assembling the positioner circuit, I was finally able to test out how the hardware worked. The basic test was `RangeChk` (Listing 12-8), which moves one servo to an extreme every couple of seconds and the other servo is left in the neutral position.

The purpose of moving one servo and keeping the other one centered was to allow me to test out the operation of the servos on the model airplane.

LISTING 12-8 Code to test the model airplane hardware.

```
;  RangeChk - Check the operating range of an R/C servo on the model
;
;  This second program moves "Servo1" to first one extreme, delays for
;   a second or so, and then moves to the other extreme.  The interrupt
;   routine from "ServCent" is used.
;
;  Each timer interrupt loop takes 18 usec.  Therefore, at the 1 msec
;   extreme, a value of 56 is used.  For the other extreme, 2 msec, a
;   value of 111 is used.  For the middle position, a value of 83 is
;   used.
;
;  Myke Predko
;  98.05.03
;
;  Hardware notes:
;  89C2051 running at 10 MHz
;  The 80C520's Timer0 running in mode 0
;  P1.6 is connected to a radio-control servo (Servo 1)
;  P1.7 is connected to a radio-control servo (Servo 2)

;  Variable declarations
Servo0 EQU 010h          ;  Position of the aileron servo (at P1.6)
Servo1 EQU 011h          ;  Position of the elevator servo (at P1.7)

 org 0
  ajmp   Mainline

 org 0Bh                 ;  Timer0 (real-time) interrupt

  mov    TH0,#191        ;  Reload the timer interrupt
  mov    TL0,#00         ;  To interrupt again in 20 msec

  push   ACC             ;  Save the accumulator
  push   PSW             ;  Save the status register

  mov    A,#0C0h         ;  Turn ON the servo's pulses
  orl    A,P1
  mov    P1,A

  mov    B,#0            ;  Use "B" as the loop counter

IntLoop:                 ;  Loop around until both servos counts are met

  mov    A,B             ;  Get the current count value
  clr    C
  subb   A,Servo0        ;  Is count past the current position?
  mov    P1.6,C          ;  Save the carry as the aileron output
```

LISTING 12-8 Code to test the model airplane hardware *Continued.*

```
  mov    A,B
  clr    C
  subb   A,Servo1
  mov    P1.7,C             ;  Save the carry as the elevator output

  inc    B                  ;  Increment the counter

  orl    C,P1.6             ;  Are both control bits low?

  jc     IntLoop            ;  Loop back, if either bit set

  pop    PSW                ;  Restore the registers pushed onto the stack
  pop    ACC

  reti

Mainline:                   ;  Program mainline

  mov    P1,#03Fh           ;  Turn off pulses to the elevators and ailerons

  mov    TMOD,#%00000001;   Timer0 - Uses internal clock
                       ;          - Run in mode 1
  mov    TCON,#%00010000;   Start Timer0 running

  mov    TH0,#191           ;  Set up 20-msec delay to the timer interrupt
  mov    TL0,#00

  mov    Servo0,#83         ;  Set up Servo0 to 1.5-msec delays
  mov    Servo1,#56         ;  Start with Servo1 at one extreme

  mov    IE,#%10000010      ;  Enable the Timer0 interrupt

;  Now, do the looping output

MainLoop:                   ;  Come back here after decrementing P3

  acall  Dlay               ;  Delay 1 second with servo in current pos'n

  mov    Servo1,#111        ;  Servo1 at the full extreme

  acall  Dlay

  mov    Servo1,#56         ;  Servo1 at the other extreme

  ajmp   MainLoop           ;  Just loop around forever


Dlay:                       ;  Delay 1 second

  mov    R2,#6
DlayLoop:
  djnz   R0,DlayLoop        ;  Use the values already in the registers
  djnz   R1,DlayLoop
  djnz   R2,DlayLoop

  ret
```

The critical portion of the code is the interrupt handler which can be modelled in C as:

```
Interrupt Timer0() {                    //  Timer0 interrupt handler
int  i;                                 //  Counter
  ElevatorServo = AileronServo = 1; //  Start the output pulse
  for ( i = 0; i < Two_msec; i++ ) {//  Loop for 2 msec
    if ( ElevatorCount < i )            //  Turn off the elevator servo?
      ElevatorServo = 0;
    if ( AileronCount < i )             //  Turn off the aileron servo?
      AileronServo = 0;
  } //  endfor
} //  endint - End the Timer0 interrupt handler
```

As I timed this code, I found that a 1-msec delay required a count of 56, while a 2 msec delay required a value of 111.

This knowledge was important because I wanted to understand what the values were that I could send to it to have the servos do a full deflection. Along with a serial interface, a scaling function was provided in ServoMov (Listing 12-9).

LISTING 12-9 **Code for ServoMov.**

```
;  ServoMov - Put servo's to specified positions.
;
;  This third servo program sets the "Servo0" and "Servo1" according
;   to the 2-byte value passed to it.  Servo0 is controlled by an "A"
;   and Servo1 is controlled by an "E."
;
;  The servo code developed earlier is used to output the two servo PWM.
;
;  Each timer interrupt loop takes 18 usec.  Therefore, at the 1 msec
;   extreme, a value of 56 is used.  For the other extreme, 2 msec, a
;   value of 111 is used.  For the middle position, a value of 83 is
;   used.
;
;  Myke Predko
;  98.05.16
;
;  Hardware notes:
;  89C2051 running at 10 MHz
;  The 80C520's Timer0 running in mode 0
;  P1.6 is connected to a radio-control servo (Servo 0)
;  P1.7 is connected to a radio-control servo (Servo 1)
;  P3.0 is connected to the position sensor
;  P3.0 is used to transmit an acknowledge character (00Dh) back

;  Variable declarations
Servo0 EQU 010h                     ;  Position of aileron servo (at P1.6)
Servo1 EQU 011h                     ;  Position of elevator servo (at P1.7)

 org 0
  ajmp   Mainline

 org 0Bh                            ;  Timer0 (real-time) interrupt

  mov    TH0,#191                   ;  Reload the timer interrupt
  mov    TL0,#00                    ;  To interrupt again in 20 msec
```

LISTING 12-9 Code for ServoMov *Continued.*

```
  push    ACC                  ;  Save the accumulator
  push    PSW                  ;  Save the status register

  mov     A,#0C0h              ;  Turn ON the servo's pulses
  orl     A,P1
  mov     P1,A

  mov     B,#0                 ;  Use "B" as the loop counter

IntLoop:                       ;  Loop around until both servos counts
                               ;   Reset the servo pins

  mov     A,B                  ;  Get the current count value
  clr     C
  subb    A,Servo0             ;  Is count past the current position?
  mov     P1.6,C               ;  Save the carry as the aileron output

  mov     A,B
  clr     C
  subb    A,Servo1
  mov     P1.7,C               ;  Save the carry as the elevator output

  inc     B                    ;  Increment the counter

  orl     C,P1.6               ;  Are both control bits low?

  jc      IntLoop              ;  Loop back, if either bit set

  pop     PSW                  ;  Restore registers pushed onto stack
  pop     ACC

  reti

Mainline:                      ;  Program mainline

  mov     P1,#03Fh             ;  Turn off the pulses to the elevators
                               ;   and ailerons

  mov     SCON,#%01010000      ;  Set the serial port to 8-bit mode
  mov     TMOD,#%00100001      ;  Timer0 - Uses internal clock
                               ;         - Run in mode 1
                               ;  Timer1 - Uses internal clock
                               ;         - Run in mode 2
                               ;         - Drive the serial port
  mov     TH1,#234             ;  Set up the Timer1 interval
  mov     TL1,#234
  mov     TCON,#%01010000      ;  Start both timers running

  mov     TH0,#191             ;  Set up the 20-msec delay to the timer
  mov     TL0,#00              ;   interrupt

  mov     Servo0,#83           ;  Set up both servos to intermediate
  mov     Servo1,#83           ;   position

  mov     IE,#%10000010        ;  Enable the Timer0 interrupt

;  Now, do the looping poll for input
```

LISTING 12-9 Code for ServoMov *Continued.*

```
MainLoop:                         ; Come back here after decrementing P3

  jnb    RI,MainLoop              ; Wait for a character to come in

  mov    R4,SBUF                  ; Save the character
  clr    RI
  mov    SBUF,#00Dh               ; Acknowledge the character coming in

  mov    R2,#6                    ; Delay 2 seconds
DlayLoop:                         ; Wait for next character (servo value)
  jb     RI,HaveChar
  djnz   R0,DlayLoop
  djnz   R1,DlayLoop
  djnz   R2,DlayLoop

HaveChar:                         ; We have the character

  mov    A,SBUF                   ; Save the incoming character
  clr    RI
  mov    SBUF,#00Dh               ; Acknowledge the incoming character

  mov    B,#5                     ; Get the range divided by 5
  div    ab
  add    A,#56                    ; 56 is the 1-msec value

  cjne   R4,#'A',SetElevators     ; Is this an "A" coming in?

  mov    Servo0,A                 ; No, set the aileron position

  ajmp   MainLoop                 ; Just loop around forever

SetElevators:                     ; Set the elevators position

  mov    Servo1,A

  ajmp   MainLoop
```

The scaling code:

```
  mov    A,SBUF                   ; Save the incoming character
     :
  mov    B,#5                     ; Get the range divided by 5
  div ab
  add    A,#56                    ; 56 is the 1-msec value
```

is not quite accurate. I originally planned for receiving a servo value from 0 to 0FFh. However, for a full range value of 55 (i.e., to get 2 msec, when added to 56), I had to divide the incoming value by 4.63. I rounded this to the next highest integer to get a simple scaling routine. Instead of the full range being 0 to 55, I actually have 0 to 51, which will result in slightly less than full travel in one direction.

With the servos working properly, I then started on reading the position of a potentiometer. This turned out to be a major amount of work because of how the 8051's I/O pins work. I ended up trying five or six different configurations using a single pin until I decided and just go ahead with a two-pin circuit (Fig. 12-24).

In this circuit, I leave the pins connected to the potentiometer low until I'm ready to read the potentiometer's position. At this time, I write a "1" to these pins (making them a high voltage) and measure the amount of time the capacitor takes to charge to the point where the pin connected to the net between the capacitor and potentiometer exceeds the logic threshold of the pin.

In Fig. 12-25, the top oscilloscope trace ("1") is the time the capacitor takes to charge. The lower trace ("2") shows the pin charging to the threshold voltage. In Fig. 12-25, I changed the lower scale to 500 mV per division (normally I use 5 V per division in the oscilloscope traces shown in this book) to make the characteristic RC rise and fall more obvious (as well as the approximately 1.4-V transition point where a "1" is read).

It is important to note that the sensor pin (the one connected between the capacitor and potentiometer) does not have an internal pull-up. If there was an internal pull-up on the sensor pin, then this circuit won't work.

The code to read two potentiometers for this circuit is also very simple (Listing 12-10).

When I created this code, I did it for a Radio Shack (Tandy) PC joystick that I was able to buy at a surplus store for a couple of bucks. This joystick turned out to be quite a bit of a problem in several respects.

The biggest problem with this joystick was that the potentiometers for the pitch and roll axis were connected together. This meant that the potentiometer value for one axis was dependent on the value of the other. The joystick was built with a copper-printed phenolic

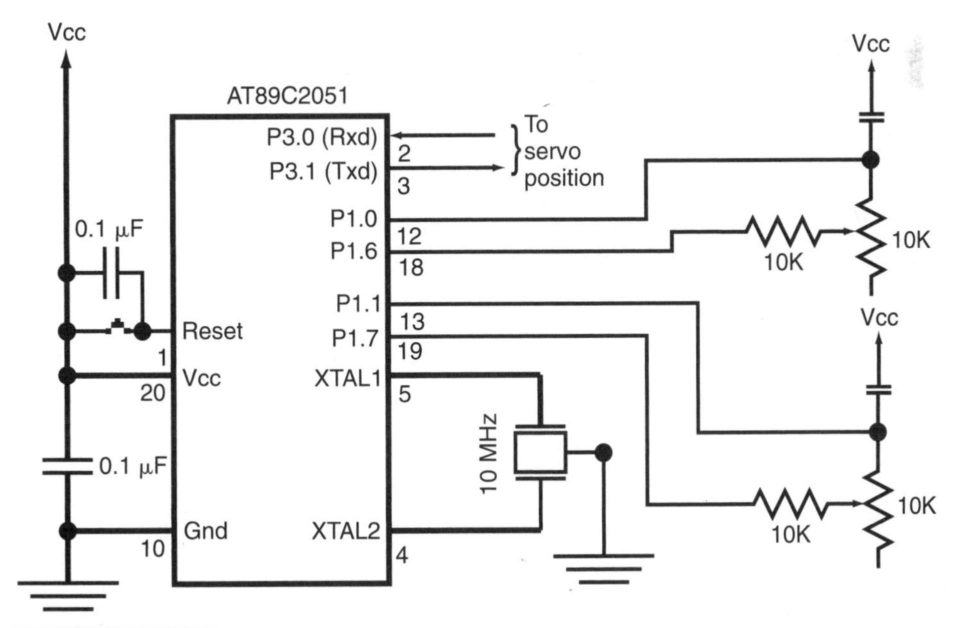

FIGURE 12-24 8051 potentiometer read.

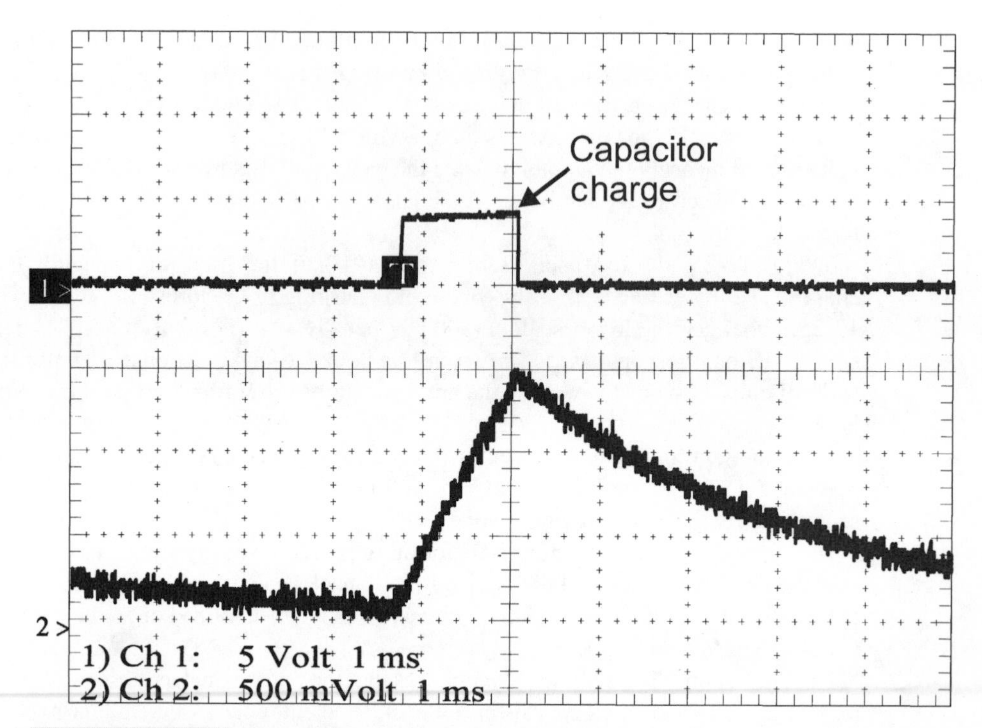

FIGURE 12-25 Potentiometer value read waveform.

LISTING 12-10 Code to read the potentiometers.

```
;  PotReadd - Read the pot values and send to the servos
;
;  This application reads resistance values by allowing a capacitor
;   wired to a pot to charge and then measure the time the cap takes
;   to discharge.
;
;  Two AT89C2051 pins are used for reading each pot.
;
;  The values of each pot are sent to the servo controller at 1200 bps
;   using a simple transmission protocol.
;
;  Myke Predko
;  98.05.16
;
;  Hardware notes:
;  89C2051 running at 10 MHz
;  P1.6 is connected to the elevator R/C charge/discharge circuit
;  P1.7 is connected to the aileron R/C charge/discharge circuit
;  P1.0 is connected to the R/C Voltage, which is the pulse state
;   between the capacitor and resistor for the elevator
;  P1.1 is connected to the R/C voltage, which is the pulse state
;   between the capacitor and resistor for the aileron
;  P3.1 is used to send data serially to the servo AT89C2051
;  P3.0 is used to receive data serially from the servo AT89C2051
```

LISTING 12-10 **Code to read the potentiometers *Continued.***

```
;  Variable declarations
;  R0 - R2: Delay counter variables
;  R3: Elevator value
;  R4: Aileron value

 org 0

  mov    SCON,#%01010000          ;  Serial port in mode 1
  mov    TMOD,#%00100000          ;  Put Timer1 in mode 2
  mov    TH1,#234
  mov    TL1,#234
  setb   TCON.6                   ;  Start the timer running

Mainline:                         ;  Program mainline

  mov    P1,#%00111111            ;  Allow the capacitors to discharge

  mov    R3,#0                    ;  Clear the elevator counter
  mov    R4,#0                    ;  Clear the aileron counter

  clr    RI                       ;  Clear the serial data receive bit

  mov    R0,#0FFh                 ;  Let the cap discharge
  mov    R1,#010h                 ;  For 10 msec
Loop1:
  djnz   R0,Loop1
  djnz   R1,Loop1

  mov    P1,#%01111111            ;  Now, wait for the elevator capacitor
                                  ;   to charge

Loop2:                            ;  Loop until the line becomes high
  inc    R3                       ;  Increment elevator position counter
  jnb    P1.0,Loop2

  mov    P1,#%11111111            ;  Now, wait for the aileron capacitor
                                  ;   to charge

Loop3:                            ;  Loop until the line becomes high
  inc    R4                       ;  Increment aileron position counter
  jnb    P1.1,Loop3

  mov    A,R3                     ;  Scale the elevator value
  clr    C
  subb   A,#4                     ;  Take 4 away for the initial value
  mov    R3,A                     ;  Multiply elevator value read by 2.5
  rl     A                        ;  Rotate left by one, multiply by 2
  xch    A,R3
  clr    C
  rrc    A                        ;  Rotate right by one, multiply by 0.5
  add    A,R3                     ;  Add 2x and 0.5x together
  mov    R3,A

  mov    A,R4                     ;  Scale the aileron value
  clr    C
  subb   A,#4
  mov    R4,A                     ;  Multiply aileron value read by 2.5
  rl     A                        ;  Rotate left by one, multiply by 2
```

LISTING 12-10 **Code to read the potentiometers *Continued.***

```
  xch    A,R4
  clr    C
  rrc    A                      ; Rotate right by one, multiply by 0.5
  add    A,R4                   ; Add 2x and 0.5x together
  mov    R4,A

  mov    SBUF,#'A'              ; Indicate we're sending aileron value
  acall  DlayCheck              ; Do we get an acknowledge?
  jnb    RI,Mainline            ; If "RI" not set, then no, start over
  clr    RI                     ; Clear the received character bit

  mov    SBUF,R4                ; Send the aileron value
  acall  DlayCheck              ; Did we get an acknowledge?
  jnb    RI,Mainline
  clr    RI

  mov    SBUF,#'E'              ; Indicate we're sending elevator value
  acall  DlayCheck              ; Do we get an acknowledge?
  jnb    RI,Mainline            ; If "RI" not set, then no, start over
  clr    RI                     ; Clear the received character bit

  mov    SBUF,R3                ; Send the elevator value
  acall  DlayCheck              ; Did we get an acknowledge?
  jnb    RI,Mainline
  clr    RI

  ajmp   Mainline

DlayCheck:                      ; Delay 2 seconds and check for "RI" to
                                ;  be set
  mov    R2,#6
DlayCheckLoop:                  ; Loop until time is up or "RI" is set
  jb     RI,DlayCheckEnd        ;  If "RI" set, return
  djnz   R0,DlayCheckLoop
  djnz   R1,DlayCheckLoop
  djnz   R2,DlayCheckLoop

DlayCheckEnd:                   ; Finished, return to the caller
  ret
```

board that had wipers running on it to provide the potentiometer functions. By cutting the copper traces, I was able to eliminate this problem.

A secondary problem was that the potentiometers were logarithmic. This meant that, for them to work properly in the application, I would have to convert the value read out to a linear value for the position. This would actually turn out to be an interesting exercise because I would have to figure out how to plot the curve to some known values.

For the final project (and what's shown previously in `PotReadd`), I decided to go with two linear potentiometers. I did this so that you could get a better idea of how the circuit works (and can build it up yourself).

When you look at Fig. 12-24, you might be surprised to see that I have included 10K resistors in line to the potentiometer's wipers. When I first wired this circuit, I found that very low resistances would result in a virtually non-existent pulse. By adding the 10K resistor, I was able to increase the RC constant of the circuit so that the resistance could be read under all circumstances.

In `PotReadd`, you'll see that I multiply the delay by 2.5. After building the circuit, I found that the range of the pot reads was from 0 to 101. In this case, I had to multiply the result by 2.52 to get the full range from 0 to 255.

This was actually pretty easy to do by using the formula:

$$OutputPosition = (PotCount << 1) + (PotCount >> 1)$$

This formula adds two times the count to one half the count to effect a multiply by 2.5.

Figure 12-25 does not show the jitter that I found when I ran this read. The actual RC charging waveforms that I produced with the potentiometer circuit were very constant, but the actual values read out could change by five or six from either side of the mean. This was caused by noise on the line and when the AT89C2051 was able to poll the line with regards to the waveform. As you will see later, this can be a problem in some applications.

With the servo move and potentiometer read completed, I could then concentrate on the data transfer between the potentiometer read AT89C2051 and the servo move AT89C2051. I decided on a very simple protocol that had the sender (the potentiometer position sensor) first sending a byte indicating which servo position was to be updated (this was either an "A" or an "E"), waiting one second for the receiver (the servo positioner) to read the byte and respond with a byte back before sending the position, and waiting for another acknowledgement byte back. This was designed to allow both AT89C2051's to run their respective applications without requiring the other to be operating.

This data was sent at 1200 bps using the built-in serial ports on the AT89C2051. In neither case, interrupts were used for the data coming in. I avoided interrupts because both applications are timed circuits and I didn't want to have any problems with them due to unexpected serial port interrupts.

With both AT89C2051s operating and the model airplane's servos connected to the circuit, I found that the servos would jitter quite a bit. This was due to the different value reads that I mentioned previously. To eliminate them, I could have experimented with averaging several position values and sending the average value or by using an external ADC. (In the latter case, it could have been connected directly up to the servo positioner.)

Fortunately for me, Joel decided that he didn't want to do this project and wanted to make a catapult using a spring (I suspect to lay siege on his brother), so I didn't end up doing the work to smooth out the position value.

This example application is really the combination of three different circuits, the servo positioner, the ADCless potentiometer read, and the two 8051 communications protocol that can be used in a variety of different applications.

PROG39: Light Sensors for Robots

One of the most basic mobile robot applications is to follow a light beam, turning left or right to keep the robot on the path to the light source. I didn't include this project in the 51Bot, but I thought I'd show how easy it is to implement in a device like the AT89C2051 that has a comparator.

The circuit that I used is really a simple modification of the basic AT89C2051 core circuit (Fig. 12-26). In this circuit, two light-dependent resistors (LDRs) have been wired in a voltage-divider configuration. On one LDR, I have placed a 4.7K pull-up and on the

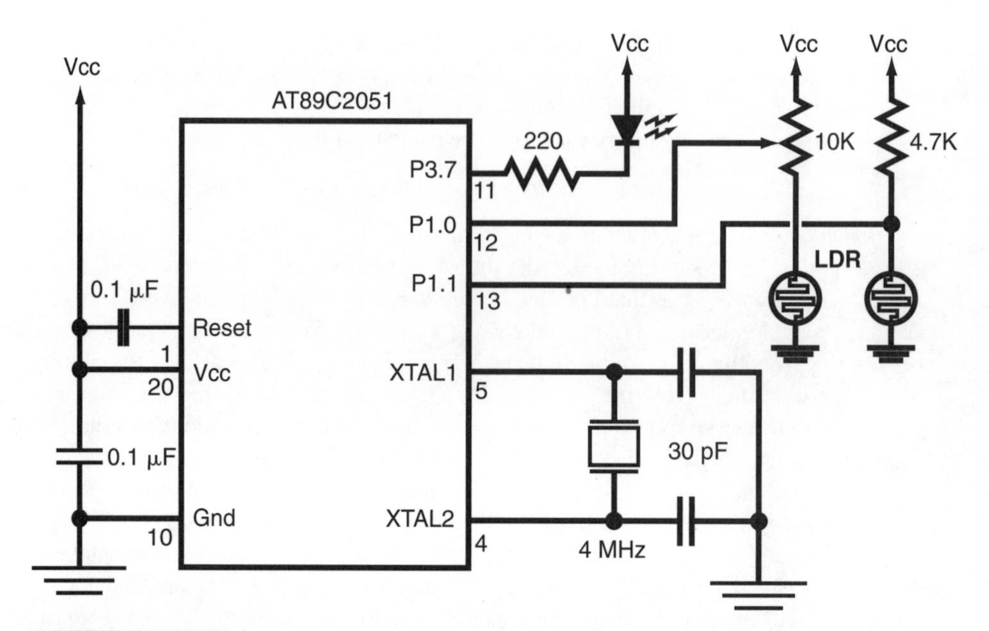

FIGURE 12-26 AT89C2051 light sensor.

other a 10K pot. The theory behind this circuit is that, by tweaking the pot, the voltage output from each of the two LDR voltage dividers is adjusted to be as close as possible. When the light level is uneven between the two LDRs, the comparator voltage changes.

In the actual application, the two LDRs would be placed at angles away from each other, each facing forward. As one LDR pointed more toward the light, its resistance would drop and the comparator would signal the change, requiring a small turn by the AT89C2051 controlling the Robot's motors.

To test out this theory, I came up with the program in Listing 12-11. Hopefully this code is not too surprising in how simple it is, but it's a wonderful example of how built-in hardware can simplify an application. I could have used the RC resistance measurement circuit shown in the previous example application to carry out the same operation, but this circuit (and software) is much simpler.

LISTING 12-11 Code to test the LDRs.

```
;  PROG39 - Comparing voltage inputs
;
;  This application compares the voltages on pins P1.0 and P1.1 and
;   lights an LED when the voltage on P1.0 is greater.
;
;  For the voltage sources, two LDRs are used, with one terminal at
;   ground and the other connected to P1.0/1.  One LDR is connected to a
;   4.7K pull-up, and the other is connected to a 10K pot pull-up.
;
;  Myke Predko
;  98.04.05
;
;  Hardware notes:
;  AT89C2051 is used as the microcontroller
```

LISTING 12-11 Code to test the LDRs *Continued.*

```
;   - Oscillator speed is 4 MHz
;  Voltage "1" is on P1.0
;  Voltage "2" is on P1.1
;   - Comparator output taken from P3.6
;  LED is connected to P3.7

;  Constant declarations

;  Variable declarations


;  Macros


;  Mainline
 org 0                              ;  Execution starts here

Loop:                               ;  Just loop around the voltage read

  mov     C,P3.6                    ;  Read the comparator
  mov     P3.7,C                    ;  Output the comparator value

  ajmp    Loop
```

In a robot application, the comparator status bit (P3.7) can simply be polled periodically (say once per second) to see if a turn is indicated.

PROG38: Ultrasonic Distance Measuring

One of the projects that I had a lot of fun with and learned a bit more about the 8051 with was creating a small device to ultrasonically measure distance using the Polaroid 6500 Module and transducer and outputting the result on an LCD. This project is just about perfect for using on a small device like the AT89C2051. The only change that I would make in the application is to use an 8051 with even less memory (only 205 bytes of control store are required for the application).

The ultrasonic transducer I chose for this application is the Polaroid 6500, which was first developed in the mid-1970s for use with autofocusing cameras.

The operation of the device is quite simple: A controller outputs a pulse to the module (which can be up to 100 msec long) and then measures the time for the echo to come back. This can be accomplished simply by two wires connected to a microcontroller.

The ultrasonic transducer both outputs the ultrasonic signal (which can be heard as a click) and receives the echo when the sound waves bounce off the first object that they come to (Fig. 12-27).

The time delay between the pulse and the echo is proportional to the distance from the transducer to the object the sound waves bounce off of. The 6500 module is specified to work for ranges from 6" to 35' with an error of 1%. My own testing seemed to confirm this.

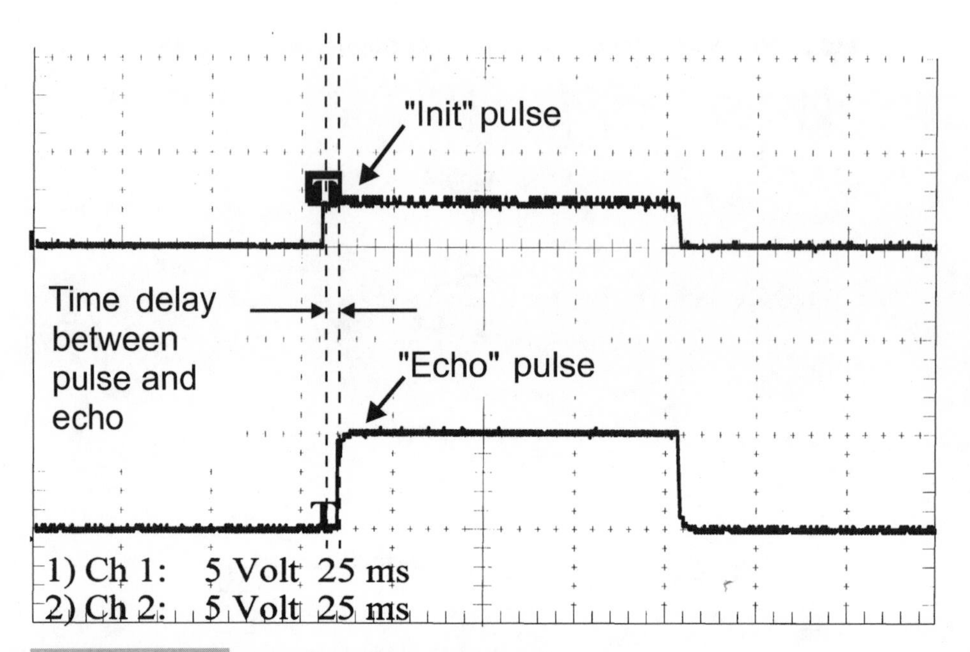

FIGURE 12-27 **Polaroid 6500 transmit/receive.**

Before going too far into this project, there are a few things that I should mention. The first is a warning. When the ultrasonic signal is sent, there is an incredible amount of energy generated. At the transducer, 400 V is produced, and about 1.2 amps is drawn from the power source. This is important to realize because nowhere in the Polaroid or any other documentation that I found was a warning about electrical shocks.

When I was first experimenting with the device and aiming it at different walls of my workshop, I inadvertently touched the two contacts on the transducer. After doing that, I discovered that the transducer was dangling by its wires and my right hand was about 4' behind me. A few seconds later the pain followed. I received a pretty good shock between my right hand's index finger and thumb.

Be very careful when handling the transducer when the 6500 is operating.

Do *not* use both hands when handling the operating transducer. When you do handle the transducer, put your left hand in your pocket to make sure there is no chance for a current path through your heart!

The voltage and current that the device produces should not be capable of severely hurting you, but you should take precautions and treat the driver module and transducer with respect when power is being applied to them. Even if you aren't hurt, you will feel the effects of a shock for some time.

On the 6500, there is a nine-conductor connector that is designed for having a flat flex cable connected to it. One of the first things that I did with the connector was to cut it off, unsolder the connections, and put wires into them. This was one of the smartest things that I did because it really made prototyping and experimenting with the 6500 very easy.

In Fig. 12-28, you can get an idea of how I added the interface wiring for the 6500. After removing the flex socket, I discovered that the connector vias were in the pattern shown in the schematic with pin 1 being on the left-hand side and pin 9 being on the right-hand side. The lower pins, halfway between the upper row, are the even pins and the upper row are the odd pins.

The pin-out of the 6500 is shown in Table 12-4.

The operation of the 6500 is very straightforward; the INIT pin is asserted for typically up to 100 msec, and the controlling processor waits for the ECHO to come back. If there is nothing for the signal to reflect off of, then ECHO will never become asserted.

The 6500 is specified to work at a minimum distance of 1.33'. This can be reduced to about 6" by asserting the BINH line. By asserting BINH, any ringing of the transducer will be masked.

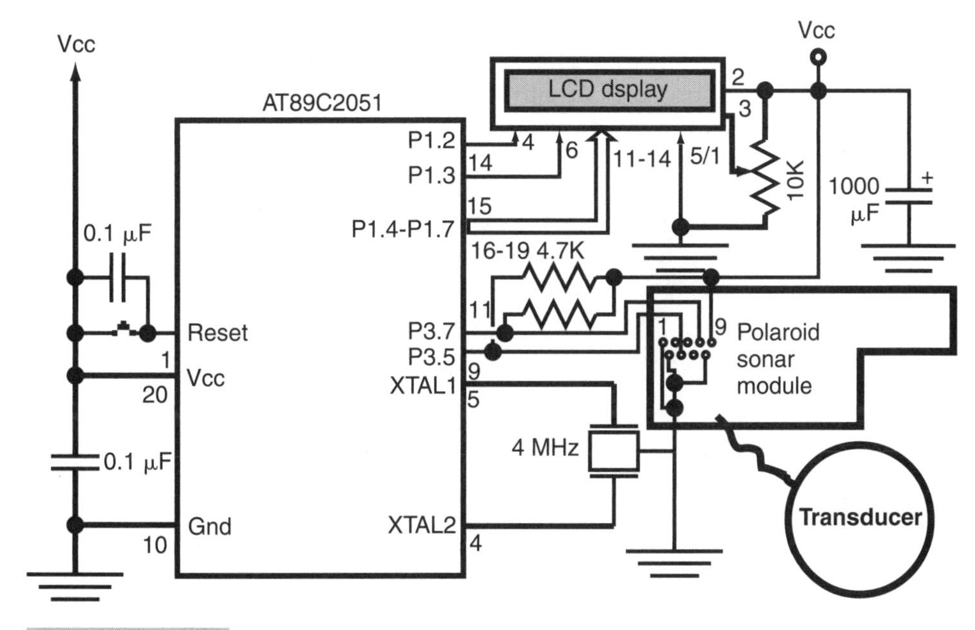

FIGURE 12-28 PROG38 schematic.

TABLE 12-4 THE PIN-OUT OF THE 6500

PIN	FUNCTION	COMMENTS
1	Ground	Connected to negative pin of 1000-μF filter cap
2	BLNK	When asserted, the 6500 will blank out any reflected signal
3	N/C	
4	INIT	When asserted, the 6500 will output the ultrasonic burst
5	N/C	
6	OSC	49.4-KHz oscillator
7	ECHO	Asserted after the received signal returned
8	BINH	When asserted, the 6500 will ignore any transducer ringing
9	Vcc	Connected to positive pin of 1000-μF filter cap

BLNK will ignore any reflected inputs at specific times after INIT is asserted. This is useful if there are multiple objects that the 6500 could detect. By blanking out the receiver when echoes are being received, a more comprehensive understanding of what is around the 6500 is possible.

When wiring the 6500 into a circuit, there are a few things to watch out for. First, power should be filtered by a 1000-μF electrolytic capacitor. As I've noted earlier, when the ultrasonic burst is output, quite a bit of energy is required by the 6500. The filter capacitor will eliminate any possibility that the 6500 will affect other components in your circuit (e.g., resetting the microcontroller) and will allow you to run the circuit from a relatively modest power source, such as a 9-V battery.

For any inputs that are going to be driven, you must pull up the lines with 4.7K resistors. This is especially important when wiring the 6500 to an 8051, which has weak MOSFET pull ups. The 8051 pins will not be able to source enough current to drive the lines high (and can be seen in the oscilloscope pictures of the INIT pulse). Even with the pull-ups, the 8051 is not able to drive the line high.

This explanation of the 6500 probably seems quite simplistic. On the CD-ROM included with this book, I have included a .PDF file of the Polaroid data sheet for your perusal. The 6500 can be purchased from Wirz Electronics (the address is given in appendix B).

When I work with devices with attached wires, I typically put on a good coating of Weldbond or hot-melt glue to provide strain relief for the solder joints. I did put Weldbond on the transducer wires and the interface wiring to make sure that, as I handled the card, I wouldn't break any joints. The transducer wires do have a hole in the card to provide some measure of strain relief.

I found that, until the Weldbond hardened (which took three days), the 6500 stopped working because the uncured Weldbond broke down and provided a relatively low-resistance path (compared to the transducer) for the 400-V output signal. If you are going to use the 6500, put the strain relief on the interface wiring and not on the transducer wiring.

After all these warnings, you're probably not feeling very good about the 6500, but what I can say about it is that it's a tough little circuit. When I first added the wires, I got Vcc and Gnd mixed up. While the chips on the card got quite hot, the circuit did work without any problems. As well, there was the problem with the uncured Weldbond. After both of this mishaps, the 6500 module worked perfectly, and I really haven't had a moment's problem with the unit I have other than the self-inflicted ones.

When I first started this project, I was planning on using the 6500 Ultrasonic Ranging Module on the 51Bot. I thought it would be interesting to have a distance sensor on the robot.

The original idea was to avoid having to use physical sensors. Instead the 6500 module would indicate when the robot was too close to an obstacle in front of it. After the shock incident, I decided that I wouldn't use the 6500 on the robot. Instead, I'd use it in its own standalone application. My major concern was how to mount it on the robot because I didn't want to alter the performance of the transducer, but I also wanted to have something that was safe for my children and with the open body of the Tamiya Robot.

So, instead, I came up with the idea of using the 6500 as a distance-measuring tool. This would be a good example of how simply something could be done with the 8051. I decided upon using the AT89C2051 and tried to keep the circuit as simple as possible. To help expedite this, I used an LCD module for outputting the current distance between the transducer and another object.

This circuit uses the standard AT89C2051 core circuit with a capacitor to help ensure that the device would start up reliably.

To simplify the operation of the 6500, I simply drove it with BINH and BLNK pulled to ground.

For outputting the distance, I used an LCD. The LCD is connected in 4-bit mode, which allowed me to control the LCD with only six pins of the AT89C2051.

The code itself is actually quite simple. The ultrasonic echo measuring code is shown in Listing 12-12.

LISTING 12-12 **Code for the ultrasonic echo measuring.**

```
  setb     P3.7                 ;  Turn on the init pulse

  clr      A                    ;  Use A to count the inches to echo

EchoLoop:                       ;  Wait for the echo to return

  acall    Dlay150              ;  Delay 148 usec/loop

  inc      A                    ;  Increment the count
  jz       NoEcho               ;  No echo if equal to zero (256)

  jnb      P3.5,EchoLoop        ;  Anything come back yet?

  mov      B,#100               ;  Figure out the distance
  div      AB

  add      A,#'0'               ;  Print out the 100s
  acall    LCDCHAR

  mov      A,B                  ;  Now, figure out the 10s
  mov      B,#10

  div      AB

  add      A,#'0'               ;  Print out the 10s
  acall    LCDCHAR

  mov      A,B                  ;  Print out the 1s
  add      A,#'0'
  acall    LCDCHAR

  clr      P3.7                 ;  Turn off the pulse

  ajmp     Loop

   :

Dlay150:                        ;  150 usec Dlay

  mov      Count,#19            ;  Delay 116 usec

Dlay150Loop:

  djnz     Count,Dlay150Loop

  ret
```

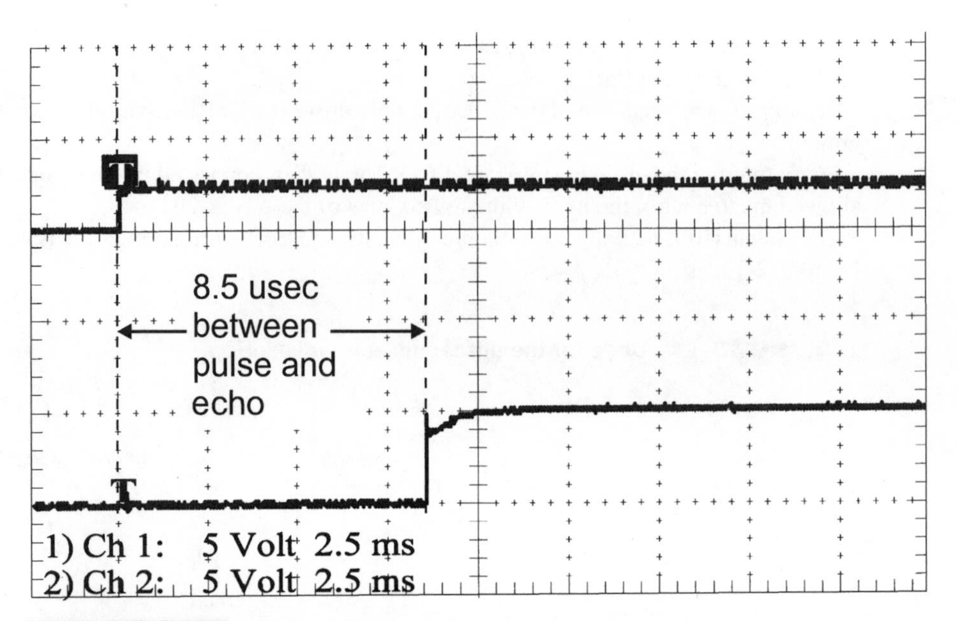

FIGURE 12-29 Polaroid 6500 transmit/receive.

This code asserts INIT and then polls ECHO once every 148 μsec (`Dlay150` is somewhat misnamed because it delays 131 μsec and not 150 μsec). When I timed this application, along with creating `Dlay150`, I wanted the loop to execute as close to 148 μsec as possible (which was achieved). So, along with the delay, I also included the time in the loop and calling the subroutine to be as accurate as possible.

Every 147.9 μsec, sound travels 2". So, in 147.9 μsec, a transmitted and reflected pulse is actually measuring 1". By polling once every 148 μsec, I am only introducing an error of 0.06%.

When the pulse's echo is received, the count of the number of 148-μsec loops is then displayed on the LCD. The 8051's `div AB` command makes the conversion to inches quite simple.

At the beginning of this application's write-up, I showed an oscilloscope picture of what the INIT and ECHO signals look like. Now, this can be examined a bit more closely.

In this oscilloscope picture, I have decreased the time interval between gradicules to see more accurately what is happening. In this example, I was using the 6500 Ultrasonic Ranging Module to measure a 60" distance between the transducer and a wall. The circuit using the ceramic resonator for the AT89C2051's clock indicated that the distance was 59".

The delay, as was measured on the oscilloscope (Fig. 12-29) is 8.5 msec. If this is divided by 147.9, you will find that the result is actually 57.5 (or 57.5"). This 4% error is somewhat worse than the specified accuracy of the 6500 (which is 1% error), and I decided to try to investigate into where the error was coming in.

When I increased the oscilloscope's sec/division display, I found that the actual time between the rising edges of the INIT and ECHO pulses was actually 8.65 msec (which works out to 58.5"). The digital oscilloscope that I was using did not have the granularity necessary at the original time base to really accurately measure the delay between the rising and falling edges for the picture shown in Fig. 12-29.

The 59" output from the circuit is accurate to within 1.5%; however, the ceramic resonator used as the AT89C2051's clock is only accurate to a few thousand PPM, which places this measurement within the expected levels of accuracy.

With a better understanding of the problem with the oscilloscope measuring time delays, I used the circuit to measure a number of different distances and found the 6500 and AT89C2051 to be quite accurate (within 2%) from about 18" to a distance of 15'.

PROG49: NTSC Composite Video Output

The TV is one of the most basic electronic output devices that we, as humans, interact with every day and is something that most people would not consider using as an output device for microcontrollers. I find that video output is often considered inappropriate for microcontrollers because of the data speeds involved as well as the fast analog voltage output required. While it is certainly challenging, it can be done quite effectively and reliably (as the photograph in Fig. 12-30 indicates).

FIGURE 12-30 **A DS87C520 driving a TV.**

In this example application, I will show you how you can use no more than a Dallas Semiconductor DS87C520 and a few resistors to create a video output circuit that is suitable for using instead of LED or LCD displays.

Before explaining the application, I should first introduce you to composite NTSC video as well as a few concepts and a warning.

The warning is, for this project, I used a $10 television that I bought at a garage sale and used a $1.50 video modulator that I bought at a surplus store to convert the composite video that the DS87C520 microcontroller produces into a signal usable by the TV (on channel 3).

I did not modify the TV in any way (such as providing a bypass to the video pre-amp from the tuner), and I don't recommend that you do this either. There are potentially lethal voltages inside a TV, stored in capacitors (not to mention voltages available at coils if the TV is still plugged in).

While the circuit here should not produce any voltages or currents that could damage a TV, I don't recommend that you hook this circuit up to the family's large-screen TV. Instead, you should look for an old 12" black-and-white TV that you can pick up for a few bucks.

The last comment is that the circuit that you are about to build is essentially a video transmitter (the modulator will convert the composite video to a frequency that can be potentially picked up by your neighbors TVs on channel 3 (which they are probably using for cable convertors or VCRs). Please be sensitive to whether or not you are interfering with their reception (or you might get a visit from your local FCC representative).

With the legal caveats out of the way, let's look at the application. After a signal has been filtered by the tuner in your TV set and demodulated, the actual video information, called *composite video*, is passed to the video drive electronics. If you were to look at the composite video signal on an oscilloscope, you would see something like Fig. 12-31.

I have identified three features that you will have to become familiar with. The first is the *vertical synch*, which is a series of specialized pulses that tells the video drivers when to start scanning the output. For each line of output, the start is identified with a *horizontal synch* pulse. During each line, data is output as an analog voltage. The description that I'm going to give here is for black-and-white composite video with no colorburst information.

Composite video gets its name from the fact that it combines three different signals (the vertical synch, horizontal synch, and video data) all in one line. Special circuitry in the TV (or CRT as I will call it for most of this section) splits this information out to control how the output is displayed.

The period of time between vertical synchs is known as the *video field*. For NTSC composite video (which this circuit creates), 59.94 fields are displayed each second. One complete frame of video consists of two fields, with the raster scan line of one field overlapping the other (which is known as *interlaced video*). The output produced by the project shown here produces the appropriate timing for the output data to repeat over two interlaced lines.

As part of the vertical synch, a number of dummy horizontal lines are passed at the same time and are known as *vertical blanking*. When the vertical synch is recognized within the TV and the raster moves to the top of the screen to begin scanning again, the CRT guns are turned off to prevent any spurious signals from being driven on the screen. (See Fig. 12-32.)

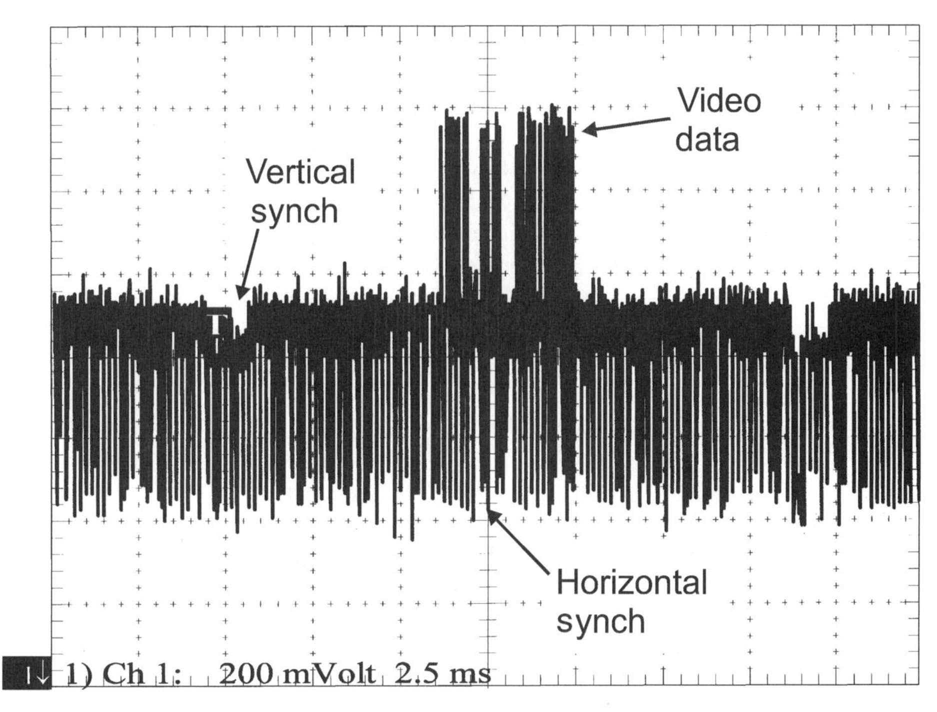

FIGURE 12-31 NTSC composite video field.

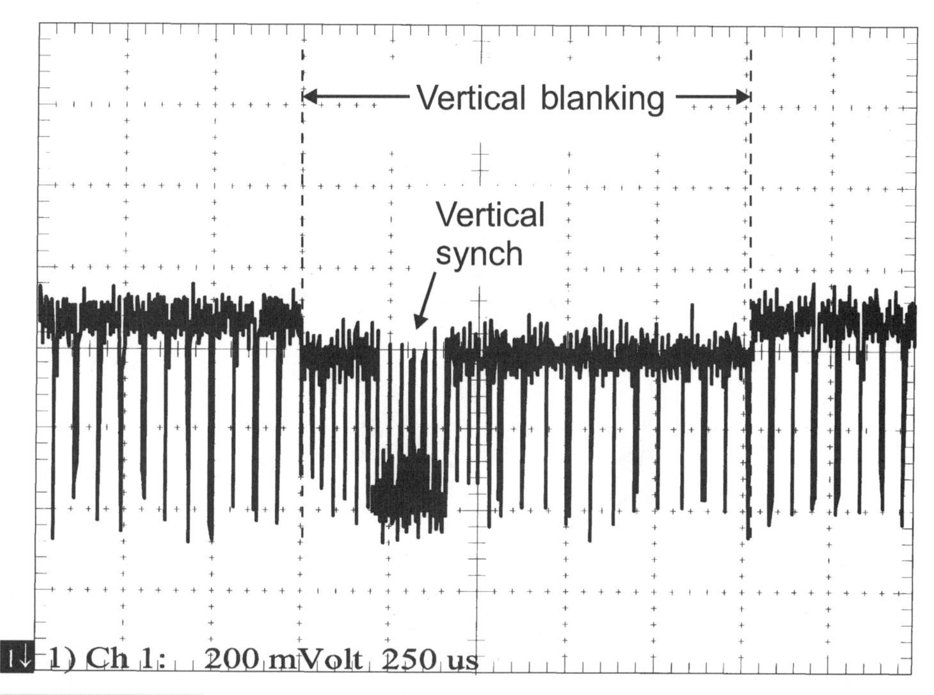

FIGURE 12-32 NTSC vertical blanking interval.

The analog voltage output used for synch pulses and CRT control is always at a level below the video data black level and is often known as "blacker than black." The normal synch level is at 0.4 V, while the active synch pulse is at 0 V.

The vertical blanking before the vertical synch consists of 6 or 7 half lines (at 31.8 μsec long), followed by 6 negative half lines (these are the vertical synch pulses), then another 6 or 7 half lines, followed by 10 or 11 full lines (at 63.5 μsec long).

With the raster gun now pointing to the top of the CRT, data can be output a line at a time.

Each line is 63.5 μsec with a horizontal synch pulse to indicate where the line starts followed by the data to be output on the line. The data output ranges from 0.48 V (black) to 1.00 V (white) with gray being in between. In Fig. 12-33, you can see the different levels (despite the noise on the line).

In Fig. 12-34, I have focussed in on the horizontal synch pulse so that you can get a better view of what it's made up of. The "Front Porch" and "Back Porch" are 1.4 μsec and 4.4 μsec in length, respectively, and are at 0.40 V. The synch pulse itself is a zero voltage active for 4.4 μsec.

For the CRT to accept the composite video, the quoted times that I have given must all be exactly on. Failure to have the same number of cycles on different lines will result in a wavy screen.

The actual circuit that I used to create the composite video is almost unbelievably simple (Fig. 12-35). In fact, it is actually less complex than many of the circuits you built in chapter 9! Now, you might think that all the science went into the code development, but I should explain how the ADC built into the circuit works.

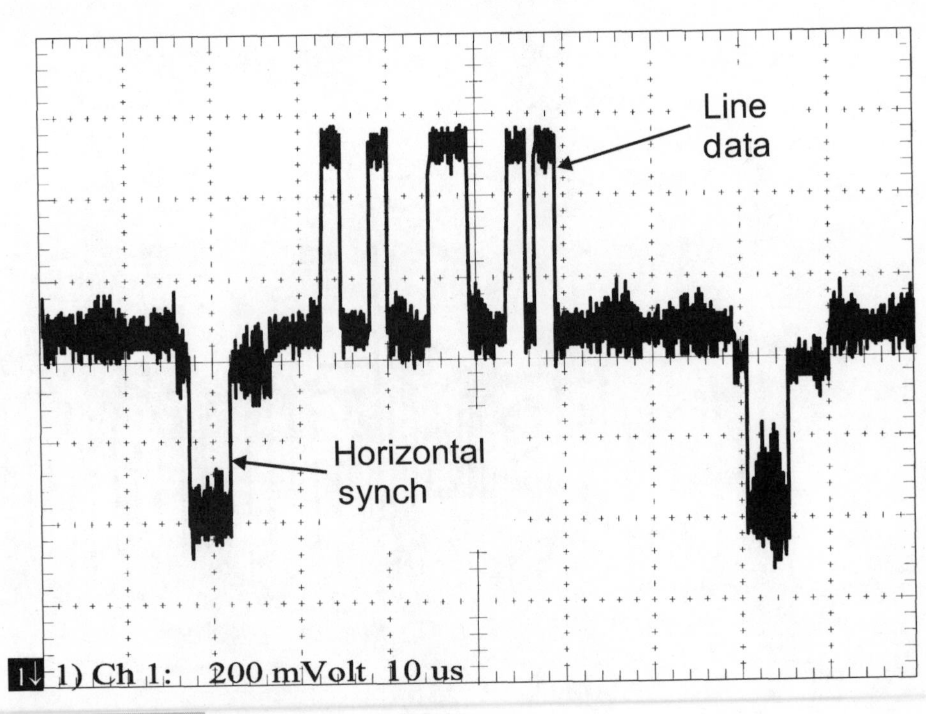

FIGURE 12-33 NTSC horizontal line.

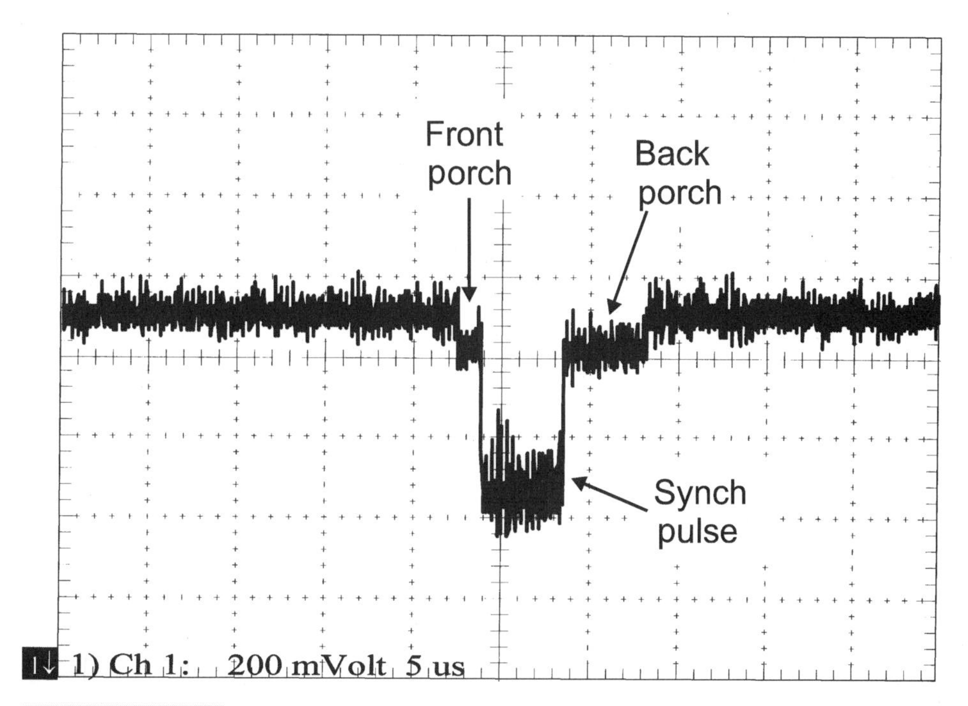

FIGURE 12-34 Horizontal synch pulse.

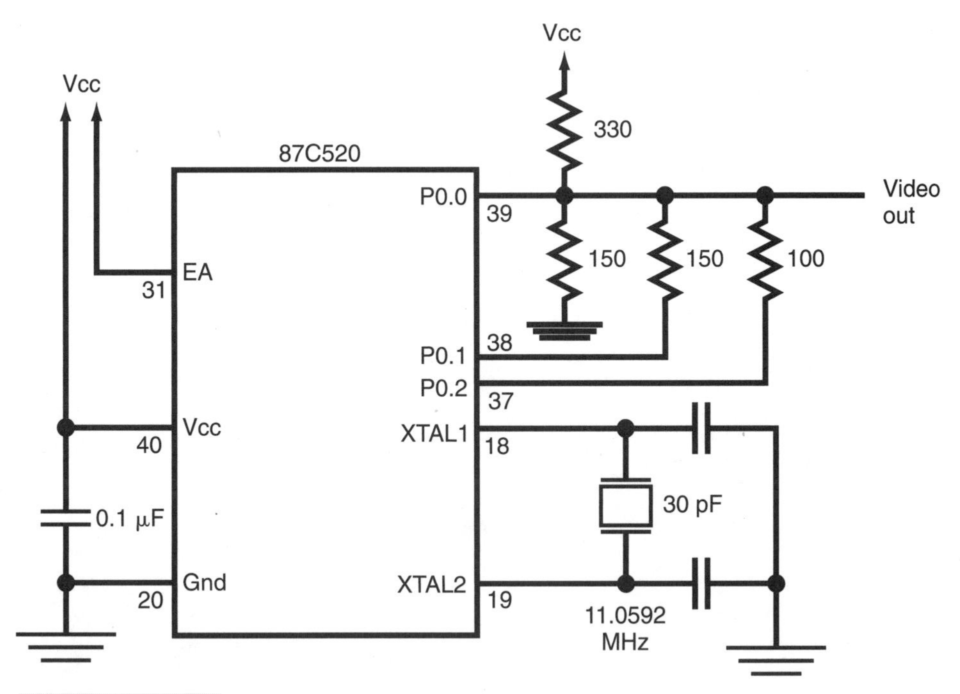

FIGURE 12-35 87C520 NTSC composite video generator.

As I've noted earlier, the composite video ranges from 0 V to 1 V. I use a programmable voltage divider to produce the voltages and select the voltages using bits of the open drain port (P0) of the DS87C520. I use the open drain port to just pull down the lines and not pull them up (which would affect the output). To output 1 V, all the pins are turned off. To output 0 V, P0.0 pulls the circuit low. Turning P0.1 or P0.2 will select 0.40 V or 0.48 V. Only one pin can be active at a time for a valid voltage to be output.

To find the resistor levels, I first calculated what the proper values would be in a circuit that connected the resistors to ground directly (and not through an open drain I/O pin). I then wired the circuit as shown Fig. 12-35 and found that the voltages had changed somewhat (actually, I just had to change the resistor always pulling to ground to 150 Ω).

With the circuit ready and tested on a protoboard, I then built it on a SimmStick prototype board. I wired it and the video modulator using point-to-point wiring. (See Figs. 12-36 and 12-37.)

In the appendices, there is more information on the SimmStick as well as its bus. For the final circuit, I used a DT003 bus board, which provides a 5-V regulator, bus sockets, and an RS-232 interface using a Maxim MAX232.

Using this circuit, I passed the composite video signal from the DS87C520 to the modulator using one of the unused SimmStick bus pins. The modulator is driven by the unregulated 12 V that is available on the SimmStick bus.

The software itself is actually quite simple, although it took a lot of work to create. The problem was in making sure that all timings would be observed and be the same all the time. If you look through PROG49B, you'll see a number of strange things to make sure that, no matter how the code executes, the same signal timing will be output from the DS87C520.

I should explain my choice of crystal. The application would have been a lot easier to time if I had used a colorburst (3.579545 MHz) or multiple crystal (ideally 14.31818 MHz). By using the 11.0592 MHz crystal, I had to make sure that I totally understood the timings for each event as it related to each feature of the signal. I chose the 11.0592-MHz crystal because I was thinking about using this circuit as a serial interface (instead of using a Hitachi 44780-based LCD).

The great idea that I had was that using a $10 TV, a DS87C520, and a surplus monitor, I could produce a serial interface more cheaply than you could buy an LCD serial inter-

FIGURE 12-36 **Top side of the SimmStick with DS87C520.**

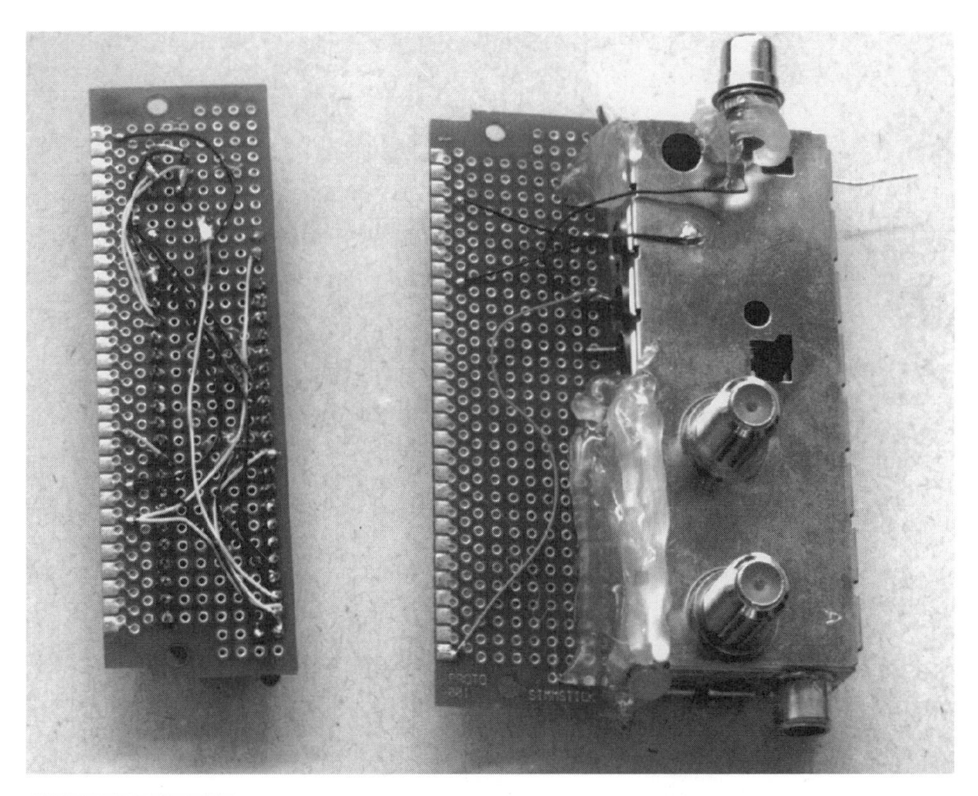

FIGURE 12-37 **Back side of the SimmStick, showing the point-to-point wiring for the DS87C520 and video modulator.**

face. If I used a 14.31818-MHz crystal, I wouldn't have been able to produce 9600 bps using the internal timer.

Actually, it's not so bad an idea.

When you look through the timing, you'll see that the screen data is essentially broken up into three parts. The first and third are a top and bottom blank (black) area with the actually lettering placed inside.

The lettering itself is read directly from the scratchpad RAM with each byte on the screen saved as 8 bytes (1 byte for each scan line of the letter) with four characters displayed on each line.

The code used reads the byte for each character line from the scratchpad RAM into the accumulator and shifts it out using the code:

```
rlc A                          ;  Load "A" with the MSB
mov  P0.1,C                    ;  Output the bit
```

After 8 bits of each byte is output, a "0" (black) is output to give time for the next byte to be loaded in.

For timing, each horizontal line is 63.5 μsec in length. With a clock speed of 11.0592 MHz, this means 176 instruction cycles must be used for each line. To output the synch information, 41 cycles are required, which means that only 135 instruction cycles are available for each line.

Now, 135 instruction cycles might seem like a lot; however, to output each bit, three instruction cycles are required for the previous `rlc A/mov P0.1,C` sequence. This means that 28 cycles (one cycle for loading the byte into the accumulator) is required or 112 instruction cycles for outputting four characters. As each character byte is output, I incremented the video memory index (R1).

This is actually quite doable. However, when I first output the data to the TV, I found that the characters were very squashed together. To make them more readable, I output each line twice. The code for doing this is:

```
cpl      NewLine                ;  2 - Do we reset R1?
jnb      NewLine,DF_Skip        ;  4 - If originally set, then "Yes"
mov      A,R1                   ;  1 - Subtract 4 to do the next line
clr      C                      ;  1
subb     A,#4                   ;  2
mov      R1,A                   ;  1
```

which uses `NewLine` as a bit to indicate whether or not the line is to be repeated (and if it is, the output index is reset to the start of the line) requires 11 instruction cycles.

This meant that I had used up 123 instruction cycles (out of 135) for displaying 4 characters on a line, with only 12 instruction cycles left for determining whether or not data was to be displayed.

When you look at PROG49B.ASM, note that I have marked how many instruction cycles each instruction takes to execute. This was turned out to be very necessary when I was timing out the application. Because of the timing restraints of this application, you should never use interrupts as they will screw up the video timings. This will manifest itself as snow or flipping of the display. I did not use interrupts for the actual timings because I could not get them precise enough.

I should also point out that, while I display the same line at 59.94 fields per second, I do interlace them to provide the best possible image on the CRT. This was done by alternating the number of half lines in the vertical blanking interval.

It might be funny to hear, but the biggest challenge I had in creating this application was finding a font for displaying ASCII characters. The solution that I came up with merits some discussion.

In my research, I found that true PC video adapters have an 8x8 font in ROM at address 0x0F000:0x0FA6E. I wrote a program (GetFont.c) that read this address and created Font.inc, which is included in PROG49B.ASM to provide the font tables for the program.

As part of this program, I had GetFont produce a subroutine that would do the table call into the font information:

```
FontLine0:
  inc A
  jnz FontLine_Skip0
  mov A,#0DFh
  nop
  ret
FontLine_Skip0:
  movc A,@A+PC
  ret
  db 000h    ;  000h
   :
```

My original intention was to have these routines called for each byte of each line, but there are obviously not enough cycles to do this. Instead, I loaded up the scratchpad RAM with the bits for the string to be displayed ("How are you Myke").

As I noted, my original thought was to use the TV as a serial interface display (i.e., a simple terminal), but the few actual instructions per line really makes that impossible. To make this application usable in actual applications, I would add external memory. When the memory was read back, I would pass the data to a hardware shift register that would be much more effective at shifting the data out (even at 3.579545 MHz, 32 characters per line could be displayed).

I would use the external memory because, without an external character generator, 8 bytes have to be stored for the bit pattern of 1 byte.

PROG47: Electronic RS-232 "Breakout Box"

The last project of the book is quite interesting. While not showing any new things that you can do with the 8051, it does show how different features can be integrated together. The project here will give you a simple LCD menu interface for your projects, as well as give you a tool with which to be able to work with RS-232 better.

If you've ever worked with RS-232, you'll know what a pain it can be with keeping signals, levels, data format, and handshaking straight. I've worked with RS-232 for a number of years. Despite my professed expertise in working with it, I still find I get requests to hook up computers using the RS-232 that can take me an hour to figure out and document what is needed.

One common tool for helping figure out how RS-232 is connected is the "breakout box" (Fig. 12-38), which consists of two connectors with a method of modifying the wiring (such as a protoboard) to try to get the two computers talking to one another. Personally, I have always found these to be a pain, with experiments taking several minutes to set up; I've always wanted to do this electronically.

Recently, I discovered the Maxim MAX214 product, which will allow a serial port to be used as either a DCE or DTE TTL/CMOS to RS-232 interface and wondered if I could use it as the interface to my electronic breakout box. The chip will swap input and output functions to allow you to use the same connector as both a DCE or DTE connection, eliminating the need to physically change the wiring.

A copy of the MAX214 data sheet is available on the CD-ROM.

Rather than trying this out, I decided to use them directly in a circuit, using a Dallas Semiconductor DS87C520, which has two serial I/O ports. The circuit I came up with is shown in Fig. 12-39. This circuit is really a rat's nest of wiring, which is why I went ahead and did my prototyping with an embedded card (Figs. 12-39 through 12-42).

Although with lots of nets, the circuit is very straightforward with the only aspect that you should note is that I placed pull ups on the LCD's E and RS bits because I used port P0 for these functions (all the other pins were used up). The circuit is designed to maximize the use of the internally pulled up pins, and the E and RS pins were the ones that I felt were least likely to affect the operation of the serial data transfer.

FIGURE 12-38 Electronic breakout box with LCD installed.

There are a few features I want to bring to your attention. The first is the use of a 11.0592-MHz crystal to drive the DS87C520. This crystal speed is a multiple of 300 Hz, which allows the internal timers to exactly run at the different data speeds. Timer1 is used in mode 2 to provide the different speeds for the application. 300, 1200, 2400, and 9600 are available. Unfortunately, this crystal will not give what I would consider a full range (from 110 to 19200 bps) of data speeds.

One feature of the project that I am particularly proud of is the menu interface that I came up with. Using the LCD and two buttons, this feature allows you to select from the second line of the LCD a number of selections. To do this, a number of subroutines had to be written.

The first was a string display routine (Listing 12-13) that records the location of each option. Upon entry into this routine, an index to the string to be displayed (its position on the LCD is assumed to be already set up by the application) is passed to the routine and then using the `movc` instruction, the correct string to output is found (because not all the strings have the same length), and the positions of blanks (ASCII 020h) are recorded.

This routine first reads through the `MsgSendTable` and decrements the index passed to the routine each time it encounters a zero (000h) character. When the index finally reaches zero, the routine goes on to read and display each character to the next zero.

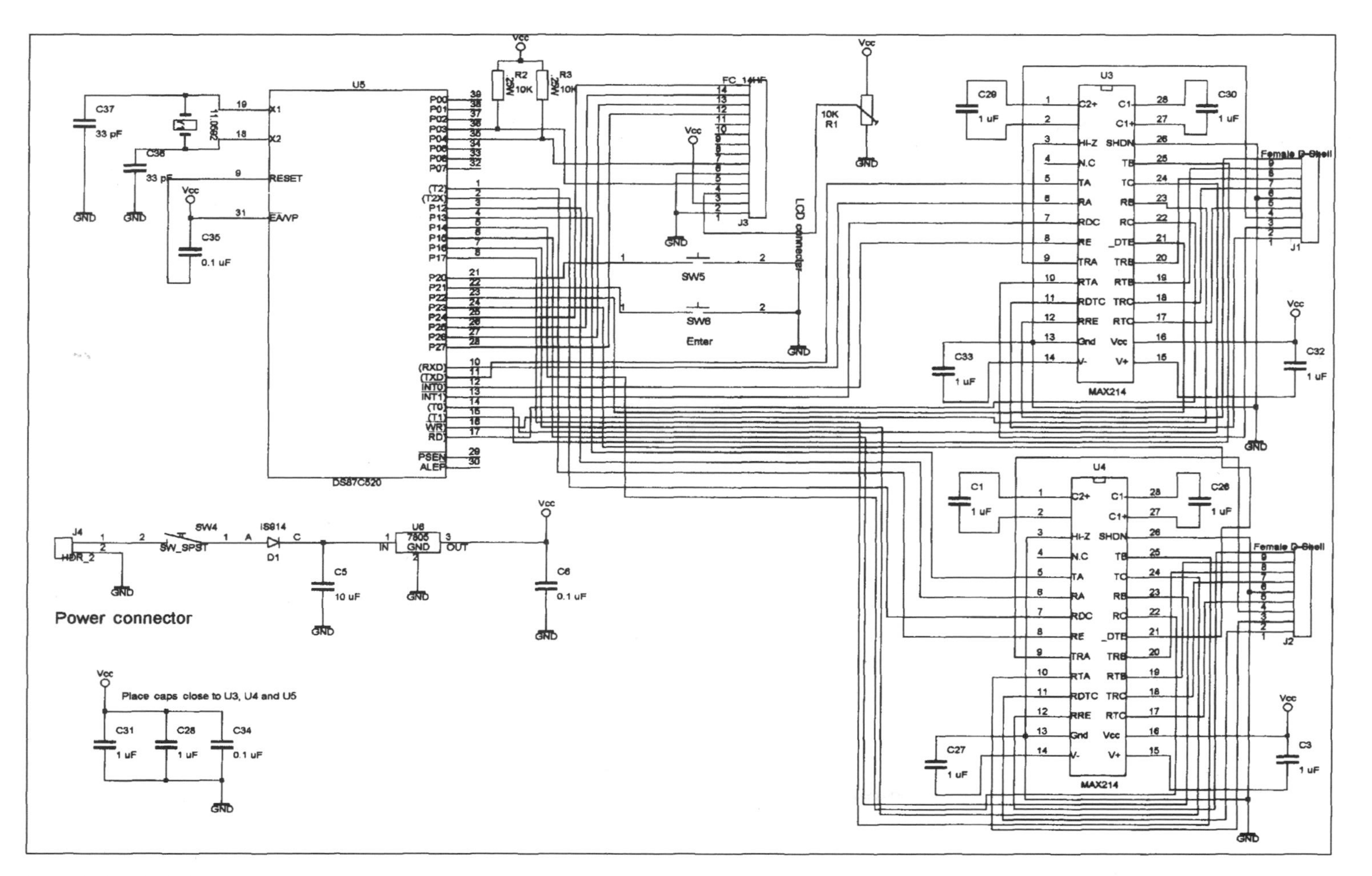

FIGURE 12-39 Breakout box raw card schematics.

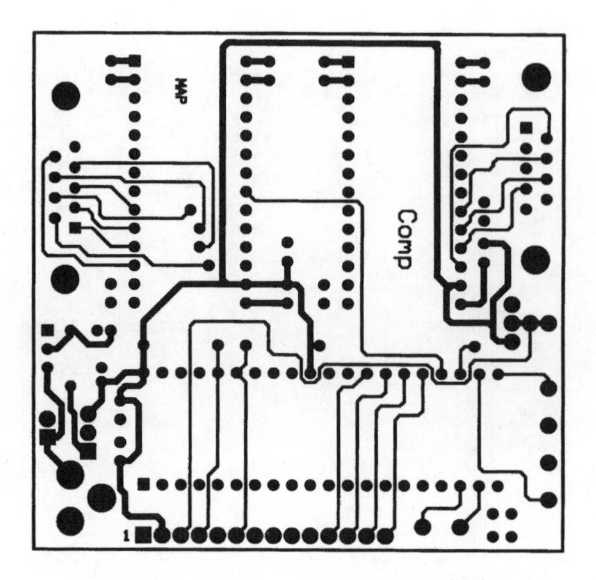

FIGURE 12-40 Top side of the breakout box raw card.

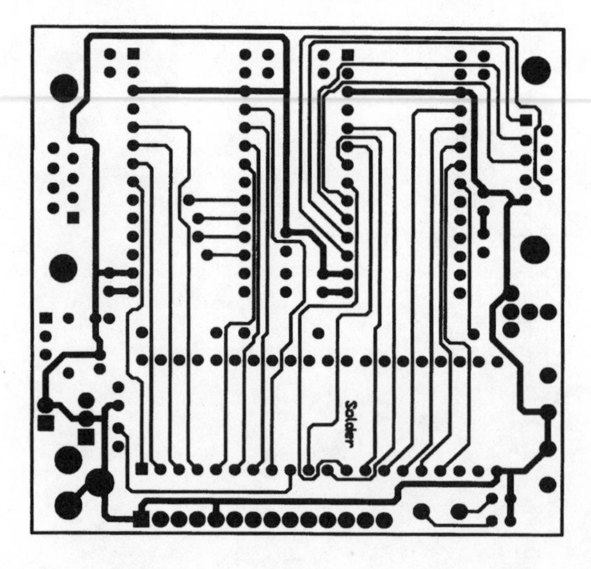

FIGURE 12-41 Bottom side of the breakout box raw card.

With the menu displayed on a 16-column, 2-row Hitachi 44780-based LCD display, the `MenuRead` routine (Listing 12-14) is called, which moves the cursor to the first blank (selection) and then moves the cursor to blanks to the right according to the user's selection.

The position of each blank is recorded in R5, R6, and R7 (from `MsgSend`), and the `MenuRead` routine takes the "select" input and moves the cursor to the next selection (or, if at the end to the beginning again). When "Enter" is pressed, the position of the cursor is returned to the calling routine. In the calling routines, I compare the returned position to the values in R5, R6, and R7.

These routines work very nicely, although there is one change I would make to them and that would be the initial position could be specified to `MenuRead` rather than having it

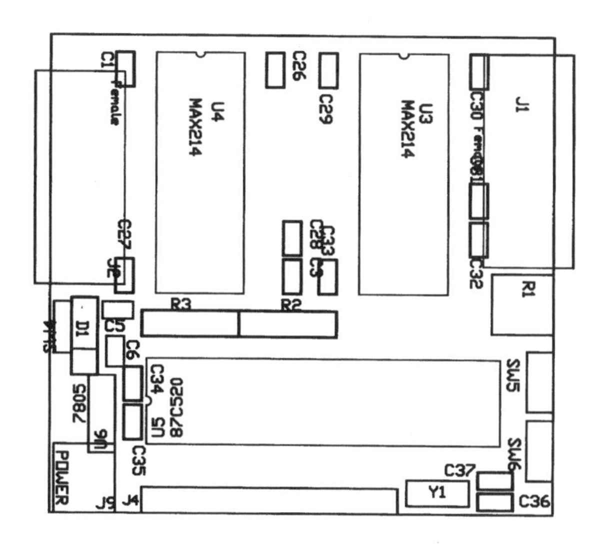

FIGURE 12-42 Overlay for the breakout box raw card.

LISTING 12-13 The string display routine.

```
MsgSend:                          ;  Send specified message to the display

  mov     R1,A                    ;  Save the message number

  mov     R0,#0                   ;  Use R0 as the offset counter

MsgSend_Find:                     ;  Look for the appropriate message

  mov     A,R1                    ;  Are we now at zero?
  jz      MsgSend_Output          ;   If count is zero

  mov     A,#LOW(MsgSendTable-MsgSendFindGet)
  add     A,R0                    ;  Get current address

  movc    A,@A+PC                 ;  Get the character in the displays
MsgSendFindGet:

  inc     R0                      ;  Point to the next character

  jnz     MsgSend_Find            ;  If NOT equal to zero,

  dec     R1                      ;  It is, decrement the message counter

  ajmp    MsgSend_Find

MsgSend_Output:                   ;  Now, output the string to '\0'

  mov     2,R0                    ;  Save the start of the message

  mov     R5,#0FFh                ;  Set the registers as invalid
  mov     R6,#0FFh
  mov     R7,#0FFh

  setb    RS                      ;  Make sure characters are going out

MsgSendOutput_Loop:               ;  Output each table character to "\0"
```

LISTING 12-13 The string display routine *Continued.*

```
  mov     A,R0
  add     A,#LOW(MsgSendTable-MsgSendOutputGet)

  movc    A,@A+PC                ;  Read the table
MsgSendOutputGet:

  inc     R0                     ;  Point to the next character

  jz      MsgSendOutput_Check    ;  If at zero, stop the reading

  mov     R1,A                   ;  Save the character
  acall   CharSend

  cjne    R1,#' ',MsgSendOutput_Loop

  mov     5,R6                   ;  Save the blank position
  mov     6,R7
  mov     A,R0                   ;  Calculate the offset
  setb    C
  subb    A,R2
  mov     R7,A

  ajmp    MsgSendOutput_Loop

MsgSendOutput_Check:             ;  Now, shift down the blanks

  mov     A,R7                   ;  Do we have no blanks?
  xrl     A,#0FFh
  jz      MsgSend_End

MsgSendOutputCheck_Loop:         ;  May only have one blank

  cjne    R5,#0FFh,MsgSend_End   ;  Everything shifted down

  mov     5,R6                   ;  Shift data down until all ' 's saved
  mov     6,R7
  mov     R7,#0FFh

  ajmp    MsgSendOutputCheck_Loop

MsgSend_End:                     ;  Message output and blanks setup

  ret

MsgSendTable:                    ;  Messages to be displayed
  db      'Left  ', 0            ;  Message  0
  db      'Right ', 0            ;  Message  1
  db      'BREAKOUT BOX',0       ;  Message  2
  db      ' Run Configure',0     ;  Message  3
  db      'Speed -',0            ;  Message  4
  db      ' 300 1200 Cont',0     ;  Message  5
  db      'Speed ¦',0            ;  Message  6
  db      ' 2400 9600 Cont', 0   ;  Message  7
  db      'Port Type',0          ;  Message  8
  db      ' DCE DTE',0           ;  Message  9
  db      'Modem Ctrl',0         ;  Message 10
  db      ' Enble Pass None',0   ;  Message 11
  db      'Display',0            ;  Message 12
  db      ' String Byte/Bit',0   ;  Message 13
  db       07Eh,'               ',0  ;  Message 14 - Point to incoming
  db      '               ',07Fh,0   ;  Message 15 - Point to incoming
```

LISTING 12-14 The MenuRead routine.

```
MenuRead:                          ;  Read the output menu

MenuRead_Skip1:
  mov     DelayCount,#0            ;  Load With a 20-msec wait
  mov     DelayCounthi,#29
MenuRead_Loop1:                    ;  Wait for "Enter" button released 20
msec
  jnb     Enter,MenuRead_Skip1
  djnz    DelayCount,MenuRead_Loop1        ;  4
  djnz    DelayCounthi,MenuRead_Loop1      ;  4

  clr     RS                       ;  Only sending commands
  mov     A,#0C0h                  ;  Move cursor to start of 2nd line
  acall   CharSend

  mov     R0,#0                    ;  Start at the first blank

MenuRead_Loop:                     ;  Wait for the buttons to be pressed

  jb      Select,MenuRead_Skip     ;  Check if "Select" pressed

MenuRead_Skip3:
  mov     DelayCount,#0            ;  Load with a 20-msec wait
  mov     DelayCounthi,#29
MenuRead_Loop3:                    ;  Wait for "Select" pressed 20 msec
  jb      Select,MenuRead_Skip3
  djnz    DelayCount,MenuRead_Loop3        ;  4
  djnz    DelayCounthi,MenuRead_Loop3      ;  4

  mov     A,R0                     ;  Get the current position
  xrl     A,R5                     ;  Are we at the first blank?
  jnz     MenuRead_CheckR6         ;   - No, check second

  mov     A,R6                     ;  Can we jump to 2nd?
  xrl     A,#0FFh
  jz      MenuRead_NewCursor       ;  No, jump to itself

  mov     0,R6                     ;  Move to the second position

  ajmp    MenuRead_NewCursor

MenuRead_CheckR6:                  ;  Are we at the 2nd value?
  mov     A,R0
  xrl     A,R6
  jnz     MenuRead_R7              ;   - No, check third

  mov     0,R5                     ;  Assume can't jump and going to roll

  mov     A,R7                     ;  Can we jump to 3rd?
  xrl     A,#0FFh
  jz      MenuRead_NewCursor       ;  No, stick with 1st

  mov     0,R7

  ajmp    MenuRead_NewCursor

MenuRead_R7:                       ;  Go from 3rd to 1st

  mov     0,R5

MenuRead_NewCursor:                ;  Go to the new cursor position
```

LISTING 12-14 The MenuRead routine *Continued.*

```
  mov     A,R0                    ;  Move the cursor
  orl     A,#0C0h                 ;  Set it with a move command
  acall   CharSend

MenuRead_Skip4:
  mov     DelayCount,#0           ;  Load with a 20-msec wait
  mov     DelayCounthi,#29
MenuRead_Loop4:                   ;  Wait "Select" button released 20 msec
  jnb     Select,MenuRead_Skip4
  djnz    DelayCount,MenuRead_Loop4        ;  4
  djnz    DelayCounthi,MenuRead_Loop4      ;  4

  ajmp    MenuRead_Loop           ;  Wait for the next button press

MenuRead_Skip:                    ;  Check to see if "Enter" pressed
  jb      Enter,MenuRead_Loop

MenuRead_Skip2:
  mov     DelayCount,#0           ;  Load with a 20-msec wait
  mov     DelayCounthi,#29
MenuRead_Loop2:                   ;  Wait for "Enter" down 20 msec
  jb      Enter,MenuRead_Skip2
  djnz    DelayCount,MenuRead_Loop2        ;  4
  djnz    DelayCounthi,MenuRead_Loop2      ;  4

  ret
```

choose the first position. By doing this, previously selected values could be retained without having to re-enter them (and potentially making a mistake).

The serial data interface in this project is quite simple. An interrupt routine passes the received data onto the other channel and saves the information to displaying on the LCD:

```
 org 023h                        ;  Serial port 0 (left serial port) int
Serial0:
  clr    SCON0.1                 ;  Clear the TX flag
  jnb    SCON0.0,Serial0_End     ;  Only handle if it was an RX interrupt
  clr    SCON0.0                 ;  Acknowledge the interrupt happened
  mov    @R0,SBUF0               ;  Save the value read
  mov    SBUF1,@R0               ;  Send it back out the other side
  inc    R0
  mov    BufferHold,R0           ;  Don't exceed the 16-byte buffer
  clr    4                       ;   This is bit 5
  mov    R0,BufferHold
Serial0_End:
  reti
```

This project was one of the last ones I worked on, and I think it shows with the interrupt handler. Note that I carry out a quite complex function, which can be modelled in C as:

```
SerInt0 Interrupt()       //  Serial interrupt handler for the
{                         //   "left" serial port
  SCON0.TI = 0;           //  Clear the serial transmit interrupt
  if ( SCON0.RI == 1 ) {        //  Handle the serial receive interrupt?
    LeftBuffer[ LeftOffset ] = SBUF0;      //  Save the incoming
                                           //   character
```

```
    SBUF1 = LeftBuffer[ LeftOffset ];    //  Output the incoming
                                         //   character to the "right"
    if ( ++LeftOffset == 16 )  //  Keep within the 16-byte buffer
      LeftOffset = 0;
  }
} //  End SerInt0
```

When you look back at my assembler code, you'll see that I was able to accomplish the functions shown in C without affecting either the microcontroller's accumulator or PSW register. One of the important points to notice is how I implemented the buffers.

In the buffers, I made sure that the bit representing its size (the buffers are 16 bits long, so this is bit 4) is always reset. By continually resetting the bit after incrementing, I have implemented a circular buffer without having to do any comparing and resetting of the index.

It probably would have been better if I had connected the receive and transmit pins of the two MAX214s directly and monitored the lines in the MAX214. This would simplify the interrupt handler and would eliminate the latency period of one character required to save the character and then send it back out. It would also eliminate the need for clearing the transmit interrupt request bit in the SCON registers as well as checking to see if a character was received.

While the breakout box is running, I display the incoming characters on the LCD display. The algorithm for doing this continually checks the buffer save index to a most recently displayed index. When they are different, the next character is displayed and the most recently displayed index pointer is incremented. (See Listing 12-15.)

LISTING 12-15 Displaying the incoming characters to the LCD.

```
  mov     A,LeftLast              ;  Get the last postion
  xrl     A,R0                    ;   Has anything changed?
  jz      RightCheck              ;  Do we have something coming in from
                                  ;   the right?
  mov     A,LeftCount
  xrl     A,#14
  jnz     DisplayLeft

  clr     RS                      ;  Redraw the top line
  mov     A,#080h
  acall   CharSend

  clr     IE.7                    ;  Disable interrupts for the message
  push    0                       ;  Save R0 and R1 which are used in
  push    1                       ;   "MsgSend"
  mov     A,#14
  acall   MsgSend
  pop     1
  pop     0
  setb    IE.7                    ;  Enable ints after message displayed

  mov     LeftCount,#0            ;  Reset the counter

DisplayLeft:                      ;  Display the character on the left

  clr     RS                      ;  Make sure the cursor is in the
  mov     A,#082h                 ;   correct spot
```

LISTING 12-15 **Displaying the incoming characters to the LCD *Continued.***

```
add     A,LeftCount
acall   CharSend

clr     IE.7                 ; Display the character
mov     A,LeftLast           ; Use "LeftLast" as index to display
xch     A,R0

mov     B,@R0                ; Save the character

xch     A,R0                 ; Restore the registers
mov     LeftLast,A
setb    IE.7                 ; Turn interrupts back on

mov     A,B                  ; Display the character
setb    RS
acall   CharSend

inc     LeftLast
mov     A,LeftLast
clr     ACC.4
mov     LeftLast,A

inc     LeftCount            ; Increment the column counter for
                             ;  initial check
ajmp    DataLoop
```

The project, as it's presented here, is pretty rudimentary although it is pretty useful. I have not implemented some of the more advanced features that I would like to. These features include:

1 Monitoring handshaking lines.
2 Passing handshaking lines between the two devices. As the project stands, all the pins are set, which means the handshaking lines are all enabled at the same time.
3 Providing a handshaking line loop back to a transmitter (i.e., tying DSR/DTS and CTS/RTS together in software on one side).
4 Displaying a previously selected value (as discussed earlier) instead of having to work through all the options.
5 Providing a serial EEPROM for saving the previously selected options when the power is removed.
6 Relaying out the board to use a specific LCD and bundle the entire project in a box.

I used this breakout box for helping to debug the emulator's RS-232 interface (which does not use the serial port), and I found it to be quite useful as a monitor of the actual data. The DS87C520 contains 16 K of control store, so additional features can be added quite easily, and I expect over time to start adding many of the features that I am listing earlier. You should check my Web site periodically to see what kind of changes I've made.

EPILOGUE

After finishing this book, I hope that I can say without sounding too corny that you are ready to begin. It has been my intention to provide you with the knowledge and confidence to create your own applications. From this book, you should have an understanding of how the 8051 hardware works, how the microcontroller is programmed, and how to create applications using the 8051.

This book (and the CD-ROM that comes with it) is not the ultimate 8051 resource. To successfully design applications, you will also need manufacturer's data sheets, application notes, and probably a few suggestions from others. I do hope that this book will be a useful reference to you as you begin to work on your own.

I'm looking forward to hearing from you, and I would be happy to answer any questions that you have. I do have one request regarding questions, please e-mail them to one of the 8051 listservers. By doing this, your question and its answer will be available to others and might help somebody else.

By joining a listserver, you are joining a community of technicians and engineers who can help you with problems or at least talk you through them. Hopefully your own comments and examples will help spur others on to working through their own applications.

Good luck on your own projects, and I hope to see what you come up with!

A

GLOSSARY

If this book is your first experience with microcontrollers, I'm sure there are a lot of terms I've thrown at you that are unfamiliar to you. I've tried to give a complete list of all the acronyms, terms, and expressions that may be unfamiliar to you. Acronyms are explained before they are described.

accumulator Register used as a temporary storage register for an operation's data source and destination.

active components Generally integrated circuits and transistors. Devices that require external power to operate.

ADC Analog-to-digital converter. Hardware devoted to converting the value of a dc voltage into a digital representation. See *DAC*.

address The location a register, RAM byte, or instruction word is located at within its specific memory space.

amps Measure of current. One amp is the movement of one coloumb of electrons in one second.

analog A quantity at a fractional value rather than a binary, one or zero. Analog voltages are the quantity most often measured.

AND Logic gate that outputs a "1" when all inputs are a "1."

ASCII American Standard Character Interchange Interface. Bit-to-character representation standard most used in computer systems.

assembler A computer program that converts assembly language source to object code. See *cross assembler*.

assembly language A set of word symbols used to represent the instructions of a processor. Along with a primary instruction, there are parameters that are used to specify values, registers, or addresses.

asynchronous serial Data sent serially to a receiver without clocking information. Instead, data synching information for the receiver is available inside the data packet or as part of each bit.

bare board See *raw card*.

BCD Binary Coded Decimal. Using 4 bits to represent a decimal number (0 to 9).

BGA Ball Grid Array. A chip solder technology that provides connection from a chip to a bare board via a two-dimensional grid of solder balls (typically 0.050" from center to center).

binary numbers Numbers represented as powers of two. Each digit is two raised to a specific power. For Example, 37 decimal is $32 + 4 + 1 = 2^4 + 2^2 + 2^0 =$ 00010101 binary. Binary can be represented in the forms: 0b0*nnnn*, B'*nnnn*' or %*nnnn* where *nnnn* is a multiple-digit binary number comprising of 1s and 0s.

bipolar logic Logic circuits made from bipolar transistors (either discrete devices or integrated onto a chip).

bit mask A bit pattern that is ANDed with a value to turn off specific bits.

burning See *programming*.

bus An electrical connection between multiple devices, each using the connection for passing data.

capacitor Device used for storing electrical charge. Often used in microcontroller circuits for filtering signals and input power by reducing transient voltages.

ceramic resonator A device used to provide timing signals to a microcontroller. More robust than a crystal but with poorer frequency accuracy.

character Series of bits used to represent an alphabetic, numeric, control, or other symbol or representation. See *ASCII*.

chip package The method used to protect a chip from the environment (usually either encased in either ceramic or plastic) with wire interconnects to external circuitry. See *PTH* and *SMT*.

CISC Complex Instruction Set Computer. A type of computer architecture that uses a large number of very complete instructions rather than a few short instructions. See *RISC*.

clock A repeating signal used to run a processor's instruction sequence.

clock cycle The operation of a microcontroller's primary oscillator going from a low voltage to a high voltage and back again. This is normally referenced as the speed that the device runs at. Multiple clock cycles might be used to make up one instruction cycle. See *instruction cycle.*

CMOS logic Logic circuits made from N-Channel and P-Channel MOSFET (Metal Oxide Silicon Field Effect Transistors) devices (either discrete devices or integrated onto a chip).

comparator A device that compares two voltages and returns a logic "1" or a "0" based on the relative values.

compiler A program that takes a high-level language source file and converts it to either assembly language code or object code for a microcontroller.

control store See *program store.*

constant Numeric value used as a parameter for an operation or instruction. This differs from a variable value that is stored in a RAM or register memory location.

CPU Central Processing Unit. What I refer to as the microcontroller's processor.

cross assembler A program written to take assembly language code for one processor and convert it to object code while working on an unrelated processor and operating system. See *assembler.*

crystal Device used for precisely timing the operation of a microcontroller.

current The measurement of the number of electrons that pass by a point in a second. The units are amps, which are coulombs per second. One coulomb of charge requires 1.6×10^{19} electrons.

DAC Digital-to-analog converter. Hardware designed to convert a digital representation of an analog dc voltage into that analog voltage. See *ADC.*

DCE Data Communications Equipment. The RS-232 standard that modems are usually wired to. See *DTE.*

debugger A program used by an application programmer to find the problems in the application. This program is normally being run on the target system that the application to run on.

decimal numbers Base 10 numbers used for constants. These values are normally converted into hex or binary numbers for the microcontroller.

digital A term used to describe a variety of logic families where values are either high ("1") or low ("0"). For most logic families, the voltage levels are either approximately 0 V or approximately 5 V with a switching level somewhere between 1.4 V and 2.5 V.

driver Any device that can force a signal onto a "net." See *receiver.*

DTE Data Terminal Equipment. The RS-232 standard that your PC's serial port is wired to. See *DCE.*

D-shell connectors A style of connector often used for RS-232 serial connections as well as other protocols. The connector is "D" shaped to provide a method of polarizing the pins and ensuring that they are connected the correct way.

duty cycle In a pulse-wave-modulated digital signal, the duty cycle is the fraction of time the signal is high over the total time of the repeating signal.

edge triggered Logic that changes based on the change of a digital logic level. See *level sensitive.*

editor Program located on your development system that is used to modify application source code.

EEPROM Electrically Erasable Programmable Read-Only Memory (a.k.a. Flash). Non-volatile memory that can be erased and reprogrammed electrically (i.e., it doesn't require the UV light of EPROM).

emulator Electrical circuit connected to the development system that will allow the application to be executed under the developer's control allowing observation of how it works and trying out some changes to the application ("what if?").

EPROM Erasable Programmable Read-Only Memory. Non-volatile memory that can be electrically programmed and erased using ultra-violet light.

external memory RAM or ROM memory attached to the 8051's P0 and P2 ports, giving the 8051 additional RAM or application space.

FIFO First In, First Out. Memory that will retrieve data in the order in which it was stored.

Flash A type of EEPROM. Memory that can be electrically erased in blocks, instead of as individual memory locations. True Flash is very unusual in microcontrollers. Many manufacturers describe their devices as having Flash, when in actuality they use EEPROM.

flip-flop A basic memory cell that can be loaded with a specific logic state and read back. The logic state will be stored as long as power is applied to the cell.

frequency The number of repetitions of a signal that can take place in a given period of time (typically one second). See *period* and *Hertz.*

FTP File Transfer Protocol. A method of transferring files to/from the Internet.

functions A subroutine that returns a parameter to the caller.

fuzzy logic A branch of computer science in which decisions are made on partially "on" data rather than "on" or "off" data like digital logic uses. These decisions are often made for controlling physical and electronic systems. See *PID.*

ground Abbreviated to "Gnd." Negative voltage to microcontroller/circuit. Also referred to as *Vss.*

GUI Graphical User Interface (often pronounced "gooey"). A GUI is used, along with a Graphical Operating System (such as Microsoft Windows), to provide a simple, consistent interface for users that consists of a screen, keyboard, and mouse.

Harvard architecture Computer processor architecture that interfaces with two memory subsystems, one for instructions (control store) memory and one for variable memory and I/O registers. See *Princeton architecture.*

Hertz A unit of measurement of frequency. One Hertz (or Hz) means that an incoming signal is oscillating once per second.

hex numbers A value from 0 to 15 that is represented using 4 bits or the numbers "0" through "9" and "A" through "F."

high-level Language A set of English (or other human language) statements that have been formatted for use as instructions for a computer. Some popular high-level languages used for microcontrollers include C, BASIC, Pascal, and Forth.

horizontal synch A pulse used to indicate the start of a scan line in a video monitor or TV set.

HSM High-Speed Microcontroller. The Dallas Semiconductor version of the 8051, which incorporates a modified processor core that has a 4-clock-cycle instruction cycle instead of the 8051's typical 12-clock-cycle instruction cycle.

Hz See *Hertz*.

index register An 8- or 16-bit register that can have its contents used to point to a location in variable storage, control store, or the microcontroller's register space. See *stack pointer*.

inductor Wire wrapped around some kind of form (metal or plastic) to provide a magnetic method of storing energy. Inductors are often used in oscillator and filtering circuits.

infrared A wavelength of light (760 nm or longer) that is invisible to the human eye. Often used for short distance communications.

interpreter A program that reads application source code and executes it directly rather than compiling it.

interrupt An event that causes the microcontroller's processor to stop what it is doing and respond.

instruction A series of bits that are executed by the microcontroller's processor to perform a basic function.

instruction cycle The minimum amount of time needed to execute a basic function in a microcontroller. One instruction cycle typically takes several clock cycles. See *clock cycles*.

in-system programming The ability to program a microcontroller's control store while the device is in the final application's circuit without having to remove it.

I/O space An address space totally devoted to providing access to I/O device control registers.

I2C Inter-InterComputer communication. A synchronous serial network protocol that allows microcontrollers to communicate with peripheral devices and each other. Only common lines are required for the network.

KHz This is an abbreviation for measuring frequency in thousands of cycles per second.

label An identifier used within a program to denote the address location of a control store or register address. See *variable*.

LCD Liquid Crystal Display. A device used for outputting information from a microcontroller. Typically controlled by a Hitachi 44780 controller, although some microcontrollers contain circuitry for interfacing to an LCD directly without an intermediate controller circuit.

LED Light-Emitting Diode. Diode (rectifier) device that will emit light of a specific frequency when current is passed through it. When used with microcontrollers, LEDs are usually wired with the anode (Positive pin) connected to Vcc and the microcontroller I/O pin sinking current (using a series 200- to 270-Ω resistor) to allow the LED to turn on. In typical LEDs in hemispherical plastic packages, the flat side (which has the shorter lead) is the cathode.

level conversion The process of converting logic signals from one family to another.

level sensitive Logic that changes based on the state of a digital logic signal. See *edge triggered*.

LIFO Last In, First Out. Type of memory in which the most recently stored data will be the first retrieved.

linker A software product that combines object files into a final program file that can be loaded into a microcontroller.

listserver An Internet server used to distribute common Interest mail to a number of individuals.

logic analyzer A tool that will graphically show the relationship of the waveforms of a number of different pins.

logic gate A circuit that outputs a logic signal based on input logic conditions.

logic probe A simple device used to test a line for either being high, low, transitioning, or in a high-impedance state.

macro A programming construct that replaces a string of characters (and parameters) into a previously specified block of code or information.

Manchester encoding A method for serially sending data that does not require a common (or particularly accurate) clock.

mask programmable ROM A method of programming a memory that takes place at final assembly of a microcontroller. When the aluminum traces of a chip are laid down, a special photographic mask is made to create wiring that will result in a specific program being read from a microcontroller's control store.

master In microcontroller and external device networking, a master is a device that initiates and optionally controls the transfer of data. See *multimaster* and *slave*.

matrix keyboard A set of push-button switches wired in an X/Y pattern to allow button states to be read easily.

MCU Acronym/abbreviation for microcontroller.

memory array A collection of flip-flops arranged in a matrix format that allows consistent addressing.

memory-mapped I/O A method of placing peripheral registers in the same memory space as RAM or variable registers.

MHz This is an abbreviation for measuring frequency in millions of cycles per second.

microwire A synchronous serial communications protocol.

MIPS Millions of Instructions Per Second. This acronym should be really be: "Misleading Indicator of Performance." It should not be a consideration when deciding which microcontroller to use for an application.

monitor A program used to control the execution of an application inside a processor.

MPU Acronym/abbreviation for microprocessor.

msec One thousandth of a second (0.001 seconds). See *nsec* and *μsec.*

multimaster A microcontroller networking philosophy that allows multiple masters on the network bus to initiate data transfers.

negative active logic A type of logic where the digital signal is said to be asserted if it is at a low ("0") value. See *positive active logic.*

nesting Placing subroutine or interrupt execution within the execution of other subroutines or interrupts.

net A technical term for the connection of device pins in a circuit. Each net consists of all the connections to one device pin in a circuit.

Net, The A colloquial term for the Internet.

NiCad Abbreviation for Nickel-Cadmium Batteries. These batteries are rechargeable, although typically provide 1.2 V per cell output compared to 1.5 V to 2.0 V for standard dry or Alkaline radio batteries.

NMOS logic Digital logic where only N-Channel MOSFET transistors are used.

NOT Logic gate that inverts the state of the input signal ("1" NOT is "0").

nsec One billionth of a second (0.000000001 seconds). See *μsec* and *msec.*

NTSC Acronym for the National Television Standards Committee. NTSC is the standards organization responsible for defining the TV signal format used in North America.

object file After assembly or high-level language compilation, a file is produced with the hex values (opcodes) that make up a processor's instructions. An object file can either be loaded directly into a microcontroller, or multiple object files can be linked together to form an executable file that is loaded into a microcontroller's control store. See *linker.*

octal numbers A method of representing numbers as the digits from "0" to "7." This method of representing numbers is not widely used, although some high-level languages, such as C, have made it available to programmers.

one's complement The result of XORing a value with 0x0FF that will invert each bit of a number. See *two's complement.*

opcodes The hex values that make up the processor instructions in an application.

open collector/drain output An output circuit consisting of a single transistor that can pull to ground the net it is connected to.

OR Basic logic gate. When any input is set to a "1", a "1" is output.

oscillator A circuit used to provide a constant frequency repeating signal for a microcontroller. This circuit can consist of a crystal, ceramic resonator, or resistor-capacitor network for providing the delay between edge transitions. The term is also used for a device that can be wired to a microcontroller to provide clocking signals without having to provide a crystal, caps, and other components to the device.

oscilloscope An instrument that is used to observe the waveform of an electrical signal. The two primary types of oscilloscopes in use today are: analog and digital. The analog oscilloscope writes the current signal onto the phosphors of a CRT. Digital storage oscilloscopes save the analog values of an incoming signal in RAM for replaying on either a built-in CRT or a computer connected to the device.

OTP One-Time Programmable. This term generally refers to a device with EPROM memory encased in a plastic package that does not allow the chip to be exposed to UV light. Note that EEPROM devices in a plastic package might also be described as OTP when they can be electrically erased and reprogrammed.

parallel Passing data between devices with all the data bits being sent at the same time on multiple lines. This is typically much faster than sending data serially.

parameter A user-specified value for a subroutine or macro. A parameter can be a numeric value, a string, or a pointer depending on the application.

passive components Generally resistors, capacitors, inductors, and diodes. Components that do not require a separate power source to operate.

PCA Printed Circuit Assembly. A bare board with components (both active and passive) soldered onto it.

PCB Printed Circuit Card. See *raw card.*

.PDF Files Files suitable for viewing with Adobe PostScript.

period The length of time that a repeating signal takes to go through one full cycle. The reciprocal of frequency.

PID Parallel Integrating Differential. A classical method of controlling physical and electronic systems. See *fuzzy logic.*

ping The operation of sending a message to a device to see if it is operating properly.

poll A programming technique in which a bit (or byte) is repeatedly checked until a specific value is found.

pop The operation of taking data off of a stack memory.

positive active logic Logic that becomes active when a signal becomes high ("1"). See *negative active logic.*

PPM Measurement of something in Parts Per Million. An easy way of calculating the PPM of a value is to divide the value by the total number of samples or opportunities and multiplying by 1,000,000. 1% is equal to 10,000 PPM, and 10% is equal to 100,000 PPM.

Princeton architecture Computer processor architecture that uses one memory subsystem for instructions (control store) memory, variable memory, and I/O registers. See *Harvard architecture* and *Von Neumann.*

program counter A counter within a computer processor that keeps track of the current program execution location. This counter can be updated by the counter and have its contents saved/restored on a stack.

programming Loading a program into a microcontroller control store, also referred to as *burning.*

program store A.k.a. *program storage.* Memory (usually non-volatile) devoted to saving the application program for when the microcontroller is powered down. Also known as *control store.*

PROM Programmable Read-Only Memory. Originally an array of fuses that were blown to load in a program. Now PROM can refer to EPROM memory in an OTP package.

PTH Pin Through Hole. Technology in which the pins of a chip are inserted into holes drilled into a FR4 printed circuit card before soldering.

pull-down A resistor (typically 100 Ω to 500 Ω) that is wired between a microcontroller pin and Ground. See *pull-up.*

pull-up A resistor (typically 1-KΩ to 10-KΩ) that is wired between a microcontroller pin and Vcc. A switch pulling the signal at the microprocessor pin can be used to provide user input. See *pull-down.*

push The operation of putting data onto a stack memory.

PWB Printed Wiring Board. See *raw card.*

PWM Pulse Width Modulation. A digital output technique where a single line is used to output analog information by varying the length of time a pulse is active on the line.

RAM Random-Access Memory. Memory that you can write to and read from. In microcontrollers, virtually all RAM is *static RAM* (SRAM), which means that data is stored within it as long as power is supplied to the circuit. *Dynamic RAM* (DRAM) is very rarely used in microcontroller applications. EEPROM can be used for non-volatile RAM storage.

raw card Fiberglass board with copper traces attached to it that allow components to be interconnected. Also known as *PCB*, *PWA*, and *bare board.*

RC Resistor/capacitor network used to provide a specific delay for a built-in oscillator or reset circuit.

receiver A device that senses the logic level in a circuit. A receiver cannot drive a signal.

recursion A programming technique where a subroutine calls itself with modified parameters to carry out a task. This technique is not recommended for microcontrollers that might have a limited stack. (An old joke for defining recursion in a glossary is: "recursion: See *recursion*."

register A memory address devoted to saving a value (like RAM) or providing a hardware interface for the processor.

relocatable Code written or compiled in such a way that it can be placed anywhere in the control store memory map after assembly and run without any problems.

resistor A device used to limit current in a circuit.

resistor ladder A circuit that consists of a number of resistors that can be selected to provide varying voltage divider circuits and output differing analog voltages.

reset Placing a microcontroller in a known state before allowing it to execute.

RISC Reduced Instruction Set Computer. This is a philosophy in which the operation of a computer is sped up by reducing the operations performed by a processor to the absolute minimum for application execution and making all resources accessible by a consistent interface. The advantages of RISC include faster execution time and a smaller instruction set. See *CISC*.

ROM Read-Only Memory. This type of memory is typically used for control store because it cannot be changed by a processor during the execution of an application. Mask programmable ROM is specified by the chip manufacturer to build devices with specific software as part of the device and cannot be programmed in the field.

rotate A method of moving bits within single or multiple registers. No matter how many times a rotate operation or instruction is carried out, the data in the registers will not be lost. See *shift*.

RS-232 An asynchronous serial communications standard. Normal logic levels for a "1" is –12 V and for a "0", +12 V.

RTOS Real-Time Operating System. A program that controls the operation of an application.

scan The act of reading through a row of matrix information for data rather than interpreting the data as a complete unit.

serial Passing multiple bits using a serial line one at a time. See *parallel*.

servo A device that converts an electrical signal into mechanical movement. Radio-control modeller's servos are often interfaced to microcontrollers. In these devices, the position is specified by a 1- to 2-msec pulse every 20 msec.

shift A method of moving bits within a single or multiple registers. After a shift operation, bits are lost. See *rotate*.

simulator A program used to debug applications by simulating the operation of the microcontroller.

slave In microcontroller networking, a device that does not initiate communications, but does respond to the instructions of a master.

SMT Surface Mount Technology (a.k.a. SMD). Technology in which the pins of a chip are soldered to the surface of a printed circuit card.

SPI A synchronous serial communications protocol.

Splat Asterisk ("*"). Easier to say, spell, and funnier than "asterisk."

SRAM Static Random-Access Memory. A memory array that will not loose its contents while power is applied.

stack LIFO memory used to store program counter and other context register information.

stack pointer An index register available within a processor that is used for storing data and updating itself to allow the next operation to be carried out with the index pointing to a new location.

state analyzer A tool used to store and display state data on several lines. Rather than requiring a separate instrument, this is often an option available in many logic analyzers.

state machine A programming technique that uses external conditions and state variables for determining how a program is to execute.

string Series of ASCII characters saved sequentially in memory. When ended with 0x000 to note the end of the string, it is known as an *ASCIIZ string.*

subroutines A small application program devoted to carrying out one task or operation. Usually called repeatedly by other subroutines or the application mainline.

synchronous serial Data transmitted serially along with a clocking signal that is used by the receiver to indicate when the incoming data is valid.

task A small, autonomous application, similar in operation to a subroutine, but can execute autonomously to other application tasks or mainline.

timer A counter incremented by either an internal or external source. Often used to time events, rather than counting instruction cycles.

traces Electrical signal paths etched in copper in a printed circuit card.

transistor An electronic device by which current flow can be controlled.

two's complement A method for representing positive and negative numbers in a digital system. To convert a number to a two's complement negative, it is complemented (converted to one's complement) and incremented.

UART Universal Asynchronous Receiver/Transmitter. Peripheral hardware inside a microcontroller used to asynchronously communicate with external devices. See *USART* and *asynchronous serial.*

USART Universal Synchronous/Asynchronous Receiver/Transmitter. Peripheral hardware inside a microcontroller used to synchronously (using a clock signal either produced by the microcontroller or provided externally) or asynchronously communicate with external devices. See *UART* and *synchronous serial.*

μsec One millionth of a second (0.000001 seconds). See *nsec* and *msec.*

UV light Ultra-violet light. Light at shorter wavelengths than the human eye can see. UV light sources are often used with windowed microcontrollers with EPROM control store for erasing the contents of the control store.

variable A label used in an application program that represents an address that contains the actual value to be used by the operation or instruction. Variables are normally located in RAM and can be read from or written to by a program.

Vcc Positive power voltage applied to a microcontroller/circuit. Generally 2.0 V to 6.0 V, depending on the application. Also known as *Vdd*.

Vdd See *Vcc*.

vertical synch A signal used by a monitor or TV set to determine when to start displaying a new screen (field) of data.

vias Holes in a printed circuit card.

volatile RAM is considered to be volatile because, when power is removed, the contents are lost. EPROM, EEPROM, and PROM are considered to be non-volatile because the values stored in the memory are saved, even if power is removed.

voltage The amount of electrical force placed on a charge.

voltage regulators A circuit used to convert a supply voltage into a level useful for a circuit or microcontroller.

volts Unit of voltage.

Von Neumann Chief scientist responsible for the Princeton architecture.

Vss See *ground*.

wait states Extra time added to an external memory read or write.

watchdog timer Timer used to reset a microcontroller upon overflow. The purpose of the watchdog timer is to return the microcontroller to a known state if the program begins to run errantly (or amok).

wattage Measure of power consumed. If a device requires 1 Amp of current with a 1-V drop, 1 Watt of power is being consumed.

word The basic data size used by a processor. In the 8051, the word size is 8 bits.

XA An enhanced 8051 architecture made available by Philips Semiconductor.

XOR A logic gate that outputs a "1" when the inputs are at different logic levels.

ZIF Zero Insertion Force. ZIF sockets will allow the plugging/unplugging of devices without placing stress upon the device's pins.

.ZIP files Files combined together and compressed into a single file using the PKZIP program by PKWARE, Inc.

B

8051 RESOURCES

This book is not intended to be the ultimate reference on the 8051 and its derivatives. It is meant to be an introduction to the 8051 as well as a resource to help you with your applications and give you a few ideas of what's possible with the 8051. In this book, I have really only presented two versions of the 8051 to you, which is actually a minuscule fraction of the total devices available. In this appendix, I have tried to list a number of resources that you can use for getting additional information on parts and applications.

You should note that I have tried to use the Internet as a primary source for information wherever possible because it is becoming a very comprehensive technical library for electronic information. If you don't currently have an Internet account with both e-mail and World Wide Web access, I highly recommend that you do so immediately. The Internet will give you access to a substantial amount of information in a very short period of time.

While I have done everything to ensure that addresses, phone numbers, and Internet URLs are accurate, this information can change without notice. If you are unable to locate a company, you can use search engines such as AltaVista to find the new information.

Contacting the Author

I look forward to hearing from you, and I can be reached either through e-mail (emailme@myke.com) or through my Web page at:

```
http://www.myke.com
```

However, for technical questions or suggestions, please contact me through the 8051 listservers rather than e-mailing me directly. By using the listservers, you can get a reply back very quickly (often in a matter of minutes). More importantly, others that might need the same information will get it distributed to them automatically.

8051 Suppliers

The following six companies currently supply 8051s. I have included the company name along with their Web page address and how to contact them using more traditional means:

Intel Corporation, Santa Clara
2200 Mission College Blvd.
Santa Clara, CA 95052-8119
USA
Tel: 408-765-8080
Fax: 408-765-9904
Web: http://developer.intel.com

Atmel Corporate Headquarters
2325 Orchard Parkway
San Jose, CA 95131
Tel: 408-441-0311
BBS: 408-436-4309
Web: http://www.atmel.com

Dallas Semiconductor
4100 Spring Valley Road
Suite 302
Dallas, TX 75244
Tel: 972-788-2197
Fax: 972-980-4290
Web: http://www.dalsemi.com

Integrated Silicon Solution, Inc.
2231 Lawson Lane
Santa Clara, CA 95054
Tel: 408-588-0800
Fax: 408-588-0805
Web: http://www.issiusa.com/month.html

Philips Semiconductors
811 East Arques Avenue
P.O. Box 3409
Sunnyvale, CA 94088-3409
Tel: 800-234-7381
Fax: 800-943-0087
Web: http://www.philipsmcu.com

Siemens Microelectronics
10950 North Tantau Avenue
Cupertino, CA 95014
Web: http://www.siemens.de/Semiconductor/products/ICs/34/mc_home.html

Part Suppliers

The following companies supplied components that were used in this book:

DIGI-KEY

Digi-Key is an excellent source for a wide range of electronic parts. Their parts are reasonably priced, and most orders will be delivered the next day. They are real lifesavers when you're on a deadline.

Digi-Key Corporation
701 Brooks Avenue South
P.O. Box 677
Thief River Falls, MN 56701-0677
Tel: 800-344-4539 (800-DIGI-KEY)
Fax: 218-681-3380
Web: http://www.digi-key.com/

AP CIRCUITS

AP Circuits will build prototype bare boards from your gerber files. Boards are available within three days. I have been a customer of theirs for several years, and they have always produced excellent quality and been helpful in providing direction to learning how to develop my own bare boards. Their Web site contains all the tools necessary to develop your own gerber files.

Alberta Printed Circuits Ltd.
#3, 1112-40th Avenue N.E.
Calgary, Alberta
T2E 5T8
Tel: 403-250-3406
BBS: 403-291-9342
Web: http://www.apcircuits.com/
E-mail: staff@apcircuits.com

WIRZ ELECTRONICS

Wirz Electronics is a full service microcontroller component and development system supplier. Wirz Electronics is the main distributor for projects contained in this book (the AT89C*x*051 programmer, the 51Bot board, and others) and will sell packaged kits of the projects. Wirz Electronics also carries the SimmStick prototyping systems as well as their own line of motor and robot controllers. (The toll-free number is for USA and Canada only.)

Wirz Electronics
P.O. Box 457
Littleton, MA 01460-0457
Tel: 888-289-9479 (888-BUY-WIRZ)
Web: http://www.wirz.com/
E-mail: sales@wirz.com

TOWER HOBBIES

Tower Hobbies is an excellent source for servos and R/C parts useful in home-built robots. (The toll-free numbers are for USA and Canada only.)

Tower Hobbies
P.O. Box 9078
Champaign, IL 61826-9078
Tel: 217-398-3636
Fax: 217-356-6608
800-637-7303
Ordering: 800-637-4989
Support: 800-637-6050
Web: http://www.towerhobbies.com/
E-mail: orders@towerhobbies.com

JDR

JDR supplies components, PC parts/accessories, and hard-to-find connectors. (The toll-free numbers are for USA and Canada only.)

JDR Microdevices
1850 South 10th St.
San Jose, CA 95112-4108
Tel: 408-494-1400
800-538-500
Fax: 800-538-5005
BBS: (408)494-1430
Web: http://www.jdr.com/JDR
e-mail: techsupport@jdr.com
Compuserve: 70007,1561

NEWARK

Newark sells components, including the Dallas line of semiconductors (the DS87C520, 051820, and DS275 presented in this book). (The toll-free number is for USA and Canada only.)

Tel: 800-463-9275 (800-4-NEWARK)
Web: http://www.newark.com/

MARSHALL INDUSTRIES

Marshall is a full-service distributor of Philips microcontrollers as well as other parts.

Marshall Industries
9320 Telstar Avenue
El Monte, CA 91731
Tel: 800-833-9910
Web: http://www.marshall.com

MONDO-TRONICS ROBOTICS STORE

Mondo-Tronics Robotics Stores is the self-proclaimed "World's Biggest Collection of Miniature Robots and Supplies." I have to agree with them. This is a great source for servos, tracked vehicles, and robot arms. (The toll-free number is for USA and Canada only.)

Order Desk
Mondo-Tronics Inc.
524 San Anselmo Ave #107-13
San Anselmo, CA 94960
Tel: 800-374-5764
Fax: 415-455-9333
Web: http://www.robotstore.com/

Periodicals

Here are a number of magazines that do give a lot of information and projects on microcontrollers. Every month, each magazine has a better than 90% chance of presenting at least one microcontroller application.

CIRCUIT CELLAR INK

Subscriptions:
P.O. Box 698
Holmes, PA 19043-9613
Tel: 800-269-6301
BBS: 860-871-1988
Web: http://www.circellar.com/

GERNSBACK PUBLICATIONS

- Electronics Now
- Popular Electronics

Subscriptions:
Subscription Department
P.O. Box 55115
Boulder, CO 80323
Tel: 800-999-7139
Web: http://www.gernsback.com

MICROCONTROLLER JOURNAL

Web: http://www.mcjournal.com/
This is published on the Web.

NUTS & VOLTS

Subscriptions:
430 Princeland Court
Corona, CA 91719
Tel: 800-783-4624
Web: http://www.nutsvolts.com

EVERYDAY PRACTICAL ELECTRONICS

Subscriptions:
EPE Subscriptions Dept.
Allen House, East Borough
Wimborne, Dorset
BH21 1PF
United Kingdom
Tel: +44 (0)1202 881749
Web: http://www.epemag.wimborne.co.uk

Web Resources

The World Wide Web is becoming more and more pervasive in our society, especially for engineers and technicians who rely upon the Web for data sheets, FAQs (answers to "Frequently Asked Questions"), and example applications. Unfortunately, due to the transitory nature of the Internet and the time between when I write this and when this book is published, some of these sites might no longer exist when you are looking for them.

If you have trouble finding any of these sites or are unable to locate information you are looking for, I highly recommend using AltaVista at:

```
http://www.altavista.digital.com
```

This site is capable of accessing and searching literally millions of Web sites all over the world for your search parameters.

MY FAVORITE 8051 WEB SITES

One of the first things that I should present on the 8051 is Russ Hersch's excellent 8051 FAQ (Frequently Asked Questions) which is available at:

```
ftp://rtfm.mit.edu/pub/usenet/comp.answers/microcontroller-faq/8051
```

This 50+-page document is an excellent source for all kinds of information about the 8051, including device and tool vendors.

For general information Web sites, here are my top nine sites:

1. `http://www.8052.com`—8051 Web site provided by Vault Information Service. An excellent 8051 tutorial is provided along with a Dallas Semiconductor HSM microcontroller tutorial and a code library.
2. `http://www.labyrinth.net.au/steve/8051.html`—Good basic resource page. Has PaulMon on it along with the AT89C2051 programmer
3. `http://www.labyrinth.net.au/steve/8051.html`—Good resource page. Was somebody asking about `dec DPTR`?
4. `http://www.ece.orst.edu/paul/8051-goodies/goodies-index.html`—Home of PaulMon and other 8051 resources.
5. `http://www.eg3.com/embe/gatox805.htm`—Data book references and page references.
6. `http://www.iotasys.com/`—Software library for assemblers/disassemblers/compilers.
7. `http://www.keil.com/c51/index.html`—Software development house with a free evaluation kit.
8. `http://www.tu-bs.de/studenten/akafunk/pr8051/`—8051 tools/software for Packet Radio.
9. `http://www.ece.orst.edu/serv/8051/`—Another site with PaulMon tools along with other hobbyist tools.

SOME WEB SITES OF INTEREST

While none of these are 8051-specific, they are a good source of ideas, information, and products that will make your life a bit more interesting and maybe give you some ideas for projects for the 8051.

SEATTLE ROBOTICS SOCIETY

The Seattle Robotics Society (http://www.hhhh.org/srs/) has a lot of information on interfacing digital devices to such real-world devices as motors, sensors, and servos. They also do a lot of neat things. Most of the applications use the Motorola 68HC11.

LIST OF STAMP APPLICATIONS (L.O.S.A.)

The List Of Parallax Basic Stamp Applications (http://www.hth.com/losa.htm) will give you an idea of what can be done with the Basic Stamp (and other microcontrollers, such as the 8051). The list contains projects ranging from using a Basic Stamp to giving a cat medication to providing a simple telemetry system for model rockets.

ADOBE PDF VIEWERS

Adobe (http://www.adobe.com) .PDF file format is used for virtually all vendor data sheets, including the devices presented in this book (and their data sheets on the CD-ROM).

HARDWARE FAQS

A set of FAQs (Frequently Asked Questions) about the PC and other hardware platforms (http:paranoia.com/filipg/HTML/LINK/LINK_IN.html) will come in handy when interfacing the 8051 to a Host.

LIST SERVERS

If you're not familiar with listservers, I highly recommend that you join the ones that I have listed in this Appendix. A listserver is an e-mail address that takes mail sent to it and redistributes it to a number of people. This means that a large amount of mail can be sent and received in a short period of time to a large number of people.

Listservers are really wonderful things that make it possible to get answers to questions literally within minutes after posing them (although hours afterward is probably more typical). There are a great deal of very knowledgeable people who can answer questions on a variety of subjects (and have opinions on more than just the subject at hand).

Having a great deal of people available makes the list essentially a community. This is a worldwide community, and you have to try to be sensitive to different people's feelings and cultures. To this end, I have created the following set of guidelines for listservers. The genesis of these guidelines came about after MIT's PICList listserver went through a period of time of problems with the same type of messages coming through over and over.

I think these are pretty good guidelines for any listserver, and I suggest that you try to follow them as much as possible to avoid getting into embarrassing situations (and having yourself berated or removed from a list because you made a gaff).

1. Don't subscribe to a list and then immediately start sending questions to the list. Instead, wait a day or so to get the hang of how messages are sent and replied to on the list and get a feel for the best way of asking questions.
2. Some lists resend a note sent to it (while others do not). If you receive a copy of your first message, don't automatically think that it is a bounce (wrong address) and resend it. In this case, you might want to wait a day or so to see if any replies show up before trying to resend it. Once you've been on the list for a while, you should get an idea of how long it takes to show up on the list and how long it takes to get a reply.
3. If you don't get a reply to a request and don't see it referenced in the list, don't get angry or frustrated and send off a reply demanding help. There's a chance that nobody on the list knows exactly how to solve your problem, or there's a problem with your mail system or the list's and your note didn't get distributed. In this case, try to break down the problem, ask the question a different way, and ask for somebody to send you a private note indicating that they have read your posting.
4. I've talked about being able to get replies within minutes in this appendix, please don't feel that this is something that you can count on. Nobody on any of the lists that I've given in this book are paid to reply to your questions. The majority of people who reply are doing so to help others. Please respect that and don't badger. Help out in anyway that you can.
5. If you are changing the "Subject" line of a post, please reference the previous topic (i.e., put in "was: ..."). This will help others keep track of the conversation.
6. When replying to a previous post, try to minimize how much of the previous note is copied in your note and maximize the relevance to your reply. This is not to say that

none of the message should be copied or referenced. There is a very fine balance between having too much and too little. The sender of the note that you are replying to should be referenced (with their name or ID).

My rule of thumb is, if the original question is less than 10 lines, I copy it all. If it is longer, then I cut it down (identifying what was cut out with a "SNIP" message), leaving just the question and any relevant information as quoted. Most mail programs will mark the quoted text with a ">" character. Please use this convention to make it easier for others to follow your reply.

7 If you have a program that doesn't work, please don't copy the entire source into a note and then post it to a list. As soon as I see a note like this, I just delete it and go on to the next one (and I suspect that I'm not the only one). Also, some lists might have a message size limit (anything above this limit is thrown out), and you will not receive any kind of confirmation.

If you are going to post source code, keep it short. People on the list are more than happy and willing to answer specific questions, but simply copying the complete source code in the note and asking a question like "Why won't the LCD display anything?" really isn't useful for anybody. Instead, try to isolate the failing code and describe what is actually happening along with what you want to happen. If you do this, chances are that you will get a helpful answer quickly.

A good thing to remember when asking why something won't work is to make sure that you discuss the hardware that you are using. If you are asking about support hardware (i.e., a programmer or emulator), make sure you describe your PC (or workstation) setup. If your application isn't working as expected, describe the hardware that you are using and what you have observed (e.g., if the clock lines are wiggling, or the application works normally when you put a scope probe on a pin).

8 You might find a totally awesome and appropriate Web page and want to share it with the list. Please make it easier on the people in the list to cut and paste the URL by putting it on a line all by itself in the format:

```
http://www.awesome-8051-page.com
```

9 If you have a new application, graphic, or whatever that takes up more than 1K that you would like to share with everyone on the list, please don't send it as an attachment in a note to the list. Instead, either indicate that you are have this amazing piece of work and tell people that you have it and where to request it (either to you directly or to a Web server address). If a large file is received, many listservers might automatically delete it (throw it into the "bit bucket"), and you might or might not get a message telling you what happened.

If you don't have a Web page of your own or one you can access, requesting somebody to put it on their Web page or FTP server is a good alternative.

10 Many of these listservers are made available, maintained, and/or moderated by the device's manufacturer. Keep this in mind if you are going to advertise your own product, and understand what the company's policy on this is before sending out an advertisement.

11 Putting job postings or employment requests *might* be appropriate for a list (like the previous point, check with the list's maintainer). However, I don't recommend that the rate of pay or conditions of employment should be included in the note (unless you want to be characterized as cheap, greedy, unreasonable, or exploitive).

12 Spams are sent to every listserver occasionally. Please do not reply to the note even if the message says that to get off the spammer's mailing list just reply. This will send a message to everyone in the list. If you must send a note detailing your disgust, send it to the spam originator (although to their ISP will probably get better results).

Note: There are a number of companies sending out bogus spams to collect the originating addresses of replying messages and sell them to other companies or distributors of addresses on CD-ROM. When receiving a spam, see if it has been sent to you personally or to the list before replying. However, beware if you are replying to the spam. You might just be sending your e-mail address to some company to sell to other spammers.

I know it's frustrating and, like everyone else, I'm sure you would like to have all spammers eviscerated and then taken out and shot. However, if you want to minimize how much you are bothered in by spams in the future, you just have to ignore all the spams that are sent to you. Eventually, if you don't reply, you will be dropped off from the lists that are bought, sold, and traded.

13 Off-topic messages, while tolerated, will probably bring lots of abuse upon you. If you feel it is appropriate to send an off-topic message, some lists request that you put "[OT]" in the subject line. Some members of the list use mail filters, and this will allow them to ignore the off-topic posts automagically.

Eventually a discussion (this usually happens with off-topic discussions) will get so strung out that there are only two people left arguing with each other. At this point, stop the discussion entirely or go private. You can obtain the other person's e-mail address from the header of the message. Send your message to him/her and not to the entire list. Everyone else on the list would have lost interest a long time ago and probably would like the discussion to just go away (so oblige them).

14 Posts referencing pirate sites and sources for cracked or hacked software is not appropriate in any case and may be illegal. If you are not sure if it is okay to post the latest software that you've found on the 'Net, then *don't* until you have checked with the owners of the software and gotten their permission. It would also be a good idea to indicate in your post that you have the owner's permission to distribute cracked software.

A variety of different microcontrollers are used in smart cards (such as used with cable and satellite scrambling), and asking how they work will probably result in abusive replies or having your questions ignored. If you have a legitimate reason for asking about smart cards, make sure you state it in your e-mail to the list.

Offering to read protected microcontrollers is also inappropriate for listservers. Many of the members of the list earn their livings from developing microcontroller applications; few will favorably respond to offers to read data from protected devices.

15 When you first subscribe to a list, you will get a reply telling you how to unsubscribe from the list. *Don't lose this note.* In the past, in some lists, people having trouble unsubscribing have sent questions to the list asking how and sometimes getting angry when their requests go unheeded. If you are trying to unsubscribe from a list and need help from others on the list, explain what you are trying to do and how you've tried to accomplish it.

16 If you're like me and just log on once or twice a day, read all the notes regarding a specific thread before replying. When replying to a question that has already been answered look for what you can add to the discussion, not reiterate what's already been said.

17 Lastly, please try to be courteous to all on the list. Others might not have *your* knowledge and experience, or they might be sensitive about different issues. There is a very high level of professionalism on all lists; please help maintain it.

Being insulting or rude will only get the same back and probably have your posts and legitimate questions ignored in the future by others on the list who don't want to have anything to do with you.

To put this succinctly: Don't be offensive or easily offended.

Philips has provided a moderated listserver for discussions on 8051s in general and the Philips versions in particular. Subscribing to the list is accomplished by filling out the form at:

```
http://www.philipsmcu.com/join.html
```

Intel also has a newsgroup set up for the MCS-51 at:

```
http://www.intel.com/newsgroups/mcontrol.htm
```

Interface Products has made an Atmel listserver available. At the time of writing, this listserver has about 5 to 10 messages per day on average and focuses on all Atmel microcontroller products (both the 8051 and AVR).

To subscribe, send a note to:

```
atmel-request@pic.co.za
```

with the word "JOIN" in the body of the message.

Consultants and Product Suppliers

Here are a number of companies that sell 8051 products. I have broken up the products/services into specific categories with full company information at the end of this section. Please note that information about companies might have changed since this was written.

- Programmers
  ~IBERCOMP S.A.
  ~Programmed Scientific Instruments, Inc.
  ~Systronix Inc.
  ~WF AUTOMACAO IND. COM. SERV. LTDA M.E
- Development tools
  ~IBERCOMP S.A.
  ~Programmed Scientific Instruments, Inc.
  ~Systronix Inc.
  ~Signum Systems Corp.
  ~Advanced Graphic Systems
  ~J & M Microtek, Inc.

- Compilers
  - ~Byte Craft Limited
  - ~IBERCOMP S.A.
  - ~J & M Microtek, Inc.
  - ~Programmed Scientific Instruments, Inc.
  - ~Signum Systems Corp.
  - ~Systronix Inc.
- Emulators
  - ~Display Electronics
  - ~Programmed Scientific Instruments, Inc.
  - ~Signum Systems Corp.
- Publications
  - ~ *Byte Craft Limited*—Quarterly newsletter
  - ~Systronx Inc. X Publications—*Micro-controller Idea Book* and *BASIC-52 Programming*, including example disks for each. Many online data sheets also, including mechanical drawings.
- Consultants
  - ~Doss Development Corporation
  - ~J & M Microtek, Inc.
  - ~Programmed Scientific Instruments, Inc.
  - ~Radix, Inc.
  - ~Spectrum Engineering
  - ~Steward Electronics Ltd.
  - ~Systronix Inc.
  - ~TLA Microsystems Ltd.
  - ~Universal Solution Technology

ADVANCED GRAPHIC SYSTEMS

Advanced Graphic Systems has provided high-quality industrial controls to the newspaper printing business for over 20 years, specializing in tension control and paster timing mechanisms. They have used the 8051 and or one of its derivatives in many an application and are currently using them in several systems. They have recently developed a single-board microcontroller using an 87C52 programmed with a Basic-52 interpreter for use as a learning/development tool. See their Web site for a more detailed description.

4055 Grass Valley Hwy.
Ste. #103
Auburn, CA 95602
USA
Tel: 530-268-3291
Fax: 530-268-0116
Web: http://www.ags-gv.com
e-mail: ags@gv.net

BYTE CRAFT LIMITED

Byte Craft Limited specializes in embedded systems software development tools for single-chip microcontrollers. The company excels in product design for diverse architectures, providing innovative solutions for developers, consultants, and manufacturers in various industries around the world. The success of the Byte Craft product line stems both from experience within the 8-bit market and commitment to customers' needs.

Byte Craft Limited products are developed to minimize product time to market. Their products include:

- C8051 Code Development System
- C Cross Compiler
- BCLINK linker
- BCLIDE Integrated Development Environment and the built-in macro assembler

The C8051 is offered for Windows 95/NT, DOS, HP-UX and Sun Solaris.

421 King Street North
Waterloo, Ontario
N2J 4E4 Canada
Tel: 519-888-6911
Fax: 519-746-6751
Web: http://www.bytecraft.com E
e-mail: info@bytecraft.com

CMX COMPANY

CMX Company develops and sells real-time multitasking operating system (RTOSs) products supporting *most* 8-, 16-, 32-, and 64-bit embedded microcontrollers and microprocessors. Supplied with *all* source code and *no* royalties. Over 25 compiler vendors are supported.

CMX enhances its RTOS with optional networking packages, such as Control Area Network (CAN) and TCP/IP.

CMX is also a distributor of compilers/assembler/IDEs, simulators, and debuggers, providing a complete solution for the engineer.

680 Worcester Road
Framingham, MA 01702
USA
Tel: 508-872-7675
Fax: 508-620-6828
Web: http://www.cmx.com
e-mail: cmx@cmx.com

COMTEC

Comtec sells electronic components (Atmel, PIC, and Intel microcontrollers) for retail and wholesale.

Manyar Keretoarjo 8/53
Surabaya
Jawa-Timur
60285 Indonesia
Tel: (62)31-5946720
Fax: (62)31-5940881
e-mail: comtec@indo.net.id

DISPLAY ELECTRONICS

Display Electronics specializes in surplus electronic equipment and components and always has good stocks of emulators, programmers, and many semiconductor devices, including the 8051 family and derivatives.

A full mail-order service and retail outlet, Display Electronics offers competitive prices to the hobbyist and low-volume user.

32 Biggin Way
Upper Norwood
London, England
SE19 3XF
United Kingdom
Tel: [44] 0181 653 3333
Fax: [44] 0181 653 8888
Web: http://www.distel.co.uk
e-mail: sales@distel.co.uk

DOSS DEVELOPMENT CORPORATION

Doss Development Corporation (established in 1990) is an electronic R&D firm that specializes in embedded control applications. Previous projects include products in process control, scientific instrumentation, and high-speed miniature data collection devices. Services include new product design, prototyping, PCB design (Pads PowerPCB), industrial packaging (SolidWorks), software/firmware development, and production manufacturing services.

11 Walnut Springs Dr,
Defiance, MI 63341-2815
USA
Tel: 314-987-3100
Fax: 314-987-3014
Web: http://www.doss.com
e-mail: info@doss.com

IBERCOMP S.A.

Ibercomp is an engineering company, specializing in developing custom microcontroller solutions. They also produce their own evaluation boards based on 32, 320, 535, and 535 (currently are developing a 552 board) and two Basic programmable PLCs, one based on the Zilog Z86E08 and one in 87C520. They have 10 years experience and are well accepted in industry and universities. They are studying the possibility of translating all of their documentation to English.

Currently, they distribute their products in Spain, Portugal, Andorra, Argentina, Chile, Peru, Uruguay, and Colombia.

C/Parc #8 (Bajos)
Palma De Mallorca
Islas Baleares
E07014
Spain
Tel: 971-456-642
Fax: 971-456-758
Web: http://www.ibercomp.es
e-mail: ibercomp@atlas-iap.es

J & M MICROTEK, INC.

J & M Microtek, Inc. has been manufacturing 8051-related products since 1990. Their 80C31 and 80C552 (80C51 core processor) development systems come complete with assembler, disassembler, simulator, and onboard debugger. The whole development environment for these two systems aims to ease the debugging cycles for the design engineers. The training courses included for the 80C552 development system provide the beginners the best way to learn 8051.

83 Seaman Road
West Orange, NJ 07052
USA
Tel: 973-325-1892
Fax: 973-736-4567
Web: http://www.jm-micro.com
e-mail: sales@jm-micro.com

PROGRAMMED SCIENTIFIC INSTRUMENTS, INC.

PSI is a consulting firm that has been in business since 1978, providing 8051 family hardware and software solutions, and has developed very high performance software and hardware tools to aid in the development of embedded applications. The ACORN system implements structured assemblers for the target processor that gives the programmer all the advantages of assembly language programming coupled with the advantages of high-level language program flow structures (`if`, `loop`, and `case` statements). The ACORN emulators connect to the PROM socket of the target system and provide breakpoints as well as full visibility into the target.

1031 Monte Verde Drive
Arcadia, CA 91007-6004
USA
Tel: 626-446-6315
Fax: 626-446-9561
e-mail: suchter@alumni.caltech.edu

RADIX, INC.

Radix designs industrial instrumentation products for manufacture. Specialties include mixed analog/digital circuit design, high-level language usage, real-time operating systems, and Windows CE integration, both hardware and software. Radix can also help with complete end-to-end design solutions, including packaging and PCB layout. They are extremely well acquainted with the 80c31, numerous variants, and the Keil C51 compiler and RTX51 real-time kernel.

P.O. Box 897
Clovis, CA 93613
USA
Tel: 209-297-9000
Fax: 209-297-9400
Web: http://rdx.com
e-mail: james@radixgroup.com

SIGNUM SYSTEMS CORP.

Signum Systems (est. 1980) is a leading manufacturer of in-circuit emulators (ICE) and supplier of other development tools for embedded markets. The products range from development kits, assemblers, and C compilers to ICEs with complex breakpoints, real-time trace buffers, and performance analyzers. Currently, Signum Systems supports the 8051, 80186, 80196, 8085, Z8, Z80, TMS320C1*x*, TMS320C2*x*, TMS320C5*x*, HPC, CompactRISC, and PIC families.

11992 Challenger Court
Moorpark, CA 93021
USA
Tel: 805-523-9774
Fax: 805-523-9776
Web: http://www.signum.com
e-mail: sales@signum.com

SPECTRUM ENGINEERING

Spectrum Engineering has more than 10 years of experience designing and programming embedded systems based on the 8051 and/or its derivatives. Their specialties include interfacing point-of-sale equipment (bar code scanners, keyboards, and printers) to various point-of-sale terminals and PCs. Other areas of interest include small dc motor control and temperature measurement.

3715 140th St Ct NW
Gig Harbor, WA 98332
USA
Tel: 253-853-5523
Fax: 253-853-5524
Web: http://www.speceng.com
e-mail: info@speceng.com

STEWARD ELECTRONICS LTD.

Steward Electronics are consultants capable of hardware and software development. They have experience in industrial control and telemetry, industrial fieldbus interfaces including DeviceNet, and high-performance automotive electronics. They are capable of complete microprocessor-based products, including assembler programming for 8051, 8086, 1802, and 6800 processors along with digital circuit design, analog circuit design, switching supply design, etc. PCB Layout PTH and SMD boards design is also done with arranging for production and electronic testing.

25 Treeway
Pakuranga, Auckland 1706
New Zealand
Tel: 64(25)960-278
Fax: 64(9)577-2249
Web: http://www.voyager.co.nz/dsteward/
e-mail: steward.electronics@clear.net.nz

SYSTRONIX INC.

Systronix offers off-the-shelf systems solutions for rapid prototyping of embedded control systems. They are a one-stop shop for development boards and all the accessories that plug into them: keypads, LCDs, data converters, enclosures, etc. They provide the I/O drivers for all peripheral devices and sample BASIC programs so that you can get a headstart on your application.

555 South 300 East
Salt Lake City, UT 84111
USA
Tel: 801-534-1017
Fax: 801-534-1019
Web: http://www.systronix.com
e-mail: info@systronix.com

TLA MICROSYSTEMS LTD.

TLA Microsystems is a small consultancy providing hardware and software design for microcontroller-based products, both locally and internationally (50% of TLA's work is via the Internet). Their president, Steve Baldwin, spent four years as an Applications and Design Engineer for Philips, working on development systems for the 8051 family. So he truly knows them both inside and out. Another eight years in the manufacturing sector means that TLA can provide a cost-effective and manufacturable solution for small-, medium-, or large-volume products.

16 Ulster Road
Blockhouse Bay, Aukland, 1007
New Zealand
Tel: +64 9 820-2221
Fax: +64 9 820-1929
Web: http://www.tla.co.nz
e-mail: steveb@tla.co.nz

UNIVERSAL SOLUTION TECHNOLOGY

UST is a company dedicated to develop special equipment upon other large distribution companies request. They mostly develop under the 8051 platform. Their common products includes industrial high-accuracy calibrators, voice modules, monitoring, phone voice modules for remote access, and so on. As a small company, UST is able to deal with low quantities, small fees, and strange requests (one was developed to monitor pH and Orp level of swimming pools). More than 25 years of experience in microprocessors and instrumentation make UST as a reliable resource.

13438 Mallard Cove Blvd
Orlando, FL 32837-5314
USA
Tel: 407-816-1080
Fax: 408-648-2206
e-mail: wagnerl@ix.netcom.com

WF AUTOMACAO IND. COM. SERV. LTDA M.E.

WF AUTOMACAO develops educational products, using the 8032 and 80196. They also develop a low-cost programmer for the Atmel family: 89C1051, 89C2051, 89C51, 89C52, and 89C8252. WF AUTOMACAO has also developed a page for access to remote experiments with microcontrollers (specifically the 89C51).

Rua 2 De Setembro, 733
Blumenau, S.C 89052-000
Brasil
Tel: (55)47-3233598 R32
Fax: (55)47-3233710
Web: http://www.ambiente.com.br/bs/wf.htm
e-mail: wf@ambiente.com.br

C

16-BIT OPERATIONS

When I first got out of school and started working, a wag at the office declared, "The only problem with eight big microprocessors is that they can't handle 16-bit numbers." While not being terrifically funny, there is a lot of truth to this statement. 8-bit numbers are quite limiting (especially when they are used in two's complement format) for many applications. I've found that, for most applications, I have to go almost exclusively with 16-bit variables to give me the ranges that I require.

The 8051, having a 16-bit program counter and 16-bit data pointer (the DPTR register) interfaces with 16-bit address buses very easily. This makes the requirements for handling 16-bit integers even more important. If you've gone through chapters 10, 11, and 12, you'll see that I often use 16-bit variables in my applications, and they really aren't that hard to work with.

I felt it would be useful for you to concentrate exclusively on 16-bit variables and provide an appendix that you could use as a reference for declaring them and some examples of how operations can be written to manipulate them.

Declaring 16-bit Variables

Like 8-bit variables, 16-bit variables have to have their addresses declared in the application. As I have said elsewhere, variable declarations for the 8051 are a manual process, with the designer specifying where different variables are going to be located. Before you can start accessing these variables, you have to have a clear plan of what storage they will use that will not affect other memory (or hardware registers) in the 8051.

For 16-bit variables, I recommend placing them at scratchpad RAM address 0x030 or above simply to avoid any conflicts with the bank registers and variable flags. Each 16-bit variable will take up 2 bytes, so by placing it outside of where any single or 8-bit variables can be located, you will avoid any possible contention between these features of the 8051's processor.

I should point out that I do not use 16-bit flag variables. If I require more than eight flags (or addressable bits) for an application, I use two 8-bit flag variables in the 0x020 to 0x02F data range. Placing a 16-bit variable in the flag data range can lead to confusion later when individual bits of the variables are accessed along with the variables themselves.

This is one area where you should quickly decide upon a convention that you are going to use and stick with it. I personally like to just declare the high and low bytes explicitly with suffixes to help specify the two values:

```
SixteenBitLo EQU 030h    ; Define the "SixteenBit" variable
SixteenBitHi EQU 031h
```

I don't recommend defining 16-bit variables as a single variable:

```
SixteenBit EQU 030h      ; Define the "SixteenBit" variable
```

The reason for not recommending this is that you might accidentally define variables at the second byte location such as:

```
SixteenBit EQU 030h      ; Define the "SixteenBit" variable
EightBit EQU 031h        ; 8-bit variable declared in second byte of
                         ; "SixteenBit" variable
```

Note that I store the data in what I call Intel format (which is to say the low byte at the lowest address and the high byte at the high address). Another way of doing this is to save the data in Motorola format (which is high byte at low address and low byte at the high address). The reason for doing this is simply keeping 16-bit data consistent throughout the 8051 (special-function registers, such as the DPTR index register, put data in this format).

Intel format is a bit harder to read than the Motorola format. For example, if the 16-bit variable declared earlier had the contents 01234h in Intel format, when you looked at the data using UMPS, this would look like:

```
030h 34 12
```

while in Motorola Format, the data would look like:

```
030h 12 34
```

which is obviously easier to read when you are first starting out.

Personally, I don't see any tangible reasons for picking one data storage format over the other. I have worked with the PC's 80x86 processor for so long that using the Intel format is really second nature to me and I am able to convert the data in my head without any problems. If you are just starting out with the 8051 and doing 16-bit variables for the first time, you might want to start out with the Motorola format because it is much easier to understand what is being displayed.

The important thing with declaring 16-bit variables is to stay consistent. This will make it easier for you in the long run by avoiding having to remember how different variables are defined. When you're starting out, some 16-bit operations can get quite confusing, so keeping the variables in a format that you can easily follow which variables are being used is an advantage.

Incrementing and Decrementing

Incrementing a 16-bit number is accomplished by incrementing an 8-bit number and, if the result is equal to zero or causes the carry flag to be set, then incrementing another 8-bit number.

In C, this could be represented as:

```
if ( ++Low == 0 )        // Increment the low 8 bits
 High++;                 // If ++Low == 0 then increment high 8 bits
```

Directly translating this into 8051 assembler code, this would be:

```
 inc    Low             ; Increment the low 8 bits
 mov    A,Low           ; If ++Low != 0
 jnz    IncSkip         ; then skip over
 inc    High            ; Increment the high byte
IncSkip:
```

The problem with this code is that the accumulator is changed (and an extra instruction is put into load the accumulator with the `Low` value).

If this causes a problem in your application—by changing the accumulator, taking up one more byte than you have, or by taking one more instruction cycle that you can afford—you could put the 16-bit variable into the bank registers and execute:

```
 cjne  Rlow, #0FFh, IncSkip
 inc    Rhigh           ; If Low == 0FFh, then increment high
IncSkip:
 inc    RLow            ; Increment the low 8 bits
```

I do not recommend doing a 16-bit increment as an add of one to the low byte and using the carry flag to determine whether or not to increment the high byte:

```
 mov    A, Low          ; Add 1 to the low byte
 add    A, #1
 mov    Low, A
 jnc    IncSkip         ; If carry not set,
 inc    High            ;  jump over the increment the high byte
IncSkip:
```

In this code, not only does it take more cycles than either of the previous two and it changes the accumulator, it also changes the carry flag.

Decrementing a 16-bit value would be carried out somewhat differently from incrementing:

```
 mov    A, Low
 jnz    DecSkip         ; If "Low" != 0, then jump to low decrement
 dec    High            ;  Decrement the high byte
DecSkip:
 dec    Low
```

In this case, I check to see if the low byte is actually equal to 0. If it is (which means the result will be –1), I decrement the high byte.

If either time, space, or saving accumulator contents are critical, this code can be simplified the same way incrementing was, by using two bank registers:

```
 cjne  Rlow,#0,DecSkip   ; Are we going to carry decrement to the
 dec   Rhigh             ; high byte?
DecSkip:
 dec   Rlow              ; Decrement the low byte
```

Addition, Subtraction, and Bit Operations

Sixteen-bit adds and subtracts are a natural progression from 16-bit increments and decrements. Doing a 16-bit add or subtract operation is very straightforward in the 8051 due to its addition and subtraction instructions, which use the carry flag for allowing you to easily pass carry or borrow values to more significant bytes in multiple byte operations (such as 16-bit numbers).

I tend to think of the 16-bit add operation in very mechanical terms. The operation can be expressed as:

```
Result = VarA + VarB
```

Where all three variables are 16 bits, the base code, which can be used for all cases, is:

```
mov  A,VarA              ; Add the lower 8 bits first
add  A,VarB
mov  Result,A
mov  A,VarA+1            ; Add the upper 8 bits with carry
adc  A,VarB+1
mov  Result+1,A
```

The `add` instruction adds the contents of the accumulator with the instruction's parameter (`VarB`). The lower 8 bits of the result of this operation are stored back into the accumulator. If the result is greater than 0FFh (255 decimal), the carry flag is set.

After the lower 8-bit result is stored, the sum of the upper 8 bits is calculated using the `adc` instruction, which takes into account the contents of the accumulator, the instruction's parameter, and the carry flag and puts the result back into the accumulator and the carry flag. Using the add-with-carry instruction eliminates the need for jumping over an increment if the lower 8 bits of the sum *don't* result in a carry.

This six instruction base for 16-bit addition can be used for all types of 16-bit additions. For example, a constant to a variable:

```
VarA = VarA + Constant
```

would use the instructions:

```
mov  A,VarA                  ; Add the low 8 bits together
add  A,#(Constant & 0FFh)
mov  VarA,A
mov  A,VarA+1                ; Add the upper 8 bits together
```

```
adc  A,#(Constant SHR 8 )
mov  VarA+1,A
```

By simply substituting into the six base instructions, any 16-bit addition can be implemented.

Subtraction is handled in a very similar manner using the carry as a borrow flag:

```
Result = VarA - VarB
```

The 16-bit subtraction base code is:

```
mov  A,VarA                 ; Do the lower 8 bits
clr  C
subb A,VarB
mov  Result,A
mov  A,VarA+1               ; Do the upper 8 bits
subb A,VarB+1
mov  Result+1,A
```

Because the 8051 does not have a subtract without carry (borrow) instruction, you have to make sure the carry flag is reset before executing the first `subb` instruction.

The subtract base code can be modified in the same way as addition base code.

The 16-bit addition and subtraction base code will handle two's complement numbers (with bit 15 of the value being the *polarity flag*). This means that the numbers will be in the range of +32,677 to -32,678. Values larger than this will require more storage space (three or four bytes).

Bitwise logical operations are included in this section and can actually be a lot simpler due to the extra capabilities of the `anl`, `orl`, and `xrl` instructions. These instructions can be operated on without loading the accumulator before the operation or saving the result as is required in the `add` (and `adc`) and `subb` instructions.

For example:

```
VarA = VarA & 01234h
```

is simply coded as:

```
anl  VarA,#034h
anl  VarA+1,#012h
```

Multiplication

Multiplication and division are two operations that probably seem like they are going to be unreasonably complex and take a large number of cycles. Fortunately, this is not the case. By applying mathematical tricks you learned in junior high school, developing these functions for 16-bit numbers is quite easy for the 8051.

Multiplication of two 16-bit numbers probably seems frustrating because there is an 8-bit by 8-bit multiply instruction (which produces a 16-bit result) built in, but no way to multiply two 16-bit numbers. However, by applying some simple relationships, the multiply instruction *can* be used to make quite an efficient 16-bit multiply.

A 16-bit number can be thought of as an 8-bit low byte added to an 8-bit high byte:

```
SixteenBitNumber = Low8Bits + (0100h * High8Bits)
```

Looking at the numbers this way, we can change a 16-bit multiply from:

```
Product = VarA * VarB
```

to:

```
Product = (VarALow + (0100h * VarAHigh)) *
          (VarBLow + (0100h * VarBHigh))
```

In your early high school years, when you were presented with calculating the product of two compound statements, you were introduced to "FOIL" (which stands for "first, outside, inside, last") as the order of operations for the multiplication.

This can be applied to the previous equation:

```
Product = (VarALow * VarBLow) + (VarALow * (0100h * VarBHigh)) +
          ((0100h * VarAHigh) * VarBLow)) +
          ((0100h * VarAHigh) * (0100h * VarBHigh))
```

Multiplying a number by 256 (0100h) or 65,536 (010000h) is accomplished by storing the product in a higher order byte. For example, to multiply by 256, the least significant byte of the product is put into is byte 1, not byte 0. Using this, the multiplies by 0100h simply become storing the product in a higher order byte.

In understanding how the multiply is going to work, we can now write out the code (Listing C-1).

LISTING C-1 16-bit multiplication.

```
VarA    EQU MulStart              ;  Two 16-bit variables to multiply
VarB    EQU MulStart+2            ;   together

Product       EQU MulStart+4      ;  The 32-bit product

   :

  mov  Product+3,#0               ;  Clear the most significant byte

  mov  A,VarA                     ;  Do "First" multiply
  mov  B,VarB
  mul ab
  mov  Product,A                  ;  Store the 16-bit product
  mov  Product+1,B

  mov  A,VarA                     ;  Do "Outside" multiply
  mov  B,VarB+1
  mul ab
  mov  Product+2,B                ;  Store the high byte of the product
  add  A,Product+1                ;  Now, add the low byte to the previous
  mov  Product+1,A                ;   product
  jnc  MulInside
```

LISTING C-1 **16-bit multiplication *Continued.***

```
  inc  Product+2                ;  If result is greater than 0FFh,
  mov  A,Product+2              ;   increment third byte
  jnz  MulInside
  inc  Product+3                ;  If third byte == 0 after increment,
                                ;   then increment high byte

MulInside:                      ;  Do "Inside" multiply
  mov  A,VarA+1
  mov  B,VarB
  mul ab
  add  A,Product+1              ;  Add the lower byte
  mov  Product+1,A
  mov  A,B                      ;  Add the upper byte
  adc  A,Product+2
  mov  Product+2,A
  jnc  MulLast                  ;  If no carry, then don't increment the
  inc  Product+3                ;   high value

MulLast:                        ;  Multiply the last two values
  mov  A,VarA+1
  mov  B,VarB+1
  mul ab
  add  A,Product+2              ;  Add the low 8 bits of the product to
  mov  Product+2                ;   the result
  mov  A,B                      ;  Add the high byte finally
  adc  A,Product+3
  mov  Product+3,A
```

Division

The first time that you did division in school, it was probably by repeated subtraction. The divisor was repeatedly taken away from the dividend until the dividend was less than the divisor. This method of division is simple and easy to understand.

It's also easy to code in the 8051 (Listing C-2).

LISTING C-2 **16-bit division.**

```
Dividend EQU DivStart           ;  16-bit number to be divided/remainder
Divisor  EQU DivStart+2         ;  16-bit number divided into divisor

Quotient EQU DivStart+4         ;  16-bit result of division

    :

  mov  Quotient,#0              ;  Clear the quotient (nothing taken away
  mov  Quotient+1,#0            ;   yet)

DivLoop:                        ;  Loop here until dividend is less than
                                ;   divisor
  mov  A,Dividend               ;  Subtract the low 8 bits first
  clr  C
  subb A,Divisor
  mov  B,A                      ;  Save the result value
```

LISTING C-2 **16-bit division *Continued.***

```
  mov  A,Dividend+1            ;  Subtract the high 8 bits (with carry)
  subb A,Divisor+1

  jc   DivEnd                  ;  If carry set, then "Quotient" is
                               ;   correct
  mov  Dividend,B              ;  Store the "new" Quotient
  mov  Dividend+1,A

  inc  Quotient                ;  Increment the Quotient
  mov  A,Quotient
  jnz  DivLoop
  inc  Quotient+1

  ajmp       DivLoop

DivEnd:                        ;  The Quotient is correct

  xch  A,B                     ;  Put the correct remainder into
  add  A,Divisor               ;   "Dividend"
  mov  Dividend,A
  xch  A,B
  adc  A,Divisor+1
  mov  Dividend+1,A
```

The problem with this method of division is that it can take a very long time. The worst case scenario is dividing 0FFFFh by 1. In this case, 65,535 loops will be executed. In the other extreme, if the dividend is less than the divisor, then the code will not loop at all. These extremes can make it very difficult to time an application or predict how it will operate.

A better method of division is to use the shift functions to first establish what is the highest power of two that the divisor can by multiplied by and still have a positive result after having this value subtracted from the dividend. Once this is done, the shifted divisor is used to find the quotient by repeated subtracting, followed by shifting of the dividend.

I've worked for half an hour trying to write out exactly how this algorithm works and the previous paragraph is the best I can do. A much better way to understand how this method works is to see it in C code (Listing C-3).

Converting this to 8051 assembler, I came up with the code shown in Listing C-4.

LISTING C-3 **C code for a better method of 16-bit division.**

```
Count = 0;                    //  Keep track of the shifting bits
Quotient = 0;

while ( Dividend > Divisor ) {
  Divisor = Divisor << 1;     //  Find the shifted divisor that is
  Count++;                    //   greater than the dividend
}

if ( Count != 0 ) {           //  Did we actually shift something?

  Divisor = Divisor >> 1;     //  Move the divisor back down to where it
  Count–;                     //   can be used
```

LISTING C-3 **C code for a better method of 16-bit division *Continued.***

```
  while ( Count != 0 ) {     //  Now, find the Quotient
    if ( Dividend >= Divisor ) {  //  Subtract Divisor from Dividend?
      Dividend = Dividend - Divisor;
      Quotient += ( 1 << Count ); //  Update the Quotient with the
    }                        //   Divisor shift value
    Divisor = Divisor >> 1; //  Take down the Divisor by a power of 2
    Count-;
  }
}
```

LISTING C-4 **Assembler code for a better method of 16-bit division.**

```
Dividend EQU DivStart          ;  16-bit number to be divided/remainder
Divisor  EQU DivStart+2        ;  16-bit number divided into divisor

Quotient EQU DivStart+4        ;  16-bit result of division

Count    EQU DivStart+6        ;  8-bit shift counter

   :

  mov  Count,#1                ;  Keep track of the current bit in the
  mov  Count+1,#0              ;   "Count" variable
  mov  Quotient,#0             ;  Clear the Quotient (nothing taken away
  mov  Quotient+1,#0           ;   yet)

DivLoop1:                      ;  First loop - See how far to shift the
                               ;   divisor
  mov  A,Dividend              ;  Subtract the Divisor and check for
  clr  C                       ;   the carry flag set
  subb A,Divisor
  mov  A,Dividend+1            ;  Now, see what the high value results in
  subb A,Divisor+1
  jc   DivEnd1                 ;  If carry set - Stop shifting Divisor

  mov  A,Divisor               ;  Shift Divisor up by 1
  rlc a
  mov  Divisor,A
  mov  A,Divisor+1
  rlc a
  mov  Divisor+1,A

  mov  A,Count                 ;  "Increment" count by shifting up
  clr  C
  rlc  Count
  mov  Count,A
  mov  A,Count+1
  rlc  Count
  mov  Count+1,A

  ajmp DivLoop1                ;  Repeat the operation

DivEnd1:                       ;  Have shifted the Divisor to a value
                               ;   greater than the Dividend, take it down

  mov  A,Count                     ;  Did we actually shift?
  dec  A
```

LISTING C-4 **Assembler code for a better method of 16-bit division *Continued.***

```
  jz   DivEnd                    ;  If count is still at the first bit,
no

  clr  C
  mov  A,Divisor+1
  rrc a
  mov  Divisor+1,A
  mov  A,Divisor
  rrc a
  mov  Divisor,A

  mov  A,Count+1               ;  "Decrement" the count
  clr  C
  rrc a
  mov  Count+1,A
  mov  A,Count
  rrc  a
  mov  Count,A

DivLoop2:                      ;  Now, get the count

  mov  A,Dividend              ;  Can we take away the Divisor?
  clr  C
  subb A,Divisor
  mov  B,A                     ;  Save the low 8 bits for later
  mov  A,Dividend+1            ;  Take away the high 8 bits
  subb A,Divisor+1
  jc   DivSkip2                ;  If carry, Dividend is less than Divisor

  mov  Dividend,B              ;  Save the new Divisor
  mov  Dividend+1,A

  mov  A,Quotient              ;  Update the Quotient
  add  A,Count
  mov  Quotient,A
  mov  A,Quotient+1
  add  A,Count+1
  mov  Quotient+1,A

DivSkip2:                      ;  Now, take down Divisor and count

  mov  A,Divisor+1             ;  Shift the Divisor by 1 bit
  clr  C
  rrc a
  mov  Divisor+1,A
  mov  A,Divisor
  rrc a
  mov  Divisor,A

  mov  A,Count+1               ;  Take down the bit value
  clr  C
  rrc a
  mov  Count+1,A
  mov  A,Count
  rrc a
  mov  Count,A

  jnc  DivLoop2                ;  If carry not loaded, loop around again
```

This code has a few features that I should mention.

If you look back at the C code, you'll see that the `Count` value is simply a counter. In the 8051 assembler, I stored "1" initially into `Count` and shifted it up. By doing this, I avoided having to figure out the actual shifted value to add to the `Quotient`. Looking through the code, it might look like it's actually more complex. However, in reality, I've made it quite a bit simpler (and shorter), and I do not need any temporary values for shifting a "1" for the quotient add.

You should also note that I save the lower eight bits in the B register in the division comparison code before determining whether or not I want to subtract the current divisor from the dividend (i.e., the dividend is equal to or greater than the divisor). By saving the results of this comparison, I can save them as the new dividend without having to divide again. By having the temporary register for saving this result, additional space is saved as well.

One way of making sure different values can be used with this code is by checking to see whether or not the divisor could be taken away from the dividend. What I haven't shown is how to handle dividends or divisors that are two's complement negative.

If either the dividend or divisor could be negative, I recommend recording which are negative, making them positive, and then using the chart in Fig. C-1 after the division for determining whether the quotient and remainder is positive or negative. The X-Axis of the chart specifies whether or not the dividend is positive or negative, while the Y-Axis of the chart is for the divisor. "Q" is the quotient, and "R" is the remainder.

Divisor \ Dividend	Negative	Positive
Negative	Q = Pos R = Neg	Q = Neg R = Pos
Positive	Q = Neg R = Neg	Q = Pos R = Pos

FIGURE C-1 **Determine whether the quotient and remainder are positive or negative.**

D

USEFUL ROUTINES

Through the book, I have used a number of different peripheral interfaces as well as some other routines that are useful to keep in your hip pocket when you are creating an application. I try to keep a "snippets" file with different and interesting routines that allow me to develop code while keeping the thinking to a minimum.

I have tried to choose a number of routines that can be dropped into applications as required. I have also included a few routines that should help you out in different programming situations as well.

These routines primarily focus on interfaces, but I think you'll find them useful. While working through them, you should learn more about the 8051.

Timing Delays

In many applications, you will have to come up with specific time delays or measure the time between events. My first recommendation is to use the timers built into the 8051, but sometimes event timing can be so critical that the timers cannot be used because of the uncertainty of the exact cycle that the program will execute at. Understanding how these methods of timing work is important to determine which method is most appropriate for an application.

There are actually two different ways that an 8051 timer can be used for specific delays. The first is to allow a timer to interrupt execution of the code. For example, in a system with a 1-MHz clock, a 1-msec delay could be handled with the following code:

```
 org   0Bh
Tmr0:                      ; Timer0 interrupt handler
  clr  TCON.4              ; Turn off the timer input
```

```
  setb flag             ; Indicate that the operation is complete
  reti
   :
  mov  TMOD, #%00000010 ; Run Timer0 in mode 2
  mov  TH0,#173         ; Set up Timer0 to overflow in 1.0 msec
  mov  TL0
  mov  IE, #%10000010   ; Enable Timer0 interrupt
  clr  flag             ; Clear the timer overflow flag
  setb TCON.4           ; Start the timer executing
Loop:                   ; Wait for the delay
 jnb  Flag, Loop
```

This code could also be run without the interrupt timer by just waiting for the Timer0 overflow flag to be set:

```
  mov  TMOD, #%00000010 ; Run Timer0 in mode 2
  mov  TH1, #173
  mov  TL1, #173
  setb      TCON.4      ; Start the timer executing
Loop:                   ; Wait for the timer to overflow
  jnb  TCON.5, Loop
```

To get the 1-msec delay, I used the timer mode 2 delay formula:

$$TimeDelay = (12 * (256 - Dlay)) / Freq$$

which can be rearranged to find *Dlay*:

$$\begin{aligned} Dlay &= 256 - ((TimeDelay * Freq) / 12) \\ &= 256 - ((1 \text{ msec} * 1 \text{ MHz}) / 12) \\ &= 256 - (1000 / 12) \\ &= 256 - 83.3 \\ &= 173 \end{aligned}$$

The interrupt version of this code allows something to execute in the foreground while the timer is counting out the delay. This could be an advantage in some applications that require simple multitasking (i.e., there is a background and foreground task, and when the background task is ready to run, it takes priority).

The other way of using the timer for delays is to let it run free and poll it to find out when an external event happened or when the desired delay has passed:

```
  mov  TMOD, #%00000001 ; Run Timer0 in mode 1
   :
EventStart:             ; Get the initial timer values
  mov  StartHi, TH0
  mov  StartLo, TL0
   :
EventEnd:               ; Get the final timer values
  mov  EndHi, TH0
  mov  EndLo, TL0
```

With having the timer value at both ends of the event, understanding the total time could simply be the start values subtracted from the end values. The total time delay would be:

$$TimeDelay = (12 * (End - Start)) / Freq$$

These operations could be reversed to get a specific time delay. After finding the time after starting the event, the number of cycles to delay could be added to this value, and then the timer could be polled for the end of the event:

```
  clr  c                   ; Clear the carry flag for following
                           ;  subtractions
EndLoopHi:                 ; Wait for the high 8 bits to become
  mov  A,EndHi             ;  the expected value
  subb A,TH0
  jnz  EndLoopHi
EndLoopLo:                 ; Wait for the low 8 bits to exceed
  mov  A,EndLo             ;  the expected
  subb A,TL0
  jnc  EndLoopLo
```

The delay check execution falls out of `EndLoopLo` when the low 8 bits of the timer are greater than the expected value. Because I require three instructions to loop (and up to six instruction cycles), there is a very good chance I won't poll at the exact cycle TL0 is at the value I'm waiting for. When the carry is set after the subtraction, then the time delay will have been achieved.

Using a timer for producing and timing delays is quite easy to do but has one drawback: The timer resources that are used could be applied elsewhere in the application (although the second method that I've shown uses a timer that is already running for the application). Using a timer does have a significant advantage over code that counts cycles. It's processor independent and somewhat insensitive to interrupts occurring during the delay.

A small loop can be used as well to provide a specific delay:

```
  mov  Rn, Delay           ; Load the delay value
Loop:
  djnz Rn, Loop
```

In a true 8051, the `mov` and `djnz` instructions (operating on the bank register) each take one instruction cycle to execute. This results in the number of instruction cycles for the delay being defined as:

$$DelayCycles = 1 + Delay$$

or, in terms of time:

$$TimeDelay = (12 * (1 + Delay)) \,/\, Freq$$

This loop can delay up to 257 instruction cycles (with `Delay` equal to zero) or 256 instruction cycles (with `Delay` equal to 255, 0FFh) and can be easily expanded to give longer delays:

```
  mov  Ra, DelayLo
  mov  Rb, DelayHi
Loop:
  djnz Ra, Loop
  djnz Rb, Loop
```

In this example, the time delay works out to:

$$TimeDelay = (12 * (DelayLo + (257 * DelayHi) - 254)) \,/\, Freq$$

To determine the correct values for the two 8-bit `Delay` values, I first find the highest `DelayHi` value that does not exceed the desired delay.

For example, if I wanted to delay 10.0 msec in a 20-MHz 8051, I would assume `DelayLo` was equal to 0 and find the `DelayHi` that did not result in 10 msec to be exceeded:

$$\begin{aligned} TimeDelay &= (12 * (0 + (257 * DelayHi) - 254)) \,/\, Freq \\ 10\text{ msec} &= (12 * ((257 * DelayHi) - 254)) \,/\, 20\text{ MHz} \\ 10\text{ msec} * 20\text{ MHz} &= 12 * ((257 * DelayHi) - 254) \\ 200{,}000 &= 12 * ((257 * DelayHi) - 254) \\ 200{,}000 \,/\, 12 &= (257 * DelayHi) - 254 \\ 16{,}667 &= (257 * DelayHi) - 254 \\ 16{,}667 + 254 &= 257 * DelayHi \\ DelayHi &= 16{,}921 \,/\, 257 \\ &= 65.8 \end{aligned}$$

To get the value less than the specified delay, I will use 65 as my `DelayHi` value.

Plugging 65 into the previous formula for the `TimeDelay`, I can now calculate the correct value for `DelayLo`:

$$\begin{aligned} TimeDelay &= (12 * (DelayLo + (257 * DelayHi) - 254)) \,/\, Freq \\ 10\text{ msec} &= (12 * (DelayLo + (257 * 65) - 254)) \,/\, 20\text{ MHz} \\ 10\text{ msec} * 20\text{ MHz} &= 12 * (DelayLo + 16{,}705 - 254) \\ 200{,}000 &= 12 * (DelayLo + 16{,}451) \\ 200{,}000 \,/\, 12 &= DelayLo + 16{,}451 \\ DelayLo &= 16{,}667 - 16{,}451 \\ &= 215 \end{aligned}$$

So, for this loop to delay 10 msec in a 20-MHz 8051, `DelayLo` and `DelayHi` must be set to 215 and 65, respectively.

Delay loops like this can be expanded by adding more counter bytes with corresponding `djnz` instructions to the loop.

A simple way of doubling the number of cycles executed by the delay loops is by using direct addressing. Both the `mov` and `djnz` instructions take two instruction cycles when a scratchpad RAM byte is used instead of a bank register.

There are two concerns with using these code loops. The first is that these loops should not be used for critical timing when interrupts are enabled. As an interrupt handler executes, the cycles required will not be available to the loop, making it actually take longer. For many applications, this is not a problem because the delays are approximate and usually the *minimum* delay is the issue.

The other concern with this approach to timed delays is what happens if the application code is ported to an enhanced 8051 processor like the Dallas Semiconductor HSM? If this code is run in an HSM processor, even at the same clock speed, you'll notice that the delay will be radically less (due to the fewer clock cycles per instruction cycle of the HSM).

For this reason, I recommend that all delays use one of the timer algorithms presented previously, and the HSM timers always use the divide-by-12 option so that the timer delays will be exactly the same as what was used in the stock 8051s.

Table Operations

In chapter 9, I introduced you to some of the aspects of tables. As I worked through the projects, I have found a few aspects that I felt I should expand upon.

As I presented in chapter 9, a typical table read subroutine could be:

```
Read_Table:                          ; Read the table entry from "A"
  add  A,#(InitTable-InitGetTable) ; Point to offset in table
  movc A,@A+PC
InitGetTable:                        ; Table offset to display
  ret
InitTable:                           ; The control store table for
  db      'Hello',0                  ;  initializing "Array"
```

This subroutine can only access a maximum of 255 entries, but occasionally, you have to access a full 256 for data conversion.

To return the corresponding table value for a byte (up to 256 different values), the following subroutine could be used:

```
Read_Table:                          ; Read the table entry from "A"
  inc  A                             ; Point to offset in table/get
                                     ;  test for 0FFh
  jnz  Read_Table_Get
  mov  A,#Entry_0FFh                 ; Return the 0FFh entry
  ret
InitGetTable:                        ; Table offset to display
  movc A,@A+PC
  ret
InitTable:                           ; The control store table for
  db      ...                        ;  initializing "Array" - 255 entries
```

This code takes advantage of the knowledge that the space between the `movc` and `ret` instructions is only 1 byte, which means that, instead of adding a calculation as I do in the first `Read_Table`, I can just increment the contents of the accumulator. If the result is equal to 0, then I know I started with 0FFh, and I can return an explicit value; otherwise, I can simply read from the table for the remaining 255 values.

Often, you will want to output strings to an output device for the operator. Instead of creating a number of table routines (like the previous ones), you might want to have a single subroutine that outputs a specified message (Listing D-1).

LISTING D-1 Subroutine to output a specified message.

```
SendMsg:                             ;  Send the specified Message to display

  mov      R1,A                      ;  Save the message number

  mov      R0,#0                     ;  Use R0 as the offset counter

SendMsg_Find:                        ;  Look for the appropriate message

  mov      A,R1                      ;  Are we now at zero?
  jz       SendMsg_Output            ;   If count is zero
```

LISTING D-1 **Subroutine to output a specified message *Continued.***

```
  mov     A,#LOW(SendMsgTable - SendMsgFindGet)
  add     A,R0                 ;  Get current address

  movc    A,@A+PC              ;  Get the character in the displays
SendMsgFindGet:

  inc     R0                   ;  Point to the next character

  jnz     SendMsg_Find         ;  If NOT equal to zero

  dec     R1                   ;  It is, decrement the message counter

  sjmp    SendMsg_Find

SendMsg_Output:                ;  Now, output the string to '\0'

  mov     A,R0
  add     A,#LOW(SendMsgTable - SendMsgOutputGet)

  movc    A,@A+PC              ;  Read the table
SendMsgOutputGet:

  inc     R0                   ;  Point to the next character

  jz      SendMsg_End          ;  If at zero, stop reading/displaying

  lcall   SendChar             ;  Output the character

  sjmp    SendMsg_Output

SendMsg_End:                   ;  Message output

  ret

SendMsgTable:                  ;  Messages to be Displayed
  db      CR,LF,'AT89C2051 Emulator',CR,LF,0      ;  Message  0
  db      'Invalid Command',CR,LF,0               ;  Message  1
  db      'Enter New Value ',0                    ;  Message  2
  db      'Invalid Address',CR,LF,0               ;  Message  3
  db      'No Address Specified',CR,LF,0          ;  Message  4
```

By passing a string number in "A," the routine will read through each message, decrementing the original count each time 000h is encountered until it is pointing to the specified message. Once it is pointing to the message, it can output it until the message's 000h character is encountered.

LCD Interfaces

While LEDs are useful in indicating that an application is running, is connected, or is waiting for input, they are not capable of providing the range of output options of a liquid crystal display (LCD). An LCD allows your application to output a very specific message (or prompt) to the user, making the application much more user friendly and impressive. I also find LCDs to be invaluable for displaying status messages and information during application debug.

ASCII-input LCDs, even though they have these advantages, have a reputation of being difficult to hook up and get to work. In this section, I want to go through how LCDs work and how they can be wired to microcontroller, to show you that it isn't that difficult.

Most alphanumeric LCDs use a common controller chip, the Hitachi 44780, and a common connector interface. Both of these factors have resulted in alphanumeric LCDs that range in size from 8 characters to 80 (arranged as 40 by 2 or 20 by 4) and are all interchangeable, without requiring hardware or software changes.

The most common connector used for the 44780 based LCDs is 14 pins in a row, with pin centers 0.100" apart. The pins are wired as shown in Table D-1.

As you would probably guess from this description, the interface is a parallel bus, allowing simple and fast reading and writing of data to and from the LCD. (See Fig. D-1.)

This waveform will write an ASCII byte code out to the LCD's screen. I call it an ASCII code, but it's actually a modified ASCII character set with a few changes and additions for Japanese characters. The most notable difference is the lack of the backslash ("\" or 05Bh) character and the first eight characters (000h to 007h), which are user definable. (See Fig. D-2.)

TABLE D-1 HOW TO WIRE AN LCD

PIN	DESCRIPTION
1	Ground
2	Vcc
3	Contrast voltage
4	"R/S" instruction/register select
5	"R/W" read/write select
6	"E" clock
7–14	Data I/O pins

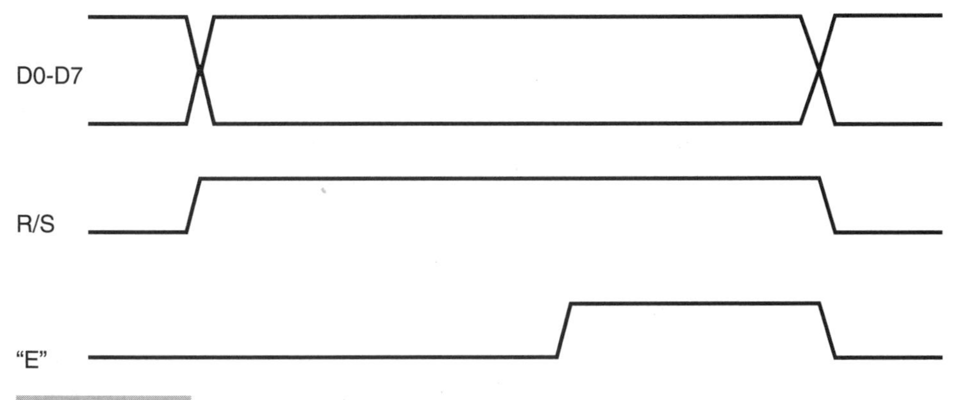

FIGURE D-1 Waveform for writing to an LCD.

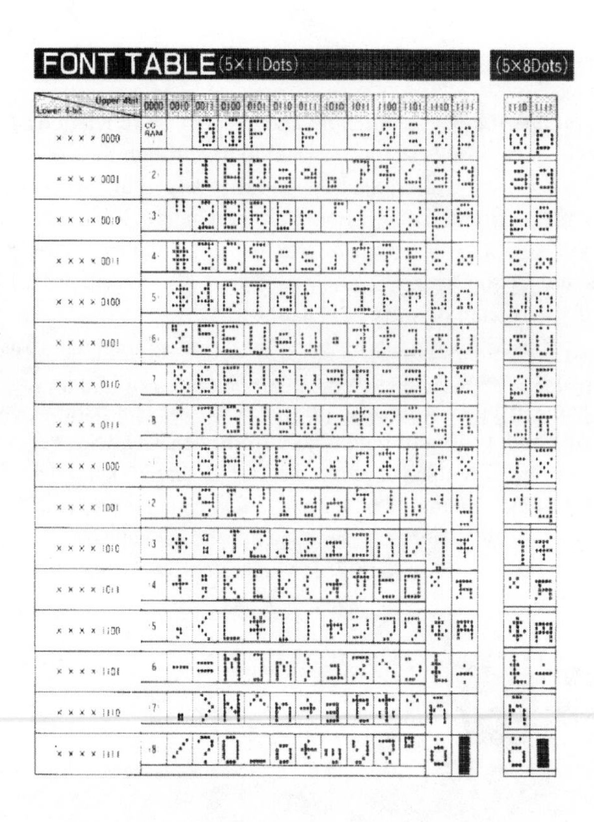

FIGURE D-2 Hitachi 44780 built-in font.

The ASCII code to be displayed is 8 bits long and is sent to the LCD either 4 or 8 bits at a time. If the 4-bit mode is used, two nybbles of data (sent high 4 bits then low 4 bits with an "E" clock pulse with each nybble) are sent to make up a full 8-bit transfer. The "E" clock is used to initiate the data transfer within the LCD.

If you've done any console programming, working with an LCD will probably be very familiar to you. In the 44780, there is a cursor, which specifies where the next data character is to be written. This cursor can be moved or be made invisible or to blink. The blinking function is very rarely used because it is pretty obnoxious.

Sending parallel data as either 4 or 8 bits are the two primary modes of operation. While there are secondary considerations and modes, deciding how to send the data to the LCD is most critical decision to be made for an LCD interface application.

8-bit mode is best used when speed is required in an application and at least 10 I/O pins are available. 4-bit mode requires a minimum of 6 bits. To wire a microcontroller to an LCD in 4-bit mode, just the top 4 bits (DB4-7) are written to (Fig. D-3).

This can be further reduced by using a shift register so that a minimum of 3 I/O pins are required (Fig. D-4).

For this type of application, I use something like a 74*x*174 (where "*x*" is "HC" or "LS") wired up as a shift register. 8-bit mode could be used with a shift register, but a ninth bit (which will be used as R/S will be required).

The R/S bit is used to select whether data or an instruction is being transferred between the microcontroller and the LCD. If the bit is set, then the byte at the current LCD cursor position can be read or written. When the bit is reset, either an instruction is being sent to the LCD or the execution status of the last instruction is read back (whether or not it has completed).

Eight programmable characters are available and use codes 0x000 to 0x007. They are programmed by pointing the LCD's cursor to the character generator RAM (CGRAM) area at eight times the character address. The next eight characters written to the RAM are each line of the programmable character, starting at the top.

Each 44780 instruction, which controls the operation of the LCD and what is displayed on it, is defined by the position of the most significant "1" in the instruction. The easiest way to list the different instructions is to use Table D-2.

The busy flag is set as long as the instruction is executing within the LCD. For creating an application that runs as quickly as possible, this line can be polled (eliminating the requirement to delay the worst-case amount of time). This is typically not a big point because putting in an explicit software delay is not that difficult. All instructions take a

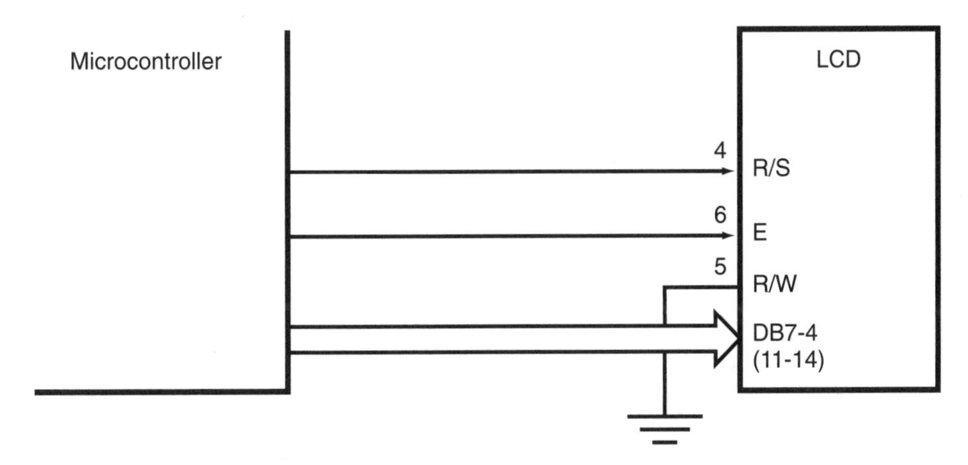

FIGURE D-3 LCD 4-bit microcontroller interface.

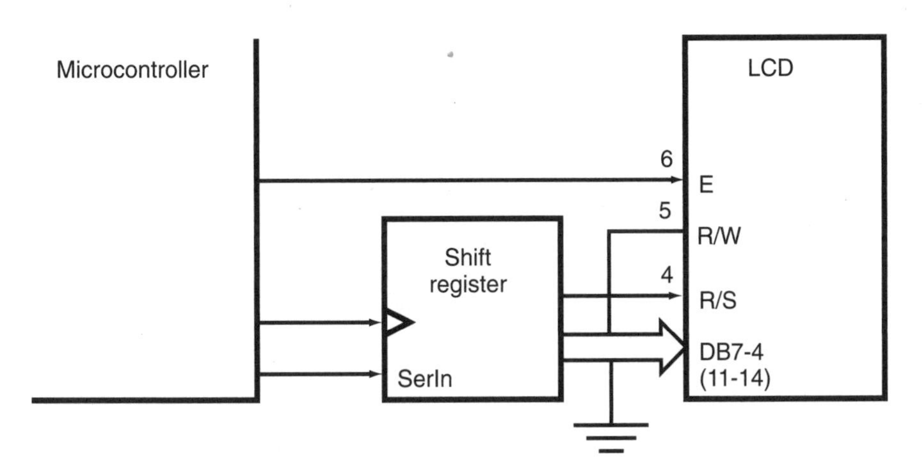

FIGURE D-4 LCD 4-bit shift register interface.

TABLE D-2 THE 4470 INSTRUCTIONS

R/S	R/W	D7	D6	D5	D4	D3	D2	D1	D0	INSTRUCTION/DESCRIPTION
0	0	0	0	0	0	0	0	0	1	Clear display
0	0	0	0	0	0	0	0	1	*	Return cursor to home
0	0	0	0	0	0	0	1	ID	S	Set cursor move direction
0	0	0	0	0	0	1	D	C	B	Enable display/cursor
0	0	0	0	0	1	SC	RL	*	*	Move cursor/shift display
0	0	0	0	1	DL	N	F	*	*	Set interface length
0	0	0	1	A	A	A	A	A	A	Move cursor to CGRAM
0	0	1	A	A	A	A	A	A	A	Move cursor to display
0	1	BF	*	*	*	*	*	*	*	Read the busy flag
1	0	D	D	D	D	D	D	D	D	Write ASCII to the display
1	1	D	D	D	D	D	D	D	D	Read ASCII from the display

Optional bit descriptions:

Set cursor move direction:

ID—Increment the cursor after each byte written to display if set
S—Shift display when byte written to display

Enable display/cursor:

D—Turn display on(1)/off(0)
C—Turn cursor on(1)/off(0)
B—Cursor blink on(1)/off(0)

Move cursor/shift display:

SC—Display shift on(1)/off(0)
RL—Direction of shift right(1)/left(0)

Set interface length:

DL—Set data interface length 8(1)/4(0)
N—Number of display lines 1(0)/2(1)
F—Character font 5x10(1)/5x7(0)

Move cursor to CGRAM/display:

A—Address

Read/write ASCII to the display:

D—Data

maximum of 160 μsec except for Clear Display and Return Cursor to Home, which take a maximum of 4.1 msec (although I always delay 5 msec just to be on the safe side).

Different LCDs execute instructions at different rates. As I've pointed out, the delays quoted earlier are maximums, some LCDs will execute in less time. However, unless the busy flag is polled, I recommend that the maximum delay is always used.

For 8-bit mode, I use this subroutine to send data or instructions to the LCD. The carry flag contains the R/S bit value while the accumulator contains the value to output:

```
LCD8Out:                    ; Send 8-bit command to the LCD.
  clr  RS                   ; Wait for the previous write to complete
  setb RW
  setb E                    ; Poll bit 7 of the data until low
```

```
LCD8Out_Loop:
  jb    Pn.7,LCD8Out_Loop
  clr   E                        ; Not polling any more - "E" is low
  mov   RS,C                     ; Put the carry flag as "RS" for data
                                 ;  or instructions
  mov   Pn,A                     ; Output the data bits
  clr   RW                       ; Make sure we're writing
  setb  E                        ; Toggle the data in
  clr   E
  ret
```

In this subroutine, before the byte is to be output to the LCD, I poll the busy flag until it goes low, indicating the operation has completed. By polling for the bit *before* I write to the LCD, rather than after, I can write at the maximum possible speed, with most of the delay being taken up by instructions between the writes.

For many applications, I don't bother with polling the busy flag. Instead, I put in the 160 µsec and 5 msec delays mentioned previously.

To write to a 4-bit interface, I would use the code:

```
LCD4Out:                         ; Send 8-bit command to the LCD as
                                 ;  two 4-bit nybbles.
                                 ; Assume "RW" is tied to ground
  mov    RS,C                    ; Set the instruction/data flag
  push   ACC                     ; Save the accumulator for later
  swap   A                       ; Output the high 8 bits
  anl    A,#00Fh
  mov    Pn,A
  setb   E                       ; Toggle out the bits
  clr    E
  pop    ACC                     ; Ouput the low 8 bits
  anl    A,#00Fh
  mov    Pn,A
  setb   E
  clr    E
  acall Dlay160                  ; Delay for the LCD to output
  ret
```

If a shift register was used in 4-bit mode, then this code would include the shifting out code instead of the `mov Pn,A`.

Reading data back from the LCD RAM is best used in applications that require data to be moved back and forth on the LCD (such as in applications that scroll data between lines). In most applications, I just tie the R/W line to ground because I don't read anything back. This greatly simplifies the application because, when data is read back, the microcontroller I/O pins have to be alternated between input and output modes. Data is read from the current cursor position.

An 8-bit interface read could be accomplished by the code:

```
LCD4Read:                        ; Read the 8 bits at the cursor
  setb   RS                      ; This is a data read
  mov    Pn,0FFh                 ; Make sure the 8051 isn't driving data out.
  setb   RW
  setb   E
  mov    A,Pn                    ; Read the data
  clr    E
  ret
```

For reading, there is no need to worry about the busy flag interface. The operation takes place immediately. Like the data write command, a data read increments the cursor (if appropriate).

In terms of options, I have never seen a 5×10 LCD display. This means that the F bit in the "Set Interface" instruction should always be reset (equal to "0").

Before you can send commands or data to the LCD module, the module must be initialized. This is done using the following series of operations for 8-bit mode:

1 Wait more than 15 msec after power is applied.
2 Write 0x030 to LCD and wait 5 msec for the instruction to complete.
3 Write 0x030 to LCD and wait 160 μsec for instruction to complete.
4 Write 0x030 *again* to LCD and wait 160 μsec or poll the busy flag.
5 Set the operating characteristics of the LCD
  a. Write "Set Interface Length."
  b. Write 0x010 to turn off the display.
  c. Write 0x001 to clear the display.
  d. Write "Set Cursor Move Direction" setting cursor behavior bits.
  e. Write "Enable Display/Cursor" and enable the display and optional cursor.

In describing how the LCD should be initialized in 4-bit mode, I will specify writing to the LCD in terms of nybbles. This is because just single nybbles are sent initially (and not two nybbles, which make up a byte and a full instruction). As I mentioned earlier, when a byte is sent, the high nybble is sent before the low nybble and the E pin is toggled each time 4 bits are sent to the LCD.

The following series of operations is for 4-bit mode:

1 Wait more than 15 msec after power is applied.
2 Write 0x03 to LCD and wait 5 msec for the instruction to complete.
3 Write 0x03 to LCD and wait 160 μsec for instruction to complete.
4 Write 0x03 *again* to LCD and wait 160 μsec (or poll the busy flag).
5 Set the Operating Characteristics of the LCD
  a. Write 0x02 to the LCD to enable 4-bit mode.
  b. Write "Set Interface Length."
  c. Write 0x01/0x00 to turn off the display.
  d. Write 0x00/0x01 to clear the display.
  e. Write "Set Cursor Move Direction" setting cursor behavior bits.
  f. Write "Enable Display/Cursor" and enable the display and optional cursor.

Note: Instructions/data writes 5.b through 5.f require two nybble writes.

Once the initialization is complete, the LCD can be written to with data or instructions as required.

The last aspect of the LCD to discuss is how to specify a contrast value for the display. I typically use a potentiometer wired as a voltage divider (Fig. D-5). This will provide an easily variable voltage between Ground and Vcc, which will be used to specify the contrast (or darkness) of the characters on the LCD screen.

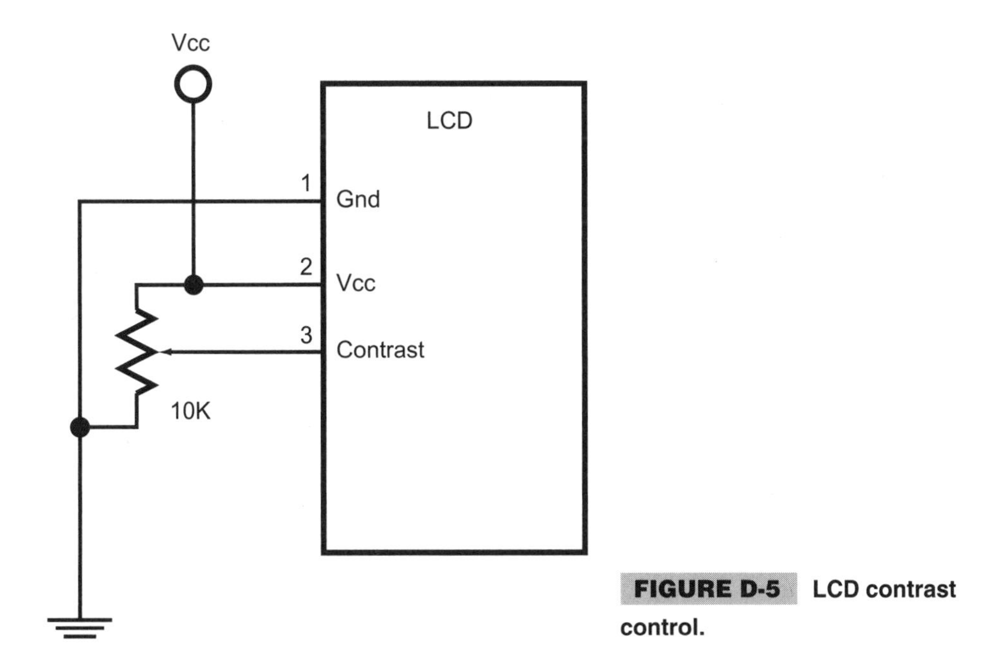

FIGURE D-5 LCD contrast control.

At the start of this section, I said that LCDs have a reputation for being hard to get working. The biggest problem with using LCDs is not properly initializing them. If you follow the instructions given previously, you shouldn't have any problems. As well, the minimum E pulse width is 450 nsec. If this pulse is shorter in duration, you might have intermittent problems with the LCD.

I2C Bus Master Interface

I2C is an interesting protocol that is very well suited for use with the 8051. This two wire protocol can be easily implemented to provide extra I/O capabilities to your application with very little software overhead or circuitry. The only thing I would caution you on using the code presented here is to make sure that the specified timings are not violated.

To achieve the required signals, only two pins of the 8051 are required. Ideally, these lines should be pulled up with 1-KΩ resistors. (See Fig. D-6.) Even though the 8051 does have internal pull-ups, using the 1-KΩ resistors will make sure that the rise times meet the I2C specifications under all conditions.

The routines in Listing D-2 are all that are required to implement a single master bus. In the code, I have assumed that `SDA` and `SCL` are I/O pins that have been defined with these labels.

What I haven't shown in these routines are the delays between the `SCL` and `SDA` transitions to make sure the timing isn't violated. To make sure that no timings are violated, refer to Fig. D-7, and make sure that the 1.3 µsec between clock pulses, the 0.6-µsec clock pulse width, and the 0.6 µsec time between `SCA` and `SCL` transitions aren't violated.

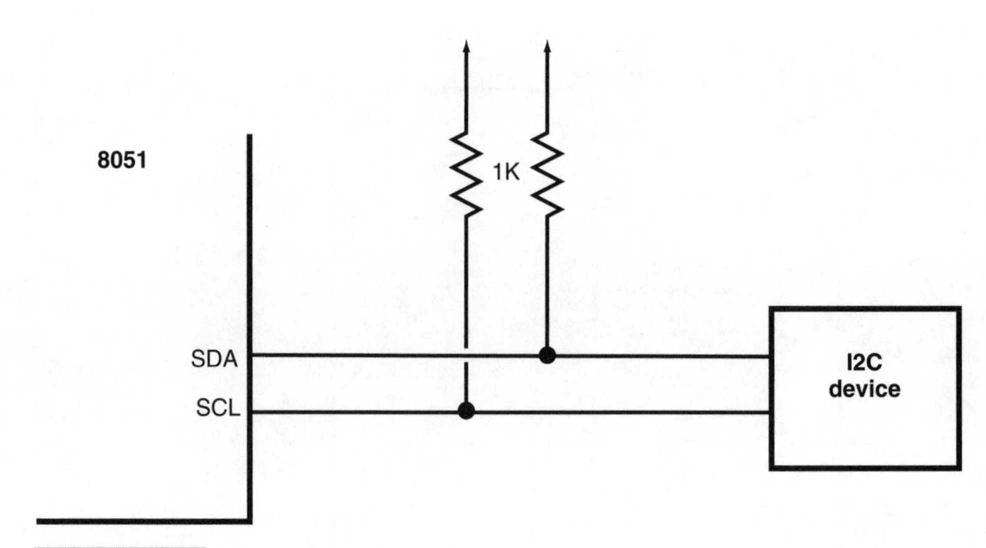

FIGURE D-6 8051 I2C interface.

LISTING D-2 Code to implement a single master bus.

```
I2CStart:                         ;  "Claim" the bus with a start waveform

  setb SDA                        ; Make sure the lines are active
  clr  SCL
  setb SCL

  clr  SDA                        ;  Change "SDA" while "SCL" is high

  clr  SCL

  ret

I2CStop:                          ;  Release the I2C bus

  setb SCL

  setb SDA                        ;  Raise the "SDA" while "SCL" is high

  ret

I2CSend:                          ;  Send 8 bits to the I2C receiver

  mov  R0,#8

I2CS_Loop:

  rrc  A                          ;  Output the next bit to send
  mov  SDA,C

  setb SCL                        ;  Clock it out
  clr  SCL

  djnz R0,I2CS_Loop               ;  Do 8x

  setb SDA                        ;  Now, get the "ACK" bit
```

LISTING D-2 **Code to implement a single master bus *Continued.***

```
  setb SCL
  mov  C,SDA                 ;  Save "ACK" in carry
  clr  SCL
  clr  SDA

  ret

I2CGet:                      ;  Get 8 bits and return "ACK" from carry

  rrc  A                     ;  Save the carry flag

  mov  R0,#8

  setb SDA                   ;  SDA is high throughout this routine

I2CG_Loop:

  setb SCL                   ;  Read while the clock is high
  mov  C,SDA                 ;  Read the bit and save it
  rrc  A
  clr  SCL

  djnz R0,I2CG_Loop

  mov  SDA,C                 ;  Output carry flag (passed "ACK" bit)
  setb SCL
  clr  SCL
  clr  SDA

  ret
```

"Bit Banging" Asynchronous Serial Interface

Providing a hardware-independent asynchronous serial interface might seem redundant for the 8051 because of the infrared serial port built into the device, but there are some simple 8051 devices, like the AT89C1051, that do not have a built-in serial port or applications like the AT89C*x*051 emulator that work best without the serial port. You probably think that software routines for providing asynchronous serial interfaces are quite complex, but they're actually quite simple to understand and code.

Understanding how these routines work is also useful from the perspective of understanding how to work through time critical applications.

As I discussed earlier in the book, an asynchronous serial data line is normally high. When data is sent, the line will go low for one bit (the start bit). Data is then sent, followed by a high stop bit to allow the receiver to handle the character.

In Fig. D-8, only 5 bits are shown along with a parity bit. In most communications, 8 bits are sent with no parity. This is known as "8-N-1," which means that 8 data bits are sent with no parity bit and one stop bit.

For the examples shown here, I will deviate from my normal format to show the instruction cycles required by each instruction for the critically timed sections of the code. This is needed to figure out how to make the different delays.

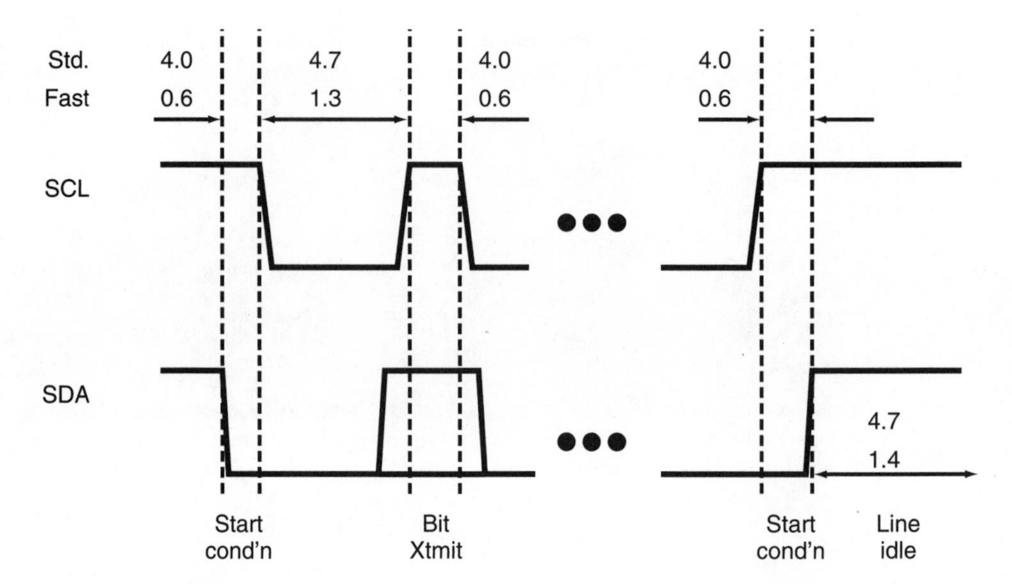

FIGURE D-7 I2C signal timing.

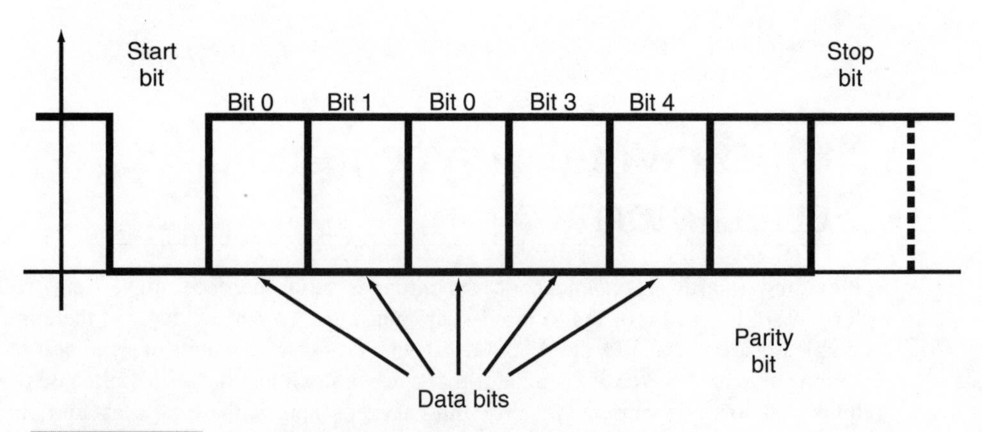

FIGURE D-8 Asynchronous serial data stream.

For all the examples, I am assuming that the target device is a Dallas Semiconductor HSM running at 4 MHz with data being transmitted and received at 9600 bps. At this data speed, each bit period is 104.167 μsec. I'll treat the bit period as just 104 μsec. With a 4-MHz clock in the HSM, each instruction cycle takes 1 μsec.

Transmitting data is quite easy to do. The routine in Listing D-3 will transmit a start bit followed by 8 data bits, and a stop bit.

In this code, the contents of the carry flag are output. By resetting the carry flag and shifting after the start bit is output, I avoid having to create a routine for sending the start bit explicitly. The same can be said for the stop bit. By loading the accumulator with ones as I shift out the data, I can guarantee that the tenth bit sent (which is the carry bit) is always "1," and I don't have to come up with code to output a stop bit.

In `TXLoop` in the previous code, to get the `TXLoop1` count value, I have counted all the instruction cycles inside the loop with the exception of the `djnz R0,TXLoop1` to find the intrinsic loop timing. This value is the number of instruction cycles required to do the data output and bit count, but not the actual delay.

To get the number of instruction cycles that `TXLoop1` has to be timed up, I use the formula:

$$\begin{aligned} \textit{Delay Cycles} &= \textit{Bit Period} - \textit{Intrinsic Delay} \\ &= 104\ \mu\text{sec} - 11\ \mu\text{sec} \\ &= 93\ \mu\text{sec} \end{aligned}$$

Because each `djnz R0,TXLoop1` instruction requires three instruction cycles, a `djnz` has to execute 31 times for `TXLoop` to take 104 μsec for each iteration.

Implementing a bit-banging receive is very similar (Listing D-4).

In `GetRX`, when the input line goes low, I wait a half a bit before checking to see if the line is still low and not a glitch. This is timed the same way as I did with the transmit delay loop, but a total of 52 instruction cycles are delayed, which places the read halfway through each bit, minimizing the opportunity for invalid data to be read.

The `RXLoop` code is timed exactly the same way as `TXLoop`. Looking at these two loops, the relationship between the `TXLoop` delay counter initial value and `RXLoop`'s is that `RXLoop`'s value is one greater than `TXLoop`. After calculating the value for one, the initial delay counter can be easily calculated for the other.

These routines work well; they're the same routines I used in the AT89C*x*051 emulator. However, there are a few drawbacks to using them. The obvious one is that interrupts must

LISTING D-3 **Code to transmit the data.**

```
Instruction
Cycles

      TXSend:                 ;  Send the character in ACC

        mov   R0,#10          ;  We're actually sending 10 bits

        clr   C               ;  First bit is a "0"

      TXLoop:                 ;  Loop here for each bit

2       mov   TXOut,C         ;  Output the bit in the carry flag

2       mov   R1,#31          ;  Delay to 104 cycles per loop
      TXLoop1:
3       djnz R1,TXLoop1
1       nop
1       nop

1       setb C                ;  Shift the next bit to carry
1       rrc  A                ;   and load in a "1" for stop bit

3       djnz R0,TXLoop

        ret
```

LISTING D-4 **Code to implement a bit-banging receive.**

```
Instruction
Cycles

         GetRX:                       ; Wait for a byte to come in and
                                      ;  return it in "ACC"
           mov  R0,#9                 ; Want to read 9 bits

         RXLoop1:                     ; Wait for a start bit
4          RXBit,RXLoop1

2          mov  R1,#14                ; Delay 1/2 bit
         RXLoop2:
3          djnz R1,RXLoop2

4          jb   RXBit,RXLoop1         ; If "glitch", keep waiting

         RXLoop:                      ; Now, read 9 bits

2          mov  C,RXBit               ; Read the input bit
1          rrc  A

2          mov  R1,#32                ; Delay loop for 104 usec per bit
         RXLoop3:
3          djnz R1,RXLoop3

3          djnz R0,RXLoop

           ret
```

be disabled while these routines are running. If they are interrupted during execution, the data will not be sent or received properly.

A more sophisticated way of implementing an asynchronous serial receiver/transmitter in software is to use a timer, running in mode 2 and overflowing at three times the bit rate, driven by an interrupt routine. By interrupting at three times the bit rate, the input serial line can be polled for the start bit. If it is encountered, the routine will poll in the middle of the next bit by delaying a bit and a third period. Once the "middle" has been established, polling can be accomplished every three timer overflows.

This method (Listing D-5), from the mainline's perspective, operates very similarly to hardware built into the device.

The delays after the transmit function are used to make sure that the receive line is always polled at the same point in a bit (between 30% and 60% of a bit period from the start of the bit).

To send a byte, the `TXChar` and count information is loaded as:

```
clr  EA                   ; Disable interrupts during operation
mov  A,Char               ; Set up the character for transmit
clr  C                    ;  shifting up by 1 and setting a 0 start bit
rlc  A
mov  TXChar,A
mov  A,#0FFh
mov  A.0,C
mov  TXChar+1,A
mov  TXBitCount,#3        ; Send the first character for three cycles
mov  TXCount,#10          ; Sending 10 bits
setb EA                   ; Turn interrupts back on
```

LISTING D-5 This code operates very similarly to hardware built into the device.

```
Instruction
Cycles
          TimerInt:                      ;  Timer interrupt routine

            push ACC                     ;  Save changed registers
            push PSW

            djnz TXBitCount,RXBitDelay   ;  Send out a bit?

2           clr  TXDone                  ;  "TXDone" indicates bit sent

1           setb C                       ;  Get the Next bit to send
2           mov  A,TXChar+1              ;  Shift down from the top
1           rrc  A
2           mov  TXChar+1,A
2           mov  A,TXChar
1           rrc  A
2           mov  TXChar,A

2           mov  TXOut,C                 ;  Output the bit

3           mov  TXBitCount,#3           ;  Reset the bit counter to delay 3
                                         ;   cycles
4           djnz TXCount,RXBitDelay2       ;  At end of the transmission?

2           setb TXDone

3           mov  TXBitCount,#1           ;  Send out a "1" each cycle

2           inc  TXCount

3           ajmp RXBit

          RXBitDlay:                     ;  Delay for the first period

2           mov  A,#5
          Dlay1:
1           dec  A
3           jnz  Dlay1

          RXBitDlay2:                    ;  Delay for the send data end code

2           mov  A,#2
          Dlay2:
1           dec  A
3           jnz  Dlay2

          TXBit:                         ;  Read incoming

            mov  C,RXBit                 ;  Get line state no matter what

            jnb  RXBitRead,RXCheckBit    ;  Is ReadBit set?

            djnz RXBitCount,IntEnd       ;  Okay to save the bit?

            mov  A,RXChar                ;  Store the bit
            rrc  A
            mov  RXChar,A

            mov  RXBitCount,#3           ;  Wait for the next bit
```

LISTING D-5 **This code operates very similarly to hardware built into the device *Continued.***

```
        djnz RXCount,IntEnd      ;  Last bit?

        clr  RXBitRead           ;  Yes

        mov  RXBitCount,#4       ;  Set up for next incoming char
        mov  RXCount,#8

        ajmp IntEnd

      RXCheckBit:                ;  Nothing being read, something to
                                 ;   read?
        jc       IntEnd          ;  Line high - Nothing to read

        setb        RXBitRead    ;  Else, a data bit to read

      IntEnd:                    ;  End of the interrupt handler

        pop      PSW             ;  Restore the changed registers
        pop      ACC

        ret
```

For receiving a character, the `RXBitRead` flag can be polled. If it is reset, then a character has been received.

There are two potential problems with this code. The first is that it does not check the received bit for a glitch. After polling the line and reading a "0," it is not checked again after one overflow to see if it was an actual "0."

The second potential problem has to do with the number of cycles required for the asynchronous communications interrupt handler. For reasonably fast data rates at relatively slow clock speeds, the interrupt handler might take more cycles than are available between timer overflow interrupts.

Hex to ASCII Conversion

After getting through the basic programming and interfacing of the 8051, you're going to want to do more and more complex applications. One of the most potentially frustrating things that you will have to do is convert a hex byte to its 20 byte ASCII equivalent or visa versa.

This is actually a pretty good test of your capabilities. How would you convert the byte 042h into the two byte string "42"?

The first thing that you would probably do is break the byte up into two 4-bit nybbles and call a conversion routine:

```
ByteToASCII:                   ; Convert the byte in "A" to two ASCII
                               ;  characters in "B" and "A"
 mov     B,A                   ; Save the byte for low nybble conversion
 swap    A                     ; Do high nybble conversion
 anl     A,#00Fh
 acall   NybbleToASCII
 xch     A,B                   ; Save the ASCII of the high byte
```

```
 anl    A,#00Fh              ; Do conversion on the low nybble
 acall  NybbleToASCII
 ret
```

This looks pretty good. Now all we have to do is figure out the conversion.

If you were programming in a high-level language, like C, you could code the Nybble-ToASCII subroutine as:

```
char NybbleToASCII( int Nybble ) { // Convert "Nybble" to an ASCII
                                   //  character
 if ( Nybble > 9 )
  return Nybble + 'A' - 10;        // Return "A" to "F"
 else
  return Nybble + '0';             // Return "0" to "9"
} // End NybbleToASCII
```

However, this is fairly difficult to do in assembler. Instead, you might consider using a table:

```
NybbleToASCII:                 ; Convert the Contents of "ACC" to an
                               ;  ASCII Character
  add  A,NToATable-GNToA       ; Get the offset into the table
  movc A,@A+PC                 ;  Read the table
GNToA:
  ret
NToATable:                     ; Conversion table
  db   "0123456789ABCDEF"
```

This isn't bad, but it takes up quite a bit of control store space.

A better way of doing this is algorithmically by noting that ASCII "A" is ASCII "9" with 7 added to it:

```
NybbleToASCII:                 ; Convert the contents of "ACC" to an
                               ;  ASCII character
  add  A,#036h                 ; If ( A & 0x00Fh ) + 6 > 0x010h
  jnb  AC,Skip
  add  A,#7                    ;  Add 7 to jump to "A" to "F"
Skip:
  subb A,#6                    ; Take away flag set
  ret
```

This method is almost as fast as the table look up, and it is much smaller and, to my eye, more elegant. The subroutine adds a 6 to the nybble. If a digit has been carried to the upper nybble, then it is in the range of "A" to "F" and 7 is added so that, when the 6 used to set the flag is taken away, the result will be an accurate ASCII value for the nybble.

There is one thing in this subroutine that I want to bring to your attention. In virtually all other cases where I use a `subb` instruction, I have a `clr C` instruction to make sure the carry/borrow flag is reset before executing the subtraction. In this case, I don't have to because I know that there is no way that the one or two adds to the nybble can cause the carry flag to be set and they actually reset the flag.

Converting an ASCII character to a nybble can be done in a similar manner:

```
ASCIIToNybble:                 ; Convert an ASCII character to a nybble
  clr  C
  jnb  ACC.6,Skip              ; If bit 6 of the ASCII value is set
```

```
  subb A,#7                    ;  subtract difference between "A" & "9"
Skip:
  subb A,#030h                 ; Take away "0" to get the nybble
  ret
```

In the last two subroutines, I have tried to take advantage of the properties of the ASCII characters to simplify the conversions. The real lesson in this section is that, if you spend some time examining data and looking for relationships that you can exploit, you can often come up with better routines to implement functions that seem to require many cycles and complex operations.

Sorting Numbers

When I was in college taking computer courses, sorting data was something I found unbelievably boring and not that applicable in real life (at that time). This was probably the worst attitude I could have had because I find that I have to add a sorting routine to an application I'm working on about once every two years or so. While I wish that I had spent more time paying attention to the professor, I think I have been able to develop enough knowledge within myself to be able to write efficient sorting routines as well as understand which routines are best used for different applications.

Maybe I should qualify what I've said. I have discovered that I can meet pretty well all my requirements with two different types of sorting. *Bubble sorting* consists of reading through an array and comparing adjacent values and exchanging them to put the two in order. *Quick sort* is a much more sophisticated algorithm that splits data up into upper and lower halves and then repeats the process on each half until the array is sorted. These two algorithms represent a choice between two algorithms, which are best suited for simple and short sort arrays as well as large data structures.

To demonstrate these two different sorting routines I first had to come up with a random number routine. I guess I could have created the data on my own, but I find that's a real pain and, to quote Dr. Hannibal Lecter, "desperately random." I also wanted to come up with a small routine that would execute quickly, would not have repeating data, and could be easily changed to produce different amounts of data.

To meet these requirements, I decided upon a simple routine where I add a prime number repeatedly to itself to get a non-repeating value. I used the routine:

```
Randomize:                 ; Store random values in memory
 mov    R0,#ArrayStart     ; Start of the random buffer
 mov    A,#181             ; Load the accumulator with the prime value
RandomLoop:                ; Loop here until the memory array is loaded.
 mov    @R0,A              ; Save the random value
 add    A,#181             ; Create a new random value
 inc    R0                 ; Point to the next array element
 cjne   R0,#LOW(ArrayEnd+1),RandomLoop
 ret
```

I choose the value of 181 (or 0B5h) as the random value because it gives a good variety of values while not repeating for 256 iterations. When I ran the routine to load the high 128 bytes of a DS80C320s scratchpad RAM, I came up with the following data:

```
0080 B5 6A 1F D4 89 3E F3 A8 5D 12 C7 7C 31 E6 9B 50
0090 05 BA 6F 24 D9 8E 43 F8 AD 62 17 CC 81 36 EB A0
00A0 55 0A BF 74 29 DE 93 48 FD B2 67 1C D1 86 3B F0
00B0 A5 5A 0F C4 79 2E 3E 98 4D 02 B7 6C 21 D6 8B 40
00C0 F5 AA 5F 14 C9 7E 33 E8 9D 52 07 BC 71 26 DB 90
00D0 45 FA AF 64 19 CE 83 38 ED A2 57 0C C1 76 2B E0
00E0 95 4A FF B4 69 1E D3 88 3D F2 A7 5C 11 C6 7B 30
00F0 E5 9A 4F 04 B9 6E 23 D8 8D 42 F7 AC 61 16 CB 80
```

When you look at these values, they probably don't look that random, because the least significant digit is the same for each column of numbers, but the most significant digit is quite random and certainly acceptable for the purposes of the bubble sort and quick sort programs.

The bubble sort algorithm compares one byte to the next and exchanges them if the one at the lower address is greater than the other. This starts at the beginning of the list and goes to the end and repeats this until there are no bytes are out of order.

I was able to write a simple bubble sort subroutine in only 28 bytes of 8051 control store (Listing D-6).

LISTING D-6 **The bubble sort routine.**

```
Sort:                                ;  Sort the value

  mov     EndLoc,#LOW(ArrayEnd+1)

SortLoop1:                           ;  Setup the array information

  mov     ArrayIndex,#LOW(ArrayStart+1)  ;  Current pointer is in R0
  mov     R1,#ArrayStart

SortLoop2:                           ;  Loop here Until ArrayIndex == EndLoc

  mov     A,@R0                      ;  Load the next value to check
  clr     C
  subb    A,@R1                      ;  Do we have a higher value

  jnc     SortLoopSkip               ;  Yes, leave as is

  mov     A,@R1                      ;  Swap the values
  xch     A,@R0
  mov     @R1,A

SortLoopSkip:

  inc     R0                         ;  Point to the next test value
  inc     R1                         ;  Point to the next previous value

  mov     A,R0                       ;  Are we at the end?
  xrl     A,EndLoc

  jnz     SortLoop2                  ;  Nope...

  dec     EndLoc

  mov     A,EndLoc                   ;  Are we at the start of the array?
  clr     C
  subb    A,#ArrayStart+1            ;  Stop at one after the first

  jnz     SortLoop1

  ret                                ;  Everything is sorted
```

TABLE D-3 BUBBLE SORT PREFORMANCE	
SORT ARRAY SIZE	BUBBLE SORT CYCLES REQUIRED
16	2,077
32	7,816
48	17,270
64	30,619
80	47,488
96	68,063
112	92,389
128	120,394
144	152,096
160	187,990
176	227,098
192	269,894
208	316,423

This code is very efficient and very frugal in terms of resources (only the A and B registers and R0 and "R1" are required for its operation). However, it can take an unreasonably long amount of time to execute when sorting large amounts of data.

When the bubble sort algorithm executes, on average, it will move each byte to be sorted half the size of the array (some bytes will be left where they are and some bytes will be moved from one end to the other). Because there are "n" bytes, the time required to run the program can be expressed as:

$$\begin{aligned} Time &= k * (n / 2) * n \\ &= (k / 2) * n ** 2 \end{aligned}$$

where k is the time required to move one byte from one position to one position in the data to be sorted.

Because the length of time needed to sort data is proportional to the square of the number of pieces of data (or elements) to sort, the bubble sort is known to be a sort algorithm that is of the order "n-squared." This means that, if you have determined how long it will take to sort one amount of data, it will take four times to sort twice as much, or nine times to sort three times as much.

Once I had created the `Sort` subroutine that I have outlined earlier, I ran it on different sized random value arrays to see how long it took. In Table D-3, I have listed the results of the operation. I have not removed the number of cycles that it takes to create the random numbers, but it really is only a small fraction of the number of cycles actually used.

As I show actual cycle counts for the sorting routines, you'll notice that they deviate somewhat from the expected value. This is due to how the unsorted data is arranged. Depending on how often individual bytes have to be moved, the number of cycles required to sort the array may change. The deviations are typically less than 2% or 3% and as such should not affect the overall results. However, when you try to plot curves, you might discover that they are somewhat "lumpy."

I used UMPS to run the two sort routines, and you might want to do the same thing by selecting `View`, *Internal RAM*, then running the programs. Watching the Internal RAM window should give you an idea of how the program works (although it's pretty boring for bubble sort).

If you go through the previous bubble sort data, you'll see that the number of cycles required to sort increases non-linearly. If you look at the number of cycles required to sort different sized data arrays, such as 64 bytes and 128 bytes, you'll see that 128 bytes is approximately four times 64.

Quick sort, as I said earlier, splits the data into two halves, with one half greater than the mean and one half lower than the mean and then calls itself to sort each half. The algorithm could be written in C as shown in Listing D-7.

LISTING D-7 **The quick sort routine in C.**

```
QuickSort( int Bottom, int Top )      //  Sort the array from
{                                     //   bottom to top

int i, j;
int mean;
int temp;

  if ( Top == ( Bottom + 1 )){        //  Do a sort on last two elements

    if ( Array[ Bottom ] > Array[ Top ] ) {
      temp = Array[ Bottom ];         //  Swap the two elements
      Array[ Bottom ] = Array[ Top ];
      Array[ Top ] = temp;
    }

  } else {                            //  Sort > 2 elements in to two halves

    for ( i = Bottom; i < ( Top + 1 ); i++ )      //  Get array total
      mean += Array[ i ];

    mean = mean / ( Top + 1 - Bottom )      //  Get array mean

    i = Bottom;  j = Top;
    while ( i != j ) {              //  Split the data into two halves

      while (( Array[ i ] < mean ) && ( i != j ))
        i++;                        //  Find an array element above the mean

      while (( Array[ j ] < mean ) && ( i != j ))
        j++;                        //  Find an array element below the mean

      if ( i != j ) {               //  Swap the two value positions
        temp = Array[ i ];          //   - Lower half is less than mean
        Array[ i ] = Array[ j ];
        Array[ j ] = temp;          //   - Upper half is >= mean
      }
    }                               //  Finished splitting the data

    QuickSort( j, Top );            //  Sort the top half of the data
    if ( i > Bottom )               //  Sort the bottom half of the data
      QuickSort( Bottom, i );

  } //  Finished sorting

} //  End QuickSort
```

This is a good example of a recursive subroutine. After the data is split into two halves, `QuickSort` calls itself and passes the addresses of the two halves until all the data is sorted. When there are just two elements left to sort, the subroutine swaps them if the one at the lower address is greater than the one at the higher address.

At the end of the routine, you'll see that I check to see if there is only one element to be passed to the subroutine for the lower half of the data. I do this because, depending on how the averaging routine works, I could end up with no data values that are less than it (this is very possible in the case where there is only one or two elements passed to `QuickSort`).

I was able to implement the C code quick sort algorithm in 8051 assembler in 153 bytes, and it can be found on the CD-ROM that comes with this book in the PROG40 subdirectory as PROG40Q. I have omitted reproducing it here because it does take up a lot of space and the C pseudo-code in Listing D-7 is a lot easier to understand.

To find the average of the elements, I have had to add an 8-bit to 16-bit addition routine and a 16-bit divided by 8-bit division routine (not using the `div AB` instruction) to find the mean of the values passed to `QuickSort`. This code actually adds quite a bit of overhead to the software. If there were instructions capable of operating on 16-bit integers, the execution time of each instance of `QuickSort` would be reduced considerably.

The code itself executes in remarkably few cycles when large amounts of data are passed to the sort routine. This is definitely something that you would want to watch executing because, for large data tables, it executes quite quickly and you can watch the data move in a very purposeful (at least compared to the bubble sort) fashion.

Repeating the experiment where I plotted the results for different values, I came up with the results shown in Table D-4.

To compare this with the bubble sort results, I graphed both algorithms against the number of bytes to sort (Fig. D-9).

In the graphs, you can see the square relationship to the number of bytes to sort in the bubble sort's curve. However, in the quick sort's curve, you can see that it's much gentler

TABLE D-4 QUICK SORT PERFORMANCE

SORT ARRAY SIZE	QUICK SORT CYCLES REQUIRED
16	4,332
32	9,977
48	18,817
64	23,407
80	32,373
96	40,602
112	48,382
128	58,454
144	65,276
160	72,998
176	81,123
192	89,184
208	99,978

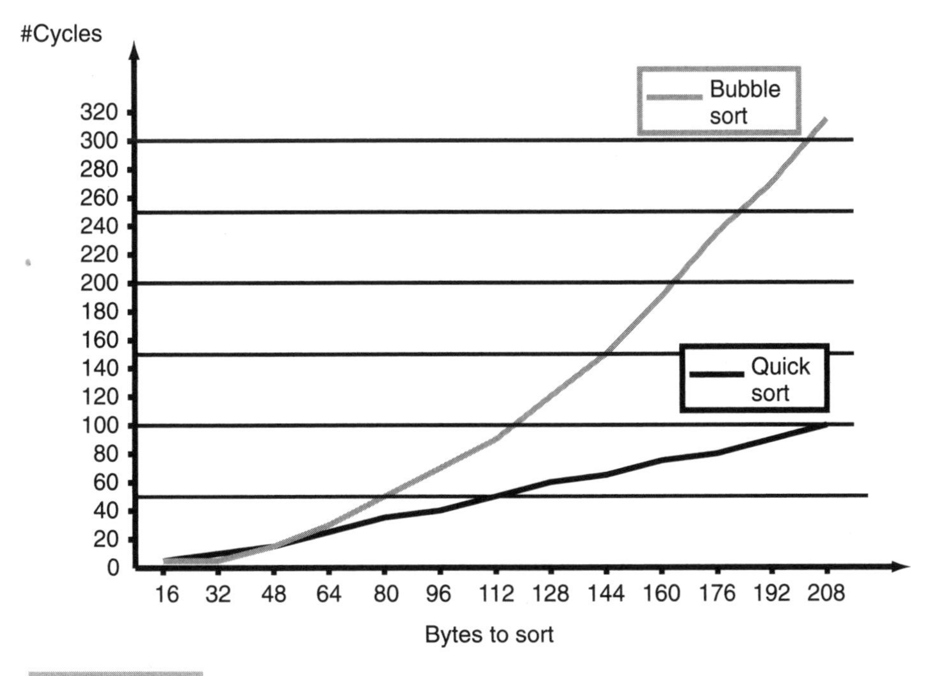

FIGURE D-9 **8051 sorting comparison.**

and almost linear as the amount of data to sort increases. This difference is due to the fewer times data has to be moved in the quick sort algorithm.

In quick sort, each element can be moved once during each time the `QuickSort` subroutine is called. Because the `QuickSort` subroutine halves the data each time it is invoked, the number of times it is invoked is the base 2 logarithm of the number of bytes to sort. Because each time the subroutine is invoked, each byte in the routine could be moved, the total number of cycles can be stated mathematically as:

$$Time = k * n * \log_2 n$$

Quick sort and some other algorithms are known as being on the order of "n log n." This means that, as the size of data increases, the number of cycles required to sort the code increases at this rate. The best possible rate of increase for a sorting routine would be on the order "n," which means that, as more elements are added to the array to sort, the time required to sort them will increase linearly. There is no algorithm that will sort unknown input array elements on the order "n".

There is one important aspect of the recursive quick sort algorithm that I've shown that I haven't discussed yet: the stack required to execute the subroutine. Each time the subroutine is called, 4 bytes of stack space are used. These 4 bytes consist of 2 bytes for the calling return address and 2 bytes used for saving the `Top` and `Bottom` indices for each invocation of `QuickSort`.

The stack space for quick sort can be defined as:

$$Stack\ Space = (4 * \log_2 n) + 4$$

which means that you have to understand how large the array is that you are going to sort. As I increased the number of bytes that I was sorting, I could very clearly see the stack growing to require more memory.

The previous stack space formula means that, for each power of two the size of the array of data to be sorted grows to, four more bytes will be required by the stack. Using the quick sort algorithm to sort an array of 2 bytes will require 12 bytes of stack space. For an array of 4 bytes, 16 bytes of stack is required. At 208 bytes (the maximum size that I could try out with this code), 32 bytes of stack was required. The quick sort stack requirements were the primary reason why I only went up to a 208 byte test array to sort.

There is one interesting thing to note about Fig. D-9: The bubble sort is actually more efficient than quick sort for processing small arrays. For arrays up to 48 bytes, the bubble sort algorithm is more efficient, with the best advantage somewhere around 16 array elements.

With this information, if I were going to use the quick sort algorithm in an actual application, I would check to see if the array received by `QuickSort` is less than 30 elements. If this were the case, I would probably execute the bubble sort algorithm for this data because it would be faster than the quick sort code and would save me a few bytes on the stack. If I was to avoid using quick sort on all 16-byte arrays passed to `QuickSort`, I would save 16 or 20 bytes of stack space, which could be crucial in the application.

Neither sorting algorithm is tremendously expensive in terms of control store or memory used. The big differentiator between the two is their performance for different sized arrays. Bubble sort is faster and more efficient with small data sizes, while quick sort is the obvious choice for large data arrays to sort. When implementing a quick sort algorithm, it's important to make sure that there is sufficient variable and stack space and that it won't be overrun when the data is loaded (this is of primary concern when an array is loaded into the 8051's scratchpad RAM).

Both algorithms can be used in the 8051 for sorting individual fields in large data structures in external memory. In this case, the data will have to be pointed to by the DPTR index register, which will require some changing of it when looking for data to move as well as when moving the actual data. If you do plan on implementing a sort routine on large data structures in external memory, you should make sure you reserved some memory in the scratchpad RAM for making the transfers easier.

Earlier, I said that order "n" sort algorithms weren't possible. This isn't quite true. You can implement an order "n" sort routine if you know where incoming data is going to go, but you don't know in what order it's going to show up in. Actually, the case for the order "n" sort is even more restrictive with their having to be all the possible data points being received, with no missing ones or duplicates.

If these conditions are met, by simply using the incoming data as the index to the address in memory where it's going to be stored, an order "n" sort algorithm can be implemented. Admittedly, this is a pretty unusual situation (I've never seen it). However, to avoid e-mail later saying that there was the possibility of an order "n" sort algorithm for some applications, I felt I should point out that this is possible.

Loading the "Encryption Array"

Earlier in the book, I told you that, if any unused EPROM or the encryption array was left unprogrammed, somebody could figure out what the EPROM or encryption array contents were by reading out the EPROM contents and checking to see if what's read back can be executed directly or, if the encryption array is programmed, looking for a repeating pattern.

To prevent this, a non-repeating pattern should be put into both the encryption array as well as the unused EPROM.

The following conditional assembler code could be used:

```
ProgEnd EQU $                  ; Identify the end address of the code
 while ( $ < DEVEND )          ; Repeat to the end address of the device
 db     ( ProgEnd + (( $ - ProgEnd ) * 3 ) & 0FFh
 wend
```

If this code was put at the end of an 8051's program, the remainder of control store memory, no matter how large, would be filled with a pattern increasing by three for each byte and starting with the least significant byte of the address after the end of the program. By incrementing by three, the pattern won't repeat at the same address for 768 bytes. This can be increased by using a larger prime number.

Along with incrementing by three, the algorithm presented earlier that I used for creating random numbers (simply repeating an add of decimal 181 to a sum) for testing the sort routines could also be used. It repeats every 255 addresses and not every 256, which would mean that the starting address of the pattern would not be repeated for roughly 65,000 bytes of control store.

Using two different algorithms (or the same one with different primes) to produce pseudo-random data to be loaded into control store and the encryption array will provide a very high level of security for your code. You could use a much more complex algorithm or random data from some source (such as copying the source file from the end of the program to the end of the control store's EPROM). However, using something that will not repeat at the same address for a large number of iterations will make it very difficult for a pirate to figure out what the contents of the 8051's EPROM are loaded with.

Circular Buffers

In some applications that require incoming data to be processed serially and where data might come in before the microcontroller can devote the cycles to processing it, the incoming data has to be *buffered* (saved) for some period of time. Ideally, the algorithms used for saving and retrieving this data, as well as maintaining the buffer, should be as simple and execute as fast as possible. Like most problems, there are many ways of solving it, with one of the best being the *circular buffer*.

A circular buffer is a memory area that has been set aside for an index to store data and another index to retrieve the data. Once an index has reached the end of the buffer, it is reset to the start of the buffer. In this way, the buffer is circular in that it never ends.

An application that has an interrupt handler that stores incoming serial data that is processed by mainline code would probably use a circular buffer.

To put data into such a buffer, the following code could be used:

```
Push( Data ) {                    // Push "Data" into the buffer
  Buffer[ PushIndex ] = Data;     // Save the data in the buffer
  PushIndex++;                    // Increment the buffer
  if ( PushIndex == BufferEnd )
    PushIndex = BufferStart;      // Reset the buffer if necessary
} // End Push
```

In 8051 assembler, this could be:

```
Push:                    ; Push the contents of the accumulator
                         ;  into the circular buffer
 mov  @R0,A              ; Save the contents of A into buffer pointed
                         ;  to by R0
 inc  R0                 ; Point to the next location in the buffer
 cjne R0,#BufferEnd+1,PushEnd
 mov  R0,#BufferStart    ; If at the end of the buffer, point to the
                         ;  start
PushEnd:
 ret
```

Pop would use almost exactly the same code. The only difference would be in the first (mov) instruction. In Pop, it would be mov A,@R0. In most applications, the pop code would use R1 as its index register.

After Push has gone through each address, the data pointer (R0) is reset to the start of the buffer. I used R0 as the index because it can be compared to and updated without affecting the contents of the accumulator or PSW register.

This can be improved by sizing and locating the buffer in such a way that its starting address has a "0" at the least significant bit before the bits used to address within the buffer. For a 16-byte circular buffer, having bit 4 of the address always reset will allow the Push routine to be simplified to:

```
Push:                    ; Push the contents of the accumulator into the
                         ;  circular buffer
  mov  @R0,A             ; Save the contents of A into buffer pointed to
                         ;  by R0
  inc  R0                ; Point to the next location in the buffer
  anl  R0,#(BufferStart+(BufferSize-1))
  ret
```

If the 16-byte buffer was located in memory from 020h to 02Fh, then by ANDing the incremented address by 02Fh (start address plus the buffer size minus one), R0 will always remain pointing to an address within the buffer boundaries.

In this example, if a push was done with R0 pointing to address 02Dh, R0 would be incremented to 02Eh, which would not change. The next push would cause R0 to be incremented to 02Fh (which is the end of the buffer). The third push would increment R0 to 030h, but the AND instruction would set R0 to 020h, the start of the buffer.

The advantages of this method over the previously shown code include taking fewer control store bytes and, each time Push was executed, a constant number of instruction cycles would be required. There could be problems with timing applications in the first example if the circular buffer push takes a different number of instruction cycles depending on what R0 is set to when Push is called.

The disadvantage of this method is that the buffer must be a power of two (to allow the least significant buffer address bit to be reset). This limits the buffer size to 2, 4, 8, 16, and 32 bytes in the standard 8051 scratchpad RAM (of 128 bytes).

E

UMPS

As I discussed in chapter 8, finding good tools for 8051 program development is difficult. I have been very lucky to discover a real treasure in the UMPS (Universal Microprocessor Program Simulator) integrated development environment (IDE). This program (Fig. E-1), which runs under Microsoft Windows, is able to handle the development for a large number of different microcontrollers (basically everything presented in this book). Along with an editor and assembler, UMPS contains the ability to simulate devices with external hardware, which really sets UMPS apart from other tools and eliminates the need for stimulus files. This capability of connecting UMPS to virtual devices gives you a real "what if" capability in your application design.

UMPS runs under Microsoft Windows (3.11, 95, and NT) PC operating systems. Along with its own internal assembler, UMPS can initiate the operation of the Microchip PIC Assembler (MPASMWIN) as well as the Cosmic Assembler and C Compiler and use their symbolic output for the UMPS simulator. More compilers will be added over time.

UMPS will run from the CD-ROM directly without any problems. But, source code must be copied from the CD-ROM onto your PC's hard file before attempting to edit, assemble, or simulate them under UMPS.

The most powerful and useful feature of UMPS is its graphical "resources" that can be "wired" to the device. At the time of writing, the virtual hardware for the simulated microcontroller includes:

- LEDs
- Push buttons
- Logic functions (AND, OR, XOR, NOT)
- 7-segment LEDs

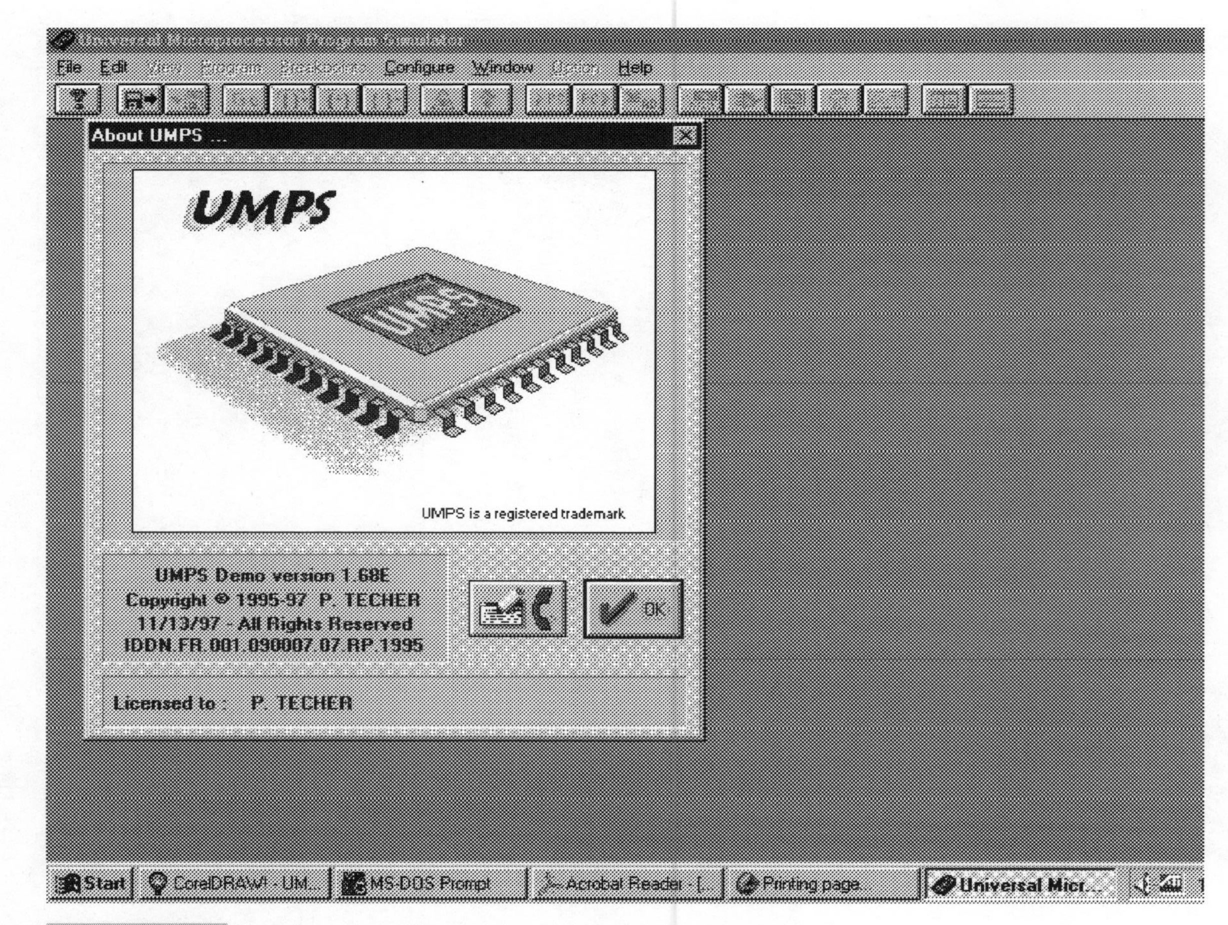

FIGURE E-1 The Universal Microprocessor Program Simulator.

- A/D convertor
- A/D "slider"
- Serial receivers/transmitters
- I2C memory (RAM and EEPROM)
- I2C peripherals (LED displays, real-time clocks, etc.)
- HD44780 compatible LCDs
- Phase-locked loops
- PWM monitors
- PWM generators
- Pull-up resistors
- Serial A/D convertors
- 74LS138, 74LS139, 74LS373, and 74LS374 TTL chip simulations
- CD4017 and CD4094 CMOS chip simulations

More simulations will be added as time goes on, but you should see that there really is a critical mass of different devices available, which should make simulating many different applications possible. These simulated devices (which are known as *resources*) are very full featured. I was able to simulate all of the 8051s experiments and many aspects of the example applications using UMPs.

When setting up the resources, you can move them around the resource window to arrange them as they make sense to you or how they will be laid out in the final application. You can also modify many of the I/O devices with different options. (For example, the LCD display can be one or two rows and 8 to 20 columns or, along with the pull-up resistor, the I/O pins can be hard wired to Vcc or Ground.) In chapter 9, I went through a simple tutorial on setting up UMPS and show how resources can be wired to the virtual processor.

UMPS currently supports the following devices (with more being added all the time):

- Dallas Semiconductor: DS87C310, DS87C320
- Intel: 8031, 8032, 8051
- Atmel: AT89C1051, AT89C2051
- Microchip: PIC12C5*xx*, PIC16C5*x*, PIC16C84, PIC16F84, PIC16F83, PIC16C554, PIC16C556, PIC16C558, PIC16C671, PIC16C672, PIC16C71, PIC16C710, PIC16C711
- Motorola: 68HC705, 68HC705J1A, 68HC705P9, 68HC705B16, 68HC705B32, 68HC11
- SGS Thomson: ST6210, ST6215, ST6220, ST6225, ST6252, ST6253, ST6260, ST6262, ST6263, ST6265
- National Semiconductor: COP820C

The assembler and simulator available in UMPS are very fast. I saw 50,000 instruction cycles per second on my 133-MHz Pentium PC.

With the consistent interface, constant improvements, and additional microcontrollers and resources being added all the time, UMPS is an investment that you will use for a very long time.

The copy of UMPS on the CD-ROM included with this book is the latest demo version. This is a full version of UMPS that will run for three months after being installed. The only restriction is that projects cannot be saved (which means that you will have to recreate your graphical simulator each time you use it).

Additional features in this version of UMPS include:

1. Back trace as you single step. You can go back up to 16 instructions to see what happened.
2. A register can be incremented or decremented by selecting it and using the "+" or "-" keys.
3. In the UMPS assembler source, a binary word can be written in the following formats:
   ~%010010110
   ~10010110B
   ~B'10010110'

To find out more about UMPS contact:

Virtual Micro Design
I.D.L.S.
Technopole Izarbel
64210 Bidart
France
Phone: 011-33(0)559.438.458
Fax: 011-33(0)559.438.401
email: p.techer@idls.izarbel.tm.fr
http://idls.izarbel.tm.fr/entp/techer/index.htm

UMPS is available in North America from:

Wirz Electronics
P.O. Box 457
Littleton, MA 01460-0457
email: sales@wirz.com
http://www.wirz.com

F

SIMMSTICK

One of the most useful products to come along for developing fast microcontroller applications (either for hobbyists or industrial applications) is the SiStudio/DonTronics SimmStick cards and bus. This system of cards will provide you with a basis for electrically wiring a microcontroller into an application. I have used SimmSticks for a number of applications, and I can usually wire an application in under an hour.

The SimmStick was originally built around the PICStic (by Micromint). This device, using a PIC16C84, could functionally simulate a BASIC Stamp (using microEngineering Lab's PBASIC Compiler) or be used as an application unique to itself by deleting or adding components to the various boards. The original PICStic was just designed for the PIC16C84; the SimmSticks were originally designed to provide prototype cards for the full range of PICMicro products. Now, along with the PICMicros, the SimmSticks can also support the Atmel 8051 and AVR devices as well.

The SimmStick can be loaded into a standard 30-pin memory SIMM socket.

The SimmStick bus consists of the pins listed in Table F-1.

I should discuss a few things about this bus. The first being the number of pins available for digital I/O. The 23 I/O pins were originally specified to provide I/O capabilities for a variety of microcontrollers. If the microcontroller that you are using does not have up to 23 I/O pins, then the extra pins can be used for other purposes (the composite video output application places a voltage divider D/A convertor in the prototyping area to provide the composite video). For each microcontroller, you will have to understand how the I/O lines are wired to the SimmStick bus.

Serial data can be transmitted or received at TTL (CMOS) or RS-232 levels. The RS-232 interface uses a Maxim MAX-232 chip that has vias and connections built into the SimmStick card. These pins could also be used for digital signal I/O. The RS- 232 communications

TABLE F-1 THE PINS OF THE SIMMSTICK BUS

PIN	LABEL	DESCRIPTION/COMMENTS
1	A1/Tx	RS-232 transmit line from the SimmStick
2	A2/Rx	RS-232 receive line to the SimmStick
3	A3	Microcontroller general purpose I/O
4	PWR	Unregulated power in
5	CI	Line to microcontroller's XTAL pins
6	CO	Line to microcontroller's XTAL pins
7	+5V	Regulated +5 V
8	Reset	Microcontroller negative active reset
9	Ground	System ground
10	SCL	I2C clock line/microcontroller general purpose I/O
11	SDA	I2C data line/microcontroller general purpose I/O
12	SI	CMOS logic serial line in/MCU general purpose I/O
13	SO	CMOS logic serial line out/MCU general purpose I/O
14	IO	Microcontroller general purpose I/O and Timer In
15	D0	Microcontroller general purpose I/O
16	D1	Microcontroller general purpose I/O
17	D2	Microcontroller general purpose I/O
18	D3	Microcontroller general purpose I/O
19	D4	Microcontroller general purpose I/O
20	D5	Microcontroller general purpose I/O
21	D6	Microcontroller general purpose I/O
22	D7	Microcontroller general purpose I/O
23	D8	Microcontroller general purpose I/O
24	D9	Microcontroller general purpose I/O
25	D10	Microcontroller general purpose I/O
26	D11	Microcontroller general purpose I/O
27	D12	Microcontroller general purpose I/O
28	D13	Microcontroller general purpose I/O
29	D14	Microcontroller general purpose I/O
30	D15	Microcontroller general purpose I/O

feature should really be wired in such a way that the connector on a bus card or on a prototype card is used and the appropriate bus lines are used for transmitting the data.

Actually, this brings up a good point. I typically only use prototype cards for I/O with signals passed through the backplane. This method of wiring will allow the microcontroller to be reused for new applications while the prototype I/O card can be hacked to bits.

The SimmStick cards are capable of taking unregulated power, regulating it, and distributing it throughout the application. Personally, I don't like to do this, instead I use a SimmStick backplane (either the DT001 or DT003) for power (and RS-232 connections) and eliminate the need for me to wire the backplane myself. (See Fig. F-1.)

The MCU's clock lines are also available on the bus. Either a crystal or ceramic resonator can be mounted on the SimmStick board with optional capacitors using the built-in vias and traces. The XTAL lines can also be distributed using the bus (a clock is mounted on a single card and distributed to microcontroller SimmSticks on the bus). (See Fig. F-2.)

I routinely cut the traces leading from the bus to the microcontroller on the SimmStick if I am using an onboard crystal. These traces will cause extra capacitances on the crystal lines and could cause application execution problems later. If the clock is to be distributed on the bus, I would recommend redriving the crystal signal on the SimmStick using a buffer redriving it on an unused pin or other devices on the card.

Microcontroller reset is negative active only. On the DT104, if an Atmel 8051 20-pin microcontroller (which has the same pin-out as the 20-pin AVR) is put on the card, a single transistor inverter can be wired in to provide the positive active reset the device requires. Actually, I usually replace R1 with a 0.1-μF capacitor and R6 with a jumper (0-Ω resistor).

The current range of SimmStick products includes bus and development platforms, microcontroller interface boards, and a number of add on interface boards for the bus. (See Tables F-2 through F-4.)

FIGURE F-1 DT003 SimmStick bus and power supply.

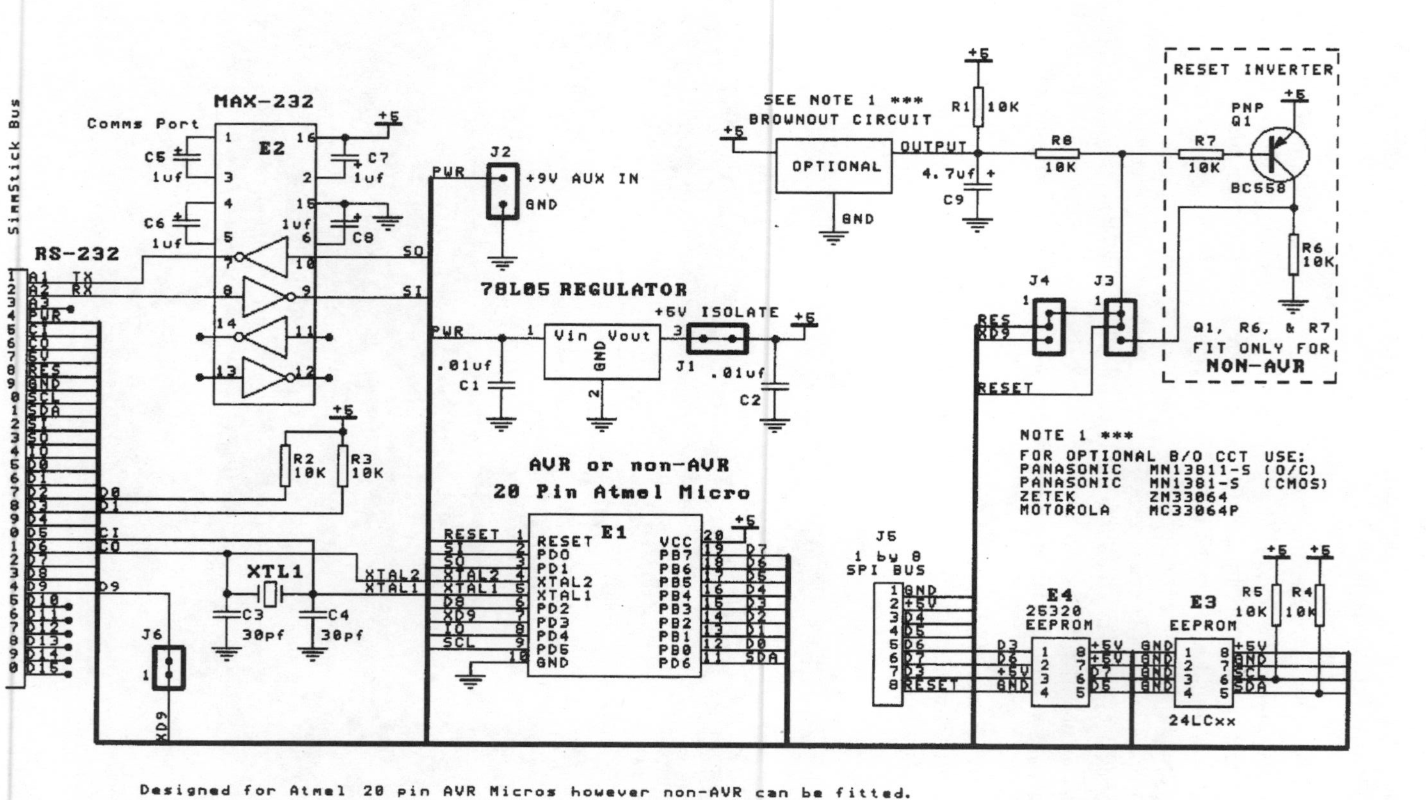

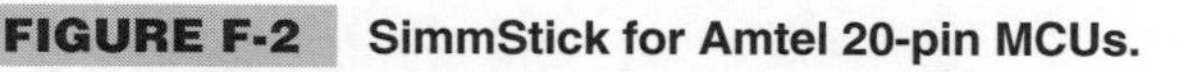
FIGURE F-2 SimmStick for Amtel 20-pin MCUs.

TABLE F-2 BACKPLANES AND DEVELOPMENT SYSTEMS

PART NUMBER	FEATURES
DT001	PICMicro programmer/SimmStick backplane
DT003	Backplane with integrated power and RS-232 interface

TABLE F-3 MICROCONTROLLER INTERFACE BOARDS

PART NUMBER	FEATURES
DT101	18-pin PICMicro SimmStick board with full RS-232 and I2C/SPI EEPROM access and control
DT103	40-pin AVR Atmel microcontroller SimmStick board with RS-232 access
DT104	20-pin Atmel microcontroller SimmStick board with RS-232 access and I2C/SPI EEPROM access and control

TABLE F-4 INTERFACE SIMMSTICK BOARDS

PART NUMBER	FEATURES
DT203	24 LED, 4/8 switch interface board
DT204	4-slot expansion board for SimmStick bus
DT205	4-relay control card with 4 LED outputs

Along with these boards, there are two prototype boards for the SimmStick bus.

With the SimmStick, there is one advantage I wonder if you've noticed. The SimmStick product line, at the time of writing, can support five of the most popular microcontroller architectures available today: the PICMicro, Atmel's AVR and 20-pin 8051, the Zilog Z8 (put backwards into the Atmel DT104), and Parallax's BS1 design. This gives you an interesting amount of capabilities because the devices can be interchanged in a single application to evaluate the performance of the different devices). This is why I like to keep the I/O peripherals on a separate SimmStick card.

What I have written here is really an introduction to the SimmStick. I highly recommend you looking at the DonTronics Web site (http://www.dontronics.com) to see the options available for the SimmStick as well as to go through the documentation to understand how the SimmStick can help you in your next application.

The SimmStick is available from:

DonTronics
P.O. Box 595
Tullamarine 3043, Australia
Phone: 011 +(613)9338-6286
Fax: 011 +(613)9338-2935
email: don@dontronics.com
http://www.dontronics.com

The SimmStick is available in North America from:

Wirz Electronics
P.O. Box 457
Littleton, MA 01460-0457
email: sales@wirz.com
http://www.wirz.com

G

REMOTE 8051 DEBUGGER

Before ending the book, I wanted to share with you one of the most interesting uses of the Internet that I have ever seen. The Remote 8051 Debugger allows you to develop 8051 applications on your PC and then test them out on an actual 8051 that is physically located in Brazil but that can be accessed through the Internet as if it were a local development system. This tool will allow you to try out applications and experiment with an 8051 without any investment in hardware. On the CD-ROM in the REMOTE subdirectory, I have included the code necessary to run the Remote 8051 Debugger and create applications for it.

The PC interface that you would use consists of a simple window that can be executed directly from the CD-ROM. To execute it, open your Web browser to the index.htm file in the root directory of the CD-ROM and then select `Internet Based 8051 Debugger`. This is shown in Fig. G-1. The Remote Debugger ("RExLab" or "CLI40.exe") can be run directly from the CD-ROM.

When the window is up and running, click on `Connect` to access the hardware in Brazil. Once you are connected (if somebody else is using the system, you might have to wait for them to finish), you can download an application and execute it.

The application itself only has a few special requirements. The three biggest are:

1. You have to start the application code at address 08000h (and not the usual 0).
2. Other than P1 and the INT0 and T0 pins, you don't have any input capability.
3. There is no "end address"; when the application code has completed, the 8051 should execute an endless loop.

To show how this tool works, I created the following example application, which polls P1 and then saves the value on P1 in bank register R0:

```
; Myke Predko's Test of the Remote 8051 Debugger
P1 EQU 090h                    ; Define the P1 register
ORG 8000H
LOOP:
 mov  A,P1                     ; Read the stimulus bytes
 mov  R0,A
 SJMP LOOP
END
```

This program is assembled using asm51 (which is in the CD-ROM's REMOTE subdirectory). Then the hex file is converted to a binary format using the following command sequence:

```
Asm51 myketest
Hexbin myketest.hex myketest.asm
```

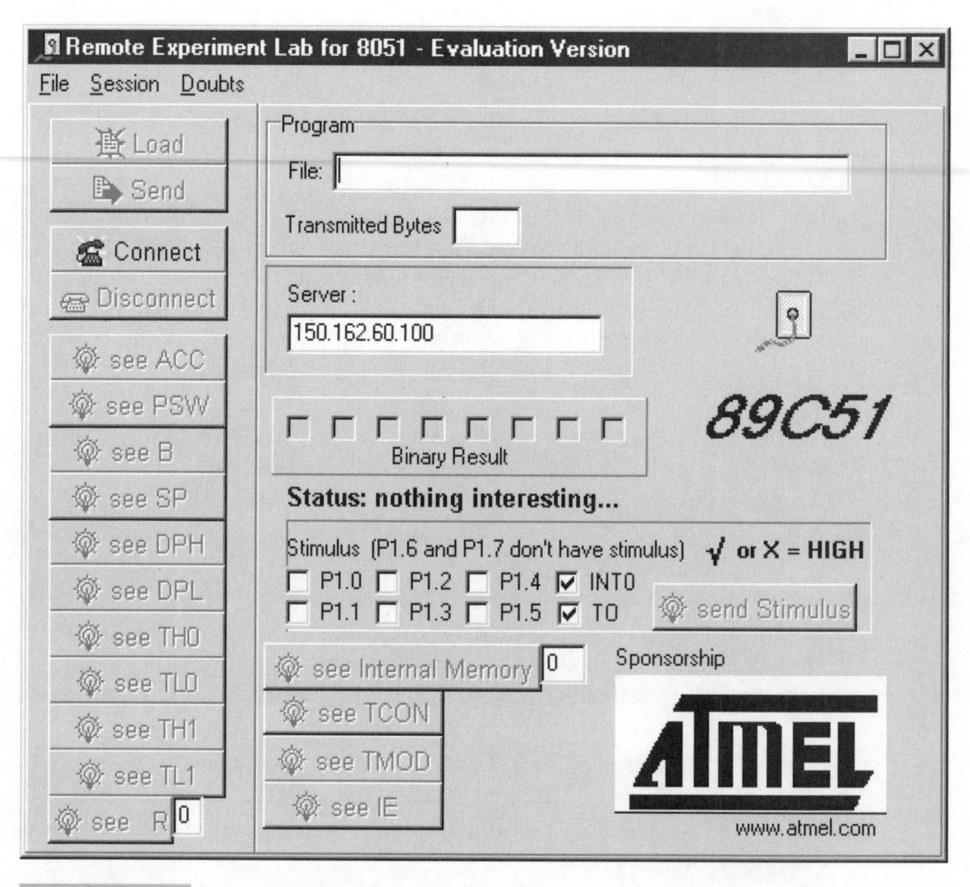

FIGURE G-1 Remote 8051 Debugger window.

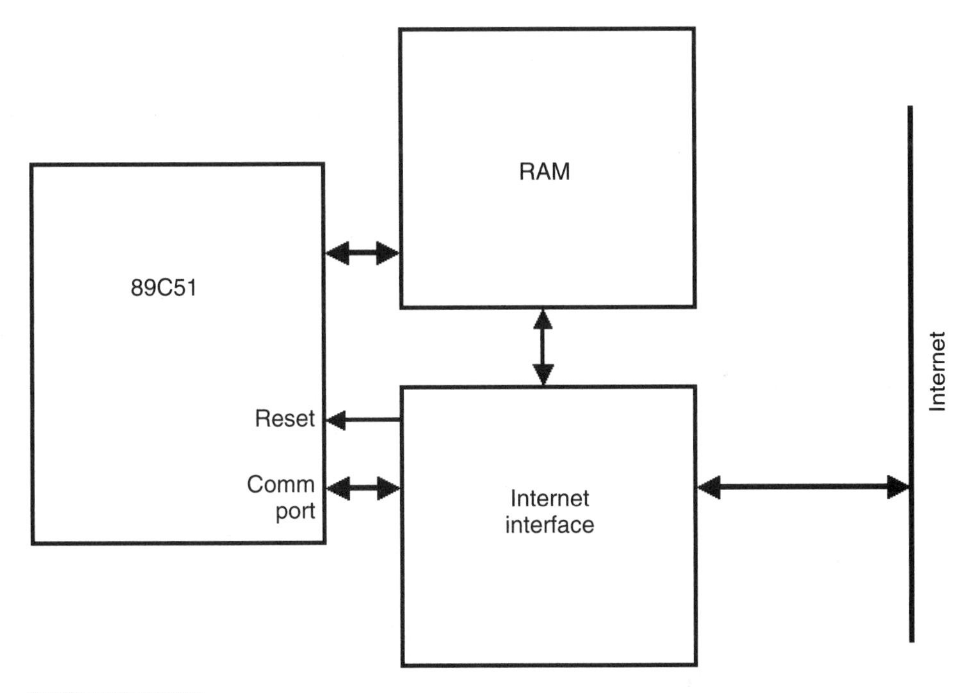

FIGURE G-2 **Remote 8051 block diagram.**

With the file ready, you execute CLI40.EXE and click `Connect`. Once the status message changes to `Connected`, you can `Load` the application, at which time it begins to run.

With myketest, you can select different stimulus bits (P1.0 to P1.5) and click on `Send Stimulus`. After an acknowledgment returns, you can see R0 and look at the P1 inputs passed to R0.

myketest is a very simple application. One of the Remote 8051 Debugger authors, Miguel Wisintainer, has tested it with the programs in chapter 9 (after changing the `org` statement to 08000h) and found that most of them run without any modifications.

The remote debugger itself consists of an Atmel 89C51 with external RAM connected to the Internet as shown in Fig. G-2. The assembled user application program code is sent over the Internet in a binary format, loaded into the RAM, and executed.

To read registers, a request is passed to the 89C51's kernel program by the Internet interface, which interrupts the 89C51 and passes data back and forth between the Internet and the microcontroller.

This appendix is just an introduction to this fascinating 8051 application. I suggest that you print out the Microsoft Word documents in the REMOTE subdirectory or contact the authors, Miguel Wisintainer (wf@ambiente.com.br) or Luis Marques (cleber@inf.ufsc.br), for more information.

As a final note: There are only eleven of these 8051 remote debuggers currently in operation. This means that you might have a wait to get on. As well, please be considerate of others and don't hog the 89C51 or leave it without disconnecting.

H

THE CD-ROM

With the latest advances in technology, computer and electronics books have really become full featured with all of the necessary files and information available in one package, and updates are a few mouse clicks away on the Internet. I'm pleased to be able to apply this technology to this book and provide with it a CD-ROM with virtually all of the information that you will require to work through all the experiments and projects in this book.

Included on the CD-ROM is all the source code for the applications presented in this book along with part data sheets, bareboard information, and a copy of the UMPS IDE for creating your own applications.

The CD-ROM consists of four primary subdirectories:

- CODE—PROG1 through PROG50 source files, each in their own subdirectory.
- DATASHT—Adobe .pdf files containing the data sheets for all of the parts used within this book.
- HTML—HTML source code for the Web browser CD-ROM interface.
- REMOTE—RexLab 8051 Internet Debugger.
- UMPS—A demonstration copy of UMPS.

To help make the CD-ROM easier to navigate and use, I have added an .html interface that will allow you to look through the data and programs much more efficiently than using the standard Windows or MS-DOS tools.

To access the .html file, start a Web browser (for this appendix, I will use Microsoft's Internet Explorer). Under `Files`, select `Open` as is shown in Fig. H-1. Now, click on `Open` (as shown in Fig. H-2), and then click on `Browse` (as shown in Fig. H- 3).

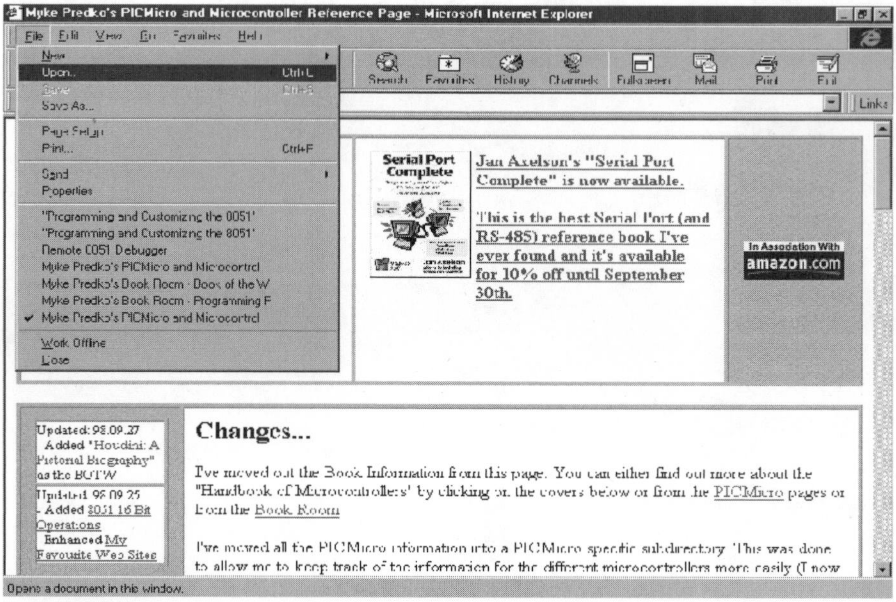

FIGURE H-1 Opening a file from Internet Explorer.

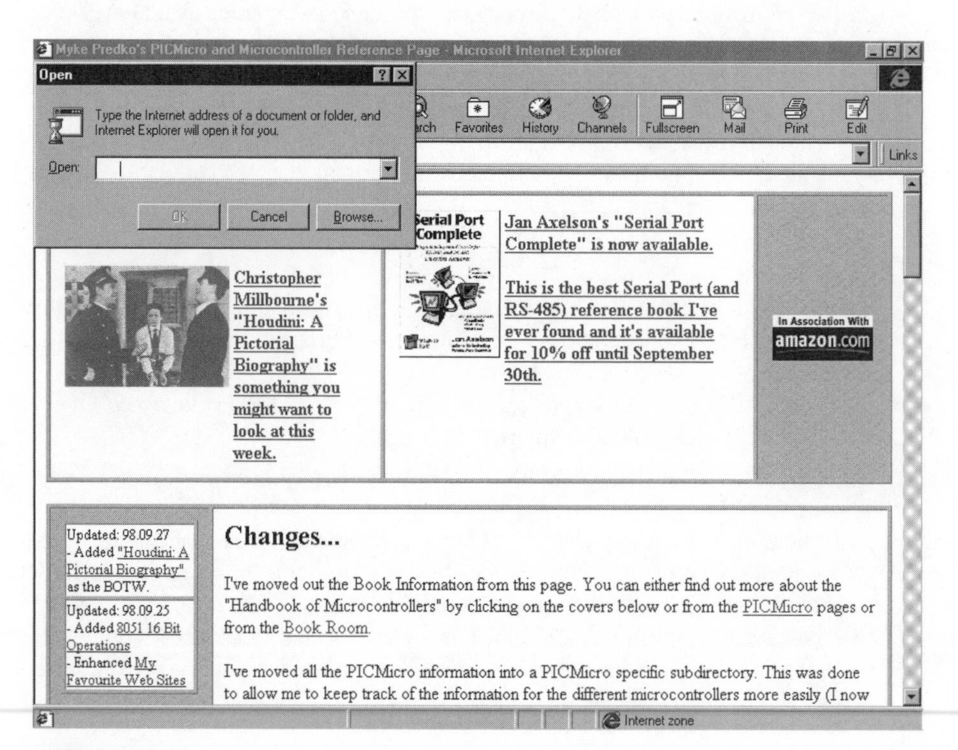

FIGURE H-2 Specifying a file to open in Internet Explorer.

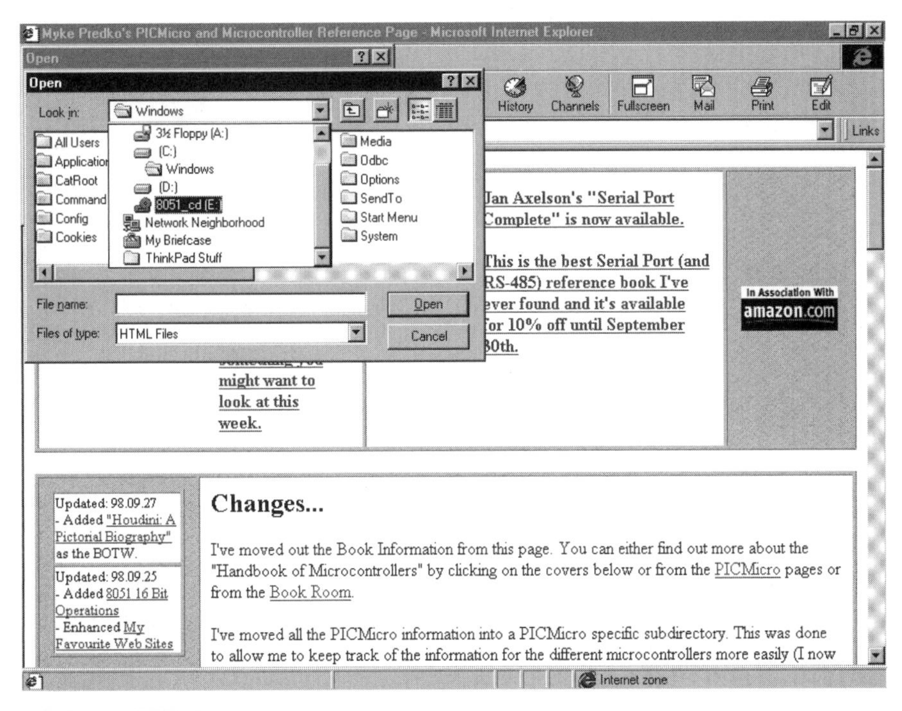

FIGURE H-3 Selecting the CD-ROM as the .html source.

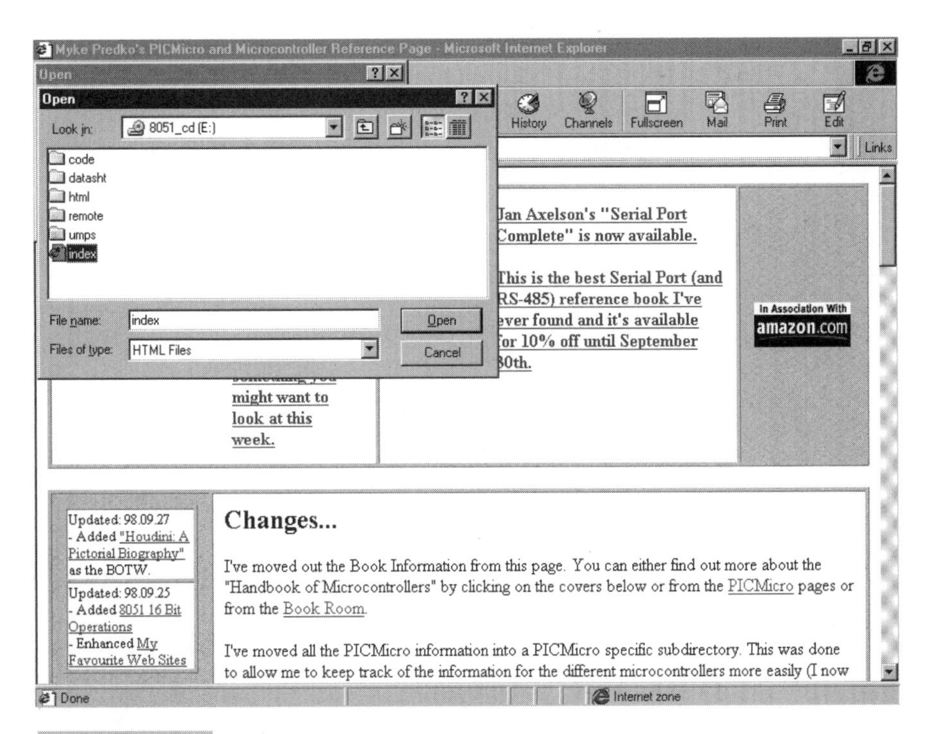

FIGURE H-4 The CD-ROM's .html selection.

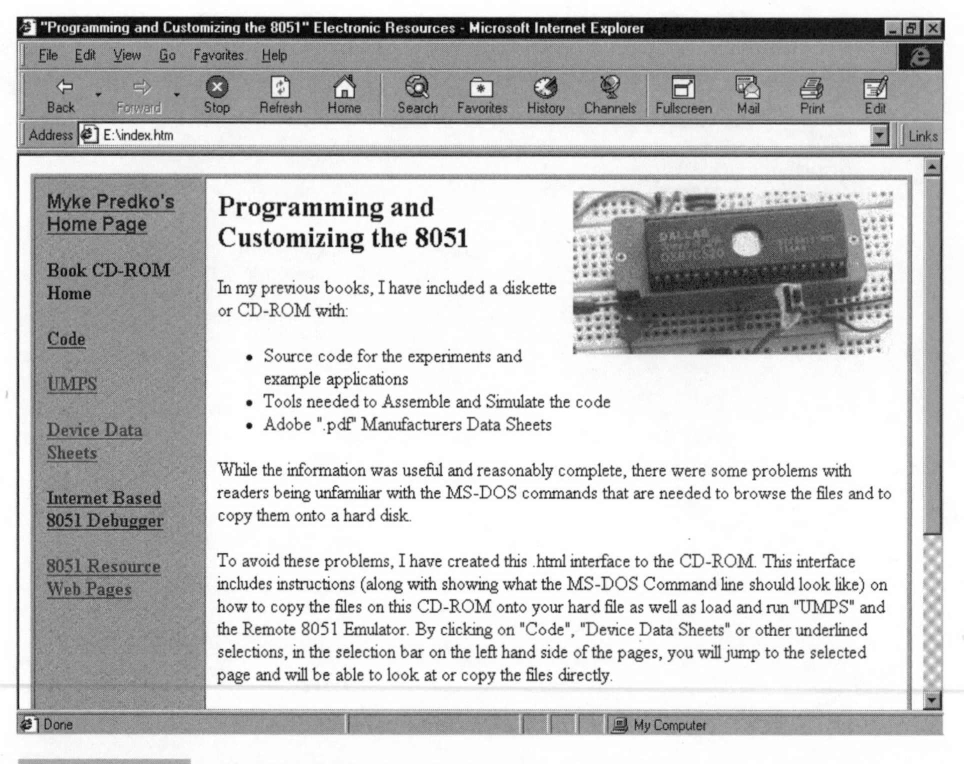

FIGURE H-5 The CD-ROM's .html interface.

Note, in Fig. H-3, that I have selected `8051_cd`, which is the CD-ROM in the drive (which has been allocated the drive letter "E" by Windows). Click on this selection, and you will get the files to execute in the browser. This can be seen in Fig. H-4.

Doubleclick on `Index` (which is shown highlighted in Fig. H-4). Click on `OK` when `Browse` gets the focus back, and you will now be in the CD-ROM's .html code. You can work through it exactly as you would a Web page. See Fig. H-5.

When working through the .html, I recommend that you have your Internet connection up and running. There are a few external pages that are reference. You should also check my Web page (http://www.myke.com) for errors and addendums.

Every effort has been made to ensure that the CD-ROM works properly and that all the files on it are correct. If you have any problems at all, please contact me via e-mail, and I'll get you the missing data files or data.

I

PROG50: ORDERING RAW CARDS AND KITS

When I wrote *Programming and Customizing the PIC Microcontroller*, I didn't realize the interest people would have in building kits out of the example applications and projects that I wrote about in the book. To help allow you to build the applications shown here that use embedded raw cards, I have provided you with three different ways of getting the cards.

The first is that the top (component), bottom (solder), and overlay diagrams have been provided in the sections that the projects appear in. Photocopies of these pages can be made, and boards can be manufactured by optical means.

On the CD-ROM in the \code\prog50 subdirectory, I have provided the gerber, aperture, and drill files in the code subdirectory. This information can be packaged up and sent to AP Circuits (following the instructions on their Web page) or another quick turn prototype PCB shop. To make this more efficient, I have laid out the AT89C*x*051 programmer, AT89C2051 emulator, 51Bot, and RS-232 Breakout Box all on the same image. (See Figs. I-1 through I-3.)

The individual cards can be cut apart using a saw or sheet metal brake.

While providing this information, I am not licensing you to reproduce these boards for sale without my written permission. As with the application design software presented in this book, the raw card images are for your own personal use only.

If you don't want to build the cards yourself, kits will be available from Wirz Electronics and their dealers. These kits will consist of all the parts necessary to build the applications, including programmed microcontrollers.

With these options, along with standard prototyping methods, you should find a way to build the applications in the manner that you are most comfortable.

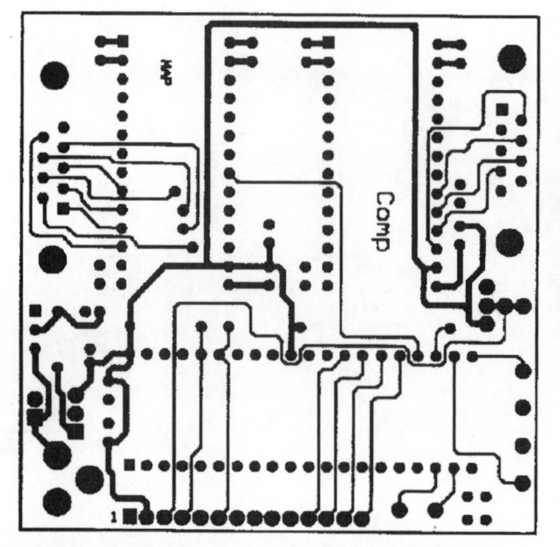

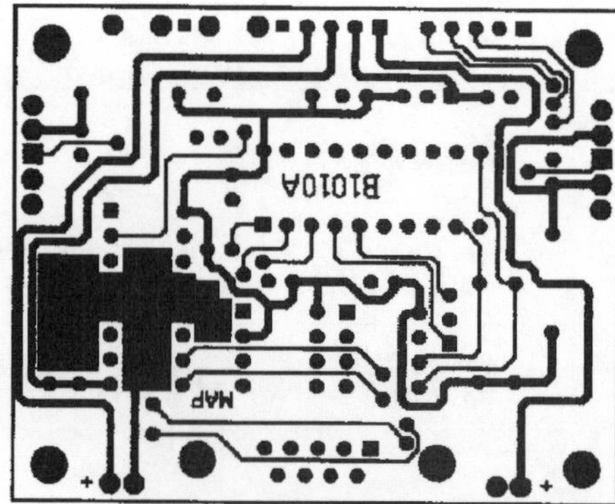

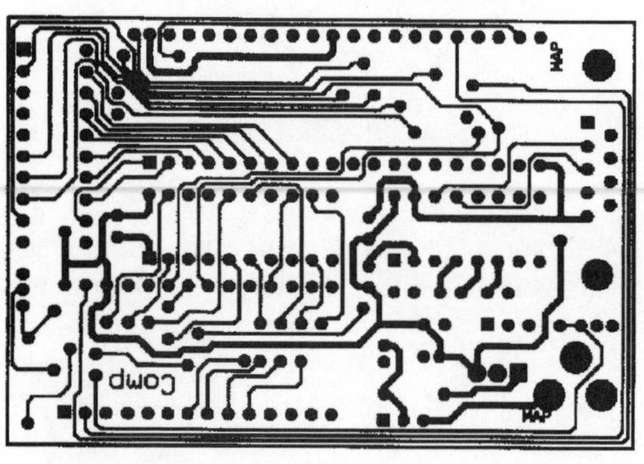

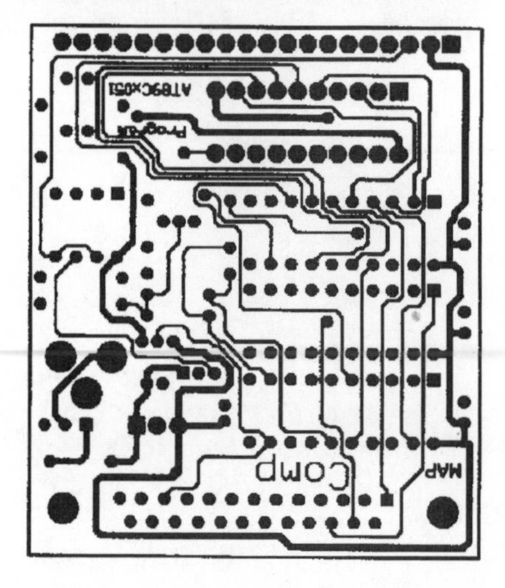

FIGURE I-1 Top side of the complete raw card.

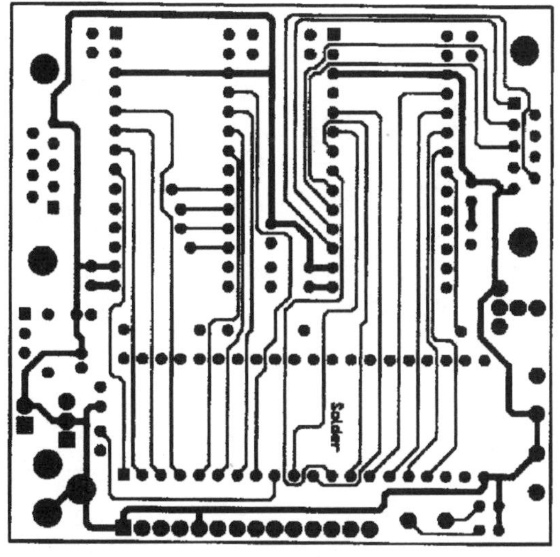

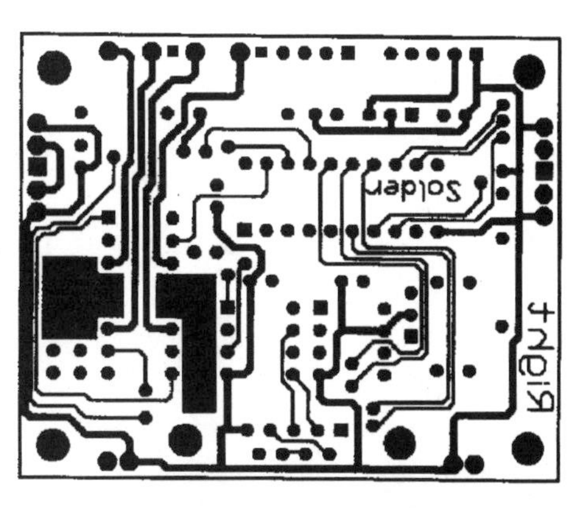

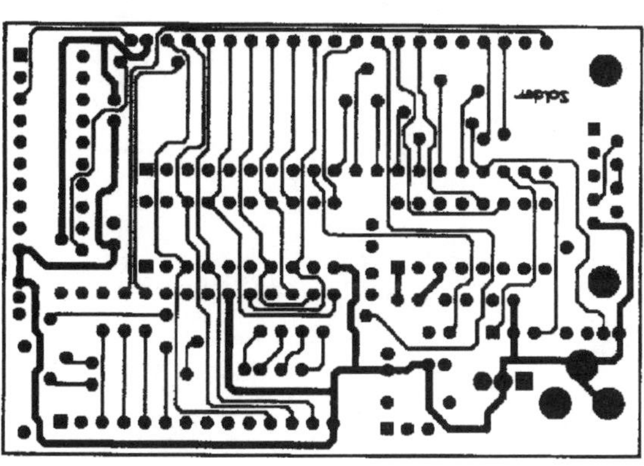

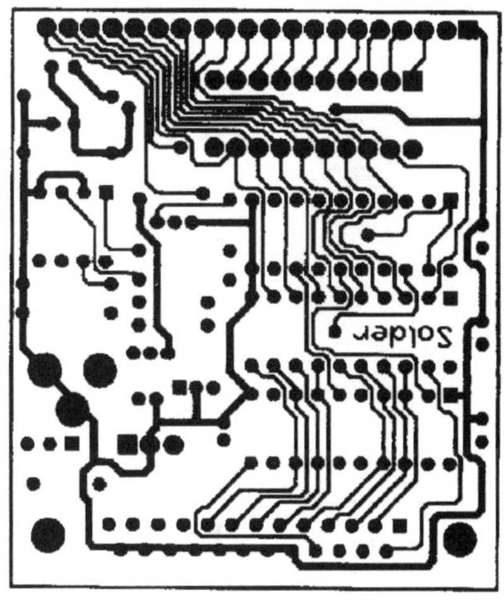

FIGURE I-2 Bottom side of the complete raw card.

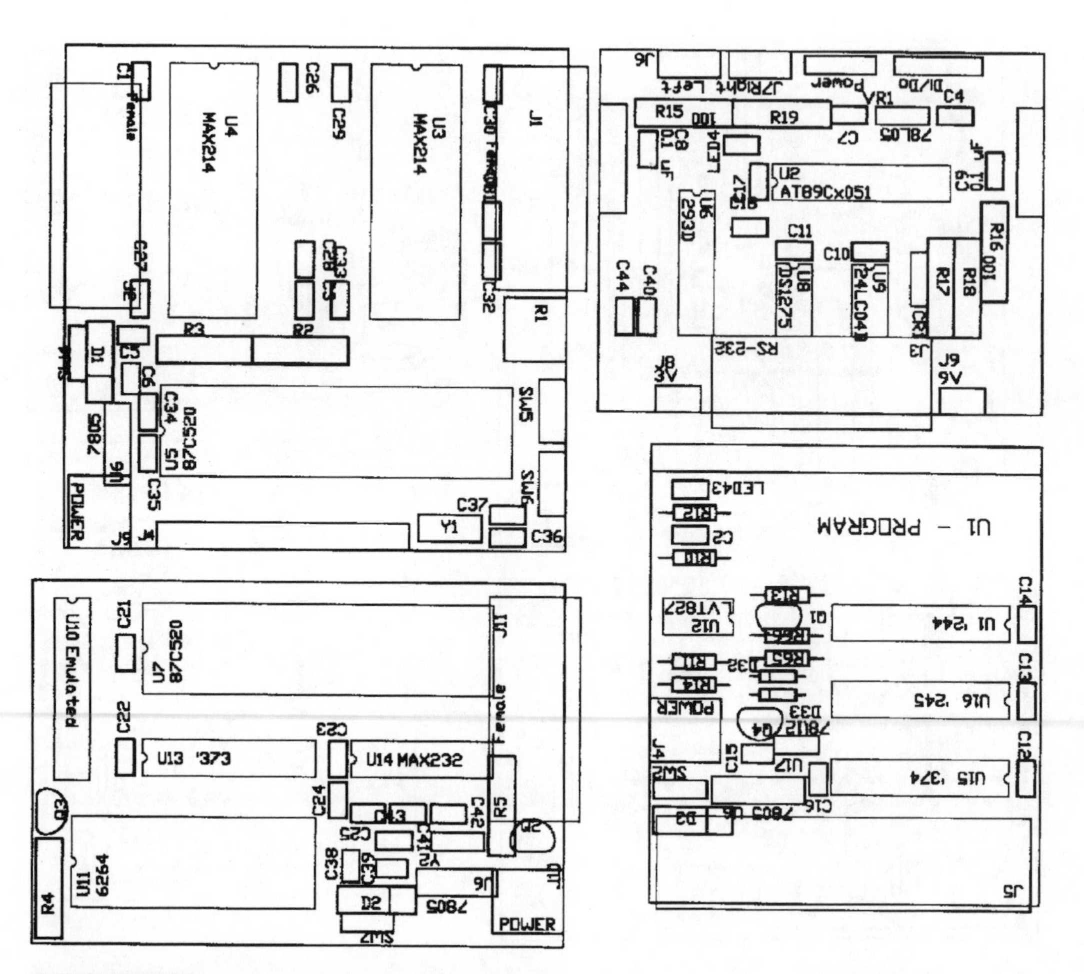

FIGURE I-3 Overlay of the complete raw card.

INDEX

Note: **Boldface** numbers indicate illustrations

C

N

O

P

T

SOFTWARE AND INFORMATION LICENSE

The software and information on this diskette (collectively referred to as the "Product") are the property of The McGraw-Hill Companies, Inc. ("McGraw-Hill") and are protected by both United States copyright law and international copyright treaty provision. You must treat this Product just like a book, except that you may copy it into a computer to be used and you may make archival copies of the Products for the sole purpose of backing up our software and protecting your investment from loss.

By saying "just like a book," McGraw-Hill means, for example, that the Product may be used by any number of people and may be freely moved from one computer location to another, so long as there is no possibility of the Product (or any part of the Product) being used at one location or on one computer while it is being used at another. Just as a book cannot be read by two different people in two different places at the same time, neither can the Product be used by two different people in two different places at the same time (unless, of course, McGraw-Hill's rights are being violated).

McGraw-Hill reserves the right to alter or modify the contents of the Product at any time.

This agreement is effective until terminated. The Agreement will terminate automatically without notice if you fail to comply with any provisions of this Agreement. In the event of termination by reason of your breach, you will destroy or erase all copies of the Product installed on any computer system or made for backup purposes and shall expunge the Product from your data storage facilities.

LIMITED WARRANTY

McGraw-Hill warrants the physical diskette(s) enclosed herein to be free of defects in materials and workmanship for a period of sixty days from the purchase date. If McGraw-Hill receives written notification within the warranty period of defects in materials or workmanship, and such notification is determined by McGraw-Hill to be correct, McGraw-Hill will replace the defective diskette(s). Send request to:

Customer Service
McGraw-Hill
Gahanna Industrial Park
860 Taylor Station Road
Blacklick, OH 43004-9615

The entire and exclusive liability and remedy for breach of this Limited Warranty shall be limited to replacement of defective diskette(s) and shall not include or extend to any claim for or right to cover any other damages, including but not limited to, loss of profit, data, or use of the software, or special, incidental, or consequential damages or other similar claims, even if McGraw-Hill has been specifically advised as to the possibility of such damages. In no event will McGraw-Hill's liability for any damages to you or any other person ever exceed the lower of suggested list price or actual price paid for the license to use the Product, regardless of any form of the claim.

THE McGRAW-HILL COMPANIES, INC. SPECIFICALLY DISCLAIMS ALL OTHER WARRANTIES, EXPRESS OR IMPLIED, INCLUDING BUT NOT LIMITED TO, ANY IMPLIED WARRANTY OF MERCHANTABILITY OR FITNESS FOR A PARTICULAR PURPOSE. Specifically, McGraw-Hill makes no representation or warranty that the Product is fit for any particular purpose and any implied warranty of merchantability is limited to the sixty day duration of the Limited Warranty covering the physical diskette(s) only (and not the software or information) and is otherwise expressly and specifically disclaimed.

This Limited Warranty gives you specific legal rights; you may have others which may vary from state to state. Some states do not allow the exclusion of incidental or consequential damages, or the limitation on how long an implied warranty lasts, so some of the above may not apply to you.

This Agreement constitutes the entire agreement between the parties relating to use of the Product. The terms of any purchase order shall have no effect on the terms of this Agreement. Failure of McGraw-Hill to insist at any time on strict compliance with this Agreement shall not constitute a waiver of any rights under this Agreement. This Agreement shall be construed and governed in accordance with the laws of New York. If any provision of this Agreement is held to be contrary to law, that provision will be enforced to the maximum extent permissible and the remaining provisions will remain in force and effect.